NEMATODE BEHAVIOUR

NEMATODE BEHAVIOUR

Edited by

Randy Gaugler and Anwar L. Bilgrami

Department of Entomology
Rutgers University
New Brunswick
New Jersey, USA

CABI Publishing

CABI Publishing is a division of CAB International

CABI Publishing
CAB International
Wallingford
Oxfordshire OX10 8DE
UK

CABI Publishing
875 Massachusetts Avenue
7th Floor
Cambridge, MA 02139
USA

Tel: +44 (0)1491 832111
Fax: +44 (0)1491 833508
E-mail: cabi@cabi.org
Web site: www.cabi-publishing.org

Tel: +1 617 395 4056
Fax: +1 617 354 6875
E-mail: cabi-nao@cabi.org

A catalogue record for this book is available from the British Library, London, UK.

Library of Congress Cataloging-in-Publication Data
Nematode behaviour / edited by Randy Gaugler and Anwar L. Bilgrami.
    p. cm
Includes bibliographical references and index.
    ISBN 0-85199-818-6 (alk. paper)
    1. Nematode --Behaviour. I. Gaugler, Randy. II. Bilgrami, Anwar L.,
1956-
QL391.N4 N37 2004
592'.5715--dc22

2003025545

ISBN 0 85199 818 6

Typeset by MRM Graphics Ltd, Winslow, Bucks
Printed and bound in the UK by Cromwell Press, Trowbridge

Contents

Contributors

Mary E. Barbercheck, Department of Entomology, Pennsylvania State University, University Park, PA 16802, USA (e-mail: meb34@psu.edu)

Maureen M. Barr, School of Pharmacy, University of Wisconsin-Madison, 777 Highland Avenue, Madison, WI 53705, USA (e-mail: mmbarr@pharmacy.wisc.edu)

Anwar L. Bilgrami, Department of Entomology, Rutgers University, New Brunswick, NJ 08901-8524, USA (e-mail: anwarbil@rci.rutgers.edu, bilgrami1956@hotmail.com); permanent address: Section of Nematology, Department of Zoology, Aligarh Muslim University, Aligarh 202002, India

Brian Boag, Scottish Crop Research Institute, Invergowrie, Dundee, DD2 5DA, Scotland, UK (e-mail: bboag@scri.sari.ac.uk)

A.H. Jay Burr, Department of Biological Sciences, Simon Fraser University, Burnaby, BC, V5A 1S6, Canada (e-mail: burr@sfu.ca)

Keith G. Davies, Nematode Interaction Unit, Rothamsted Research, Harpenden, Hertfordshire AL5 2JQ, UK (e-mail: keith.davies@bbsrc.ac.uk)

Larry Duncan, Department of Entomology and Nematology, University of Florida, IFAS, Citrus Research and Education Center, 700 Experimental Station Road, Lake Alfred, FL 33850, USA (e-mail: lwdn@lal.ufl.edu)

Randy Gaugler, Department of Entomology, Rutgers University, New Brunswick, NJ 08901-8524, USA (e-mail: gaugler@rci.rutgers.edu)

Jinghua Hu, School of Pharmacy, University of Wisconsin-Madison, 777 Highland Avenue, Madison, WI 53705, USA (e-mail: jhu@pharmacy.wisc.edu)

Robin N. Huettel, Department of Entomology and Plant Pathology, Auburn University, AL 36849, USA (e-mail: huettro@auburn.edu)

E.E. Lewis, Department of Entomology, Virginia Tech, Blacksburg, VA 24061-0319, USA (e-mail: lewise@vt.edu)

Aaron G. Maule, Parasitology Research Group, School of Biology and Biochemistry, Queen's University Belfast, Medical Biology Centre, 97 Lisburn Road, Belfast BT9 7BL, UK (e-mail: a.maule@qub.ac.uk)

E.E. Pérez, Department of Entomology, Virginia Tech, Blacksburg, VA 24061-0319, USA (e-mail: eperez@vt.edu)

Roland N. Perry, Plant Pathogen Interactions Division, Rothamsted Research, Harpenden, Hertfordshire AL5 2JQ, UK (e-mail: roland.perry@bbsrc.ac.uk)

Ekaterini Riga, Washington State University, 24106 North Bunn Road, Prosser, WA 99350, USA (e-mail: riga@wsu.edu)

A. Forest Robinson, USDA ARS, Southern Crops Research Laboratory, 2765 F & B Road, College Station, TX 77845, USA (e-mail: forestr@cpru.usda.gov)

Patricia Timper, USDA-ARS, Crop Protection and Management Research Unit, PO Box 748, Tifton, GA 31793, USA (e-mail: ptimper@tifton.usda.gov)

David A. Wharton, Department of Zoology, University of Otago, PO Box 56, Dunedin, New Zealand (e-mail: david.wharton@stonebow.otago.ac.nz)

Denis J. Wright, Department of Biological Sciences, Imperial College London, Silwood Park Campus, Ascot, Berkshire SL5 7PY, UK (e-mail: d.wright@ic.ac.uk)

Gregor W. Yeates, Landcare Research, Private Bag 11052, Palmerston North, New Zealand (e-mail: yeatesG@landcareresearch.co.nz)

Dedication

This volume is dedicated to the memory of Neil Argo Croll (1941–1981), Professor of Parasitology at McGill University and Director of McGill International, whose untimely death prevented his witnessing the profound influence his work would have on the discipline of nematology. His legacy includes 90 research papers and

seven books, including *Ecology of Parasites*, which was written even before completing his PhD in 1966, and the seminal *Behaviour of Nematodes*. In addition to the PhD, he was awarded the Doctor of Medicine degree and the prestigious Doctor of Science degree from London University. He was particularly passionate about nematode behaviour and crossed over traditional boundaries to work with animal- and plant-parasitic as well as free-living marine and freshwater species. It is remarkable that one man could accomplish so much in the mere 39 years allotted, and we can only wonder as to 'what might have been' given a full lifespan.

Introduction and Overview

Randy Gaugler and Anwar L. Bilgrami

Department of Entomology, Rutgers University, New Brunswick, NJ 08901-8524, USA

Nematodes inhabit virtually every environment and are among the most ubiquitous organisms on earth. Four of every five metazoans is a nematode, and a mere 100 g of soil typically will house 3000 individuals. With an estimated 1 million species, only insects rival nematodes in biodiversity.

Nematology, the youngest of the zoological disciplines, is splintered along taxonomic lines into plant, insect, animal and human-parasitic, and free-living nematode factions. Historically these disparate discipline camps have communicated poorly, when at all. The bewildering differences in terminology (e.g. juvenile vs. larva) among the subdisciplines are products of this separation. In recent years, however, these barriers have become more fluid due to common interest in the extraordinary fundamental advances being made by molecular biologists with the free-living nematode, *Caenorhabditis elegans*.

Studies in nematology have traditionally concentrated on understanding structure, function, ecology, systematics and control. Nevertheless, an enormous body of behavioural research has accumulated on nematode responses, movement, taxis, search, feeding, swarming, dispersal, hatching, aggregation, moulting, and mating of parasitic and free-living species. The data, however, are as scattered and fragmented as the discipline itself. Little effort has been made to assemble and analyse these studies. It has been more than 30 years since Neil Croll's (1970) *The Behaviour of Nematodes* was published. Croll built a framework for those who followed, but noted that 'some may think I have moved prematurely into a subject which is still in its infancy'.

A seminal work, Croll's book was by necessity narrow in scope, focusing heavily on nematode responses to abiotic stimuli; chapters were devoted to light, temperature, chemicals, electrical and magnetic currents, gravity, and mechanical responses. Subsequent reviews were rare and were devoted to single aspects of nematode behaviour such as feeding or orientation. The present effort is far broader in scope and attempts to consolidate the disparate studies on nematode behaviour into a comprehensive, up-to-date volume.

Fuelled by unprecedented levels of research funding, *C. elegans* reigns as the model species of choice for students of nematode behaviour. Interest in behaviour was spurred in particular by the publication of *Nematodes as Biological Models*, Vol. I, *Behavioural and Developmental Models* by Zuckerman (1980), which advocated a focus on *C. elegans*. Research into *C. elegans* behaviour continues to expand, but comparative studies with other nematode species, chiefly parasitic species of economic importance, are growing in parallel. The present book cuts across taxonomic lines, encompassing the full range of nematode behaviour as it pertains to plant, insect, animal and free-living species, often within the context of significant insights gained from *C. elegans*.

We have similarly endeavoured to treat behaviour within a broad context, considering every facet of behaviour relevant to nematode biology, sometimes at both individual and population levels, including the mechanisms and evolutionary causes of behaviour. Examples are drawn from across the full breadth of nematology. Tools and methods useful to the study of nematode behaviour are highlighted whenever possible, and practical aspects are considered when appropriate. Research gaps, of which there are many, are identified and promising approaches suggested.

Defining Behaviour

Behaviour is the integration of sensory and motor inputs, which are assessed in terms of locomotion, movement, feeding, mating, migration, penetration, etc. Behaviour is a set of patterns that involves processes such as perception; for example, nematodes sense variations in their internal and external surroundings and respond to the changes they perceive. This in essence forms the total behavioural component in lower invertebrates including nematodes. Insects with their complex reactions and social interactions enjoy hierarchical dominance in terms of behavioural patterns. Still higher are vertebrates, which display instinctive behaviour that is increasingly modified by learning, and finally human beings who can look upon themselves, become the object of their own thoughts, distinguish the world they know from themselves and whose behaviour is altered by changes in culture.

Behaviour depends upon potentialities of the nervous system, which vary between individuals, groups, and controls the organism's changing relations with the environment. The nervous system does not operate in a vacuum, since it is affected by limitations and biases imprinted during the development and long evolution of the organism.

The nervous system is influenced by developing stages of individuals, changes imprinted upon it, repeated actions, and types and intensities of internal or external stimuli. Although the nervous system plays the central role in controlling activities, our aspiration to understand behaviour can not end with it alone, since glands and muscles also play significant roles. Nematode behaviour may therefore be defined as a set of activities and responses translated by the nervous system in response to internal and external stimuli.

Describing Types of Nematode Behaviour

In the absence of well-defined criteria, behavioural classification of nematodes based upon their activities and adaptations becomes a difficult task. At what point do nematodes sense physical and chemical stimuli? Is the ingestion of food by nematodes passive or active? What type of behaviour do nematodes show when sperm are released? Many similar questions could be addressed if a classification system of nematode activity based on endogenous and exogenous, and biotic and abiotic components was adopted.

Nematodes display behaviours as coordinated and interacted responses. This is why behaviour in such deceptively simple organisms as nematodes is complex. Nematodes use receptors, the central nervous system, and somatic musculature to perform exogenous activities (e.g. locomotion), whereas the sympathetic nervous system, changes in body turgor pressure and somatic muscles are used to accomplish endogenous activities, such as ingestion and defaecation.

Behaviours may be classified into two basic types: operational (voluntary) and consequential (influenced by stimuli). Behaviour by operation refers to what nematodes do and describes exogenous or endogenous activities such as movement, hatching, vulval contractions, stylet movement, etc. Classification by operation brings behavioural patterns of similar spatio-temporal organizations into a single group, including body postures, wave patterns, movements in egg, muscular movements during defecation, swarming, nictation, orientation and penetration. It is simple in some cases (e.g. determining the number of muscles involved in coordinating a particular activity) but difficult where patterns are mixed and complicated, for instance, the different movements involved in single types of behaviour such as orientation, feeding or copulation. Various subcomponents such as crawling, forward and backward movements, coiling, sideways movements, lateral head movements, and head and tail movements may be considered to simplify such classification.

Behaviour by consequence is extremely varied, since nematodes are governed by various endogenous activities and sensory stimuli of physical and chemical natures. Changes in behaviour due to chemical (e.g. kairomones, allomones, sex pheromones) or physical (e.g. temperature, moisture, electricity, evasive actions taken to avoid predation) stimuli are examples of behaviour by consequence. Description of behaviour by consequence brings clarity in experiments, distinguishes different types and reduces experimental error. Behaviour by consequence can take several forms. First, chemical and physical stimuli are information-generating patterns (e.g. movement patterns). Next, modes of triggering stimuli and actions of the nervous system constitute muscular, neurological, and/or physiological mechanisms. These mechanisms present displays which constitute behaviours.

Overview

Information on behavioural adaptations to diverse ecological conditions is assembled and discussed by Gregor Yeates in Chapter 1 ('Behavioural and Ecological

Adaptations'). The author has analysed the significance of ongoing adaptive processes with reference to nematode behaviour. Nematodes are classified depending upon their adaptations. For example, Yeates considers marine, freshwater or phreatic nematodes that inhabit permanently water-saturated habitats as *interstitial* nematodes. Conversely, nematodes using meniscus forces directly for efficient locomotion, and indirectly benefiting from the more rapid gaseous exchange of thin films, are classified as *pellicole* nematodes. Diversity in nematode assemblages is discussed over a wide range of habitats with particular reference to adaptation. Behavioural adaptations are apparent in cephalic specialization, body length, number of juvenile stages (Yeates and Boag, 2002), differences in geographical distribution (Bloemers *et al.*, 1997), migration and aggregation (Moens *et al.*, 1999), tolerance to environmental stress (Jacquiet *et al.*, 1996), etc.

Locomotion is fundamental to nematode behaviour and impacts feeding, food finding, mating and migration. Nematodes move by undulatory propulsions as described in detail by Burr and Gans (1998) and Alexander (2002). Locomotion includes crawling or swimming through leaf surfaces, stomata, root hairs and tissues, rotting plant matter, water, excreta, intestinal microvilli, animal tissues, blood vessels and insect tracheae. Nematodes may also accelerate, stop, reverse, turn, omega-turn, probe, orient, swim, burrow, penetrate, poke, lace, climb, bridge, roll, graze, cruise, nictate, aggregate, swarm, ambush, hitchhike, loop or somersault. Although nematodes cannot fly, infective juveniles of some entomopathogenic species nearly accomplish this feat by leaping distances of nine body lengths when in the presence of an insect host (Campbell and Kaya, 1999a,b). These fascinating aspects of nematode locomotion are explored by Jay Burr and Forest Robinson in Chapter 2 ('Locomotion Behaviour'). This chapter details the various types of locomotion, the role of the hydrostatic skeleton and neuromuscular control in locomotion. The authors explain how movements are achieved and controlled, how propulsive forces are generated against external substrates, what role neuromuscular system plays, how locomotion is adapted in different environments, and what alternate means are involved in locomotion when undulatory propulsion is absent. The authors elucidate functions of neuromuscular structures associated with nematode locomotion. They discuss the fine points of body wall structure, elastic properties, lateral connection to cuticle, tonus and development of bending muscles. Transmission of forces during locomotion, the functions of synapses, neurotransmitters, and neuromodulators, the propagation of waves, the mechanism of orientation and factors generating frictional resistance are discussed. The authors emphasize the need to study mechanical properties and putative functions of the hydrostatic skeleton, inflation pressures, change in dimensions, crossing angle, elasticity, cuticle ultrastructure and bending. There is a strong need for more integrative investigations on locomotion based on comparative structural, functional and behavioural diversities with reference to nematode adaptations. Of special interest is the promise of novel methodologies, particularly video capture and editing (VCE) with microscopy (De Ley and Bert, 2002) to archive nematode locomotion for in-depth analysis of the mechanisms involved.

Nematodes receive and interpret signals from the environment and from each other that allow them to find hosts, mates, develop and survive. Ekaterina Riga uses Chapter 3 ('Orientation Behaviour') to describe various facets of nematode orientation behaviour by discussing types of stimuli and mechanisms involved in nematode chemo-, mechano-, photo-, thigmo- and thermo-tactic responses. The functional aspects of receptors are discussed in relation to their role in orientation. Riga suggests that novel control methodologies could develop by disrupting certain phases of the nematode life cycle (e.g. phases during search for food or mates) (Perry, 1994). As hypothesized by Bone and Shorey (1977), such disruptions might be achievable by saturating the nematode's soil environment with artificial pheromones or host cues. Bone (1987) and Perry (1994) have specifically suggested that nematode reproduction could be disrupted using the sex pheromone confusant method. Further studies are needed to understand the role and functions of receptors in nematode orientation.

Comprehensive knowledge about food and feeding habits is fundamental to understanding aetiology. Anwar Bilgrami and Randy Gaugler's Chapter 4 ('Feeding Behaviour') focuses on patterns of feeding types and mechanisms, muscular movements, food and feeding habits, extracorporeal digestion, host recognition, host tissue penetration, prey catching, cannibalism, ingestion and defecation. A central theme is that despite remarkable structural and functional similarities, nematodes show great diversity in food and feeding habits, obtaining nutrients from bacteria, protozoa, fungi, algae, other nematodes, or plant and animal tissues. They may be monophagous or polyphagous with some species even showing dual or biphasic feeding habits (Yeates *et al.*, 1993), switching food resources at different stages of the life cycle. This diversity, coupled with the disparate nature of the disciplines that work with plant, animal, insect and free-living nematodes, has complicated nematode characterization into different feeding groups.

The structure of feeding organs predicts a nematode's food and feeding habits (Bird and Bird, 1991). These structures may differ between nematode groups but they perform the same functions of feeding and ingestion. The feeding apparatus can be categorized into engulfing (e.g. predatory mononchs), piercing (e.g. plant-parasitic, predatory and fungal feeding), cutting (e.g. predatory diplogasterids, some marine and animal-parasitic nematodes) and sucking types (e.g. bacterial feeding and some animal-parasitic nematodes). Plant and fungal feeders and some predatory nematodes possess a needle-like stylet for piercing tissues. Plant-parasitic triplonchids and predatory nygolaims wield a solid pointed tooth, whereas predatory mononchs, diplogasterids and monohysterids have buccal cavities with wide openings and specialized structures such as dorsal tooth, ventral teeth and denticles. The subventral lancet, cutting plates, dorsal gutter and dorsal cone are structures associated with animal-parasitic nematode feeding. These structures perform similar functions, i.e. piercing, penetration, cutting food resource and ingestion of nutrients.

Feeding mechanisms vary with the type of nematode feeding apparatus and habitats. Food capture and feeding mechanisms of plant-parasitic (Doncaster and Seymour, 1973), predatory (Bilgrami, 1992, 1993, 1997), insect-parasitic (Grewal *et al.*, 1993; Bilgrami *et al.*, 2001), and microbivore nematodes (Avery and Horvitz,

1990) are reviewed by Bilgrami and Gaugler in Chapter 4. The authors divide nematode feeding into six major components – structure and function of feeding apparatus, feeding types, food search, food capture, post-feeding activities and food preference – by adopting a scheme of classification based on nematode food and feeding habits. Explanations on the role of extracorporeal digestion of nutrients prior to ingestion, formation of tubes and plugs during feeding, and extra-intestinal food absorption provide a better understanding of adaptive processes during nematode feeding.

There is greater variation in nematode reproductive biology than any other aspect of their behaviour. Robin Huettel discusses this diversity in Chapter 5 ('Reproductive Behaviour'), including evolutionary and ecological aspects. How nematodes respond to sex pheromones and what behavioural mechanisms are adopted during mating are among the subjects treated. Sensory habituation, influence of age, sexual status, physical and chemical factors on orientation, recognition and copulation are also considered. Sex pheromone activity in *Heterodera glycines* females has been attributed to vanillic acid (Jaffe *et al.*, 1989). Bioassays such as lactin binding are used to study chemotactic responses of *H. schachtii* males to female sex pheromones (Aumann *et al.*, 1998). Aumann and Hashem (1993) extracted attractive substances from females of *H. schachtii* which possess pheromone activity for males. Greet and Perry (1992) reviewed patterns and evolution of sexuality, genetic basis of sex determination, sexual behaviour and differentiation, the role of sex attractants, spicule function and copulation behaviour in nematodes. The cellular basis of chemotaxis, thermotaxis and developmental switching may be studied in nematodes by killing selected neurones with laser ablation and assaying effects on behaviour. Ablation is a powerful but underutilized tool that may be applied to study diverse functions of the nervous system and the cellular basis of specific nematode behaviours.

Ageing has been defined as a time-dependent series of cumulative, progressive, intrinsic and deleterious functional and structural changes that begin to manifest at reproductive maturity, eventually culminating in death (Arking, 1999). Ageing affects locomotion, fecundity, oviposition, vulval contractions, sexual attraction, copulation, osmotic fragility and feeding activities of nematodes (Zuckerman *et al.*, 1972; Gems, 2002; Herndon *et al.*, 2002). Ed Lewis and E.E. Pèrèz in Chapter 6 ('Ageing and Developmental Behaviour') describe age-dependent nematode behaviour during development and after reaching adulthood. Learning processes in nematodes are non-associative but are involved in the modification of behaviour due to repeated exposure to a single or multiple cues, e.g. habituation and sensitization (Bernhard and van der Kooy, 2000). Chapter 6 is made more interesting when the authors explain learning processes in nematodes and correlate this with ageing. Our understanding of ageing behaviour has made a deep breach at the genetic and molecular levels using *C. elegans* as a model organism (Gershon and Gershon, 2001). However, studies on ageing behaviour of other groups of nematodes have been neglected. In this chapter, the authors suggest several lines of research on ageing behaviour in other important nematode species based upon accomplishments with *C. elegans*. These could prove useful in the development of novel management practices, and in monitoring environmental pollution using nematode behavioural parameters as bioindicators.

Denis Wright brings together the scattered information on behaviour as it relates to waste and ionic regulation in free-living and parasitic nematodes in Chapter 7 ('Osmoregulatory and Excretory Behaviour'). Earlier reviews emphasized the molecular cell biology of animal-parasitic species (Thompson and Geary, 2002). Nematodes regulate water content to adapt to changing osmotic conditions (Wright, 1998). Failure to do so may disrupt locomotion because of the anisometric nature of the cuticle (Wright and Newall, 1980). Osmotic factors also govern nematode survival (Glazer and Salame, 2000), freezing tolerance (Wharton and To, 1996), hatching (Perry, 1986), reproduction (Gysels and Tavernier-Bracke, 1975) and feeding (Raispere, 1989). The nematode excretory system has been described as a 'secretory–excretory' system (Wright, 1998) because of its dual role in osmoregulation and excretion. Chapter 7 integrates the two mechanisms so as to explain changes in nematode behaviour occurring due to osmotic stresses and excretion of nitrogenous waste products.

Behavioural responses are the result of intrinsic and extrinsic stimulations involving various physiological and biochemical activities. Roland Perry and Aaron Maule in Chapter 8 ('Physiological and Biochemical Basis of Behaviour') describe how these activities play a huge role in regulating nematode behavioural responses. Physiology and biochemistry influence functions of sense organs, cuticle, muscles, glands, digestive and excretory organs associated with nematode behaviours. Correlations with physiological and biochemical factors have been established with sensory responses (Perry and Aumann, 1998), female sex pheromones (Aumann *et al.*, 1998) and starvation in nematodes (Reversat, 1981). Results on chemosensory responses suggest each receptor cell in the amphids detects different chemicals. A promising approach to investigate chemosensory responses would be to identify and analyse attractants and repellents in the natural environment, and test them for independence. Chapter 8 further describes the role of hormones and enzymes during behaviour including ecdysis, extracorporeal digestion, salivation, hunger, chemoattraction, nerve conduction, dauer formation, population regulation and sex attraction. Chemical neurotransmitters play an important role in nematode behaviour (Sulston *et al.*, 1975; Wright and Awan, 1976). Willet *et al.* (1980) considered the nervous system in nematodes as cholinergic, with acetylcholine and γ-aminobutyric acid as excitatory and inhibitory transmitters respectively. Vaginal and vulval movement in *C. elegans*, *Aphelenchus avenae* and *Panagrellus redivivus* are controlled by serotonin, 5 hydroxytryptophan and adrenaline (Croll, 1975). It is not difficult to study nematode behaviour in relation to neurones as their numbers do not exceed 250 (Willet *et al.*, 1980). Neurones that mediate behavioural responses to different chemicals have been identified through laser ablation (Troemel, 1999). Genetic and molecular studies on nematode behaviour have indicated involvement of G protein signalling pathways in chemotransduction. Nematodes (e.g. *C. elegans*) are estimated to use approximately 500 chemosensory receptors to detect a large spectrum of chemicals in the environment. Unfortunately, the physiological and biochemical bases of nematode behaviour are less developed than the rapidly expanding knowledge on the molecular aspects of behaviour, yet Chapter 8 takes a molecular approach to securing a broader perspective on biochemical and physiological aspects of nematode behaviour.

Genes control behaviour and molecular genetics has become a powerful new tool to study nematode behaviour. Maureen Barr and Jinghua Hu in Chapter 9 ('Molecular Basis for Behaviour') make a novel effort to describe various aspects of behaviour at the molecular level using *C. elegans* as their model. Laser ablation experiments have revealed important similarities in *C. elegans* with visual and olfactory transductions in vertebrates (Mori and Ohshima, 1997). Movement and migration have been analysed theoretically (Anderson *et al.*, 1997a) and experimentally (Anderson *et al.*, 1997b). The role of ageing, acetylcholinesterase, motor neurone M3 and genes in regulating *C. elegans* behaviour has been established. Studies on neural G protein signals show that EGL-10, RGS-1 and RGS-10 proteins alter signals in *C. elegans* to induce nematode behavioural responses. Oviposition is regulated by the FLP-1 peptide (Waggoner *et al.*, 2000) and reversals in foraging movements are controlled by small subsets of neurones (Zheng *et al.*, 1999).

Most nematodes live in the soil, an environment heavily colonized by other organisms. Nematodes interact with this biotic community and adapt to survive, adaptations that often have behavioural components. How nematodes interact with beneficial and antagonistic organisms including intraspecific interactions, what causes interactions to occur, and how these interactions affect nematode behaviour constitute the subject of Chapter 10 ('Biotic Interactions') by Patricia Timper and Keith Davies. This chapter describes different types of interactions, e.g. phoresy, antagonism, mutualism, commensalism and amensalism, with reference to behaviour. The authors draw particular attention to the extraordinary behaviours shown by nematodes in phoretic associations. Phoretic hosts provide transport to fresh resources and protection from unfavourable biotic and abiotic environments. Many species of Rhabditida, Diplogasterida and Aphelenchida develop phoretic relationships but few species of Tylenchida (Massey, 1974) and Strongylida are phoretic (Robinson, 1962). Formation of dauer juveniles (Sudhaus, 1976; Maggenti, 1981) and the ability to nictate are a few examples of phoretic adaptations discussed in this chapter. Such interactions are either facultative (e.g. synchronization of *Bursaphelenchus seani* with helictid wasps, Giblin and Kaya, 1983) or obligate (e.g. synchronization of *Bursaphelenchus cocophilus* with *Rhynchophorus*, Giblin-Davis, 1993) depending upon behavioural adaptations. Antagonism, a varied but interactive behaviour leading to nematode predation and parasitism, is discussed in association with important natural enemies such as fungi, bacteria, insects and nematodes (Stirling, 1991). Chapter 10 indicates that competition is the basis for nematode responses such as co-existence, population fluctuation, migration, spatial displacement, aggregation and sharing food resources.

Abiotic factors including temperature, microwaves, electromagnetic waves, radiation, gravity, pH and water potential (Sambongi *et al.*, 2000; Saunders *et al.*, 2000; Soriano *et al.*, 2000; de Pomerai *et al.*, 2002) also influence nematode behaviour and have been studied with special intensity. Chapter 11 ('Abiotic Factors') by

Mary Barbercheck and Larry Duncan reviews nematode responses to these chemical and physical challenges with emphasis on updating and expanding topics treated by Croll (1970). The influence of many abiotic factors on nematode activities has adaptive significance, as these influences elicit responses such as acclimation and orientation. The reasons why such adaptations are necessary in nematodes exposed to adverse physical conditions are discussed. The authors emphasize that nematode sensitivity, tolerance and response to the abiotic environment are keys to exploiting nematode behaviour for the management of economically important species.

Gregorich *et al.* (2001) defined population dynamics as 'the numerical changes in population within a period of time,' whereas Lawrence (1995) described it as 'changes in population structure over a period of time'. The former encompasses seasonal variations in nematode populations (Kendall and Bluckland, 1971) whereas the latter includes population cycles (Goodman and Payne, 1979). Chapter 12 ('Population Dynamics'), contributed by Brian Boag and Gregor Yeates, reflects both components in an expansive examination of behaviour at the nematode population rather than at the individual level by focusing on temporal and spatial patterns of migration. The authors explain non-uniform distribution behaviour of nematodes in soils, sediments, and plant and animal tissues. They also describe changes in behaviour during migration and environmental conditions responsible for variations in populations and migrations. For example, variation in vertical migratory behaviour (Boag, 1981) is attributed to root distribution, soil type, moisture, temperature, etc. (Yeates, 1980; Rawsthorne and Brodie, 1986; Young *et al.*, 1998). Horizontal migratory behaviour is suggested (Taylor, 1979) as a useful tool to detect and estimate the size of nematode populations (Been and Schomaker, 1996), as well as mechanisms of nematode aggregation (Goodell and Ferris, 1981). Migratory behaviour deserves special attention with reference to infective soil stages because these nematodes respond to host cues. The distance migrated may be a function of some nematodes waiting passively for a host while others are dispersers.

Like other organisms, nematodes have strategies to resist adverse environmental conditions. Such strategies may have behavioural, biochemical or morphological components but all three have strong associations with each other. In Chapter 13 ('Survival Strategies'), David Wharton describes various behavioural approaches nematodes employ to avoid or mitigate stress, including synchronizing parasite life cycles with host availability or migration (e.g. phoresy) to escape environmental extremes until favourable conditions return. Other species develop a degree of tolerance or resistance. These adaptive strategies enhance survival under biological (food inadequacy, predation, pathogens and competition), physical (temperature, desiccation, pressure and radiation) or chemical (pH, osmotic stress, anoxia) stress. Wharton explains how various tactics and mechanisms govern nematode behaviour in challenging environments. The reasons for and mechanisms of attaining resting, infective and dauer stages, diapause and egg diapause, arrested development and delays in the life cycles are elaborated in this chapter.

Future Prospects

Nematodes have been known since biblical times (the Bible's 'fiery serpent' almost certainly was the guinea worm, *Dracunculus medinensis*) and have been the subject of scientific inquiry since the 18th century, yet the singular scientific event in this long history occurred in 2000 with sequencing of the *C. elegans* genome. With a sequencing capacity of more than 25 billion base pairs a year and new organismal sequences being added almost weekly, additional nematodes species can soon be expected to join *C. elegans*, including four more *Caenorhabditis* species. Consortia have also been assembled with aspirations to sequence economically relevant nematode species including the plant-parasitic species *Meloidogyne hapla*, the entomopathogen *Heterorhabditis bacteriophora* and the filariasis nematode *Brugia malayi*. These epic tasks will prove to be the easy part relative to what follows: functional genomics. The use of powerful new tools for functional genomics, particularly interference RNA which permits gene function to be blocked with relative ease, is enabling us to study nematode behaviour in ways not previously imagined.

Nematode behaviour has tended to be studied as a secondary issue to physiological, biochemical or ecological questions, and consequently has lagged far behind its animal behaviour counterparts: bird, insect and mammalian systems. This is about to change. The new linkages establishing between functional genomics and behaviour will elevate nematode behaviour to the next level, enabling nematodes to serve as a broader model for animal behaviour studies. Moreover, behaviour is poised to move from a minor subject of nematological research, to becoming a unifying focus that enhances communication among the narrow subdisciplines of nematology. This book documents steady progress made in the study of nematode behaviour since Croll (1970), but the next 30 years promise dazzling advances.

Acknowledgements

We thank the contributors for their time, expertise, patience, and dedication. Our wives, Cheryl and Tabassum, receive our deepest appreciation for their long years of support.

References

Alexander, R. McN. (2002) Locomotion. In: Lee, D.L. (ed.) *The Biology of Nematodes.* Taylor and Francis, London, pp. 345–352.

Anderson, A.R.A., Young, I.M., Sleeman, B.D., Griffiths, B.S. and Robertson, W.M. (1997a) Nematode movement along a chemical gradient in a structurally heterogeneous environment. 1. Experiment. *Fundamental and Applied Nematology* 20, 157–163.

Anderson, A.R.A., Sleeman, B.D., Young, I.M. and Griffiths, B.S. (1997b) Nematode movement along a chemical gradient in a structurally heterogeneous environment. 2. Theory. *Fundamental and Applied Nematology* 20, 165–172.

Arking, R. (1999) *Biology of Aging*, 2nd edn. Sinauer Associates, Sunderland, Massachusetts, 570 pp.

Aumann, J. and Hashem, M. (1993) Effect of the formamidine acaricide chlordimeform on mobility and chemoreception of *Heterodera schachtii* males (Nematoda: Heteroderidae). *Nematologica* 39, 75–80.

Aumann, J., Dietsche, E., Rutencrantz, S. and Ladehoff, H. (1998) Physiochemical properties of the female sex pheromone of *Heterodera schachtii* (Nematoda: Heteroderidae). *International Journal for Parasitology* 28, 1691–1694.

Avery, L. and Horvitz, H.R. (1990) Effects of starvation and neuroactive drugs on feeding in *Caenorhabditis elegans*. *Journal of Experimental Zoology* 253, 263–270.

Been, T.H. and Schomaker, C.H. (1996) A new sampling method for the detection of low population densities of potato cyst nematodes (*Globodera pallida* and *G. rostochiensis*). *Crop Protection* 15, 375–382.

Bernhard, N. and van der Kooy, D. (2000) A behavioural and genetic dissection of two forms of olfactory plasticity in *Caenorhabditis elegans*: adaptation and habituation. *Learning and Memory* 7, 199–212.

Bilgrami, A.L. (1992) Resistance and susceptibility of prey nematodes to predation and strike rate of the predators, *Mononchus aquaticus*, *Dorylaimus stagnalis* and *Aquatides thornei*. *Fundamental and Applied Nematology* 15, 265–270.

Bilgrami, A.L. (1993) Analyses of relationships between predation by *Aporcelaimellus nivalis* and different prey trophic categories. *Nematologica* 39, 356–365.

Bilgrami, A.L. (1997) *Nematode Biopesticides*. Aligarh Muslim University Press, Aligarh, 262 pp.

Bilgrami, A.L., Kondo, E. and Yoshiga, T. (2001) Host searching and attraction behaviour of *Steinernema glaseri* using *Galleria mellonella* as its host. *International Journal of Nematology* 11, 168–176.

Bird, A.F. and Bird, J. (1991) *The Structure of Nematodes*. Academic Press, London, 316 pp.

Bloemers, G.F., Hodda, M., Lambshead, P.J.D., Lawton, J.H. and Wanless, F.R. (1997) The effects of forest disturbance on diversity of tropical soil nematodes. *Oecologia* 111, 575–582.

Boag, B. (1981) Observations on the population dynamics and vertical distribution of trichodorid nematodes in a Scottish forest nursery. *Annals of Applied Biology* 98, 463–469.

Bone, L.W. (1987) Pheromone communication in nematodes. In: Veech, J.A. and Dickson, D.W. (eds) *Vistas in Nematology*. Society of Nematologists, Hyattsville, Maryland, pp. 147–152.

Bone, L.W. and Shorey, H.H. (1977) Disruption of sex pheromone communication in a nematode. *Science* 197, 694–695.

Burr, A.H.J. and Gans, C. (1998) Mechanical significance of obliquely striated architecture in nematode muscle. *Biological Bulletin* 194, 1–6.

Campbell, J.F. and Kaya, H.K. (1999a) How and why a parasitic nematode jumps. *Nature* 397, 485–486.

Campbell, J.F. and Kaya, H.K. (1999b) Mechanism, kinematic performance, and fitness consequences of jumping behavior in entomopathogenic nematodes (*Steinernema* spp.). *Canadian Journal of Zoology* 77, 1947–1955.

Croll, N.A. (1970) *The Behaviour of Nematodes: Their Activity, Senses, and Responses*. Edward Arnold, London, 117 pp.

Croll, N.A. (1975) Indolealkylamines in the coordination of nematode behavioral activities. *Canadian Journal of Zoology* 53, 894–903.

De Ley, P. and Bert, W. (2002) Video capture and editing as a tool for the storage, distribution and illustration of morphological characters of nematodes. *Journal of Nematology* 34, 296–302.

de Pomerai, D.I., Dawe, A., Djerbib, L., Allan, J., Brunt, G. and Daniells, C. (2002) Growth and maturation of the nematode *Caenorhabditis elegans* following exposure to weak microwave fields. *Enzyme and Microbial Technology* 30, 73–79.

Doncaster, C.C. and Seymour, M.K. (1973) Exploration and selection of penetration site by Tylenchida. *Nematologica* 19, 137–145.

Gems, D. (2002) Ageing. In: Lee, D.L. (ed.) *The Biology of Nematodes*. Taylor and Frances, London, pp. 413–456.

Gershon, H. and Gershon, D. (2001) *Caenorhabditis elegans* – a paradigm for aging research: advantages and limitations. *Mechanisms of Ageing and Development* 123, 261–274.

Giblin-Davis, R.M. (1993) Interactions of nematodes with insects. In: Khan, A.W. (ed.) *Nematode Interactions*. Chapman and Hall, London, pp. 303–344.

Giblin, R.M. and Kaya, H.K. (1983) Field observations on the association of *Anthophora bomboides stanfordiana* (Hymenoptera: Anthophoridae) with the nematode *Bursaphelenchus seani* (Aphelenchida: Aphelenchoididae). *Annals of the Entomological Society of America* 76, 228–231.

Glazer, I. and Salame, I. (2000) Osmotic survival of the entomopathogenic nematode *Steinernema carpocapsae*. *Biological Control* 18, 251–257.

Goodell, P.B. and Ferris, H. (1981) Sampling optimization for five plant parasitic nematodes. *Journal of Nematology* 13, 304–313.

Goodman, A. and Payne, E.M.F. (1979) *Longman Dictionary of Scientific Usage*. Longman Group Ltd, Harlow, 684 pp.

Greet, D.N. and Perry, R.N. (1992) Nematoda and Nematomorpha. In: Adiyodi, K.G. and Adiyodi, R.G. (eds) *Reproductive Biology of Invertebrates*, Vol. V, *Sexual Differentiation and Behaviour*. John Wiley & Sons, Chichester, UK, pp. 147–173.

Grewal, P., Gaugler, R. and Lewis, E.E. (1993) Host recognition behaviour by entomopathogenic nematodes during contact with insect gut contents. *Journal of Parasitology* 79, 495–503.

Gregorich, E.G., Turchenek, L.W., Carter, M.R. and Angers, D.A. (2001) *Soil and Environmental Science*. CRC Press, Boca Raton, Florida, 577 pp.

Gysels, H. and Tavernier-Bracke, E. (1975) The influence of physiological stress situations as a consequence of changing osmotic pressures, upon the development and growth of the free-living nematode *Panagrellus silusiae* (De Man, 1913) Goodey, 1945. *Natuurwetenschappelijk Tijdschrift* 57, 15–37.

Herndon, L.A., Schmeissner, P.J., Dudaronek, J.M., Brown, P.A., Listner, K.M., Sakano, Y., Paupard, M.C., Hall, D.H. and Driscoll, M. (2002) Stochastic and genetic factors influence tissue-specific decline in ageing *C. elegans*. *Nature* 419, 808–814.

Jacquiet, P., Cabaret, J., Dia, M., Cheikh, D. and Thiam, E. (1996) Adaptation to arid environment: *Haemonchus longistipes* in dromedaries of Saharo-Sahelian areas of Mauritania. *Veterinary Parasitology* 66, 193–204.

Jaffe, H., Huettel, R.N., Demilo, A.B., Hays, D.K. and Rebois, R.V. (1989) Isolation and identification of a compound from soybean cyst nematode, *Heterodera glycines*, with sex pheromone activity. *Journal of Chemical Ecology* 15, 2031–2043.

Kendall, S.M.G. and Buckland, W.R. (1971) *A Dictionary of Statistical Terms*. Longman Group Ltd, London, 166 pp.

Lawrence, E. (1995) *Henderson's Dictionary of Biological Terms*. Addison Wesley Longman Ltd, Harlow, 693 pp.

Maggenti, A. (1981) *General Nematology*. Springer-Verlag, New York, 372 pp.

Massey, C.L. (1974) *Biology and Taxonomy of Nematode Parasites and Associates of Bark Beetles in the United States.* US Government Printing Office, Washington, DC, 233 pp.

Moens, T., Verbeeck, L., de Maeyer, A., Swings, J. and Vincx, M. (1999) Selective attraction of marine bacterivorous nematodes to their bacterial food. *Marine Ecology – Progress Series* 176, 165–178.

Mori, I. and Ohshima, Y. (1997) Molecular neurogenetics of chemotaxis and thermotaxis in the nematode *Caenorhabditis elegans. Bioassays* 19, 1055–1064.

Perry, R.N. (1994) Studies on nematode sensory perception as a basis for novel control strategies. *Fundamental and Applied Nematology* 17, 199–202.

Perry, R.N. (1996). Chemoreception in plant parasitic nematodes. *Annual Review of Phytopathology* 34, 181–199.

Perry, R.N. and Aumann, J. (1998) Behavior and sensory responses. In: Perry, R.N. and Wright, D.J. (eds) *Physiology and Biochemistry of Free-living and plant-parasitic Nematodes.* CAB International, Wallingford, UK, pp. 75–102.

Raispere, A. (1989) Influence of the water regime of the plant-host on the development of the potato gall nematode. *Eesti NSV Teaduste Akadeemia Toimetised, Bioloogia* 38, 274–284.

Rawsthorne, D. and Brodie, B.B. (1986) Root growth of susceptible and resistant potato cultivars and population dynamics of *Globodera rostochiensis* in the field. *Journal of Nematology* 18, 501–504.

Reversat, G. (1981) Effects of ageing and starvation on respiration and food reserve contents in adult *Hirschmanniella spinicaudata. Revue de Nematologie* 4, 125–130.

Robinson, J. (1962) *Pilobolus* spp. and the translocation of the infective larvae of *Dictyocaulus viviparus* from faeces to pasture. *Nature* 193, 353–354.

Sambongi, Y., Takeda, K., Wakabayashi, T., Ueda, I., Wada, Y. and Futai, M. (2000) *Caenorhabditis elegans* senses protons through amphid chemosensory neurons: proton signals elicit avoidance behavior. *Neuroreport* 11, 2229–2232.

Saunders, L.M., Tompkins, D.M. and Hudson, P.J. (2000) The role of oxygen availability in the embryonation of *Heterakis gallinarum* eggs. *International Journal for Parasitology* 30, 1481–1485.

Soriano, I.R.S., Prot, J.C. and Matias, D.M. (2000) Expression of tolerance for *Meloidogyne graminicola* in rice cultivars as affected by soil type and flooding. *Journal of Nematology* 32, 309–317.

Stirling, G. (1991) *Biological Control of Plant Parasitic Nematodes.* CAB International, Wallingford, UK, 282 pp.

Sudhaus, W. (1976) Vergleichende untersuchungen zur phylogenie, systematik, okalogie, biologie und ethologie der Rhabditidae (Nematoda). *Zoologica* 43, 1–229.

Sulston, J., Dew, M. and Brenner, S. (1975) Dopaminergic neurons in the nematode *Caenorhabditis elegans. Journal of Comparative Neurology* 163, 215–226.

Taylor, L.R. (1979) Aggregation as a species characteristic. In: Patil, G.P., Pielon, E.C. and Waters, W.E. (eds) *Statistical Ecology.* Pennsylvania Press, University Park, pp. 357–377.

Thompson, D.P. and Geary, T.G. (2002) Excretion/secretion, ionic and osmotic regulation. In: Lee, D.L. (ed.) *The Biology of Nematodes.* Taylor and Francis, London, pp. 291–320.

Troemel, E.R. (1999) Chemosensory signaling in *C. elegans. Bioassays* 21, 1011–1020.

Waggoner, L.E., Hardaker, L.A., Golik, S. and Schafer, W.R. (2000) Effect of a neuropeptide gene on behavioral states in *Caenorhabditis elegans* egg-laying. *Genetics* 154, 1181–1192.

Wharton, D.A. and To, N.B. (1996) Osmotic stress effects on the freezing tolerance of the

Antarctic nematode *Panagrolaimus davidi*. *Journal of Comparative Physiology – B, Biochemical, Systemic and Environmental Physiology* 166, 344–349.

Willet, D.J., Rahim, I., Geist, M. and Zuckerman, B.M. (1980) Cyclic nucleotide exudation by nematodes and the effects on nematode growth, development and longevity. *Age* 3, 82–87.

Wright, D.J. (1998) Respiratory physiology, nitrogen excretion and osmotic and ionic regulation. In: Perry, R.N. and Wright, D.J. (eds) *The Physiology and Biochemistry of Free-living and Plant-parasitic Nematodes*. CAB International, Wallingford, UK, pp. 103–131.

Wright, D.J. and Awan, F.A. (1976) Catecholaminergic structures in the nervous system of three nematode species, with observations on related enzymes. *Journal of Zoology* 185, 477–489.

Wright, D.J. and Newall, D.R. (1976) Nitrogen excretion, osmotic and ionic regulation in nematodes. In: Croll, N.A. (ed.) *The Organization of Nematodes*. Academic Press, London, pp. 163–210.

Yeates, G.W. (1980) Populations of nematode genera in soils under pasture III. Vertical distribution at eleven sites. *New Zealand Journal of Agricultural Research* 23, 117–128.

Yeates, G.W. and Boag, B. (2002) Post-embryonic growth of longidorid nematodes. *Nematology* 4, 883–889.

Yeates, G.W., Bongers, T., de Goede, R.G.M., Freckman, D.W. and Georgieva, S.S. (1993) Feeding habits in soil nematode families and genera – an outline for soil ecologists. *Journal of Nematology* 25, 315–331.

Young, I.M., Griffiths, B.S., Robertson, W.M. and McNicol, J.W. (1998) Nematode (*Caenorhabditis elegans*) movement in sand as affected by particle size, moisture and the presence of bacteria (*Escherichia coli*). *European Journal of Soil Science* 49, 237–241.

Zheng Yi., Brockie, P.J., Mellem, J.E., Madsen, D.M. and Maricq, A.V. (1999) Neuronal control of locomotion in *C. elegans* is modified by a dominant mutation in the GLR-1 ionotropic glutamate receptor. *Neuron* 24, 347–361.

Zuckerman, B.M. (1980) *Nematodes as Biological Models,* Vol. I, *Behavioral and Developmental Models*. Academic Press, New York, 306 pp.

Zuckerman, B.M., Nelson, B. and Kisiel, M. (1972) Specific gravity increase of *Caenorhabditis briggsae* with age. *Journal of Nematology* 4, 261–262.

1 Ecological and Behavioural Adaptations

GREGOR W. YEATES

Landcare Research, Private Bag 11052, Palmerston North, New Zealand

1.1. Introduction

As early as 1915, N.A. Cobb recognized the diversity and abundance of nematodes when he wrote:

> They occur in arid deserts and at the bottom of lakes and rivers, in the waters of hot springs and in polar seas where the temperature is constantly below the freezing point

of fresh water. They were thawed out alive from Antarctic ice. A thimbleful of mud from the bottom of river or ocean may contain hundreds of specimens. A lump of soil no larger than the end of one's finger may contain hundreds, even thousands of nematodes.

Nematodes are adapted to both living and non-living substrates (Perry and Wright, 1998; Anderson, 2000). In New Zealand, 29 nematode species parasitize sheep and 27 parasitize cattle (McKenna, 1997), while globally 68 parasitize potatoes (Jensen *et al.*, 1979). However, most nematodes are free-living and occur in non-living substrates. At two estuarine sites, 98–123 species were found, with maximum abundances of 650,000–800,000/m^2 (Eskin and Coull, 1987). In deep-sea trenches, abundances up to 930,000 nematodes/m^2 have been reported (Tietjen *et al.*, 1989). Nematode abundance in terrestrial soil may reach 10 million/m^2. There are few full inventories of soil nematode assemblages, but 154 species have been recorded from an English grassland, 182 species from a protected landscape in Slovakia, 228 species from a Kansas prairie and 431 species from a tropical forest (Orr and Dickerson, 1967; Šály, 1985; Hodda and Wanless, 1994; Bloemers *et al.*, 1997).

That such diverse and abundant nematode assemblages occur in such a wide range of environments shows the adaptability of nematodes. That one host species or non-living substrate can provide resources for so many species of nematodes demonstrates specificity or resource partitioning. Adaptations to permit this occur in a variety of ways, including:

1. A range of morphological specialization of the cephalic region to allow feeding in a variety of ways on a variety of food resources.
2. A range of feeding behaviours and responses to environmental conditions and cues.
3. A range in adult length of 0.3 mm to over 8 m.
4. An egg and four juvenile stages typically precede maturity. Each may have distinct morphological and behavioural adaptations, and utilize distinct environments and food resources, facilitating its contribution to the life history and maintenance of the species.
5. Differences in geographical distribution.

As we shall see, nematode species that occur in a given host or non-living substrate may, as adults, occupy different organs (e.g. plant roots, stems or florets; mammalian gastro-intestinal, vascular or respiratory systems) or use the same organ but differ in very subtle biological or behavioural adaptations.

This chapter discusses nematode behaviour adaptations in relation to various ecological and environmental conditions. The scattered information on behavioural adaptations of various nematode factions is gathered and analysed to emphasize the significance of ongoing adaptive processes in nematode behaviour and their role in practical applications.

1.2. Ecological Adaptations

1.2.1. Habitat

The only habitats that nematodes do not use directly are water and air; nematodes neither swim in open water nor fly in air. However, they use these spaces indirectly as parasites of hosts such as fish, arthropods and birds. The three primary factors that determine habitats occupied by nematodes, and to which they show adaptations, are food, water and temperature.

1.2.1.1. Food

Nematodes require sufficient, appropriate food resources for completion of their life cycle and the production of sufficient propagules to perpetuate the species. An adequate food resource may occur in extreme habitats (e.g. the vinegar eelworm *Turbatrix aceti* in vinegar; near deep-sea hydrothermal vents; in the mammalian intestine). Changes in substrate or environmental conditions (e.g. changes in soil pH affecting the plants on which nematodes feed) also affect nematodes and their behaviour, albeit indirectly. Further, complex life cycles in which successive developmental stages may use differing food resources, including alternative cycles in non-living (free-living species) and living (parasitic species) substrates, demonstrate the flexibility which enables some nematodes to link and exploit otherwise unconnected food resources.

1.2.1.2. Water

Like most organisms, nematodes require free water to maintain activity. Fundamental chemical processes such as osmosis are important, and the inability of active nematodes to control their water content excludes them from habitats of high ionic concentrations. However, the simple, unregulated nature of the nematode body allows the same body plan to be used by freshwater nematodes at a pressure of 1 atmosphere and by abyssal marine nematodes at a depth of 5000 m, corresponding to over 400 atmospheres.

For over a century, a distinction has been made between soil and freshwater or marine nematodes on a habitat basis without any consideration of the ecological implications of the differences between these habitats. Populations of bacterial-feeding soil nematodes (e.g. *Rhabditis*, *Cephalobus* and *Pristionchus*) continue to reproduce in soil even when moisture approaches the wilting point (−1500 kPa) and water films are thin (1 μm) (Yeates *et al.*, 2002). The migration of foliar nematodes (e.g. *Aphelenchoides*) and infective juveniles of some animal parasites (e.g. *Teladorsagia*) through water films on leaves, as well as negatively geotropic females of insect-parasitic *Mermis* that migrate up thin water films on plants so that herbivorous hosts may ingest their eggs, have long been known. In truly aquatic situations interstitial pores are continually water-filled. As not only the efficacy of nematode locomotion but also gaseous diffusion rates differ markedly between thin film and saturated situations, the distinction may provide a sound biological basis for discriminating between classes of nematodes or stages of particular nematodes. Nematodes that inhabit permanently water-saturated habitats, be they marine, freshwater or phreatic, can be regarded as interstitial nematodes. In contrast, those occurring in situations where they may directly use meniscus forces for efficient locomotion and indirectly benefit from the more rapid gaseous exchange of thin films can be regarded as pellicole nematodes (from the Italian, *pellicola*, a film or membrane).

1.2.1.3. Temperature

Temperature influences the metabolic rate of both nematodes and their food resources. Activity–temperature relationships are similar to those of other organisms. At low temperatures some nematodes may survive through effective dehydration or biochemical mechanisms (see Wharton, Chapter 13, this volume). High temperatures may be lethal to nematodes, and yet species have been recorded from hot springs at temperatures of up to 61°C (Winslow, 1960). Nematodes near oceanic volcanoes use resources in zones which are warm (15–25°C) rather than extremely hot (Luther *et al.*, 2001).

Temperature limits have been demonstrated in a range of bacterial- (Venette and Ferris, 1997), fungal- (Okada and Ferris, 2001) and plant-feeding (Trudgill, 1995) and predacious (Jones, 1977) nematodes. Among vertebrate parasites there are many cases in which developmental temperature thresholds are adaptations to survival outside the primary host (Evans and Perry, 1976; Schjetlein and Skorping, 1995). Infective stages may be positively thermotactic, and chilling is required by some eggs of *Nematodirus battus* before hatching (Ashton *et al.*, 1999). These are adaptations linking the activity of host and nematode, serving to increase the number of infective juveniles reaching the final host and thus completing the life cycle.

1.2.2. Oxygen

Nematodes clearly require oxygen, but the absence of oxygen does not always preclude nematode activity. The marine *Oncholaimus campylocercoides* is adapted to sulphidic (i.e. essentially anoxic) sediments by the development of S-8 rings and polysulphur chains in the epidermis; on return to normal oxygen levels these disappear (Thiermann *et al.*, 2000). The adaptation of *Stilbonema* and *Laxus* to sulphide-rich sediments involves bacterial ectosymbionts capable of respiratory reduction of nitrate to nitrite, permitting migration into deeper anoxic sediments (Hentschel *et al.*, 1999). Undoubtedly other adaptation mechanisms remain undescribed.

1.2.3. Substrates

Nematodes have traditionally been categorized as free-living (inhabiting soils, freshwater, marine, brackish or estuarine environments) or parasitic (inhabiting plant, vertebrate or invertebrate hosts). However, in its life history a given nematode species may occupy more than one of these habitat types. Yeates and Boag (2004) regarded nematodes as occupying either non-living (mineral soil, marine sediments, freshwater sediments, organic layers, plant residues, animal cadavers) or living substrates (plant tissues and tissue and body cavities of invertebrates and vertebrates). As discussed in Section 1.3, different stages of a given species may be adapted to use different substrates, both non-living and living.

1.2.4. Habitat-specific behaviour

Stimulation, orientation and migration of infective juvenile nematodes in response to host signals are primarily dependent on the host (Anderson, 2000; Lewis, 2002). However, many nematodes show a range of behaviours necessary for completion of their life cycles that are independent of proximal host stimuli. Such behavioural adaptations include:

1. Attachment of aquatic nematodes to their substrate, apparently using the three caudal glands, and showing draconian movement to maintain their position on the substrate.
2. Movement of infective juveniles of parasitic species to plant extremities to increase the chance of contacting a, sometimes phoretic, host. This tends to involve movement in a water film.
3. Infective juveniles of the entomopathogenic *Steinernema* may overcome meniscus forces, standing vertically out from the substrate, and attach to a passing insect, a behaviour termed nictation. Some forms merely crawl, while in others host cues stimulate nictating individuals to jump towards passing hosts (Lewis, 2002).
4. Nematodes associated with plant florets and seeds (e.g. *Anguina, Ditylenchus*) may migrate to germinating seeds and be carried up the growing plant rather than migrating up the plant, in a negative geotropic manner, to the floret itself.
5. A photic response has been reported for *Chromadorina bioculata* but no other definite responses have been related to nematode pigment spots or eye-spots (Croll *et al.*, 1972).

1.2.5. Biogeographical behaviour

Natural processes such as plate tectonics, water currents, winds and animal host dispersal, coupled with human-mediated movement of animals, plants and associated soil and detritus, have all contributed to the present distribution of nematodes. These factors operate on differing time and geographical scales in differing nematode groups. For nematodes inhabiting non-living substrates, the impact of differing distributions of plants, roots and their products on the small-scale distribution patterns of soil nematodes in forest and adjacent pasture has been shown by Ettema and Yeates (2003). Vanreusel *et al.* (1997) have discussed the impact of limited nematode dispersal ability on the assemblages around hydrothermal deep-sea vents.

Assessing the global relationships of soil nematodes is difficult as their study has been unevenly distributed. For example, among terrestrial rhabditids, bacterial-feeding *Rhabditis spiculigera* and *Protorhabditis oxyuroides* were reported only from Europe, *Rhabditis tripartita* from Europe, North America and South America, and *Protorhabditis wirthi* only from New Zealand. In contrast, *Rhabditis marina* was recorded from littoral seaweed deposits in the North Atlantic, South Atlantic and Pacific, while *Rhabditis bengalensis* occurred only in such deposits in the Indian Ocean (Sudhaus, 1974a,b).

Global records of aquatic nematodes have been compiled (Gerlach and Riemann, 1973, 1974) but no comprehensive analysis of their distribution patterns has been made. A more robust systematic base is a prerequisite for such an analysis.

The distribution of cyst-forming heteroderid nematodes is generally related to the evolutionary history of the various taxa, their host plants and the geographical setting. *Cryphodera* occurs only in New Zealand and Australia; *Atalodera* and *Thecavermiculatus* occur in North and South America. Superimposed on such patterns are the effects of human-mediated dispersal, particularly of agricultural pests. *Globodera rostochiensis* and *Globodera pallida* originated in the Andes and have been dispersed with potatoes; soybean cyst nematode, *Heterodera glycines*, was introduced to the USA from Japan (Nickle, 1991). Of 28 species of migratory, root-feeding *Radopholus* most have been recorded only from natural habitats in Australia (19 species) and New Zealand (two species). Some workers regard Australasia as the centre of radiation for this genus (Nickle, 1991). *Radopholus similis* has been widely distributed by humans – especially with banana and citrus; the status of the banana and citrus races is discussed below. Most species of *Xiphinema* have been described from areas formerly comprising Gondwanaland. Speciation appears to have been greatest in Africa and may partly be due to a drier climatic phase in southern Africa following the last ice age, leading to fragmentation of once continuous populations (Hunt, 1993).

Animal-parasitic nematode distribution reflects the distribution of their hosts. The combination of nematode clades, hosts and life histories have led to a wide range of adaptations, including:

1. Drilonematoidea are parasites of the coelomic cavity of annelids. While their distribution is poorly known (Africa, Asia, Australia, New Zealand) it seems that they do not occur in European earthworms.

2. The Camallanidae are parasites of the stomach and intestine of lower, predacious vertebrates, especially freshwater fish. The group is most diverse in the Tropics. Their distribution pattern suggests an Old World Tropics origin, followed by spreading to the New World Tropics and undergoing species radiation (Stromberg and Crites, 1974).

3. Hookworms (Ancylostomatidae) occur as adults in the small intestine of mammals. For dispersal they utilize two bacterial-feeding J1 and J2 stages; the infective J3 is adapted to penetrate the host skin. Fourth-stage and adult hookworms suck blood. The distribution of *Ancylostoma* has been modified by human migration (*A. duodenale*) and by human-mediated transfer of hosts (e.g. *A. caninum*). Adults of the genus *Placoconus* are restricted to racoons, mustelids and bears of the New World (Anderson, 2000).

4. There has been recent radiation of *Cloacina* (Strongyloidea), parasitic in the stomachs of Australian kangaroos and wallabies (Beveridge *et al.*, 2002). Most were found to show a high degree of specificity, occurring in a single or two closely related host species. Most hosts harboured two to four species of *Cloacina* but there were up to 22 nematode species per host. Within a recent macropod clade there was no correlation between the evolutionary age of the host and the number of parasite species present. The degree of parasite diversity provides scope for future studies of the adaptations permitting the coexistence of multiple congeners in parasitic nematodes.

5. The Metastrongylidae are diverse, with about 180 species; adults are confined to mammals. However, in cattle and horses they are replaced by *Dictyocaulus* (Anderson, 2000).

6. The oxyuroids have a simple life history, being transmitted as eggs. The two principal groups, Thelastomatidae and Oxyuridae, develop in invertebrates and vertebrates, respectively.

7. Eggs of Ascaridoidea transmitted in aquatic habitats are generally thin-shelled, whereas those transmitted in terrestrial habitats show the adaptation of thick shells and usually do not hatch until ingested by an intermediate or definitive host (Anderson, 2000).

8. The Crassicaudinae parasitize cetaceans. *Crassicauda boopis* has been recorded from the renal system of fin, humpback and blue whales (Anderson, 2000). The largest known nematode, *Placentonema gigantissima* (females 6.7–8.4 m long, 1.5–2.5 cm wide), occurs in the sperm whale (*Physter catodon*). The size reflects its adaptation to living in, and support from, tissue of the nutrient-rich placenta (Gubanov, 1951).

Global distribution patterns reflect the influence of food, water and temperature on the primeval nematode fauna. Natural and human-mediated processes have produced the present distribution patterns and, within the basic abilities of nematodes, the overwhelming significance of food resources is clear. Nematode distribution changes as the distribution of their food changes.

1.3. Biological Adaptations

1.3.1. Effective number of stages

Most nematodes have four juvenile stages. While most studies focus on the biology and adaptation of post-hatching stages, even the nematode egg shows great adaptation. Eggs are fairly uniform in size, most measuring 20–50 μm by 50–100 μm, despite females ranging in length from 0.3 mm (e.g. *Bunonema, Paratylenchus*) to

8.4 m (*P. gigantissima*). The nematode eggshell is complex, with multiple layers of differing permeability (Bird and Bird, 1991). Changes in eggshell permeability (e.g. *Globodera*) and the presence of an operculum (e.g. *Trichuris*) are important adaptations related to J2 hatching and emergence from the egg. Eggs of animal-parasitic nematodes in particular have adaptive features that aid identification (Chitwood and Chitwood, 1974; Soulsby, 1982), including a single polar operculum (e.g. *Oxyuris*), two polar opercula (e.g. *Trichuris*), an abyss for adhesion to plants (e.g. *Mermis*) and a thick, mamillated uterine layer (e.g. *Safaris*). In contrast, embryos of *Wuchereria* develop with a thin shell or sheath that may be retained as the microfilariae move in the blood or host tissue (Anderson, 2000). Microfilariae develop in the uteri of female nematodes and live in the host blood or skin where they are available to bloodsucking vectors. Such nematodes have, in effect, five juvenile stages. This represents a highly significant adaptation of nematode development and it facilitates the transmission of this group of nematodes, whose 70–80 genera have defined host ranges (Anderson, 2000).

Certain species of *Xiphinema*, *Longidorus*, *Seinura* and *Pristionchus* have an adaptation that involves the loss of a stage, with sexual maturity being achieved after only three juvenile stages (Hunt, 1993; Félix *et al.*, 1999; Yeates and Boag, 2002; 2003). A more common adaptation is moulting in the egg before hatching, again effectively reducing the number of stages. That is, the first moult usually occurs in the egg, with the J2 hatching. Throughout the Oxyuroidea two moults occur in the egg, and there are reports of *Ascaris lumbricioides* and *Ascaris suum* also undergoing two moults in the egg (Anderson, 2000).

Nematode growth curves suggest continuous growth during development, with saltatorial (stepwise) growth of the stylet or buccal cavity between stages (Yeates and Boag, 2003). This saltatorial growth has a parallel in Dyar's rule (Hutchinson *et al.*, 1997), commonly applied to insects in which the number of instars may vary both between and within species (Koning and Jamieson, 2001). Within the phylum Nematoda the occurrence of the putative four juvenile stages is variable. As we have seen above, in some nematode species a stage is omitted while in others it occurs before hatching. Such reduction may occur uniformly in a group (possibly in Tylenchina, Aphelenchina and Oxyuroidea), or in some species of the genus (e.g. *Xiphinema*, *Longidorus*) and, in life history terms, may indicate that sufficient resources may be assimilated in the shorter duration, without larger size, or within the abilities of fewer stages. The addition of sheathed, motile embryos of Onchocercidae (microfilariae) to the life history is apparently an ecological adaptation to allow a more complex life history and utilization of host resources not otherwise available to nematodes. This is an emerging area worthy of further study.

1.3.2. Ongoing adaptation

The adaptation of nematodes to their environment is a continuing process. As biochemical and molecular techniques develop, the impact of human-mediated selection on nematode populations and natural variability is becoming better appreciated. Wild populations of many nematode species inhabiting living tissues may have subtle genetic differences that become apparent when hosts resistant to one nematode population are challenged by another population. A classic example among plant-feeding nematodes is the potato cyst nematode, in which differences between *G. rostochiensis* and *G. pallida* or their pathotypes have now been resolved (Marks and Brodie, 1998). Further, repeated culture of early potatoes led to selection of a rapidly maturing population of *G. rostochiensis* in Scotland (Hominick, 1982). Populations of potato cyst nematode can exist wherever their food resource (potato) can exist. Studies of the burrowing nematode of banana and citrus suggest that the putative species *Radopholus similis* and *R. citrophilus*, while differing genetically in their preferred host plants, are not reproductively isolated (Kaplan *et al.*, 1997). They can be regarded as 'races' of *R. similis*.

Among vertebrate nematodes, a range of Trichostrongylidae species has been distributed with their mammalian hosts and transferred to other domesticated mammals. Such relocation may be associated with changes in reproductive behaviour such as the increased temperature requirements of bacterial-feeding and infective stages of *Teladorsagia circumcincta* in warmer climates (Crofton and Whitlock, 1965). Electrophoresis of isoenzymes has shown the existence of sheep and goat strains of *T. circumcincta* and, subsequently, morphological differences in the dorsal bursal rays have been found (Gasiner and Cabaret, 1996; Gasiner *et al.*, 1997). Species groups, such as *A. lumbricoides* and *A. suum* (from humans and pigs, respectively) and *Haemonchus contortus* and *Haemonchus placei* (from sheep and cattle, respectively), have been discussed by Anderson (2000). Eggs of *N. battus* were regarded as requiring winter chilling to overcome diapause before hatching (Evans and Perry, 1976). However, recent studies indicate that some populations are evolving rapidly. There is evidence that eggs may hatch in the autumn, without diapause, and that *N. battus* may cause clinical nematodiriasis in calves (Armour *et al.*, 1988).

Only after 5 years of culturing bacterial-feeding *Panagrolaimus* on agar plates did Sohlenius (1988) discover that his initial field isolate contained two forms (*P. superbus* and *P. detritophagus*). These apparently coexisted because *P. superbus* dispersed rapidly over the plate whereas *P. detritophagus* tended to aggregate in fresh cultures. Recent molecular and morphological studies have shown that five molecular operational taxonomic units (MOTUs) of *Panagrolaimus* isolated from a single field and indistinguishable by traditional morphological techniques belonged to two reproductively isolated species (Eyualem and Blaxter, 2003).

1.3.3. Populations

Nematode abundance is measured in various ways. Soil and aquatic nematode populations are usually expressed as total numbers of individuals per sample unit, per 100 g substrate or on an area basis (e.g. per m^2). In the case of parasites, estimates of density need to be related to the nematode life cycle, preferably using a quantitative model to link fecundity (egg production) to fertility (mature adults produced) as these provide net changes in abundance from generation to generation. While population models have been developed for some parasitic nematodes (e.g. *G. rostochiensis*, *Trichostrongylus tenuis*) they are not necessarily based on stage-specific recruitment and mortality data for the nematodes themselves. Some models concern nematode impact on host population dynamics and others on host exposure to infective juveniles (Fenton *et al.*, 2002; Nodtvedt *et al.*, 2002; see Boag and Yeates, Chapter 12, this volume). Adaptations include:

1. Large females which produce large numbers of eggs are often associated with high-risk life cycles including factors such as the need to locate, invade and successfully infect widely dispersed hosts. The extreme example may be the 8.4 m long female *P. gigantissima* (32 ovaries), whose eggs must infect another whale.
2. Many nematodes have highly developed eggshells. In some species the outer layers are uterine in origin. At least in the Trichostrongyloidea, Oxyuroidea and Ascaridoidea these layers are considered to be an adaptation providing desiccation protection for the embryo within the egg. Filaments are an adaptation in some marine forms by which eggs are attached to the substrate. Maternal investment in complex eggshells has a metabolic cost; this is often considered to be traded off by production of fewer eggs. Mamillation or numerous small spines occur on the eggshell in many groups of nematodes (e.g. *Ascaris, Plectus, Prionchulus, Diplenteron, Acrobeles*) – whether this is a functional modification or a display of latent ability is unknown.
3. Some females develop into resistant cysts, which provide either protection or a means of dispersal for the next generation (e.g. Heteroderidae).
4. In genera such as *Criconemoides, Syngamus* and *Tetradonema* adaptation of the female for egg production has led to males being significantly smaller than females. In several groups of plant-parasitic nematodes (e.g. Meloidogynidae, Criconematidae) the males of some species appear to be non-feeding (Yeates and Boag, 2003).

A hitherto unexplored dimension of population dynamics is how the rates of population increase are related to food resource distribution in time and space. Rates of population increase have been expressed in terms of the r–K strategy continuum. The concept was tentatively applied to plant-parasitic nematodes (Jones, 1980), and indiscriminately applied across plant and soil nematode families in the maturity index (Bongers, 1990). Differences have been demonstrated within genera (e.g. *Rhabditis mariannae* vs. *Rhabditis synpapillata*) and families (e.g. Trichodoridae, Longidoridae) (Alphey, 1985; Sudhaus, 1985; Hunt, 1993), and assumed relationships have been found to be erroneous for bacterial-feeding soil nematodes (Yeates *et al.*, 2002). Ecologists and animal parasitologists regard the r–K concept as representing a suite of characters that covary across life-cycle stage within (genus level) groups (Skorping *et al.*, 1991; Gems, 2002; Reznick *et al.*, 2002).

Examples of variation in behaviour with food resource in different substrates include:

1. In abyssal, marine sediments occupied by microbial-feeding and predacious nematodes, abiotic (e.g. temperature, sea-water composition) and biotic (e.g. rain of organic detritus, predation) conditions tend to be relatively static over time. Similarly, constant reproductive output should give a stable population; any disturbance of the sediment is likely to lead to passive distribution of nematodes without significant mortality.

2. In temperate natural forest or grasslands there are continuous changes in some abiotic (e.g. temperature, moisture, soil cracking) and many biotic (e.g. root growth, root exudation, leaf litter input, predation) conditions. Under these conditions, species whose life-cycle strategy provides positive responses to changing conditions (e.g. rapid reproduction, resistant stages, dormancy) may be able to exploit more of the resource than other species and thus be at a competitive advantage.

3. Parasites of terrestrial arthropods typically have both free-living and parasitic stages in their life histories to facilitate transfer between hosts. In some species there are separate free-living and parasitic generations, and either may be repeated without recourse to the other.

4. Animal parasites show a diversity of adaptive behaviours. Analyses of monoxenous species fail to clearly separate direct life cycles (e.g. *Enterobius*) from those with bacterial-feeding and infective juveniles (e.g. Trichostrongylidae). These infective juveniles may be ingested and mature in the intestine or infect via the skin and migrate through tissues. Heteroxenous nematodes use one or more intermediate hosts, with such behaviours being particularly common in vertebrate parasites (Anderson, 2000). Variation in strategies may: (i) increase spatial distribution of the nematode; (ii) allow growth and/or moulting in intermediate hosts to permit maturation in the final host; (iii) protect juveniles from adverse conditions; (iv) increase chances of transmission to the primary host; or (v) allow specialization such as arrested development (hypobiosis), precocious development (early growth of genital primordia), paratenic hosts (with minimal development) and autoinfection, by which a population maintains itself within a single host (Anderson, 2000; Lee, 2002).

1.3.4. Behavioural adaptation

Many, if not all, nematode behaviours are innate and it is only those that have been studied that are regarded as behavioural adaptations. As described above, many nematode activities and behaviours are stage-specific. While knowledge is too fragmentary to permit integration, the following describe some of the specific behavioural adaptations recorded in nematodes.

1.3.4.1. Migration to non-specific sites

The origin of nematode aggregations in heterogeneous non-living substrates can be difficult to resolve. However, Moens *et al.* (1999) demonstrated experimentally that, while each of the four marine Monhysterida was attracted to a bacterial strain, responses differed between live and dead bacteria and with bacterial abundance. Some form of chemotaxis clearly underlies the process.

Density-dependent predation is considered by some to demonstrate the importance of chance encounters (e.g. Yeates, 1969), but others suggest aggregations of predacious nematodes occur at sites of earlier predation (Bilgrami and Jairajpuri, 1990; Bilgrami *et al.*, 2001).

1.3.4.2. Migration in response to host stimuli

Root exudates stimulate mobility of nematodes such as *G. rostochiensis* (Clarke and Hennessy, 1984). However, the importance of specific kairomones, in contrast to a general indication of a site of biological activity, such as carbon dioxide (Perry and Aumann, 1998), in inducing behavioural adaptations in nematodes remains unclear.

The jumping response of infective *Steinernema* juveniles to volatile host cues was mentioned earlier. Such ambush behaviour contrasts with cruisers, which move in search of sedentary hosts, possibly following gradients of volatile cues. Once a potential host is contacted recognition behaviours may initiate (Lewis, 2002).

1.3.4.3. Sexual attraction

Females of *Heterodera* and *Globodera* emit sex attractants which modify male movement patterns (Green and Plumb, 1970). Dusenbery and Snell (1995) gave a theoretical body size threshold for the efficacy of such pheromones in a 0.2–5 mm water film, suggesting that pheromone use in nematodes may be marginal. While pellicole nematodes can presumably utilize pheromones in thin films, behavioural adaptations present in the ancestral nematode will affect the potential for such adaptations to be distributed across clades.

1.3.4.4. Enhancing transmission to host

Having developed through the bacterial-feeding J1 and J2 stages in faeces or soil, sheathed infective (J3) juveniles of many Trichostrongylidae crawl up herbage, thus exposing themselves to ingestion by the mammalian herbivores that are their primary hosts.

There are two strains of the filarial nematode *Loa loa*. The strain infecting humans is active in surface blood-vessels during the day, when their insect vector, *Chrysops*, is biting humans. In contrast, the strain infecting monkeys has a nocturnal periodicity, mirroring biting by those *Chrysops* species which attack monkeys in the forest canopy at night (Anderson, 2000).

There has been speculation that the insect nematodes *Iotonchium* and *Sphaerularia* have a negatively geotropic adaptation to crawl up so as to infect their hosts. Careful observations have indicated that the insects become infected when they enter soil containing infective stages – whether or not there is a fungal-feeding generation of the nematode remains undetermined (Poinar and van der Laan, 1972; Tsuda and Futai, 2000).

1.3.5. Adaptations in nematodes for the use of food resources

Developmental stages of nematode species may differ in their habitat type (non-living vs. living substrates) and in the food resource that they use. Plant-parasitic nematodes use a stomatostyle (Tylenchina) or odontostyle (Dorylaimina) to feed on plant cell contents (including any induced gall, syncytium, multinucleate feeding site). The food choices of other nematodes using living substrates (i.e. parasites) are less clear. Those with a simple, undifferentiated stoma and inhabiting the intestines of invertebrates are assumed to be bacterial-feeders. Ascaridoidea usually inhabit the stomach and intestine of their vertebrate host and consume food ingested by the host (Anderson, 2000). *Bradynema,* with no gut but with microvilli on its body wall, is presumed to absorb available nutrients from the intestinal lumen of its insect host (Riding, 1970). Nematodes inhabiting the body cavity of their host are presumably dependent on haemolymph.

A stylet is not necessarily an adaptation for plant feeding. In Tylenchoidea and Aphelenchoidea the stylet may be present in species whose adults parasitize arthropods, and some of these Aphelenchoidea may have both fungal-feeding and arthropod-parasitic stages or life cycles (Hunt, 1993; Siddiqi, 2000). There has long been uncertainty about the food resource for which the relatively delicate spear of tylenchids is adapted (Yeates *et al.*, 1993); evidence is now growing that some are fungal-feeders (Okada *et al.*, 2002), which relates well to their distribution in ecosystems. The Dorylaimina have several genera (e.g. *Tylencholaimus, Doryllium, Leptonchus*) in which fungal-feeding is probably important. Predation on other nematodes occurs in stylet-bearing nematodes in which the lumen is very narrow (0.5 μm) (e.g. *Seinura*) or broad (5 μm) (e.g. *Aporcelaimus*). Further adaptation is found in the ectoparasitic Acugutturidae in which the elongate (50–185 μm) stylet also aids attachment. The stomatostyle penetrates the cuticle of cockroaches and moths, with the nematode apparently feeding on insect haemolymph (Hunt, 1993).

Infective and migrating stages generally increase in body size and must feed. However, as in all nematodes, the most significant food requirement is that of the female for maturation and egg production. Females of *Haemonchus, Necator* and *Ancylostoma* use diverse methods to release and ingest host blood. Whipworms (*Trichuris*) induce syncytium formation and consume externally digested material.

Most adult Metastrongyloidea occupy the lungs of the host, with exceptions occurring in blood vessels (e.g. *Elaphostrongylus*) and frontal sinuses (e.g. *Skrjabingylus*) (Anderson, 2000). Although the actual food resource used is not well documented, in all such locations the nematode will be bathed in fluid relatively rich in nutrients. Oxyuroidea adults are parasites of the intestinal lumen of vertebrates or invertebrates. In this group it is the J3 which hatches and all post-hatching stages are probably bacterial-feeding. The juvenile stages of nematode genera *Ascaris, Ancylostoma* and *Strongylus* migrate through tissues of their mammalian host and yet mature in the part of the body from which migration commenced. Comparison of the growth of similar taxa with and without a juvenile tissue phase has shown that those with the adaptation of tissue migration have bigger and thus more fecund females. The mechanism responsible for the associated growth is unknown (Read and Skorping, 1995).

The first stage of *Trichinella spiralis* lives within a muscle cell and induces formation of a nurse cell; later stages are spent within a group of enterocytes in the small intestine, with the contents of these cells providing the diet (Wright, 1979; Jasmer, 1995). In *Brugia*, the intestine and microvilli seem to be poorly developed in all stages, but transcuticular uptake of nutrients has been shown in infective juveniles and adults (Chen and Howells, 1979).

1.3.6. Morphological adaptations for feeding

The nematode stoma (buccal cavity) shows a wide range of variations and in many cases teeth are an adaptation for obtaining food; smaller denticles if present may contribute to retention or breaking down of food. The simplest form of tooth is a solid, fixed cuticular protuberance in the stoma. As the stoma is triradiate this tooth may occur in any of the three sectors, the particular location being of systematic significance (e.g. *Diplogaster*, *Pareudiplogaster*, *Punctodora*, *Eurystomina*, *Onchulus*). Such teeth, whether fixed or semi-mobile, puncture the food source. Small food items (e.g. protozoa, diatoms) enter the stoma, sufficient lip or pharyngeal suction being used to bring the food against the tooth (Moens *et al.*, 2004). In nematodes with three similar, solid teeth (e.g. *Ethmolaimus*, *Oncholaimus*, *Ironus*) a relationship between the teeth and the successive elements of the stoma (labial cuticle of the cheilostom, gymnostom, pro-, meso-, meta- and telo- components of the metastom, procorpus of the pharynx) (de Ley *et al.*, 1995) is often more obvious and they may be mobile. Observations on their eversion under natural conditions are not consistent. Adult nematodes inhabiting the gut of vertebrates may have well-developed teeth, which may not only wound the host (e.g. subventral teeth or cutting plates and dorsal teeth in *Ancylostoma*) but also be an adaptation to help hold the nematode in place within the host. In *Strongylus* the teeth are set deeper in the stoma and clearly have a different function, whereas in *Haemonchus* the stoma contains a small dorsal lancet, which is an adaptation to cause haemorrhaging wounds. In the Triplonchida the onchiostyle is a dorsal tooth, often associated with complex stomal ribbing or posterior extensions. At least in the Trichodoridae the tooth is used as a pick to penetrate plant cell walls, but as the tooth lacks a pore or lumen there is no food passage (Hunt, 1993).

In two groups, Tylenchina and Enoplida, a tubular hollow stylet, through which all ingested material must pass, has evolved. Such stylets are used by nematodes that feed on fungi, on vascular plants and as predators or parasites of invertebrates and vertebrates. The tylenchid stomatostyle and dorylaimid odontostyle differ in their ontogeny (Bird and Bird, 1991). The posterior portion, which may have knobs or flanges for muscle attachment, is triradiate in origin. Likewise both have an anterior cuticular sheath that provides physical continuity from the cuticle, through the lumen to the pharyngeal lumen. Vertebrate parasites of the orders Trichinellida and Dioctophymatida and the arthropod-parasitizing Mermithida all have a stylet similar to that of the Dorylaimina in having its origin in the subventral sector of the anterior pharynx, in at least the first stage. In *G. rostochiensis*, J2 hatch is preceded by a period of stylet thrusting, making a line of perforations through the shell (Doncaster and Shepherd, 1967). Thus the stylet, a clear morphological adaptation, facilitates both hatching and feeding behaviours.

The effect of bacteria cell concentration, and other factors, on growth and reproduction of bacterial-feeding nematodes has been widely studied (Yeates, 1970; Schiemer, 1983; Venette and Ferris, 1997, 1998). However, apart from relationships between cephalic structure and bacterial intake (de Ley, 1991; de Ley and Coomans, 1997; Moens *et al.*, 2004) few reports mention behavioural adaptations to access bacterial cell contents. Indeed, in *Diplenteron colobocercus* and *Cephalobus nanus* live bacteria are present in faeces (Yeates, 1970; Bird and Ryder, 1993). The physical crushing of bacterial cells between the pharyngeal valve plates of *C. nanus* was described and illustrated by Bird and Ryder (1993). It is not known whether pharyngeal valve plates of all bacterial-feeding nematodes are a similar adaptation for crushing food, or whether it is incidental to their role in pumping.

1.4. Feedback Effects

The interaction between nematodes and their food resource is dynamic. Like all ecological processes there is a series of checks and balances. There may be mutually beneficial interactions. What is beneficial for one species may be detrimental to another. Parasitism is generally detrimental – unless the host population is regarded as a pest and then the consequences of biological control are seen as favourable.

1.4.1. Biological control

The European sawfly *Sirex noctilio*, a significant pest of *Pinus radiata* forests in Australia and New Zealand, can be controlled using the tylenchid nematode *Beddingia* (=*Deladenus*) *siridicola*. *Beddingia siridicola* has biphasic fungal-feeding and insect-parasitic life cycles (Bedding, 1967, 1984). The nematode can persist as a fungal-feeder indefinitely without the insect-parasitic cycle.

The role of entomopathogenic nematodes as biological control agents provides visible evidence of one positive feedback of nematode activity. Nematodes are applied to manage insect pests of cranberries, turf grass, artichokes, mushrooms, ornamentals and many other pests in horticultural, agricultural and urban situations (Gaugler, 2002). Entomopathogenic nematodes such as *Steinernema* and *Heterorhabditis*, do not work alone; their pathogenicity depends in large part on a mutualistic nematode–bacterial complex (with *Xenorhabdus* and *Photorhabdus*, respectively) (Gaugler, 2002). Similarly, slugs such as *Deroceras reticulatum* can be practically managed using the rhabditid nematode *Phasmarhabditis hermaphrodita* and certain associated bacteria (Wilson *et al.*, 1993, 1995).

The cause of regular, cyclic fluctuations of numerous vertebrate populations was not clearly demonstrated until Hudson *et al.* (1998) used the nematocide lev-amisole to control the caecal nematode *T. tenuis* in red grouse, preventing popula-tion crashes. This finding established the important role of a nematode parasite in controlling the population of a vertebrate. The implications are great and, while nematode infections do not underlie all vertebrate population cycles, the role of nematode parasitism in regulating similar vertebrate populations should prove a fruitful area for future research.

1.4.2. Nutrient mineralization

The effects of nematodes on nutrient mineralization result directly from the excre-tion of ingested nutrients not used in tissue production (the ecological growth effi-ciency of nematodes (10%) is far less than that of protozoa (40%)). The effects result indirectly from modification of the microbial community, inoculation of new substrates with microorganisms, accelerated turnover of microbial cells and leakage into the rhizosphere of nutrients from feeding sites (Bardgett and Griffiths, 1997; Griffiths and Bardgett, 1997; Yeates *et al.*, 1999). In a classic paper, Ingham *et al.* (1985) demonstrated that soil microcosms containing *Pelodera* or *Acrobeloides* had higher bacterial densities than similar microcosms without nematodes. Plants growing in such microcosms grew faster and initially took up more nitrogen because of increased nitrogen mineralization by bacteria, NH_4^+-N excretion by nematodes and greater initial exploitation of soil by plant roots. Mineralization due to such activity of microbial grazers was considered to be significant for increasing plant growth when mineralization by microflora alone was insufficient to meet the plant's requirements.

A laboratory study using freshwater sediment showed that bacterial activity was greater in microcosms containing bacterial-feeding nematodes (*Caenorhabditis elegans*) than in those with fungal-feeding nematodes or no nematodes (Traunspurger *et al.*, 1997). The impacts of nematodes on bacterial activity were similar to those found in the soil microcosms described above.

A range of marine nematodes, including *Diplolaimella*, *Diplolaimelloides*, *Geomonhystera*, *Daptonema*, *Sabatiera* and *Enoplus*, make sinusoidal burrows when moving in agar. This behaviour, mimicking interstitial nematodes in the seabed, stimulated bacterial and ciliate populations (Jensen, 1996).

Demonstrating the positive contribution of fungal-feeding nematodes to nutri-ent mineralization in soil was more elusive. However, when substrates with a C:N ratio of 15 were used, Chen and Ferris (1999, 2000) showed an increase in nitro-gen mineralization when fungi were grazed by *Aphelenchus* or *Aphelenchoides*. This finding is consistent with ecological situations in which fungi are known to con-tribute to soil processes.

1.5. Indices of Nematode Assemblages

The abundance of particular parasitic nematodes (e.g. infective juveniles of *G. rostochiensis* and *G. pallida* in potato fields or eggs per gram of faeces of grazing mammals) has long been used to index the condition of agroecosystems. This reflects the negative feedback traditionally associated with parasitic nematodes. Analyses of marine nematode assemblages, which are often very diverse, used a variety of indices in an effort to unravel relationships with environmental factors (Tietjen, 1989; Boucher, 1990; Vanreusel, 1990) but recent analysis has shown that straightforward species accumulation curves are valuable (Lambshead and Boucher, 2003). When it was appreciated that soil nematodes make positive contributions to ecosystem processes, a large literature developed, using both general ecological indices (e.g. Shannon–Weiner diversity (H); species richness (SR) and novel nematological indices (e.g. maturity index; channel index; enrichment index) (Yeates, 1984; Ferris *et al.*, 2001; Neher, 2001).

Recent advances in understanding nematode biology, such as the improved interpretation of feeding types (Moens *et al.*, 2004) and the relative population increases of three genera observed in real soils (Yeates *et al.*, 2002), have cast serious doubt on the utility of nematological indices. The specialized indices may, however, have validity in comparing local nematode assemblages drawn from the same species pool, in which the impacts of assigned values are self-compensating (Hohberg, 2002; Ruess, 2003).

1.6. Use of Plant and Soil Nematodes in Environmental Impact Assessment

Plant and soil nematodes, with generation times measured in days or weeks and ultimate dependence on photosynthetic inputs, integrate many aspects of ecosystem function. Their specific diversity and numerical abundance make them among favoured organisms as indicators of soil health and, consequently, for assessing the environmental impact of natural and anthropogenic changes. While the use of nematodes in such studies must be constrained by the caveats in Section 1.5, there is validity in comparing local nematode assemblages drawn from the same species pool. Thus the impact of agricultural, horticultural and forestry practices, including cultivar, nutrient and biocide use, under uniform climatic and soil conditions can be assessed by examining their effects on nematode assemblages – that is, by measuring the adaptation of the local nematode fauna (Šály, 1985; de Goede *et al.*, 1993; Freckman and Ettema, 1993; Hohberg, 2002; Wall *et al.*, 2002). It must be recognized that many impacts are indirect, acting through food webs, soil structure and both quantitative and qualitative changes in inputs. Also, indices are information-rich, incorporating predetermined rankings and ratings that may disguise adaptations within, for example, a given feeding group.

The use of a specific nematode for assessing impact gives a very narrow index and results must be interpreted in that light. Changes in populations of potato cyst nematodes *(G. rostochiensis, G. pallida)* are primarily controlled by growth, and degree of growth, of a suitable potato crop. Use of laboratory cultures of microbial-feeding nematodes such as *C. elegans* and *Cruznema tripartitum* (Donkin and Dusenbery, 1993; Lau *et al.*, 1997; Dhawan *et al.*, 2000) in bioassays assumes that only the variable being tested affects the population; many other variables, such as the ability to feed on the microbes present (Venette and Ferris, 1998; Ruess *et al.*, 2000), tend to be ignored.

1.7. Conclusions and Future Directions

Nematodes use a wide range of adaptations and behaviours to exploit diverse food resources in non-living and living substrates, with cephalic morphology indicating resource use. Availability of food resources and free water, coupled with a suitable temperature, constitute the principal factors that control nematode abundance and distribution. Some nematode groups have developed in temperate climates, in tropical climates or in association with host groups; such factors underpin the natural distribution patterns of nematodes. While many species show only size differences among developmental stages, other species show great adaptation, not only of feeding type but also of activity and tropisms, from stage to stage; these are the keys to nematode diversity. The critical importance of these stage-specific differences reduces our ability to generalize about the significance of particular factors influencing the life cycles, abundance and diversity of nematodes. In ecological terms, nematodes may have significant effects on plant nutrient cycling and in regulating the populations of both harmful and beneficial organisms (biological control and parasitism, respectively).

Perhaps the greatest areas for improved understanding of nematode adaptations and behaviours lie in considering stages of disparate species that have similar environmental requirements. Just as adult gastrointestinal and plant-feeding nematodes have been widely studied, all those stages that, for example, feed on bacterial cells, feed on fungal hyphae or resist desiccation have much in common. How their behaviours correlate with those of *C. elegans*, whose behaviours in the field are virtually unknown, may tell much about the generality of nematode biology. The more we understand nematode biology the better we shall be able to understand and manage nematode populations and their effects.

References

Alphey, T.J.W. (1985) A study of spatial distribution and population dynamics of two sympatric species of trichodorid nematodes. *Annals of Applied Biology* 107, 497–509.

Anderson, R.C. (2000) *Nematode Parasites of Vertebrates: Their Development and Transmission*, 2nd edn. CAB International, Wallingford, UK, 650 pp.

Armour, J., Bairden, K., Dalgleish, R., Ibarra-Silva, A.M. and Salman, S.K. (1988) Clinical nematodiriasis in calves due to *Nematodirus battus* infection. *Veterinary Record* 123, 230–231.

Ashton, F.T., Li, J. and Schad, G.A. (1999) Chemo- and thermosensory neurons: structure and function in animal parasitic nematodes. *Veterinary Parasitology* 84, 297–316.

Bardgett, D.R. and Griffiths, B.S. (1997) Ecology and biology of soil protozoa, nematodes, and microarthropods. In: van Elas, J.D., Trevors, J.T. and Wellington, E.M.H. (eds) *Modern Soil Microbiology*. Marcel Dekker, New York, pp. 129–163.

Bedding, R.A. (1967) Parasitic and free-living cycles in entomogenous nematodes of the genus *Deladenus*. *Nature, London* 214, 174–175.

Bedding, R.A. (1984) Nematode parasites of Hymenoptera. In: Nickle, W.R. (ed.) *Plant and Insect Nematodes*. Marcel Dekker, New York, pp. 755–795.

Beveridge, I., Chilton, N.B. and Spratt, D.M. (2002) The occurrence of species flocks in the nematode genus *Cloacina* (Strongyloidea: Cloacininae), parasitic in the stomachs of kangaroos and wallabies. *Australian Journal of Zoology* 50, 597–620.

Bilgrami, A.L. and Jairajpuri, M.S. (1990) Aggregation of *Mononchoides longicaudatus* and *M. fortidens* (Nematoda: Diplogasterida) at feeding sites. *Nematologica* 34, 119–121.

Bilgrami, A.L. Pervez, R., Yoshiga, T. and Kondo, E. (2001) Attraction and aggregation behaviours of predatory nematodes, *Mesodorylaimus bastiani* and *Aquatides thornei* (Nematoda: Dorylaimida). *Applied Entomology and Zoology* 36, 243–249.

Bird, A.F. and Bird, J. (1991) *The Structure of Nematodes*, 2nd edn. Academic Press, San Diego, 316 pp.

Bird, A.F. and Ryder, M.H. (1993) Feeding of the nematode *Cephalobus nanus* on bacteria. *Journal of Nematology* 25, 493–499.

Bloemers, G.F., Hodda, M., Lambshead, P.J.D., Lawton, J.H. and Wanless, F.R. (1997) The effects of forest disturbance on diversity of tropical soil nematodes. *Oecologia* 111, 575–582.

Bongers, T. (1990) The maturity index: an ecological measure of environmental disturbance based on nematode species composition. *Oecologia* 83, 14–19.

Boucher, G. (1990) Pattern of nematode species diversity in temperate and tropical subtidal sediments. *Marine Ecology* 11, 133–146.

Chen, J. and Ferris, H. (1999) The effects of nematode grazing on nitrogen mineralization during fungal decomposition of organic matter. *Soil Biology and Biochemistry* 31, 1265–1279.

Chen, J. and Ferris, H. (2000) Growth and nitrogen mineralization of selected fungi and fungal-feeding nematodes on sand amended with organic matter. *Plant and Soil* 218, 91–101.

Chen, S.H. and Howells, R.E. (1979) The uptake *in vitro* of dyes, monosaccharide and amino acids by the filarial worm *Brugia pahangi*. *Parasitology* 78, 343–354.

Chitwood, B.G. and Chitwood, M.B. (1974) *Introduction to Nematology*. University Park Press, Baltimore, Maryland, 334 pp.

Clarke, A.J. and Hennessy, J. (1984) Movement of *Globodera rostochiensis* (Wollenweber) juveniles stimulated by potato-root diffusate. *Nematologica* 30, 206–212.

Cobb, N.A. (1915) Nematodes and their relationships. *Yearbook of the United States Department of Agriculture* 1914, 457–490.

Crofton, H.D. and Whitlock, J.H. (1965) Ecology and biological plasticity of sheep nematodes. III. Studies on *Ostertagia circumcincta* (Stadelman, 1894). *Cornell Veterinarian* 55, 259–262.

Croll, N.A., Riding, I.L. and Smith, M.J. (1972) A nematode photoreceptor. *Comparative Biochemistry and Physiology* 40A, 999–1009.

de Goede, R.G.M., Verschoor, B.C. and Georgieva, S.S. (1993) Nematode distribution, trophic structure and biomass in a primary succession of blown-out areas in a drift sand landscape. *Fundamental and Applied Nematology* 16, 525–538.

de Ley, P. (1991) The nematode community of a marginal soil at Cambérène, Senegal, with special attention to functional morphology and niche partitioning in the family Cephalobidae. *Academiae Analecta Mededelingen van de Konnklijke Acadamie voor Wetenschappen, Letteren en Schone Kunsten van België* 53, 108–153.

de Ley, P. and Coomans, A. (1997) Terrestrial nematodes from the Galápagos Archipelago. 7. Description of *Tylocephalus nimius* sp. n. and new data on the morphology, development and behaviour of *T. auriculatus* (Bütschli, 1873) Anderson, 1996 (Leptolaimida: Plectidae). *Fundamental and Applied Nematology* 20, 213–228.

de Ley, P., van de Velde, M.C., Mounport, D., Baujard, P. and Coomans, A. (1995) Ultrastructure of the stoma in Cephalobidae, Panagrolaimidae and Rhabditidae, with a proposal for a revised stoma terminology in Rhabditida (Nematoda). *Nematologica* 41, 153–182.

Dhawan, R., Dusenbery, D.B. and Williams, P.L. (2000) A comparison of metal-induced lethality and behavioural responses in the nematode *Caenorhabditis elegans*. *Environmental Toxicology and Chemistry* 19, 3061–3067.

Doncaster, C.C. and Shepherd, A.M. (1967) The behaviour of second-stage *Heterodera rostochiensis* larvae leading to their emergence from the egg. *Nematologica* 13, 476–478.

Donkin, S.G. and Dusenbery, D.B. (1993) A soil toxicity test using the nematode *Caenorhabditis elegans* and an effective method of recovery. *Archives of Environmental Contamination and Toxicology* 25, 145–151.

Dusenbery, D.B. and Snell, T.W. (1995) A critical body size for use of pheromones in mate location. *Journal of Chemical Ecology* 21, 427–438.

Eskin, R.A. and Coull, B.C. (1987) Seasonal and three-year variability of meiobenthic nematode populations at two estuarine sites. *Marine Ecology – Progress Series* 41, 295–303.

Ettema, C.H. and Yeates, G.W. (2003) Nested spatial biodiversity patterns in a New Zealand forest and pasture soil. *Soil Biology and Biochemistry* 35, 339–342.

Evans, A.A.F. and Perry, R.N. (1976) Survival strategies in nematodes. In: Croll, N.A. (ed.) *The Organisation of Nematodes*. Academic Press, London, pp. 383–424.

Eyualem, A. and Blaxter, M. (2003) Comparison of biological, molecular and morphological methods of species identification in a set of cultured *Panagrolaimus* isolates. *Journal of Nematology* 35, 119–128.

Félix, M.-A., Hill, R.A., Schwarz, H., Sternberg, P.W., Sudhaus, W. and Sommer, R.J. (1999) *Pristionchus pacificus*, a nematode with only three juvenile stages, displays major heterochronic changes relative to *Caenorhabditis elegans*. *Proceedings of the Royal Society, London B* 266, 1617–1621.

Fenton, A., Gwynn, R.L., Gupta, A., Norman, R., Fairbairn, J.P. and Hudson, P.J. (2002) Optimum application strategies for entomopathogenic nematodes: integrating theoretical and empirical approaches. *Journal of Applied Ecology* 39, 481–492.

Ferris, H., Bongers, T. and de Goede, R.G.M. (2001) A framework for soil food web diagnostics: extension of the nematode faunal analysis concept. *Applied Soil Ecology* 18, 13–29.

Freckman, D.W. and Ettema, C.H. (1993) Assessing nematode communities in agroecosystems of varying human intervention. *Agriculture, Ecosystems and Environment* 45, 239–261.

Gasiner, N. and Cabaret, J. (1996) Evidence for the existence of a sheep and a goat line of *Teladorsagia circumcincta* (Nematoda). *Parasitology Research* 82, 546–550.

Gasiner, N., Cabaret, J. and Durette-Desset, M.C. (1997) Sheep and goat lines of *Teladorsagia circumcincta* (Nematoda): from isoenzyme to morphological identification. *Journal of Parasitology* 83, 527–529.

Gaugler, R. (ed.) (2002) *Entomopathogenic Nematology.* CAB International, Wallingford, UK, 388 pp.

Gems, D. (2002) Ageing. In: Lee, D.L. (ed.) *The Biology of Nematodes.* Taylor and Francis, London, pp. 413–455.

Gerlach, S.A. and Riemann, F. (1973) The Bremerhaven checklist of aquatic nematodes. *Veröffentlichungen des Instituts für Meeresforschung in Bremerhaven, Supplement* 4, 1–404.

Gerlach, S.A. and Riemann, F. (1974) The Bremerhaven checklist of aquatic nematodes. *Veröffentlichungen des Instituts für Meeresforschung in Bremerhaven, Supplement* 4, 405–736.

Green, C.C. and Plumb, S.C. (1970) The interrelationships of some *Heterodera* spp. indicated by the specificity of the male attractants emitted by their males. *Nematologica* 16, 39–46.

Griffiths, B.S. and Bardgett, R.D. (1997) Interactions between microbe-feeding invertebrates and soil micro-organisms. In: van Elas, J.D., Trevors, J.T. and Wellington, E.M.H. (eds) *Modern Soil Microbiology.* Marcel Dekker, New York, pp. 165–182.

Gubanov, N.M. (1951) [A giant nematode from the placenta of cetaceans – *Placentonema gigantissima* n.g., n.sp.] *Doklady Akademii Nauk USSR* 77, 1123–1125.

Hentschel, U., Berger, E.C., Bright, M., Felbeck, H. and Ott, J.A. (1999) Metabolism of nitrogen and sulfur in ectosymbiotic bacteria of marine nematodes (Nematoda, Stilbonematinae). *Marine Ecology – Progress Series* 183, 149–158.

Hodda, M. and Wanless, F.R. (1994) Nematodes from an English chalk grassland: species distribution. *Nematologica* 40, 116–132.

Hohberg, K. (2002) Soil nematode fauna of afforested mine sites: genera distribution, trophic structure and functional guilds. *Applied Soil Ecology* 22, 113–126.

Hominick, W.M. (1982) Selection of a rapidly maturing population of *Globodera rostochiensis* by continuous cultivation of early potatoes in Ayrshire, Scotland. *Annals of Applied Biology* 100, 345–351.

Hudson, P.J., Dobson, A.P. and Newborn, D. (1998) Prevention of population cycles by parasite removal. *Science* 282, 2256–2258.

Hunt, D.J. (1993) *Aphelenchida, Longidoridae and Trichodoridae: Their Systematics and Bionomics.* CAB International, Wallingford, UK, 352 pp.

Hutchinson, J.M.C., McNamara, J.M., Houston, A.I. and Vollrath, F. (1997) Dyar's Rule and the investment principle: optimal moulting strategies if feeding rate is size dependent and growth is discontinuous. *Philosophical Transactions of the Royal Society of London, B* 352, 113–138.

Ingham, R.E., Trofymow, J.A., Ingham, E.R. and Coleman, D.C. (1985) Interactions of bacteria, fungi and their nematode grazers: effects on nutrient cycling and plant growth. *Ecological Monographs* 55, 119–140.

Jasmer, D.P. (1995) *Trichinella spiralis:* subversion of differentiated mammalian skeletal muscle cells. *Parasitology Today* 11, 185–188.

Jensen, H.J., Armstrong, J. and Jatala, P. (1979) *Annotated Bibliography of Nematode Pests of Potato.* International Potato Center, Lima, 315 pp.

Jensen, P. (1996) Burrows of marine nematodes as centres for microbial growth. *Nematologica* 42, 320–329.

Jones, F.G.W. (1977) Temperature and the development of *Mononchus aquaticus. Nematologica* 23, 123–125.

Jones, F.G.W. (1980) Some aspects of the epidemiology of plant-parasitic nematodes. In: Patti, J. and Kranz, J. (eds) *Comparative Epidemiology – a Tool for Better Disease Management.* Pudoc, Wageningen, pp. 71–92.

Kaplan, D.T., Vanderspool, M.C. and Opperman, C.H. (1997) Sequence tag site and host range assays demonstrate that *Radopholus similis* and *R. citrophilus* are not reproductively isolated. *Journal of Nematology* 29, 421–429.

Koning, J.W. and Jamieson, I.G. (2001) Variation in size of male weaponry in a harem-defence polygynous insect, the mountain stone weta *Hemideina maori* (Orthoptera: Anostostomatidae). *New Zealand Journal of Zoology* 28, 109–117.

Lambshead, P.J.D. and Boucher, G. (2003) Marine nematode deep-sea biodiversity – hyperdiverse or hype? *Journal of Biogeography* 30, 475–485.

Lau, S.S., Fuller, M.E., Ferris, H., Venette, R.C. and Scow, K.M. (1997) Development and testing of an assay for soil ecosystem health using the bacterial-feeding nematode *Cruznema tripartitum*. *Ecotoxicology and Environmental Safety* 36, 133–139.

Lee, D.L. (2002) Life cycles. In: Lee, D.L. (ed.) *The Biology of Nematodes*. Taylor and Francis, London, pp. 61–72.

Lewis, E.E. (2002) Behavioural ecology. In: Gaugler, R. (ed.) *Entomopathogenic Nematology*. CAB International, Wallingford, UK, pp. 205–223.

Luther, G.W., Rozan, T.F., Taillefert, M., Nuzzio, D.B., Di Meo, C., Shank, T.M., Lutz, R.A. and Cary, S.C. (2001) Chemical speciation drives hydrothermal vent ecology. *Nature, London* 410, 813–816.

McKenna, P.B. (1997) Checklist of helminth parasites of terrestrial mammals in New Zealand. *New Zealand Journal of Zoology* 24, 277–290.

Marks, R.J. and Brodie, B.B. (1998) *Potato Cyst Nematodes: Biology, Distribution and Control*. CAB International, Wallingford, UK, 424 pp.

Moens, T., Verbeeck, L., de Maeyer, A., Swings, J. and Vincx, M. (1999) Selective attraction of marine bacterivorous nematodes to their bacterial food. *Marine Ecology – Progress Series* 176, 165–178.

Moens, T., Yeates, G.W. and de Ley, P. (2004) Use of carbon and energy sources by nematodes. *Nematology Monographs and Perspectives* (in press).

Neher, D.A. (2001) Role of nematodes in soil health and their use as indicators. *Journal of Nematology* 33, 161–168.

Nickle, W.R. (ed.) (1991) *Manual of Agricultural Nematology*. Marcel Dekker, New York, 1035 pp.

Nodtvedt, A., Dohoo, I., Sanchez, J., Conboy, G., DesCoteaux, L., Keefe, G., Leslie, K. and Campbell, J. (2002) The use of negative binomial modelling in a longitudinal study of gastrointestinal parasite burdens in Canadian dairy cows. *Canadian Journal of Veterinary Research* 66, 249–257.

Okada, H. and Ferris, H. (2001) Effect of temperature on growth and nitrogen mineralization of fungi and fungal-feeding nematodes. *Plant and Soil* 234, 253–262.

Okada, H., Tsukiboshi, T. and Kadota, I. (2002) Mycetophagy in *Filenchus misellus* (Andrássy, 1958) Lownsbery and Lownsbery, 1985 (Nematoda: Tylenchida), with notes on its morphology. *Nematology* 4, 795–801.

Orr, C.C. and Dickerson, O.J. (1967) Nematodes in true prairie soils of Kansas. *Transactions of the Kansas Academy of Science* 68, 317–334.

Perry, R.N. and Aumann, J. (1998) Behaviour and sensory responses. In: Perry, R.N. and Wright, D.J. (eds) *The Physiology and Biochemistry of Free-living and Plant-parasitic Nematodes*. CAB International, Wallingford, UK, pp. 75–102.

Perry, R.N. and Wright, D.J. (eds) (1998) *The Physiology and Biochemistry of Free-living and Plant-parasitic Nematodes*. CAB International, Wallingford, UK, 438 pp.

Poinar, G.O. and van der Laan, P.A. (1972) Morphology and life history of *Sphaerularia bombi*. *Nematologica* 18, 239–252.

Read, A.F. and Skorping, A. (1995) The evolution of tissue migration by parasitic nematode larvae. *Parasitology* 111, 359–371.

Reznick, D., Bryant, M.J. and Bashey, F. (2002) *r*- and *K*-selection revisited: the role of population regulation in life-history evolution. *Ecology* 83, 1509–1520.

Riding, I.L. (1970) Microvilli on the outside of a nematode. *Nature, London* 226, 179–180.

Ruess, L. (2003) Nematode soil faunal analysis of decomposition pathways in different ecosystems. *Nematology* 5, 179–181.

Ruess, L., Garcia Zapata, E.J. and Dighton, J. (2000) Food preferences of a fungal-feeding *Aphelenchoides* species. *Nematology* 2, 223–230.

Šály, A. (1985) Production of free living nematodes in the protected landscape area of the Slovak Paradise. *Ekológia (ČSSR)* 4, 185–209.

Schiemer, F. (1983) Comparative aspects of food dependence and energetics of free-living nematodes. *Oikos* 4, 32–42.

Schjetlein, J. and Skorping, A. (1995) The temperature threshold for development of *Elaphostrongylus rangiferi* in the intermediate host – an adaptation to winter survival. *Parasitology* 111, 103–110.

Siddiqi, M.R. (2000) *Tylenchida: Parasites of Plants and Insects*, 2nd edn. CAB International, Wallingford, UK, 834 pp.

Skorping, A., Read, A.F. and Keymer, A.E. (1991) Life history covariation in intestinal nematodes of mammals. *Oikos* 60, 365–372.

Sohlenius, B. (1988) Interactions between two species of *Panagrolaimus* in agar cultures. *Nematologica* 34, 208–217.

Soulsby, E.J.L. (1982) *Helminths, Arthropods and Protozoa of Domesticated Animals*, 7th edn. Baillière, Tindall, London, 809 pp.

Stromberg, P.C. and Crites, J.L. (1974) Specialization, body volume and geographical distribution of Camallanidae (Nematoda). *Systematic Zoology* 23, 189–201.

Sudhaus, W. (1974a) Regarding the systematics, distribution, ecology and biology of new and lesser-known Rhabditidae (Nematoda). Part I. *Zoologische Jahrbucher Systematik* 101, 173–212.

Sudhaus, W. (1974b) Regarding the systematics, distribution, ecology and biology of new and lesser-known Rhabditidae (Nematoda). Part II. *Zoologische Jahrbucher Systematik* 101, 407–465.

Sudhaus, W. (1985) Comparative studies on life cycles and reproductive strategies of two species of the nematode genus *Rhabditis* (*Teratorhabditis*). *Zoologische Jahrbucher Systematik* 112, 455–468.

Thiermann, F., Vismann, B. and Giere, O. (2000) Sulphide tolerance of the marine nematode *Oncholaimus campylocercoides* – a result of internal sulphur formation. *Marine Ecology – Progress Series* 193, 251–259.

Tietjen, J.H. (1989) Ecology of deep-sea nematodes from the Puerto Rico Trench area and Hatteras Abyssal Plain. *Deep-Sea Research* 36, 1579–1594.

Tietjen, J.H., Demig, J.W., Rowe, G.T., Macko, S. and Wilke, R.J. (1989) Meiobenthos of the Hatteras Abyssal Plain and Puerto Rico Trench: abundance, biomass and associations with bacteria and particulate fluxes. *Deep-Sea Research* 36, 1567–1577.

Traunspurger, W., Bergtold, M. and Goedkoop, W. (1997) The effects of nematodes on bacterial activity and abundance in a profundal freshwater sediment. *Oecologia* 112, 118–122.

Trudgill, D.L. (1995) Host and temperature effects on nematode development rates and nematode ecology. *Nematologica* 41, 398–404,

Tsuda, K. and Futai, K. (2000) The insect-parasitic stage and life cycle of *Iotonchium ungulatum* (Tylenchida: Iotonchidae), the causal agent of gill-knot disease of the oyster mushroom. *Japanese Journal of Nematology* 30, 1–7.

Vanreusel, A. (1990) Ecology of the free-living marine nematodes from the Voordelta (Southern Bight of the North Sea). I. Species composition and structure of the nematode communities. *Cahiers de Biologie Marine* 31, 439–462.

Vanreusel, A., Vandenbossche, I. and Thiermann, F. (1997) Free-living marine nematodes from hydrothermal sediments – similarities with communities from diverse reduced habitats. *Marine Ecology – Progress Series* 157, 207–19.

Venette, R.C. and Ferris, H. (1997) Thermal constraints to population growth of bacterial-feeding nematodes. *Soil Biology and Biochemistry* 29, 63–74.

Venette, R.C. and Ferris, H. (1998) Influence of bacterial type and density on population growth of bacterial-feeding nematodes. *Soil Biology and Biochemistry* 30, 949–960.

Wall, J.W., Skene, K.R. and Neilson, R. (2002) Nematode community and trophic structure along a sand dune succession. *Biology and Fertility of Soils* 35, 293–301.

Wilson, M.J., Glen, D.M., George, S.K. and Butler, R.C. (1993) Mass culture and storage of the rhabditid nematode *Phasmarhabditis hermaphrodita*, a biocontrol agent for slugs. *Biocontrol Science and Technology* 3, 513–521.

Wilson, M.J., Glen, D.M., George, S.K. and Pearce, J.D. (1995) Selection of a bacterium for the mass production of *Phasmarhabditis hermaphrodita* (Nematoda, Rhabditidae) as a biocontrol agent for slugs. *Fundamental and Applied Nematology* 18, 419–425.

Winslow, R.D. (1960) Some aspects of the ecology of free-living and plant-parasitic nematodes. In: Sasser, J.N. and Jenkins, W.R. (eds) *Nematology: Fundamentals and Recent Advances with Emphasis on Plant-parasitic and Soil Forms*. University of North Carolina Press, Chapel Hill, pp. 341–415.

Wright, K.A. (1979) *Trichinella spiralis*: an intracellular parasite in the intestinal phase. *Journal of Parasitology* 65, 441–445.

Yeates, G.W. (1969) Predation by *Mononchoides potohikus* (Nematoda: Diplogasteridae) in laboratory culture. *Nematologica* 15, 1–9.

Yeates, G.W. (1970) Studies on laboratory cultures of dune sand nematodes. *Journal of Natural History, London* 4, 119–136.

Yeates, G.W. (1984) Variation in soil nematode diversity under pasture with soil and year. *Soil Biology and Biochemistry* 16, 95–102.

Yeates, G.W. and Boag, B. (2002) Post-embryonic growth of longidorid nematodes. *Nematology* 4, 883–889.

Yeates, G.W. and Boag, B. (2003) Growth and life histories in Nematoda, with particular reference to environmental factors. *Nematology* 5, 653–664.

Yeates, G.W. and Boag, B. (2004) Background for nematode ecology in the 21st century. In: Chen, Z.X., Chen, S.Y. and Dickson, D.W. (eds) *Nematology, Advances and Perspectives*, Vol. 1. Tsinghua University Press, China and CAB International, Wallingford, UK, pp. 406–437.

Yeates, G.W., Bongers. T., de Goede, R.G.M., Freckman, D.W. and Georgieva, S.S. (1993) Feeding habits in soil nematode families and genera – an outline for soil ecologists. *Journal of Nematology* 25, 315–331.

Yeates, G.W., Saggar, S., Hedley, C.B. and Mercer, C.F. (1999) Increase in ^{14}C-carbon translocation to the soil microbial biomass when five plant-parasitic nematodes infect roots of white clover. *Nematology* 1, 295–300.

Yeates, G.W., Dando, J.L. and Shepherd, T.G. (2002) Pressure plate studies to determine how moisture affects access of bacterial-feeding nematodes to food in soil. *European Journal of Soil Science* 53, 355–365.

2 Locomotion Behaviour

A.H. Jay Burr[1] and A. Forest Robinson[2]

[1]Department of Biological Sciences, Simon Fraser University, Burnaby, BC, V5A 1S6, Canada; [2]USDA ARS, Southern Crops Research Laboratory, 2765 F&B Road, College Station, TX 77845, USA

2.1. Introduction

By far the most characteristic traits of nematodes are their extremely narrow stream-lined body and undulatory style of locomotion, useful in their common burrowing habit. These traits have enabled them to be successful in an amazingly wide range of free-living and parasitic environments that is without parallel in other meiofauna. This review examines what is known of the mechanism of this locomotion and its adaptations to various environments in the light of their unique body architecture and neuromuscular system.

Nematodes typically move from place to place by an undulatory propulsion that is distinctly different from that utilized by other animals. It has been described and analysed in considerable detail (Harris and Crofton, 1957; Wallace, 1968, 1969; Croll, 1976; Lee and Atkinson, 1977; Bird and Bird, 1991; Burr and Gans, 1998; Alexander, 2002). Nematode forward undulatory locomotion involves stiff body waves that propagate backward from the anterior end and exert propulsive forces against the environment. This is achieved through dorsoventral bending rather than lateral bending, as in fishes and snakes, and without a backbone. Occasionally, nematodes move backwards by means of forwardly propagated waves. Such reversals usually stop within one body length or so, and the nematode then heads off forward in a new direction, sometimes involving an omega turn (Croll, 1970).

Undulatory propulsion in nematodes is modified or supplemented with other movements in ways that enable nematodes to crawl on or swim through diverse natural surfaces, media, channels and orifices, for example leaf surfaces, intestinal microvilli, plant hairs, water, blood, excreta, rotting plant matter, animal tissues, blood vessels, insect tracheae and stomata. Nematodes can accelerate, turn, stop, reverse, omega-turn, probe, orient, swim, burrow, penetrate, poke, lace, climb, bridge, roll, graze, cruise, nictate, aggregate, swarm, ambush, hitchhike, loop, creep or somersault. They do not fly, but some can leap nine body lengths in the presence of a potential insect host (Campbell and Kaya, 1999a,b) or crawl on to a *Pilobolus* sporangium about to be shot for distances of up to 2.5 m towards sunlit grass where ungulate hosts feed (Robinson, 1962).

Thus, locomotion in nematodes can be examined in relation to at least six questions:

1. How do nematodes achieve and control the bending of their bodies internally?
2. How does bending of the body gain purchase and provide propulsive forces against external substrates?
3. How is undulatory propulsion controlled by the neuromuscular system?
4. How is bending modified to maximize purchase and propulsion?
5. How have nematodes adapted undulatory propulsion to different environments to achieve specific goals?
6. What are the constraints and advantages of alternative means of locomotion in nematodes that do not involve undulatory propulsion?

2.2. Internal Mechanics

In a recent review on locomotion, Alexander (2002) focused on mathematical and computer models that describe nematode crawling and swimming motions. The present section takes a complementary approach, examining the internal mechanisms behind this motion. Modelling involves assumptions, model formulation and comparing predictions with observations. This approach is useful for conceptualizing complex mechanisms and usually leads to the design of new experiments.

However, this approach has not yet led to new experiments in nematode locomotion. Here we focus on the assumptions, specifically the evidence for a hydrostatic skeleton, the mechanism of bending, how the application of force to objects can propel the worm and the neural control of the body-wall muscles involved. The explanation of the hydroskeleton and bending will be mostly intuitive, and the evidence will necessarily be mostly anecdotal because of limited experimental support. We hope that the deficiencies exposed, questions raised and our radical ideas will stimulate new research.

2.2.1. Role of hydrostatic skeleton

Many invertebrate phyla include organisms whose worm-like form is maintained by a hydrostatic skeleton – an elastic body wall surrounding a constant volume of fluid and/or soft tissues under pressure (Chapman, 1958). As for a water-filled balloon, the elastic properties of the body wall determine the resting shape. In a cylindrical balloon, the membrane is stiffer in the circumferential than the axial direction, whereas in a spherical balloon the elastic moduli are isotropic.

In many nematodes, when the body wall is punctured there can be a dramatic effusion of fluid and viscera. Pseudocoelomic pressures from 16 to 225 mmHg were measured in *Ascaris* (Harris and Crofton, 1957). However, note that some nematodes appear not to have such internal pressure (Wright, 1991).

If there were no change in the amount of fluid inside the body cavity, the volume of the body cavity would remain constant, because liquids are incompressible. If more fluid is added to the body cavity, increasing its volume, the internal pressure should increase as the elastic layer is stretched to greater tension. A muscular oesophagus (= pharynx) pumps fluid and food into the intestine against the internal pressure. The energy required to pump in more fluid is stored in the elastic elements of the body wall and would be released by defecation. Defecation causes a short-term decrease in pressure because of the subsequent period of lowered volume (Harris and Crofton, 1957). Over the longer term, nematode internal volume and pressure would be governed by rates of osmoregulation, oesophageal pumping, excretion and defecation (Mapes, 1965; Davey, 1995).

The nematode body wall is composed of a cuticle and a single layer of longitudinal muscle cells (Fig. 2.1). These are separated by a thin sheet of hypodermis (= epidermis) that expands medially at intervals around the circumference to form the hypodermal cords. These divide the muscles into four to eight longitudinal bands, depending on species (Fig. 2.1). The body cavity is internal to the basal lamina that covers the muscle and hypodermal cells, and contains pseudocoelomic fluid, intestine and gonads. Since the intestine and gonads have soft, non-muscular walls, their contents are probably at the same pressure as the pseudocoelomic fluid.

If the body wall were a thin membrane, the cylindrical shape should collapse when pressure is released by puncturing the body wall. However, in nematodes the cylindrical shape of the body tube is maintained when punctured and its diameter changes only slightly. Therefore the nematode body wall must have a high

A   B

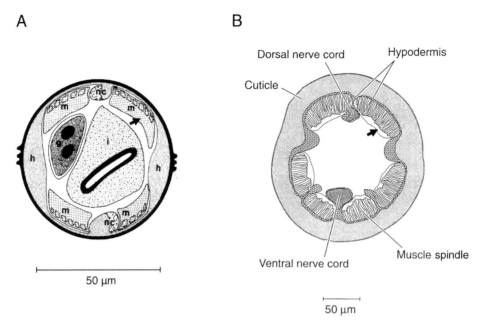

Fig. 2.1. Two examples of body-wall structure in nematodes. A. Hermaphrodite of *C. elegans*. Expansions of the hypodermis (h), called hypodermal cords, contain nerve cords (nc) and separate the muscles (m) into two dorsal and two ventral bands. A thin sheet of hypodermis extends between the muscle spindles (small rectangles) and cuticle. The intestine (i) and gonads (g) lie inside the body cavity. (From Wood, 1988.) B. Female *Mermis nigrescens*. In this much larger nematode (120 mm long) the cuticle and muscle spindles are thicker. The body cavity is usually filled with eggs and a modified gut (trophosome) containing stored food (not shown). Arrow, muscle belly. (From Gans and Burr, 1994.)

circumferential stiffness and special properties that maintain the circular cross-section in the absence of internal pressure.

Mechanically, the nematode body wall consists of the multilayered cuticle and the longitudinal strap-like contractile region (spindle) of the muscle cells (Figs 2.1 and 2.2). The sarcomeres consist of a band of thick and thin myofilaments between two cylindrical dense bodies, which are equivalent to the Z lines of vertebrate muscle (see the complete sarcomere in the tangential section of the coelomyarian example, Fig. 2.2). A series of sarcomeres are aligned in columns parallel to the body axis; however, neighbouring columns of sarcomeres are offset to form an obliquely striated pattern in sections perpendicular to the dense bodies (Fig. 2.2).

Nematode muscle is unlike any other muscle in that each sarcomere is connected perpendicularly to the site of force application, the cuticle. The myofilaments of each sarcomere are attached via the dense bodies to the basal lamina (Fig. 2.2), which in turn is fixed to the cuticle across the sheet-like extension of hypodermal cell via short intermediate filaments and half-desmosomes (Francis and Waterston, 1985, 1991; Bartnik *et al.*, 1986; Waterston, 1988; Barstead and Waterston, 1989; Moerman and Fire, 1997). This architecture transmits contrac-

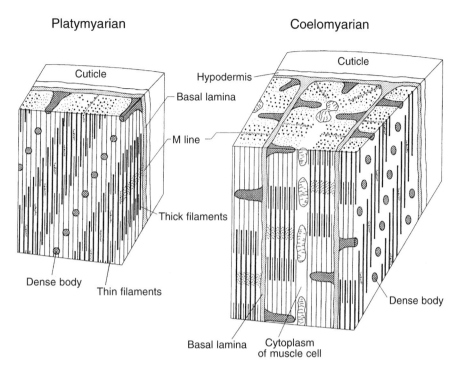

Fig. 2.2. Contractile architecture of the two main types of body-wall muscle: platymyarian (example, *Caenorhabditis elegans*) and coelomyarian (example, *Mermis nigrescens*). The sarcomeres are attached to the basal lamina by dense bodies (Z-line equivalents). In coelomyarian muscle, the basal lamina forms a ribbon between the folded plates of spindle. In both types, the basal lamina is attached to the cuticle across the thin sheet of hypodermis. For clarity, the scale is shortened in the vertical direction and only two thin filaments are shown in the tangential section of *C. elegans* and none are shown in the equivalent radial section of *Mermis*. (From Burr and Gans, 1998.)

tile force laterally to the cuticle rather than serially to the muscle ends, and thus provides for smooth bending and avoids buckling of the body tube (Burr and Gans, 1998). In the coelomyarian muscle type found in large nematodes such as *Ascaris* and *Mermis nigrescens*, the spindle is folded around an inward extension of the basal lamina (Figs 2.2 and 2.3). This ribbon of basal lamina would need to provide a stiff connection between the dense body attachment sites and the part of the basal lamina that is fastened to the cuticle.

Nematodes are unique among worm-like organisms in lacking circumferential muscles; therefore their motions are limited to what can be accomplished by applying longitudinal forces to the body wall. In other worms, contraction of circumferential muscles, thereby decreasing the circumference of one region as in squeezing a water-filled balloon, can displace a large volume from that region and expand the body elsewhere (Chapman, 1958). Thus crawling, as in earthworms, by peristaltic extension-with-thinning and contraction-with-thickening (Quillin, 1999) is not

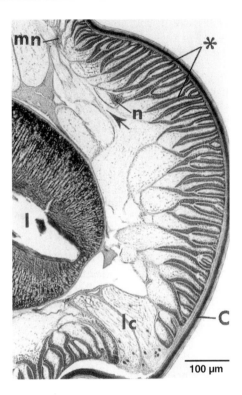

Fig. 2.3. Cross-section through mid-body of a large ascarid, *Toxascaris* sp., showing coelomyarian muscle spindles, muscle bellies and an innervation process (arrow), which fill space in the pseudocoelom not occupied by the intestine (I) and median and lateral cords (mn and lc). C, cuticle; n, nucleus; *, darkly stained myofilaments in folded muscle spindle. (From Wright, 1991.)

possible in nematodes. The peristaltic locomotion of *Criconemoides* is unusual for nematodes and involves relatively small changes in circumferential and axial dimensions (Stauffer, 1924; Streu *et al.*, 1961).

It is commonly assumed that the pressurized body cavity plays an essential role in bending motion, with muscle contraction shortening one side of the body tube while pressure maintains the length of the other. Then, when the muscle relaxes, the internal pressure is thought to straighten the tube, providing the 'restoring force'. These assumptions have been incorporated in several mathematical models of nematode locomotion (Harris and Crofton, 1957; Seymour, 1983; Alexander, 1987, 2002). However, these assumptions are based on only one experimental study, which established that the pseudocoelom is under a pressure that is uniformly distributed throughout the body length (Harris and Crofton, 1957). The role of internal pressure in bending has never been directly investigated in nematodes, nor has the alternative possibility that elastic compression of the cuticle provides all or part of the restoring force (as was assumed in the model developed by Niebur and Erdos (1991)). This lack of experimental investigation of the hydrostatic skeleton and its role in body movements is surprising in view of the keen interest in nematode function spawned by genetic, developmental and molecular studies of *Caenorhabditis elegans*. In contrast, there has been extensive work on hydroskeleton and locomotion in the leech (Miller, 1975; Stern-Tomlinson *et al.*, 1986; Wadepuhl and Beyn, 1989) and the earthworm (Quillin, 1999).

2.2.2. Role of cuticle

If one compares the body motion of a nematode with that of a fly larva or earthworm, the comparative stiffness of the nematode body is very apparent. This could be partly due to the much greater thickness of the cuticle relative to body diameter, the ratio being about 1:30, but it can be 1:5 to 1:3 in *M. nigrescens* (Fig. 2.1B) and 1:100 in adults of large ascarids (Fig. 2.3; Bird and Bird, 1991; Gans and Burr, 1994). It could be due also to special elastic properties of the cuticle with or without internal hydrostatic pressure. When stripped of muscle, basal lamina and epidermis, the cuticle of *C. elegans* maintains a cylindrical shape in aqueous suspension (Cox *et al.*, 1981a); however, it may require muscle tonus and/or internal pressure to maintain this adequately against external forces.

The nematode cuticle has been described at the electron microscope level in many species and is the subject of many reviews (Wharton, 1986; Wright, 1987, 1991; Bird and Bird, 1991; Kramer, 1997; Baldwin and Perry, 2004). A wide variation in cuticle structure, even in different juvenile stages of a species, probably represents important adaptations to the widely diverse environments in which nematodes are found (Inglis, 1964; Edgar *et al.*, 1982; O'Grady, 1983; Fetterer and Urban, 1988; Baldwin and Perry, 2004). However, the basic structure is a thin (6–40 nm) smooth epicuticle covering three thick layers (Figs 2.4 and 2.5B): the cortex, median zone and basal zone (Wright, 1991). These are constructed primarily, of protein secreted by the underlying epidermis during both moulting and growth (Bird, 1980; O'Grady, 1983; Wright, 1987). Here, we shall focus on features that may be significant for mechanical function and locomotion.

The cortex generally consists of one or two layers of relatively homogeneous and electron-dense material digestible by elastase (Fujimoto and Kanaya, 1973; Cox *et al.*, 1981a). Together with its outer location, apparent flexibility (Fig. 2.5) and fine structure, this suggests that the cortex provides a tough but flexible contact with the environment.

The median zone is mechanically very compliant (Fig. 2.5B) and may contain a fluid. In the simplest cases it consists of an electron-lucent region crossed by fibrils connecting the cortex and basal zone. In a number of nematode examples the median zone contains radial struts that join the cortex and basal zone through a fluid layer (Figs 2.4 and 2.5B; Inglis, 1964; Lee, 1965; Wright and Hope, 1968; Baldwin and Perry, 2004).

The basal zone is the layer most directly connected to the body-wall muscle and is where the circumferential stiffness of the body wall most probably resides. A loosely organized network of fibres connects the inner surface of the cuticle with a

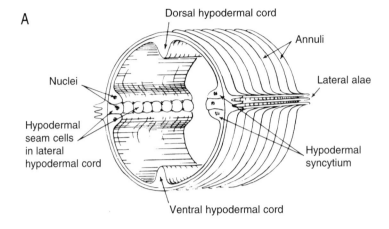

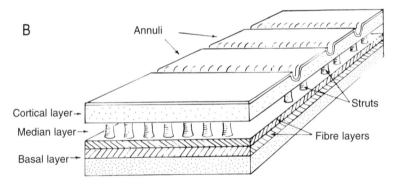

Fig. 2.4. Cuticle and hypodermis in *Caenorhabditis elegans* hermaphrodite. A. General organization of cuticle and hypodermis. B. Structure of cuticle. (From Cox *et al.*, 1981a.)

stiff layer, which occurs in three different forms, depending on species and life stage: (i) two or three layers of helical fibres that cross at characteristic angles, (ii) a layer of electron-dense striated material (Fig. 2.5B); or (iii) a multilayered, non-helical fibrous material curving in a scalloped pattern.

In addition to the basic three-layered structure covering most of the circumference of the cuticle, lateral longitudinal ridges are found in certain species or life stages. Examples include the external alae of *C. elegans* hermaphrodites or the internal ridges in mermithids (Figs 2.1A,B and 2.4). Unlike in the rest of the cuticle, positive birefringence and electron micrographs indicate that longitudinal fibres are present (Cox *et al.*, 1981a,b; Edgar *et al.*, 1982). These structures are likely to bias bending motions, stiffening the body against lateral bending while allowing dorsoventral bending.

How are the different cuticle layers affected during locomotion? The cuticle is compressed on the inside of a bend by the contraction of the sarcomeres and it may

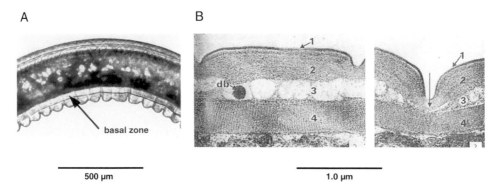

Fig. 2.5. Folding of cortex of compressed cuticle. A. Corrugations form on inside of bend of *Mermis nigrescens* female. The basal zone is thicker on compressed side. Optical section of whole mount, using differential interference contrast. B. Annuli in *Heterodera* juvenile. Right, fold along annule indents the fluid-filled medial layer (arrow). 1, epicuticle; 2, cortex; 3, median zone containing struts and fluid-filled space; 4, striated basal zone; db, dense ball. (From Wisse and Daems, 1968.)

be under tension on the outside. The cortex on the inside of the bend may fold to form large (50–150 µm) corrugations (Fig. 2.5A), or the 0.3–2.0 µm spaced annuli of the cortex (Fig. 2.4) may become indented (Fig. 2.5B). On the other hand, the cortex on the outside of the bend remains smooth and may be relatively stiff against extension (Rosenbluth, 1967; Wright and Newall, 1976; Robinson and Carter, 1986; Gans and Burr, 1994). The basal zone, on the other hand, yields to compression by becoming thicker (Fig. 2.5A; Fig. 1 in Rosenbluth, 1967). The spongy median zone appears to provide the internal compliance necessary for the two layers to react differently to mechanical stress (Fig. 2.5B). The struts or fibres of the median zone would transfer forces across the fluid-filled layer to the cortex and external objects during locomotion. The fluid in the median zone could flow circumferentially to the outside of the bend, increasing flexibility by decreasing cuticular compressive resistance on the inside of the bend. Thus the cuticle appears to be well adapted to yielding to axial compression; however, on extension, the cuticle response may be complex. The compliant basal zone and folding cortex may allow bending in the normal range, but we suspect that the cortex may strongly resist elongation at higher extensions.

The cuticle must protect against potentially harmful mechanical forces. Much as for human skin, a tough outer layer (epicuticle and cortex) would protect against abrasion, while a compliant medial zone may dissipate local stresses. Compliance of the medial zone may protect the inner layers from damage as the body pushes pass unyielding soil particles during burrowing. The stiff basal zone may resist indentation of the body cavity and provide resistance to shear forces. As noted above, its chief importance is probably provision of circumferential stiffness while allowing axial compression during bending.

2.2.3. Body stiffness and the role of muscle tonus

When a nematode stops crawling on a surface or swimming, its body remains in its undulatory bent shape. When *C. elegans* is picked up with a wire pick, its body appears quite stiff against the applied forces. This short-term resistance to bending forces is especially noticeable in *Mermis* because of its 100 mm length. *Mermis* can stiffen its body enough to cantilever more than a third of its body during head-elevation bouts (Fig. 2.6A) and when bridging gaps in vegetation (Fig. 2.7; Gans and Burr, 1994).

The nematode body wall, in addition, transmits forces axially. When nematodes burrow through agar, soil or host tissues, the propulsive forces that are developed in posterior regions (see below) must be transmitted to the anterior where the greatest resistance to motion would be experienced. This is easily observed in *Mermis* as it pokes its head into felt and pushes through fibres in the surface (Fig. 2.6A). More than 10 mm of body can intervene between the tip and the nearest site of force generation, and body stiffness appears to be essential to transmit the force (Gans and Burr, 1994). In all nematode locomotion, propulsion depends on the transmission of forces generated by muscular contraction in one region to an axially distant contact site (see below) and this would be impossible if the body were limp. Furthermore, each body region must be able to resist mild tensile and compressive stresses as it is either pulled or pushed through the medium by other body regions.

How, then, does the body provide axial stiffness while allowing locomotory bending? Dropping *Ascaris* on to a table causes a sudden straightening and stiffening, probably by stimulating muscle contraction equally in the four quadrants. In contrast, when a nematode is anaesthetized, the body becomes limp. These and other anecdotal observations suggest that muscle tonus is involved. The advantage of using muscle tonus to stiffen the body is that it can be adjusted locally on

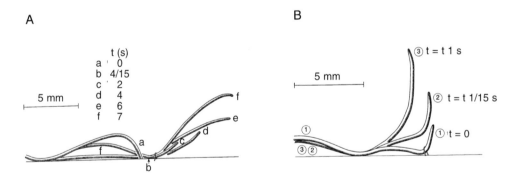

Fig. 2.6. Axial transmission of forces and body stiffness during locomotion of *Mermis nigrescens*. A. Sequence illustrating poking (a) and lacing motion (b–c), followed by application of propulsive force to a fibre loop (e–f) and a head-elevation bout (f) involving cantilevering. B. Tracings from video sequence of a fibre loop breaking due to application of force by the anterior. Position 1/15 second after the break is the result of the force applied to the fibre loop by the anterior and the force on the body transmitted from the posterior (see Fig. 2.12). (From Gans and Burr, 1994.)

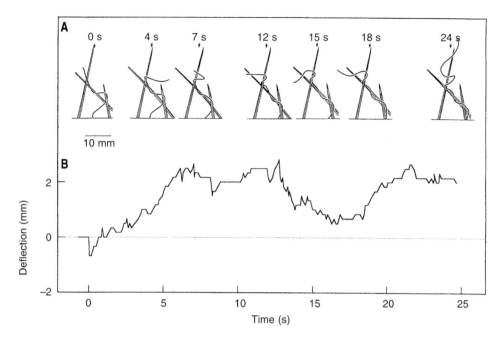

Fig. 2.7. Application of force to a blade of grass by *Mermis nigrescens* female. A. Successive body positions traced from videotaped images. Arrow, original blade position. B. Plot of the deflection of the blade as the body slides past. The deflection would be proportional to the force the blade applies to the nematode – specifically, to the component of this force in the plane of the image and perpendicular to the blade. (From Gans and Burr, 1994.)

command to provide the desired mechanical result within the cuticle's normal length range. The stiffening would be similar to that caused by increasing the tension in opposing muscles of vertebrate body trunks.

In an attempt to determine the contribution of muscle tonus to stiffness in *Ascaris suum*, one of us (AHJB) measured resistance to elongation before and after injection of γ-aminobutyric acid (GABA). This inhibitory neurotransmitter is known to relax *Ascaris* muscle (Davis and Stretton, 1989), and, when present, the slope of length/tension curves was much lower (A.H.J. Burr, unpublished results). Since the slope is proportional to stiffness it is clear that the resistance of the body wall to elongation is due primarily to muscle tonus.

The axial force due to tonus would increase internal pressure, and muscle tonus can be modulated to regulate internal pressure. A momentary higher pressure is produced by contraction of the body-wall muscles during the defecation cycle of *C. elegans* (Thomas, 1990). Harris and Crofton (1957) recorded a rhythmic variation in pressure in the posterior half of *Ascaris* that was correlated with rhythmic changes in posterior length.

Normally, muscle tension is opposed by the hydrostatic skeleton, and a decrease in pressure caused by volume reduction would lower this opposing force and cause shortening of the body tube until pressure returns to the level presumably regulated by tonus. Reducing *Ascaris* volume by 24% (1.43 ml) by withdrawing pseudocoelomic fluid caused a 15% body shortening. Under applied tension, the slope of the length/tension curve (stiffness) was the same even though body length was 15% less (A.H.J. Burr, unpublished results). This result seems to support the idea that body-wall stiffness and internal pressure may be regulated by muscle tonus.

A marine *Enoplus* responded to loss of volume in hypersaline solutions by shortening the middle part of the body to the point that the cuticle became corrugated, and they were able to locomote normally while other species could not (Inglis, 1964). Thus, shortening the body may be an important way for nematodes to compensate for fluid loss.

Apparently shortening does not compromise muscle function. The ability for nematode muscle to operate normally over a large range of lengths could be made possible by the unusually long myofilaments found in nematode muscle (Rosenbluth, 1965; Waterston, 1988). Thick and thin filaments are 10 and 6 μm long compared with 1.6 and 1.0 μm in vertebrate cross-striated muscle. Rosenbluth (1967) proposed that the large operating range is due to shearing between the offset columns of sarcomeres found in obliquely striated muscle (Fig. 2.2). However, unlike in obliquely striated smooth muscle of other organisms, where the shearing concept originated, in nematode body-wall muscle every sarcomere is attached by dense bodies to the basal lamina and cuticle, whose elastic properties would prevent shearing. Oblique striation might instead be an adaptation for distributing the force application sites of the sarcomeres more uniformly over the basal lamina and cuticle (Burr and Gans, 1998).

2.2.4. Mechanism of bending

What causes bending during undulatory locomotion and other nematode activities? With constant volume and constant elastic moduli, the shape of a hydrostatic skeleton can be changed only by an asymmetrical application of forces. For example, if the membrane on one side of a water-filled cylindrical balloon is shortened longitudinally by pinching it, this external force on the elastic membrane will bend the tube. Similarly, compression of the nematode body wall by contraction of muscle bands on one side would bend the body tube. In the presence of muscle tonus, bending can equally be caused by relaxation of muscle cells on one side, a possibility that appears not to have been considered previously. In nematodes, muscles of the body wall are controlled by inhibitory as well as excitatory neuromuscular synapses (see Section 2.4 on neural control below) and it is possible that relaxation as well as contraction, on opposite sides of the body, could be responsible for bending.

The anecdotes described in the preceding section illustrate that there may be an equilibrium between the force exerted by body-wall muscles and an opposing force due to the pressurized hydroskeleton. If muscles in all four quadrants contract, as during defecation, muscle tension would exceed the opposing force and the body tube would shorten, thus increasing pressure until the forces rebalance. Activating muscles in all four quadrants in a short segment of the body tube would compress the body tube locally, and the increase in local pressure would be transmitted through fluid inside the body cavity to other parts of the body tube. There, the hydroskeletal force would exceed muscle force and would cause elongation until opposing elastic forces build up there. When the local muscles relax, the process would reverse, the net force (the 'restoring force') stretching the local body wall, decreasing local pressure and moving fluid back from other parts of the body. During the process, some internal fluid would be translocated from one region to another. Transfer of gut contents can often be observed (Seymour, 1973; Thomas, 1990).

In the case of bending, the net force due to increased muscle tension would shorten only one side of the local body tube, while the length on the other side would be maintained by the equilibrium of forces there. The increase in local pressure and the axial volume transfer could be relatively small and may even be negligible if the local muscles on the opposite side relax at the same time. When the muscle tension is released, a restoring force would straighten the body tube.

Until this point, our development of this idea has assumed that the restoring force arises from the hydrostatic skeleton under pressure, as when the force pinching the water-filled balloon is released. In nematodes the elastic membrane is commonly assumed to lie in the cuticle and the pressurized internal fluid to be the pseudocoelomic fluid plus the fluid contents of the soft internal organs of the body cavity. However, there are two other possible origins of restoring force that have not been investigated: (i) pressurized cytoplasm of the muscle bellies, and (ii) the compressed cuticle. Note that the proportion of the body cross-section taken up by muscle bellies and cuticle thickness varies strikingly among nematode examples (Figs 2.1 and 2.3) and among different body regions of a nematode (see below), and therefore it is likely that the relative contribution of the three possible sources of restoring force could vary.

The necessity of internal pseudocoelomic pressure to bending was investigated in *Ascaris* (A.H.J. Burr and R. Davis, unpublished results). Only the anterior third of this species contributes to locomotion; the posterior regions filled with gonads appear to be dragged along by the anterior. Pseudocoelomic volume in the anterior is small, as evident in cross-sections such as Fig. 2.3 and also shown by injecting 0.1 ml dye solution into the head of intact *Ascaris* – the solution passed right through the active zone. Transecting the worm posterior to this active region or inserting a cannula into the pseudocoelom just posterior to the region did not prevent normal locomotion. The cannula evidently had successfully equalized the pressure between the pseudocoelom and the exterior, because *Ascaris* injected with Ringer's solution into the head passed through the active region and immediately exited via the cannula. Thus, a pressurized pseudocoelom is not necessary to provide a force opposing muscle tension during locomotory bending, although the results do not rule out that it may contribute if present.

This surprising result indicates that there may be some other restoring force opposing muscle tension in *Ascaris*. It would most probably arise by compression of the cytoplasm of the large muscle belly when the contractile region of the same cell shortens. Surrounded by an elastic basal lamina, the pressurized muscle bellies could serve as an intracellular hydroskeleton that could provide the necessary restoring force. Nematode basal lamina contain cross-linked type IV collagen fibres (Kramer, 1997). In the oesophagus, a basal lamina provides the elastic restoring force for the pumping action (Roggen, 1973; Saunders and Burr, 1978). Muscle bellies occupy a large proportion of the cross-section internal to the muscle spindles in the active zone of *Ascaris* (Rosenbluth, 1965) and in *Toxascaris* (Fig. 2.3) and *Nippostrongylus* (Wright, 1991). When the muscle contracts, the muscle bellies become tightly pressed together and expanded medially (Fig. 1 in Rosenbluth, 1967) and extension of muscle spindles and bellies on the outside of the bend would provide space for this locally. The pressurized cytoplasm would not be translocated out of the local region; however, the expanded muscle bellies could displace some pseudocoelomic fluid or gut contents.

Seymour (1973) suggested that the thickening of the muscle layer caused by local contraction in all four quadrants creates radial forces that can displace gut contents in small nematodes such as *Aphelenchoides blastophthorus*. When some marine species such as *Oncholaimus oxyuris* and two monhysterid species are transected, there is no explosive release of fluid and locomotion continues (Coomans *et al.*, 1988; Van de Velde and Coomans, 1989). Large vacuoles in the hypodermal cords of these species are suggested to be a local hydrostatic compartment.

The third possible elastic restoring force, provided by the compressed cuticle, was proposed by Wisse and Daems (1968), and it deserves investigation. This system would be like a bent section of thick-walled latex tubing, which tends to straighten even in the absence of internal pressure. We have noted above that the basal zone is compressed on the inside of a bend, becoming thicker (Fig. 2.5A), and that fluid in the intermediate zone could translocate to the outside of the bend, expanding the cuticle there.

2.2.5. Model of crossed helical fibres

The elastic properties of the nematode body wall – high circumferential stiffness with longitudinal flexibility – occur regardless of which type of basal zone is present in the cuticle. However, the interesting geometrical structure of the crossed-helical fibre layers in one type has attracted attention. Harris and Crofton (1957) proposed a mechanical model and provided a mathematical formula that governs the geometrical relationship between fibre crossing angle and body length, diameter, volume and pressure. They did not directly measure the relationships to crossing angle, however. By measuring the length and volume changes in the normal physiological range and applying the model to these data, they predicted a fibre crossing angle of 73°. As this turned out to be very close to that measured in sections of fixed worms, they proposed that the crossed fibre array is the prime determinant of body dimensions in *Ascaris*. The model predicts that for geometrical reasons alone, and in the absence of any elastic stiffness in the cuticle, an increase in pressure would lengthen the body tube rather than increase its circumference.

Harris and Crofton (1957) recognized that their model was an oversimplification because it considers as insignificant any elastic component that may be contributed by other components of the cuticle, including the matrix in which the fibres are embedded. In addition, the model assumes that the collagen fibres are rigid, contrary to findings in other organisms (Shadwick and Gosline, 1985; Wainwright, 1988). Further experiment on the elastic properties of nematode cuticle is required along the lines of investigation of the mechanical properties of blood vessels, in which stress/strain functions are measured at different inflation pressures (Shadwick and Gosline, 1985). Since the fibre crossing angles in *Ascaris* can be measured *in situ* with a microscope, changes in fibre crossing angles can be measured at the same time. Until such measurements are done, the Harris and Crofton model should be considered speculative.

Alexander (1987, 2002) extended the crossed-helical fibre model to include bending; however, there is no experimental evidence. Indeed, the measurement of length and volume changes that predicted a crossing angle were done by Harris and Crofton (1957) on the posterior third of *Ascaris*, which does not contribute to locomotion and where bending motion is slight. Their observations of pressure fluctuations with a 30 second periodicity cannot be related to locomotory bending, which is faster, and the phenomenon which produces this periodicity is unexplained.

Wainright (1988) suggests several other possible functions for helical fibre structures in animals: preventing bulging under pressure, providing resistance to buckling under external axial loads (such as when pushing through soil or tissue) and allowing smooth bending without kinking. These functions would be important in all nematodes, and the other two types of basal zone structure must somehow provide these as well.

2.3. Transfer of Forces and Propulsion

Locomotion in nematodes, as in snakes and spermatozoa, is usually accomplished by travelling waves of bending that apply forces to external objects or fluids as the body passes by. Force transmission from sarcomere to contact point at the surface of the cuticle traverses many structural components of the body wall. The forces would be transmitted across the compliant medial zone of the cuticle via fibres or struts. Smoothness of the epicuticle and lubrication provided by the surface coat are probably important in reducing friction. The hydrophilic surface coat must also set up the aqueous surface tension that causes adherence to a moist substrate important in transmitting propulsive forces to the substrate in the case of nematodes crawling on thin moisture films (Wallace, 1959, 1968, 1969).

It is evident that understanding undulatory propulsion has two levels: the neuromuscular mechanisms that generate the waves of forces within the body, and the relationship of these and the body form to the external forces that drive the nematode forward (Gray, 1953). The first level will be treated in the next section. The

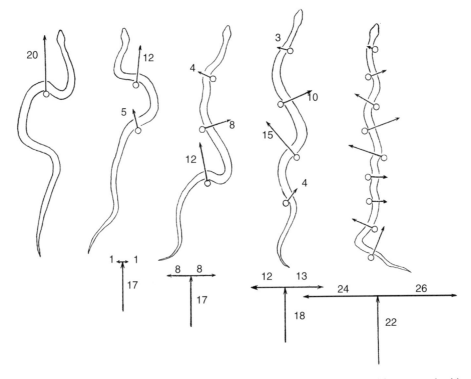

Fig. 2.8. Reactive forces (arrows) exerted by a series of objects as a result of forces applied by a grass snake. The vectors beneath each record show the sums of the forward or lateral components of the forces. (From Gray and Lissman, 1950.)

current section will connect muscle tension to the forces that propel the nematode during undulatory locomotion. The more complete treatments by Gray (1953) and Wallace (1968) would be required reading for anyone interested in investigating the topic further.

Measurement of the forces against external objects would be difficult to determine with most nematodes, though conceivable with the larger species. However, the pattern of forces would be similar to that measured on snakes. Figure 2.8 shows the reactive forces exerted by objects on to a grass snake that are equal and opposite to the forces applied by the body during undulatory propulsion. Each of the vectors can be resolved into one component in the forward direction, the propulsive thrust, and one in a lateral direction, the restraining force, and these can be summed to give the overall effect on the centre of mass (Fig. 2.8). Note that the lateral components cancel and there is a net forward propulsive force exerted on the snake. If the net propulsive force equals the net frictional force (not shown), the speed will be constant.

As the body slides past an object, each region follows the same sinuous path as that preceding it, and curvature of the region and the force applied by it to the object changes to that formerly applied by the region immediately ahead of it. This

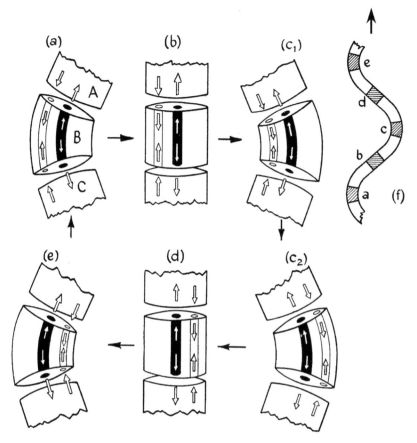

Fig. 2.9. Changes in form of an actively bending region (B) and the forces exerted on its neighbours A and C (labelled in panel a). The complete bending cycle a–e is marked along the undulatory wave in panel f. Arrow, direction of locomotion. Panels a–c_1 illustrate the bending sequence and bending couple when tension is applied on the left, and panels c_2–e illustrate the effect when applied on the right. (From Gray, 1953.)

should result in a continuous application of force to each object as the body glides by, and a continuous propulsive force applied to the body. The three-dimensional movement of an *M. nigrescens* female climbing among grass blades is more complex; however the effect of force on one of the grass blades could be observed and is seen to be continuous, though variable, as the body glides by (Fig. 2.7B). The net effect of such forces from many such contacts is the continuous upward locomotion indicated by the sequential body positions (Fig. 2.7A).

Figure 2.9 illustrates how muscle tension on one side of a local body region (thick arrows within segment) and a restoring force supplied by the hydrostatic skeleton (thin arrows in segment) act together as a 'bending couple' to change the curvature. While the muscle tension tends to shorten the cuticle on one side, the restoring force resists shortening of the centre of the tube. During forward locomotion (arrow in Fig. 2.9f), body region *a* acquires the curvature of region *b* and then *c*. For these changes in curvature to occur, the muscle tension in each region must be on the left side, as illustrated in Figs 2.9a, b and c_1. Thus, the regions along this half wave can be regarded as a single propulsive unit. Figure $2.9c_2$, d and e illustrate the bending sequence in segments of the second half wave that is generated by applying muscle tension to the right side. Note by comparison with Fig. 2.9f, that muscle tension is always on the forward side of each half wave. The same effect is achieved if the medial restoring force due to a hydrostatic skeleton is replaced by a lateral restoring force applied by a compressed cuticle.

How such internally generated local bending couples can apply forces to external objects is illustrated in Fig. 2.10. Muscle tension (shading) is on the forward side of each half wave, as in Fig. 2.9. In order to think about a half wave as an iso-

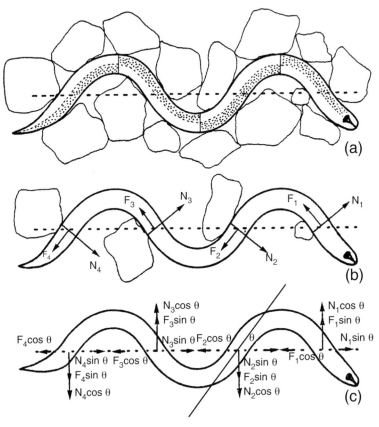

Fig. 2.10. Propulsive forces in the presence of external restraints. Each half wave marked in (a) can be regarded as a locomotory unit. Shaded areas indicate on which side muscle tension is developed. (b) A further simplification, isolating one restraining object in each half wave. Muscle tension in each half wave exerts a backward pressure on each restraint, and the normal reactive forces N_i and tangential frictional forces F_i are shown. (c) Vectors shown are the components of the N_i and F_i along axis of progression and axes perpendicular to it. (From Wallace, 1968, and based on Gray, 1953.)

lated propulsive unit, however, it must be recognized that lateral movement of both ends of the half wave is restrained by transmission of forces from preceding and following half waves. In the presence of these restraints and the external objects at the mid-point of the half waves (Fig. 2.10b), the result of shortening the forward side of each half wave causes the body to press against the objects. Note that, as in Fig. 2.8, the reactive force, N_i, exerted by each object on to the body is normal to the body surface and has a component in the forward direction. As the body slides by each object, a frictional force, F_i, develops tangential to the body. In Fig. 2.10c the forces applied by the four objects are resolved into components along the axis of progression and components transverse to it. The sum of the transverse components, $\Sigma N_i \cos \theta + \Sigma F_i \sin \theta$, is zero and forward propulsive forces, $N_i \sin \theta$, are observed to arise from the bending couples in each half wave.

When the body is moving at a constant speed, the sum of propulsive forces acting on the nematode, $\Sigma N_i \sin \theta$, is equal and opposite to the sum of frictional forces, $\Sigma F_i \cos \theta$. From this equality an important relationship can be derived between total frictional force, F, and total propulsive force, N, at constant speed: $F = N \tan \theta$. This can explain changes in wave-form that are observed when friction increases, as when the surface of agar dries (Fig. 2.11). According to the equation, increases in friction can be overcome by increasing either the amount of muscle tension in order to increase N, or by increasing θ, the angle the body makes at mid-wave with the axis of progression (Fig. 2.10c). The latter can be accomplished by decreasing wavelength and/or increasing amplitude; note that *Heterodera schachtii* decreases wavelength on dry agar (Fig. 2.11).

Thus Fig. 2.10 shows how the forces against external objects can develop from the bending couples of Fig. 2.9 during undulatory locomotion. Except on agar, which provides a uniform field of external constraints, wave-forms are not likely to be symmetrical as in Fig. 2.10. External constraints vary in position and different parts of the body contribute unequally to the propulsive force. Note also that the two examples on the left of Fig. 2.8, where propulsive forces arise from one or two objects, are not typical sinusoidal undulatory locomotion. Interestingly, the body patterns resemble what develops when *M. nigrescens* obtains propulsive force from a single fibre (Fig. 2.6). In this case, tension generated by dorsal muscles would apply forces to three restraining points (Fig. 2.12): the fibre loop, the substrate and an arbitrary point on the body through which force is transmitted from posterior regions. The change in body curvature 1/15 second after the fibre breaks (Fig. 2.6B) reflects these forces. How these restraining forces can sum to provide that propulsive force is diagrammed in Fig. 2.12.

2.4. Neuromuscular Control

Figure 2.10a is a snapshot of the body wave at one instant in time. For undulatory locomotion to occur the motor nervous system must accomplish the following:

1. Create a pattern of tension and relaxation somewhat like that in Fig. 2.10a by

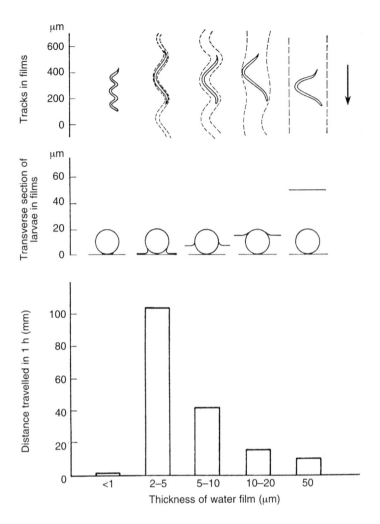

Fig. 2.11. The relationship between thickness of water films on agar, wave-form and speed in *Heterodera schachtii*. For thicknesses greater than 2 μm, speed is inversely related to slippage; however, for thicknesses below 1 μm, friction slows locomotion and wavelength decreases. (From Wallace, 1958.)

exciting dorsal or ventral muscle bands on the anterior edge of the body waves and inhibiting those on the opposite side.

2. Propagate this pattern along the body posteriorly for forward locomotion and anteriorly for backward locomotion.

3. Control the switching from forward to backward locomotion based on sensory input.

4. Provide a rhythmic pattern generator to cyclically initiate the waves.

5. Regulate wave frequency, amplitude, wavelength and rate of propagation according to environmental requirements.

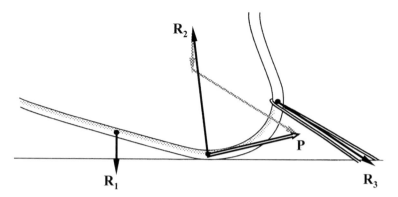

Fig. 2.12. Force diagram illustrating how a propulsive force can be generated by application of force by *Mermis nigrescens* to a loop of fibre as the body slides past, as in Fig. 2.6. Shaded area, regions of muscle tension that apply forces to the loop, substrate and posterior body. The propulsive force P is the resultant of hypothetical restraining forces R_1, R_2 and R_3. Grey vectors show how the three vectors sum. (From A.H.J. Burr, unpublished results.)

6. Utilize sensory feedback to adjust the wave-form according to the location of objects along the body.

Of the above listed required activities of the neuromuscular system, 1–3 are better understood; the others are more a matter of speculation. See Perry and Maule (Chapter 8, this volume) and previous reviews for details (Johnson and Stretton, 1980; Stretton *et al.*, 1985; Walrond *et al.*, 1985; Chalfie and White, 1988; Wicks and Rankin, 1995; Davis and Stretton, 1996, 2001; Driscoll and Kaplan, 1997; Rand and Nonet, 1997; Thomas and Lockery, 1999; Martin *et al.*, 2002). The neuromuscular anatomy and physiology is known in *C. elegans* and *Ascaris* in detail due largely to the work of White *et al.* (1976, 1986) and Stretton *et al.* (Stretton *et al.*, 1978; Angstadt *et al.*, 1989, 2001; Davis and Stretton, 1989). The number, morphology and location of the motor neurones are strikingly similar in these 1 and 300 mm long nematodes. The motor system of all nematodes may be fundamentally similar to these well-studied examples, but note that other nematodes appear to have more complex behaviour and the 120 mm long *M. nigrescens* has many more motor neurones (Table 2.1).

Anyone who has observed nematode behaviour notices a distinction between the exploring, probing and steering motions of the anterior ('head' and 'neck') of the worm and the undulatory locomotory motion of the remainder of the body which involves only dorsoventral bending. This is a reflection of the innervation of the four to eight rows of muscles that travel through both regions (Fig. 2.1). The anteriormost muscle cells in each row, in the head and neck region, are independently innervated by motor neurones in the nerve ring, providing for complex patterns of movements in both lateral and dorsoventral directions. However, in the region behind the neck, in both *C. elegans* and *Ascaris*, the muscle cells from the two subdorsal quadrants send processes to the dorsal nerve cord, and those from the two

Table 2.1. Number of motor neurones, commissures and muscle cells. The number of muscle cells per mm are equivalent across the three species; however, the number of commissures per muscle cell (and presumably motor neurones per muscle cell) vary dramatically. *Mermis*, with the most versatile locomotion, has the most commissures per mm.

Species	Length (mm)	Motor neurones	Commissures	Commissures per mm	Muscle cells	Muscle cells per mm	Commissures per muscle cell
C. elegans[a]	1	57	36	4	79	79	0.5
Ascaris[b]	300	59	36	0.12	50,000	167	0.0007
Mermis[c]	120	> 1000	1000	12	10,000	83	0.1

[a]Chalfie and White (1988).
[b]Stretton (1976); Stretton *et al.* (1978).
[c]Gans and Burr (1994).

subventral quadrants send processes to the ventral nerve cord (Fig. 2.3). In addition to making chemical synapses with motor neurones in the nerve cords, the processes are tied electrically to each other by gap junctions. This causes the dorsal muscle bands to act together separately from the ventral muscle bands, and only dorsoventral bending is possible. Reciprocal inhibition between dorsal and ventral motor nerve cords ensures that dorsal muscles relax when ventral muscles contract and vice versa.

What would cause the sinusoidal bending pattern to propagate along the body? Driscoll and Kaplan (1997) review several possibilities. The most interesting is the possibility that distal processes of the motor neurones may act as stretch receptors. The long processes extend well beyond the regions that form neuromuscular junctions and have no other apparent function. When stretched at the outside of a bend, a posterior process could increase the depolarization of the neurone, which would then more strongly excite muscles anterior to the bend. A stronger tension anterior to the outside of a bend and lower tension anterior to the inside of a bend would be appropriate for driving forward locomotion (Figs 2.9 and 2.10).

While this simple model based on the neuroanatomy of *C. elegans* and *Ascaris* would satisfactorily explain locomotion on a smooth agar surface where physical properties of the substrate are uniform, nematodes seldom encounter such conditions outside the laboratory. Changes in wave-form on different substrates are generally consistent across taxa and appear to optimize purchase and reduce slip and yawing. What causes the amplitude to increase as slippage increases (Fig. 2.11)? Why are frequency, amplitude and wave-form different when a nematode swims? Could these changes be due simply to the smaller lateral restraining forces, or is there a physiological adjustment involved? During locomotion of *Mermis*, it is clear that feedback from touch receptors along the body is needed in order for *Mermis* to maintain the force against a fibre or grass blade as the body glides past (Figs 2.6 and 2.7). Soil-inhabiting nematodes may need a similar mechanism in order to adjust to different particle sizes (Fig. 2.13). While lateral and ventral touch receptors in *C.*

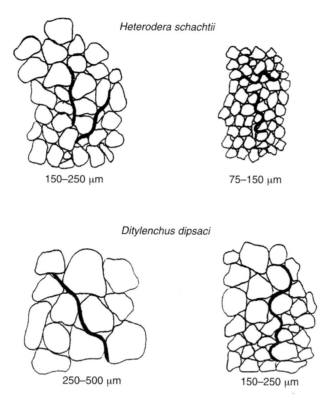

Fig. 2.13. Effect of particle size on wave-form of *Heterodera schachtii* and *Ditylenchus dipsaci.* (From Wallace, 1958.)

elegans have been shown to initiate a reversal reflex, it is not known if they can modulate muscle tension during locomotion. Could some commissures be touch-sensitive? They lie adjacent to the cuticle as they pass from nerve cord to nerve cord and there are many more of them per mm in *Mermis* than in *C. elegans* or *Ascaris* (Table 2.1).

These considerations of locomotion do not cover mechanisms of steering which are initiated by the head and neck. Marine nematodes crawling through filamentous algae display movements that suggest a highly complex sensing and steering capability. Nematodes constrained to crawling on their sides on agar cannot fully display this (Fig. 2.14). Scanning movements of the anterior end can be seen when nematodes forage for bacteria, probe roots or otherwise sample their environment for thermal, chemical or tactile cues, suggesting they thereby gather information that directs their steering. Experiments with *Mermis* demonstrate that, in order for the female to steer towards light, proprioceptive signals of the vertical and sideways head bending must be compared with the periodic signal from the photoreceptor caused by shading by the haemoglobin pigmentation in the eye (Burr and Babinszki, 1990). The innervation of head and neck muscles of *C. elegans* is known to be complex, with motor neurones controlled by interneurones that

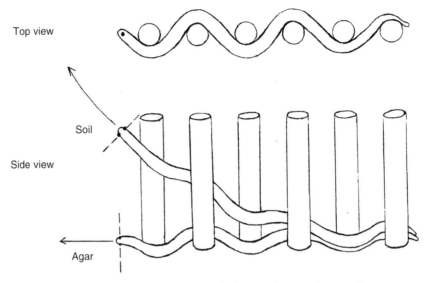

Fig. 2.14. Lateral sampling and steering is prevented when surface tension confines a nematode to a wet planar surface (side-view, lower worm, and top view), because its dorsoventral flexures force it to crawl on its side. However, this is not prevented in three-dimensional media such as a fibre network or soil (upper worm in side-view). Dots represent laterally located sensilla such as amphids. (From A.F. Robinson, unpublished results.)

integrate sensory information in the nerve ring. However, unlike the control of the dorsoventral undulatory motion, the behaviour and physiology of steering are largely unstudied.

2.5. Environmental Challenges and Adaptations

Since most nematodes are microscopic and occupy opaque habitats, their movements are seldom observed directly in nature, although there have been some observations of nematode locomotion on leaves, stems, skin, hair and excised sections of gut. Most observations have been on agar. On rigid surfaces most nematodes require some free moisture for movement. In at least some cases, progression on a moist rigid surface with the appropriate topology can be significantly faster than on agar. There are significant differences in the extent to which nematodes depend on free moisture for movement, and nematodes with tough integuments, such as the ensheathed infective juveniles of entomopathogenic nematodes, appear able to move on virtually dry surfaces.

2.5.1. Locomotion on agar

On the surface of agar wave-form depends on slippage in the surface water film (Fig. 2.11). When placed on water agar weaker than 0.5%, many nematodes will penetrate the surface and move through the agar in a fashion similar to that on the surface, but more slowly. Specific aspects of locomotion on agar can differ greatly, depending on species and goals. Bacteria-feeding rhabditids such as *C. elegans*, for example, typically move slowly on agar that supports a bacterial lawn, frequently pausing to graze with characteristic swinging motions of the head. Conversely, on food-free agar, *C. elegans* cruises for 1–2 min before reversing direction spontaneously for one or more undulations, and then heads off in a new direction. Many infective stages of parasites of vertebrates, foliage or insects move forward on agar without interruption for extended periods, stopping only when the walls of the container are encountered. Nematode tracks on agar often exhibit broad arcs or slew caused by inherent dorsoventral bias in movement, or drag from ventrally bent tails or projected male copulatory organs (Croll, 1972; Robinson, 2004).

2.5.2. Swimming

Some nematodes can swim effectively, especially aquatic species. Among other species, swimming efficiency varies greatly. In general, active nematodes abruptly become more active when transferred from agar to water, and decreasing the viscosity or agar concentration increases wave frequency, amplitude and wavelength. Bacteria-feeding rhabditids, which swim poorly, undulate in water with waves of higher frequency and greater amplitude but otherwise similar in form to those on agar (Fig. 2.11). On the other hand, the vinegar eelworm, *Turbatrix aceti*, which swims more efficiently, generates waves with amplitude increasing by a factor as great as four as they travel posteriorly, and this suppresses yawing (Fig. 2.15; Gray and Lissmann, 1964). In striking contrast, infective juveniles of *Ditylenchus* species, which progress on agar via smooth waves like other nematodes, in free water typically exhibit rapid and repeated bending and unbending in the middle with no progression of the wave along the body. The 120 mm *M. nigrescens* does not swim at all, whereas 300 mm *Ascaris* can swim in intestinal contents. It is worth noting that *Ascaris* can also crawl through a narrow tube by pressing against opposite walls and rolling its body along the opposite surfaces. With forwardly propagated waves they thus accomplish an unusual forward locomotion (Reinitz *et al.*, 2000; A.H.J. Burr, unpublished results).

2.5.3. Movement through soil

Much of our knowledge of factors governing movement of nematodes through soil comes from the classic experiments of Wallace (1968) on nematodes moving in a thin layer of soil (Wallace, 1958). Soil is a mixture of mineral particles that can be placed into three size categories: sand, silt and clay. The relative contents of these particles determine the gross soil texture. This, however, is an oversimplification because the precise distribution of sizes and the surface chemistry of soil particles vary greatly within these particle size classes and among soils. Small particles often

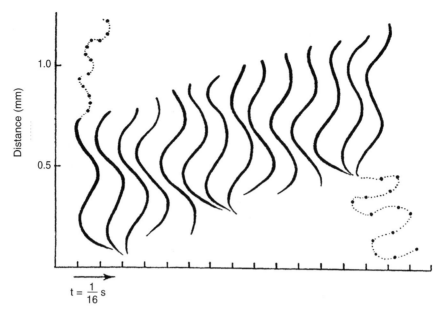

Fig. 2.15. Successive positions at 1/16 second intervals of swimming *Turbatrix aceti*. Note slippage, and amplitude of tail about four times that of head. (From Gray and Lissman, 1964.)

form aggregates or soil crumbs, and the sizes of crumbs present within the soil can be the primary determinant of the suitability of conditions for nematode movement.

The surface tension of a water film draws nematodes towards a flat surface with a force greatly exceeding the force exerted on them by gravity. In soil, a similar effect would be expected but additional forces resulting from the powerful affinity of water for electrically charged soil particle surfaces (primarily clay micelles) must also be considered. These effects are taken into account by expressing the water status of soil in terms of the sum of matric potential (Gibbs' free energy attributable to adhesive forces), osmotic potential (free energy resulting from dissolved solutes) and gravitational potential. Osmotic potential and gravity have little effect on the shape of water films. Instead, the water films in soil for the most part are shaped by surface tension and matric forces, two factors whose relative contributions vary greatly with soil clay content. As a consequence, soils with high clay content have thinner films at the same level of physiological dryness. Wallace's experiments showed that matric potential is the most reliable predictor of the suitability of soil moisture content for nematode movement across a wide range of soil types (Wallace, 1968). A mathematical model has been developed that simulates movement through soil under different controlling factors (Hunt *et al.*, 2001).

As soil dries, solutes become more concentrated. However, osmotic pressure increases little compared with the changes that occur in matric potential, and the resulting differences in osmotic pressure in most soils appear to have essentially no effect on nematodes over the range of moisture contents where surface tension permits nematode movement. This relationship was elegantly demonstrated by an experiment by Blake (1961) in which soil was wetted with a urea solution and subjected to various matric potentials. Movement was similar with water or urea and fastest as the pores of the sand began to drain (Fig. 2.16). These and other results showed that, as soil moisture is removed, reducing thickness of the water films surrounding the soil particles prevents nematode movement long before they are affected physiologically by osmotic pressure (Blake, 1961).

Fastest nematode movement has been shown to occur in soil when it is at field capacity, i.e. the point when soil pores have just partially drained (Wallace, 1968). At field capacity, moisture and gas transport are generally optimum for plant

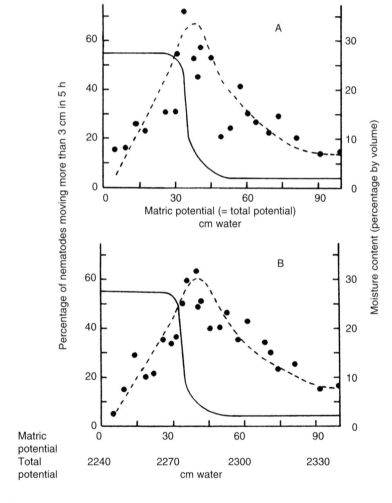

Fig. 2.16. Effect of matric potential on movement of *Ditylenchus dipsaci*. Moisture was removed from soil wetted with water (A) or 0.1 M urea (B) before stepwise drying. Solid line, moisture content (right-hand scale); filled circles and dashed line, nematode movement (left-hand scale). (From Blake, 1961.)

growth, and the gas phase is continuous, allowing free exchange of respiratory gases as well as the establishment of long-distance gradients within soil channels of volatiles that can serve as attractants and repellents for animals in the soil.

Even though surface tension draws nematodes in a thin film with a force usually exceeding gravity many-fold, a buoyancy differential between head and tail can orient microscopic nematodes in soil if soil pores are sufficiently large. Direct observations of *Heterodera rostochiensis* (Wallace, 1960) and *Rotylenchulus reniformis* (Robinson and Heald, 1993) when suspended in water showed both to point head down, and, when in soil at field capacity, live (but not dead) nematodes moved downward regardless of moisture gradient direction if soil particle sizes exceeded 325 μm.

2.5.4. Adaptations of size and shape

Nematodes typically found in soil samples appear to fall into two broad size classes with mean lengths of about 400 and 1000 μm. Those in the former may be well suited for moving through soil interstices while the others may utilize channels provided by roots and insects. Observations of juveniles of the sugar beet cyst nematode in a monolayer of soil showed them to move fastest with longest wavelength when the particle size was about 0.3 times the body length, regardless of moisture content (Wallace, 1958). Infective stages of numerous entomopathogenic and vertebrate parasites occurring in soil are similar in overall length and diameter to the bulk of other soil-inhabiting nematodes, suggesting that the soil pore size and tortuosity have selected for optimum length and diameter for movement through soil in diverse trophic groups. Conversely, the retention of a soil stage by most terrestrial vertebrate and arthropod parasites may impose a general constraint on shape that contributes significantly to the highly conserved morphology of nematodes across so many taxa and niches.

The sizes and shapes of nematodes appear attuned to requirements for movement and reproduction not just in soil but in other matrices as well. Thus, the microfilariae of filariids are among the smallest of all nematodes (*c.* 30–40 μm in length), facilitating movement into peripheral dermal capillaries and through the vector's proboscis (Casiraghi *et al.*, 2001). The saccate shape of adult females in the root-parasitic heteroderids reflects a complete trade-off between motility and reproduction. Partial compromises are apparent throughout the phylum. The sacrifice of motility for fecundity by *Ascaris*, which can lay 200,000 eggs per day, is an example. Infective juveniles of *Anguina*, *Ditylenchus*, *Aphelenchoides* and mermithids that move on to foliage as part of the life cycle are typically similar in diameter to nematodes in soil but are two to five times longer. This probably facilitates movement on foliage, where length-dependent bridging between leaf hairs, leaf folds and isolated water droplets may be critical. *Mermis nigrescens* females, with their unusual 120 mm length and 0.4 mm diameter, bridge large gaps between grass blades (Fig. 2.7A).

2.5.5. Adaptations of speed and activation

The energy cost of movement for nematodes can be considered negligible (Alexander, 2002) and so, not surprisingly, their speed often appears well tuned to ecological needs. The slowest of all nematodes are those that parasitize plant roots and fungi in the soil, i.e. hosts that grow slowly in a highly buffered environment. In comparison, the transient bacterial blooms available to bacteria feeders in and on the soil can require rapid migration to escape desiccation, hitchhike on passing animals and locate new food sources. Nematodes that crawl from soil on to the surfaces of foliage risk a harsh environment where temperature and moisture film availability change abruptly compared with soil. Nematodes that crawl on foliage, be they parasites of plants, insects or vertebrates, typically move in water much faster than soil-inhabiting nematodes. Intestinal parasites must constantly overcome the expulsive effects of peristalsis, and swallowed *Trichinella* juveniles that find themselves in the small intestine following exsheathment in the stomach enter into frenzied activity, which drives them into the intestinal wall and saves them from expulsion (Sukhdeo *et al.*, 2002).

Spontaneity of movement is also typically attuned in relation to ecological needs. For example, cessation or activation of locomotion in response to environmental cues can both result in accumulation within strategic zones that favour survival and maximize the likelihood of host encounters or phoresy. Infective stages of root and foliar parasites typically are spontaneously active and continue to move at a speed linearly related to ambient temperature (Croll, 1975; Robinson *et al.*, 1981; Robinson, 1989). Many nematodes, however, remain lethargic or entirely motionless until activated by triggering stimuli. Examples include activation of *Agamermis* infective juveniles by light (Robinson *et al.*, 1990), of infective juveniles of *Trichinella spiralis* and hundreds of other intestinal parasites on exposure to bile (Sukhdeo *et al.*, 2002), and of *Steinernema* by vibration, air movement and volatile host cues (Gaugler *et al.*, 1980; Campbell and Kaya, 1999a,b).

The wide array of plant, insect and vertebrate parasites that move in large numbers from soil on to foliage sometimes crawl up foliage more than a metre above the soil surface. Such migrations allow nematodes to distribute their eggs or themselves on foliage where their eggs may be eaten by insects or ungulates, or they may gain access to foliar terminals, leaf folds or stomata where they invade plant tissue, induce foliar or floral galls and reproduce. Careful studies of the environmental factors favouring such migrations (Rees, 1950; Adamo *et al.*, 1976; Stromberg, 1997; Robinson, 2000) have shown that rainfall, humidity, foliar pubescence and sometimes sunlight play key roles in activating nematodes when there are suitable physical conditions for locomotion. In the case of *Agamermis*, which infect orthopterans on the Australian tablelands (Baker and Poinar, 1995), orientation precisely perpendicular to light (transverse phototaxis) directs nematodes vertically at sunrise when the moisture films needed for movement on grass blades are present (Robinson *et al.*, 1990). The larger adult female *M. nigrescens* is guided by a positive phototaxis to the upper levels of vegetation, but only under wet conditions (Burr and Babinszki, 1990; Burr *et al.*, 1990; Gans and Burr, 1994).

2.5.6. Other locomotory motions

Undulatory propulsion in three dimensions (Gray, 1953) has been observed for *Hemicycliophora* and *Meloidogyne javanica* during root penetration and for juveniles of *Nippostrongylus* climbing hairs of the animal host (Wallace and Doncaster, 1964). In the case of *Nippostrongylus*, ultrastructural examination revealed a 360° twist of the body for each 360° revolution around the axis of the helix traversed (Lee and Biggs, 1990).

Many different kinds of nematodes move rhythmically and synchronously *en masse* when drawn together by surface tension within a film of water on the surface of agar, glass or plastic Petri dish lids (Gray and Lissmann, 1964; Croll, 1970). Such movement by large numbers of nematodes is called swarming and is commonly observed in crowded populations of *C. elegans*, other bacteria-feeding *Rhabditis* on mushroom beds, *Ditylenchus myceliophagus* and *Aphelenchoides composticola* migrating from foul compost, dauer juveniles of *Pelodera coarctata*, diplogasterids on nutrient plates and entomopathogenic nematodes (Croll, 1970). During swarming, large numbers of nematodes climb up each other, forming a writhing structure above the substrate.

During an additional locomotory subroutine, called nictation, many nematodes lift their anterior or even more of their body off a moist substrate and wave it in the air (Croll and Matthews, 1977). Examples include dauer juveniles of *C. elegans* (Riddle, 1988) and other bacteria feeders (Croll and Mathews, 1977), infective stages of certain hookworms (Lee, 1972; Granzer and Haas, 1991; Muller, 2002) and several entomopathogenic nematodes (Ishibashi, 2002). Nictating can be promoted by adding glass beads or sand grains to agar (Baird, 1999; J. Campbell, personal communication) and by other substrate irregularities (Ishibashi and Kondo, 1990). Nictating at the extremities of objects may improve the chance of phoresy or host encounter. In the presence of its isopod host, dauers of *Caenorhabditis remanei* nictate and crawl on to the host (Baird, 1999). Many rhabditid species associate with isopods, insects or molluscs for transport to new microenvironments, including a potential cadaver (Baird *et al.*, 1994; Baird, 1999; Kiontke *et al.*, 2002). Nictating by the dog hookworm *Ancylostoma caninum* is stimulated by carbon dioxide, warmth and humidity (Granzer and Haas, 1991). During the migration of gravid *M. nigrescens* on vegetation, nictation increases the chance of discovering new surfaces to climb on (Gans and Burr, 1994).

Steinernematid nematodes can jump – for example, the infective juveniles of *Steinernema carpocapsae*. When the nictating anterior does not contact another surface projection, it bends over into a loop, contacting a moist point on the posteriad body. The force generated by the subsequent opposing body wave abruptly breaks the anterior end free of surface tension at the point of contact. This can fling the anterior end upward with sufficient momentum to break the surface tension holding the posterior end to the substrate and jettison the entire body through the air (Reed and Wallace, 1965; Campbell and Kaya, 1999a,b). The entomopathogenic steinernematids use nictating, swarming and jumping to transfer to insects, along with a refinement that allows some species to jump seven or more nematode body lengths above the substrate towards insects. Factors influencing nictating, swarming and jumping in steinernematids have received extensive investigation because these species are important in biological control programmes. Both nictation and jumping are stimulated by carbon dioxide.

Criconemoides curvatum can propel itself forward with no dorsoventral undulatory bending by means of anteriorly propagated peristaltic contractions. Posteriorly angled cuticular ridges provide traction. Cinematography showed worms to progress very slowly (20 μm per contraction) when crawling in water on a glass slide (Streu *et al.*, 1961). A somewhat different pattern, involving non-peristaltic contraction and relaxation of the entire body, was described of *Criconemoides rusticus* as it crawled through soil particles (Stauffer, 1924). *Shaerolaimus gracilis* utilizes adhesive secretions from a caudal gland system and an anteriorly positioned renette cell to release and refasten posterior and anterior ends alternately to the substrate at sequential points in a controlled fashion within water currents (Turpenniemi and Hyvärinen, 1996).

2.6. Summary and Future Directions

Understanding the mechanism of locomotion requires knowledge of cuticle (ultra-structure and elastic properties), hydrostatic skeleton (internal pressure and elastic properties of body wall), muscle (structure, lateral connection to cuticle, tonus and development of bending couples), transmission of forces (along stiff body tube and across cuticle), motor nervous system (connectivity, function of synapses, neurotransmitters and neuromodulators, propagation of waves, mechanism of orientation and lateral steering) and the locomotory role of sensory feedback (of relative location of bends and external objects along the body tube, and of other physical factors, including frictional resistance). Separately these individual areas of investigation have little apparent meaning for locomotion. We have attempted to integrate these separate topics and explain them clearly in the context of their locomotory roles. At the same time, we have highlighted areas that desperately need further study before the basic mechanisms are satisfactorily understood.

Of the special areas listed above, those most in need of further work are: (i) cuticle – elastic properties and transfer of forces across it, including the possible mechanical role of lateral ridges and alae; (ii) muscle – tonus and tension causing stiffness and bending; (iii) proprioception – stretch, pressure and touch reception; and (iv) role of sensory feedback in locomotion. Especially needed is (v) an investigation of the mechanical properties and putative functions of the hydrostatic skeleton, relating different inflation pressures, changes in dimensions, crossing angle, elasticity, cuticle ultrastructure and bending.

Current intensive study is in other areas: the genetics, molecular composition and development of cuticle and muscle in *C. elegans*, and the effects of neuromodulators on the *Ascaris* neuromuscular system. While important and interesting for other reasons, none of these recent efforts are aimed at attaining the whole picture of locomotion. There is a strong need for more integrative investigations of nematode locomotion.

Knowledge of the mechanism in a few species, however, is not enough. Only by comparative study of the diversity of structure, function and behaviour in the context of adaptation to a varied environment can nematode locomotion be fully understood. We have ended this review by describing the results of some of these excellent studies, hoping to inspire more.

There is an amazing diversity in nematode structure and for taxonomic purposes this has been recorded at the light microscopic scale for more than a century. There is anecdotal evidence for a diversity of nematode behaviour too, and it should also be documented. Technology is now available for archiving locomotion and other behaviour as video clips. These can accompany a new method for archiving morphological information, video capture and editing (VCE) microscopy (De Ley and Bert, 2002), which provides micrographic images of nematodes as a multifocal series (see http://faculty.ucr.edu/%7epdeley/vce.html). These records will soon be provided on a website together with DNA sequences, and behavioural clips can readily be added.

References

Adamo, J.A., Madamba, C.P. and Chen, T.A. (1976) Vertical migration of the rice white-tip nematode, *Aphelenchoides besseyi*. *Journal of Nematology* 8, 146–152.

Alexander, R.M. (1987) Bending of cylindrical animals with helical fibres in their skin or cuticle. *Journal of Theoretical Biology* 124, 97–110.

Alexander, R.M. (2002) Locomotion. In: Lee, D.L. (ed.) *The Biology of Nematodes*. Taylor and Francis, London, pp. 345–352.

Angstadt, J.D., Donmoyer, J.E. and Stretton, A.O.W. (1989) Retrovesicular ganglion of the nematode *Ascaris*. *Journal of Comparative Neurology* 284, 374–388.

Angstadt, J.D., Donmoyer, J.E. and Stretton, A.O.W. (2001) The number of morphological synapses between neurons does not predict the strength of their physiological synaptic interactions: a study of dendrites in the nematode *Ascaris suum*. *Journal of Comparative Neurology* 16, 512–527.

Baird, S.E. (1999) Natural and experimental associations of *Caenorhabditis remanei* with *Trachelipus rathkii* and other terrestrial isopods. *Nematology* 1, 471–475.

Baird, S.E., Fitch, D.H. and Emmons, S.W. (1994) *Caenorhabditis vulgaris* sp.n. (Nematoda: Rhabditidae): a necromenic associate of pill bugs and snails. *Nematologica* 40, 1–11.

Baker, G.L. and Poinar, G.O. (1995) *Agamermis catadecaudata* n. sp. (Nematoda: Mermithidae), a parasitoid of Orthoptera in south-eastern Australia. *Fundamental and Applied Nematology* 18, 139–148.

Baldwin, J.G. and Perry, R.N. (2004) Nematode morphology, sensory structure and function. In: Chen, Z.X., Chen, S.Y. and Dickson, D.W. (eds) *Nematology, Advances and Perspectives*. Vol. 1, *Nematode Morphology, Physiology and Ecology*. Tsinghua University Press, Tsinghua, China, pp. 171–256.

Barstead, R.J. and Waterston, R.H. (1989) The basal component of the nematode dense-body is vinculin. *Journal of Biological Chemistry* 264, 10177–10185.

Bartnik, E., Osborn, M. and Weber K. (1986) Intermediate filaments in muscle and epithelial cells of nematodes. *Journal of Cell Biology* 102, 2033–2041.

Bird, A.F. (1980) The nematode cuticle and its surface. In: Zuckerman, B.M. (ed.) *Nematodes as Biological Models*, Vol. 2. Academic Press, New York, pp. 213–235.

Bird, A.F. and Bird, J. (1991) *The Structure of Nematodes*, 2nd edn. Academic Press, San Diego, 316 pp.

Blake, C.D. (1961) Importance of osmotic potential as a component of the total potential of the soil water on the movement of nematodes. *Nature* 192, 144–145.

Burr, A.H.J. and Babinszki, C.P.F. (1990) Scanning motion, ocellar morphology and orientation mechanism in the phototaxis of the nematode *Mermis nigrescens. Journal of Comparative Physiology A* 167, 257–268.

Burr, A.H.J. and Gans, C. (1998) Mechanical significance of obliquely striated architecture in nematode muscle. *Biological Bulletin* 194, 1–6.

Burr, A.H.J., Babinszki, C.P.F. and Ward, A.J. (1990) Components of phototaxis of the nematode *Mermis nigrescens. Journal of Comparative Physiology A* 167, 245–255.

Campbell, J.F. and Kaya, H.K. (1999a) How and why a parasitic nematode jumps. *Nature* 397, 485–486.

Campbell, J.F. and Kaya, H.K. (1999b) Mechanism, kinematic performance, and fitness consequences of jumping behavior in entomopathogenic nematodes (*Steinernema* spp.). *Canadian Journal of Zoology* 77, 1947–1955.

Casiraghi, M., Anderson, T.J., Bandi, C., Bazzocchi, C. and Genchi, C. (2001) A phylogenetic analysis of filarial nematodes: comparison with the phylogeny of *Wolbachia* endosymbionts. *Parasitology* 122, 93–103.

Chalfie, M. and White, J. (1988) The nervous system. In: Wood, W.B. (ed.) *The Nematode* Caenorhabditis elegans. Cold Spring Harbor Laboratory, Cold Spring Harbor, New York, pp. 337–391.

Chapman, G. (1958) The hydroskeleton in the invertebrates. *Biological Reviews* 33, 338–371.

Coomans, A., Verschuren, D. and Vanderhaeghen, R. (1988) The demanian system, traumatic insemination and reproductive strategy in *Oncholaimus oxyuris* Ditlevsen (Nematoda, Oncholaimina). *Zoologica Scripta* 17, 15–23.

Cox, G.N., Kusch, M. and Edgar, R.S. (1981a) Cuticle of *Caenorhabditis elegans*: its isolation and partial characterization. *Journal of Cell Biology* 90, 7–17.

Cox, G.N., Staprans, S. and Edgar, R.S. (1981b) The cuticle of *Caenorhabditis elegans*: II. Stage-specific changes in ultrastructure and protein composition during postembryonic development. *Developmental Biology* 86, 456–470.

Croll, N.A. (1970) *The Behavior of Nematodes: Their Activity, Senses, and Responses.* Edward Arnold, London, 117 pp.

Croll, N.A. (1972) Behavioral activities of nematodes. *Helminthological Abstracts A* 41 (3), 359–377.

Croll, N.A. (1975) Behavioral analysis of nematode movement. In: Dawes, B. (ed.) *Advances in Parasitology.* Academic Press, New York, pp. 71–122.

Croll, N.A. (1976) Behavioural coordination of nematodes. In: Croll, N.A. (ed.) *The Organization of Nematodes.* Academic Press, London, pp. 343–364.

Croll, N.A. and Mathews, B.E. (1977) *Biology of Nematodes.* John Wiley & Sons, New York, 201 pp.

Davey, K.G. (1995) Water, water compartments and water regulation in some nematodes parasitic in vertebrates. *Journal of Nematology* 27, 433–440.

Davis, R.E. and Stretton, A.O.W. (1989) Signaling properties of *Ascaris* motorneurons:

graded active responses, graded synaptic transmission, and tonic transmitter release. *Journal of Neuroscience* 9, 403–414.

Davis, R.E. and Stretton, A.O.W. (1996) The motor nervous system of *Ascaris*: electrophysiology and anatomy of the neurons and their control by neuromodulators. *Parasitology* 113 (Suppl.), S97–S117.

Davis, R.E. and Stretton, A.O.W. (2001) Structure–activity relationships of 18 endogenous neuropeptides on the motor nervous system of the nematode *Ascaris suum. Peptides* 22, 7–23.

De Ley, P. and Bert, W. (2002) Video capture and editing as a tool for the storage, distribution and illustration of morphological characters of nematodes. *Journal of Nematology* 34, 296–302.

Driscoll, M. and Kaplan, J. (1997) Mechanotransduction. In: Riddle, D.L., Blumenthal, T., Meyer, B.J. and Priess, J.R. (eds) *The Nematode* C. elegans *II*. Cold Spring Harbor Laboratory Press, New York, pp. 645–677.

Edgar, R.S., Cox, G.N., Kusch, M. and Politz, J.C. (1982) The cuticle of *Caenorhabditis elegans. Journal of Nematology* 14, 248–258.

Fetterer, R.H. and Urban, J.F., Jr (1988) Developmental changes in cuticular proteins of *Ascaris suum. Comparative Biochemistry and Physiology* 90B, 321–327.

Francis, G.R. and Waterston, R.H. (1985) Muscle organization in *Caenorhabditis elegans*: localization of proteins implicated in thin filament attachment and I-band organization. *Journal of Cell Biology* 101, 1532–1549.

Francis, G.R. and Waterston, R.H. (1991) Muscle cell attachment in *Caenorhabditis elegans. Journal of Cell Biology* 114, 465–479.

Fujimoto, D. and Kanaya, S. (1973) Cuticulin: a noncollagen structural protein from *Ascaris* cuticle. *Archives of Biochemistry and Biophysics* 157, 1–6.

Gans, C. and Burr, A.H.J. (1994) The unique locomotory mechanism of *Mermis nigrescens*, a large nematode which crawls over soil and climbs through vegetation. *Journal of Morphology* 222, 133–148.

Gaugler, R., LeBeck, L., Nakagaki, B. and Boush, G.M. (1980) Orientation of the entomogenous nematode *Neoaplectana carpocapsae* to carbon dioxide. *Environmental Entomology* 9, 649–652.

Granzer, M. and Haas, W. (1991) Host-finding and host recognition of infective *Ancylostoma caninum* larvae. *International Journal for Parasitology* 21, 429–440.

Gray, J. (1953) Undulatory propulsion. *Quarterly Journal of Microscopical Science* 94, 551–578.

Gray, J. and Lissmann, H.W. (1950) The kinetics of locomotion of the grass snake. *Journal of Experimental Biology* 26, 354–367.

Gray, J. and Lissmann, H.W. (1964) The locomotion of nematodes. *Journal of Experimental Biology* 41, 135–154.

Harris, J.E. and Crofton, H.D. (1957) Structure and function in the nematodes: internal pressure and cuticular structure in *Ascaris. Journal of Experimental Biology* 34, 116–130.

Hunt, H.W., Wall, D.H., DeCrappeo, N.M. and Brenner, J.S. (2001) A model for nematode locomotion in soil. *Nematology* 3, 705–716.

Inglis, W.G. (1964) The structure of the nematode cuticle. *Journal of Zoology* 143, 465–502.

Ishibashi, N. (2002) Behaviour of entomopathogenic nematodes. In: Lee, D.L. (ed.) *The Biology of Nematodes*. Taylor and Francis, London, pp. 511–520.

Ishibashi, N. and Kondo, E. (1990) Behaviour of infective juveniles. In: Gaugler, R. and Kaya, H.K. (eds) *Entomopathogenic Nematodes in Biological Control*. CRC Press, Boca Raton, Florida, pp. 139–150.

Johnson, C.D. and Stretton, A.O.W. (1980) Neural control of locomotion in *Ascaris*: Anatomy, electrophysiology and biochemistry. In: Zuckerman, B.M. (ed.) *Nematodes as Biological Models*, Vol. 1. Academic Press, New York, pp. 159–195.

Kiontke, K., Hironaka, M. and Sudhaus, W. (2002) Description of *Caenorhabditis japonica* n. sp. (Nematoda: Rhabditida) associated with the burrower bug *Parastrachia japonensis* (Heteroptera: Cydnidae) in Japan. *Nematology* 4, 933–941.

Kramer, J.M. (1997) Extracellular matrix. In: Riddle, D.L., Blumenthal, T., Meyer, B.J. and Priess, J.R. (eds) *The Nematode* C. elegans *II*. Cold Spring Harbor Laboratory Press, New York, pp. 471–500.

Lee, D.L. (1965) The cuticle of *Nippostrongylus brasiliensis*. *Parasitology* 55, 173–181.

Lee, D.L. (1972) Penetration of mammalian skin by the infective larva of *Nippostrongylus brasiliensis*. *Parasitology* 65, 499–505.

Lee, D.L. and Atkinson, H.J. (1977) *The Physiology of Nematodes*, 2nd edn. Columbia University Press, New York.

Lee, D.L. and Biggs, W.D. (1990) Two- and three-dimensional locomotion of the nematode *Nippostrongylus brasiliensis*. *Parasitology* 101, 301–308.

Mapes, C.J. (1965) Structure and function in the nematode pharynx. II. Pumping in *Panagrellus*, *Aplectana* and *Rhabditis*. *Parasitology* 55, 583–594.

Martin, R.J., Purcell, J., Robertson, A.P. and Valkanov, M.A. (2002) Neuromuscular organisation and control in nematodes. In: Lee, D.L. (ed.) *The Biology of Nematodes*. Taylor and Francis, London, pp. 321–344.

Miller, J.B. (1975) The length–tension relationship of the dorsal longitudinal muscle of a leech. *Journal of Experimental Biology* 62, 43–53.

Moerman, D.G. and Fire, A. (1997) Muscle: structure, function, and development. In: Riddle, D.L., Blumenthal, T., Meyer, B.J. and Priess, J.R. (eds) *The Nematode* C. elegans *II*. Cold Spring Harbor Laboratory Press, Cold Spring Harbor, New York, pp. 417–470.

Muller, R. (2002) *Worms and Human Disease*. CAB International, Wallingford, UK. 300 pp.

Niebur, E. and Erdos, P. (1991) Theory of the locomotion of nematodes: dynamics of undulatory progression on a surface. *Biophysical Journal* 60, 1132–1146.

O'Grady, R.T. (1983) Cuticular changes and structural dynamics in the fourth-stage larvae and adults of *Ascaris suum* Goeze, 1782 (Nematoda: Ascaridoidea) developing in swine. *Canadian Journal of Zoology* 61, 1293–1303.

Quillin, K.J. (1999) Kinematic scaling of locomotion by hydrostatic animals: ontogeny of peristaltic crawling by the earthworm *Lumbricus terrestris*. *Journal of Experimental Biology* 202, 661–674.

Rand, J.B. and Nonet, M.L. (1997) Synaptic transmission. In: Riddle, D.L., Blumenthal, T., Meyer, B.J. and Priess, J.R. (eds) *The Nematode* C. elegans *II*. Cold Spring Harbor Laboratory Press, New York, pp. 611–643.

Reed, E.M. and Wallace, H.R. (1965) Leaping locomotion in an insect parasitic nematode. *Nature* 206, 210–211.

Rees, G. (1950) Observations on the vertical migrations of the third-stage larva of *Haemonchus contortus* (Rud.) on experimental plots of *Lolium perenne* S24, in relation to meteorological and micrometeorological factors. *Parasitology* 40, 127–142.

Reinitz, C.A., Herfel, H.G., Messinger, L.A. and Stretton, A.O.W. (2000) Changes in locomotory behaviour and cAMP produced in *Ascaris suum* by neuropeptides from *Ascaris suum* or *Caenorhabditis elegans*. *Molecular and Biochemical Parasitology* 111, 185–197.

Riddle, D.L. (1988) The dauer larva. In: Wood, W.B. (ed.) *The Nematode* Caenorhabditis elegans. Cold Spring Harbor Laboratory, Cold Spring Harbor, New York, pp. 393–412.

Robinson, A.F. (1989) Thermotactic adaptation in two foliar and two root-parasitic nematodes. *Revue de Nématologie* 12, 125–131.

Robinson, A.F. (2000) Techniques for studying nematode movement and behaviour on physical and chemical gradients. In: Society of Nematologists Ecology Committee (ed.) *Methods of Nematode Ecology*. Society of Nematologists, Hyattsville, Maryland, pp. 1–20.

Robinson, A.F. (2004) Nematode behaviour and migrations through soil and host tissue. In: Chen, Z.X., Chen, S.Y. and Dickson, D.W. (eds) *Nematology, Advances and Perspectives*. Vol. 1, *Nematode Morphology, Physiology and Ecology*. Tsinghua University Press, Tsinghua, China, pp. 328–403.

Robinson, A.F. and Carter, W.W. (1986) Effects of cyanide ion and hypoxia on the volumes of second-stage juveniles of *Meloidogyne incognita* in polyethylene glycol solutions. *Journal of Nematology* 18, 563–570.

Robinson, A.F. and Heald, C.M. (1993) Movement of *Rotylenchulus reniformis* through sand and agar in response to temperature, and some observations on vertical descent. *Nematologica* 39, 92–103.

Robinson, A.F., Orr, C.C. and Heintz, C.E. (1981) Effects of oxygen and temperature on the activity and survival of *Nothanguina phyllobia*. *Journal of Nematology* 13, 528–535.

Robinson, A.F., Baker, G.L. and Heald, C.M. (1990) Transverse phototaxis by infective juveniles of *Agamermis* sp. and *Hexamermis* sp. *Journal of Parasitology* 76, 147–152.

Robinson, J. (1962) *Pilobolus* spp. and the translocation of the infective larvae of *Dictyocaulus viviparus* from faeces to pasture. *Nature* 193, 353–354.

Roggen, D.R. (1973) Functional morphology of the nematode pharynx. I. Theory of the soft-walled cylindrical pharynx. *Nematologica* 19, 349–365.

Rosenbluth, J. (1965) Ultrastructural organization of obliquely striated muscle fibers in *Ascaris lumbricoides*. *Journal of Cell Biology* 25, 495–515.

Rosenbluth, J. (1967) Obliquely striated muscle. III. Contraction mechanism of *Ascaris* body muscle. *Journal of Cell Biology* 34, 15–33.

Saunders, J.R. and Burr, A.H. (1978) The pumping mechanism of the nematode esophagus. *Biophysical Journal* 22, 349–372.

Seymour, M.K. (1973) Motion and the skeleton in small nematodes. *Nematologica* 19, 43–48.

Seymour, M.K. (1983) Some implications of helical fibers in worm cuticles. *Journal of Zoology* 199, 287–295.

Shadwick, R.E. and Gosline, J.M. (1985) Mechanical properties of the octopus artery. *Journal of Experimental Biology* 114, 259–284.

Stauffer, H. (1924) Die Lokomotion der Nematoden. *Zoologische Jahrbücher. Abteilung für Systematik. Ökologie und Geographie der Tiere* 49, 1–120.

Stern-Tomlinson, W., Nusbaum, M.P., Perez, L.E. and Kristan, W.B., Jr (1986) A kinematic study of crawling behaviour in the leech *Hirudo medicinalis*. *Journal of Comparative Physiology A* 158, 593–603.

Stretton, A.O.W. (1976) Anatomy and development of the somatic musculature of the nematode *Ascaris*. *Journal of Experimental Biology* 64, 773–788.

Stretton, A.O.W., Fishpool, R.M., Southgate, E., Donmoyer, J.E., Walrond, J.P., Moses, J.E.R. and Kass, I.S. (1978) Structure and physiological activity of the motoneurons of the nematode *Ascaris*. *Proceedings of the National Academy of Sciences, USA* 75, 3493–3497.

Stretton, A.O.W., Davis, R.E., Angstadt, J.D., Donmoyer, J.E. and Johnson, C.D. (1985) Neural control of behavior in *Ascaris*. *Trends in Neurosciences* 8, 294–300.

Streu, H.T., Jenkins, W.R. and Hutchinson, M.T. (1961) *Nematodes Associated with Carnations* Dianthus caryophyllus L. *with Special Reference to the Parasitism and Biology of* Criconemoides curvatum *Raski*. Rutgers Cooperative Extension Bulletin 800, New Jersey Agricultural Experiment Station, New Brunswick, New Jersey.

Stromberg, B.E. (1997) Environmental factors influencing transmission. *Veterinary Parasitology* 72, 247–264.

Sukhdeo, M.V.K., Sukhdeo, S.C. and Basemir, A.D. (2002) Interactions between intestinal nematodes and vertebrate hosts. In: Lewis, E.E., Campbell, J.F. and Sukhdeo, M.V.K. (eds) *The Behavioural Ecology of Parasites*. CAB International, Wallingford, UK, pp. 223–242.

Thomas, J.H. (1990) Genetic analysis of defecation in *Caenorhabditis elegans*. *Genetics* 124, 855–872.

Thomas, J.H. and Lockery, S. (1999) Neurobiology. In: Hope, I.A. (ed.) C. elegans: *A Practical Approach*. Oxford University Press, Oxford, UK, pp. 143–179.

Turpenniemi, T.A. and Hyvärinen, H. (1996) Structure and role of the renette cell and caudal glands in the nematode *Sphaerolaimus gracilis* (Monhysterida). *Journal of Nematology* 28, 318–327.

Van de Velde, M.C. and Coomans, A. (1989) A putative new hydrostatic skeletal function for the epidermis in monhysterids (Nematoda). *Tissue and Cell* 21, 525–533.

Wadepuhl, M. and Beyn, W.-J. (1989) Computer simulation of the hydrostatic skeleton: the physical equivalent, mathematics and application to worm-like forms. *Journal of Theoretical Biology* 136, 379–402.

Wainright, S.A. (1988) *Axis and Circumference – the Cylindrical Shape of Plants and Animals*. Harvard University Press, Cambridge, Massachusetts, 132 pp.

Wallace, H.R. (1958) The movement of eelworms. I. The influence of pore size and moisture content of the soil on the migration of larvae of the beet eelworm, *Heterodera schachtii* Schmidt. *Annals of Applied Biology* 46, 74–85.

Wallace, H.R. (1959) The movement of eelworms in water films. *Annals of Applied Biology* 47, 366–370.

Wallace, H.R. (1960) Movement of eelworms. VI. The influences of soil type, moisture gradients and host plant roots on the migration of the potato-root eelworm *Heterodera rostochiensis* Wollenweber. *Annals of Applied Biology* 48, 107–120.

Wallace, H.R. (1968) The dynamics of nematode movement. *Annual Review of Phytopathology* 6, 91–114.

Wallace, H.R. (1969) Wave formation by infective larvae of the plant-parasitic nematode *Meloidogyne javanica*. *Nematologica* 15, 65–75.

Wallace, H.R. and Doncaster, C.C. (1964) A comparative study of the movement of some microphagous, plant-parasitic and animal-parasitic nematodes. *Parasitology* 54, 313.

Walrond, J.P., Kass, I.S., Stretton, A.O.W. and Donmoyer, J.E. (1985) Identification of excitatory and inhibitory motoneurons in the nematode *Ascaris* by electrophysiological techniques. *Journal of Neuroscience* 5, 1–8.

Waterston, R.H. (1988) Muscle. In: Wood, W.B. (ed.) *The Nematode* Caenorhabditis elegans. Cold Spring Harbor Laboratory, Cold Spring Harbor, New York, pp. 281–335.

Wharton, D.A. (1986) *A Functional Biology of Nematodes*. Croom Helm, London, 192 pp.

White, J.G., Southgate, E., Thomson, J.N. and Brenner, S. (1976) The structure of the ventral nerve cord of *Caenorhabditis elegans*. *Philosophical Transactions of the Royal Society of London, B* 275, 327–348.

White, J.G., Southgate, E., Thomson, J.N. and Brenner, S. (1986) The structure of the nervous system of *C. elegans*. *Philosophical Transactions of the Royal Society of London, B.* 314, 1– 340.

Wicks, S.R. and Rankin, C.H. (1995) Integration of mechanosensory stimuli in *Caenorhabditis elegans*. *Journal of Neuroscience* 15, 2434–2444.

Wisse, E. and Daems, W.T. (1968) Electron microscopic observations on second-stage larvae of the potato root eelworm *Heterodera rostochiensis*. *Journal of Ultrastructure Research* 24, 210–231.

Wood, W.B. (1988) Introduction to *C. elegans* biology. In: Wood, W.B. (ed.) *The Nematode* Caenorhabditis elegans. Cold Spring Harbor Laboratory, Cold Spring Harbor, New York, pp. 1–16.

Wright, D.J. and Newall, D.R. (1976) Nitrogen excretion, osmotic and ionic regulation in nematodes. In: Croll, N.A. (ed.) *The Organization of Nematodes*. Academic Press, London, pp. 163–210.

Wright, K.A. (1987) The nematode's cuticle – its surface and the epidermis: function, homology, analogy – a current consensus. *Journal of Parasitology* 73, 1077–1083.

Wright, K.A. (1991) Nematoda. In: Harrison, F.W. and Ruppert, E.E. (eds) *Microscopic Anatomy of Invertebrates*. Vol. 4, *Aschelminthes*. Wiley-Liss, New York, pp. 111–195.

Wright, K.A. and Hope, W.D. (1968) Elaboration of the cuticle of *Acanthonchus duplicatus* Weiser, 1959 (Nematoda: Cyatholaimidae) as revealed by light and electron microscopy. *Canadian Journal of Zoology* 46, 1005–1011.

3 Orientation Behaviour

EKATERINI RIGA

Washington State University, 24106 North Bunn Road, Prosser, WA 99350, USA

3.1. Introduction

Nematodes resemble other animals in terms of perception, response and orientation to stimuli emitted from their environment, their host and each other. Nematodes that move randomly, once activated, display oriented responses. There are two types of oriented responses to changes in the environment: kinesis and taxis. Kinesis depends on the intensity of the stimulus but it is random and undirected; that is, there are neither positive nor negative responses to a stimulus. In taxis, the nematode displays a directed orientation response by moving either towards or away from a stimulus. Nematodes detect their environment with sensory receptors, which mediate a variety of behavioural responses, including chemoreception, thermoreception, photoreception, mechanoreception and orientation to magnetic fields and electric currents (Lee, 2002).

Orthokinesis and klinokinesis are two types of kinesis. In orthokinesis, the speed of the organism corresponds to the intensity of the stimulus and is non-directional. Similarly, klinotaxis is a non-oriented response during which the rates of changes in the turning of the animal are according to the intensity of the stimulus. Depending on the type of stimulus, orthokinesis and klinokinesis can be further categorized to photokinesis or photoklinokinesis if the response is due to light, and to thigmokinesis or orthokinesis if the response is due to contact. These types of behaviours result in aggregation where environmental conditions are optimum (Lee, 2002).

Klinotaxis and tropotaxis are two types of taxis. In klinotaxis, which requires only a single sensor, the nematode moves along the concentration gradient by moving the anterior part of its body from side to side to compare the stimulus intensity. In tropotaxis, the nematode responds by using two paired organs to simultaneously compare the stimulus.

Both taxis and kinesis responses can be studied by analysing nematode tracks on the surface of agar gels in the presence of a stimulus. The responses of nematodes to stimuli from chemicals, light, touch, electric currents and temperature are discussed in detail in this chapter.

3.2. Receptor Types

Nematodes have a variety of sensory organs (see review by Jones, 2002) with which to perceive cues from their environment. The main sense organs are located anteriorly and consist of paired amphids, six inner and six outer labial papillae and four cephalic papillae. The amphids and other cephalic sensilla are open to the environment and function mostly in chemoreception. The amphids and inner labial sensilla consist of a sheath cell, a socket cell and dendritic processes (Ward et al., 1975). Some of the dendritic processes are exposed to the external environment (Jones, 2002). *Caenorhabditis elegans* mutants with defective dendritic processes in the cephalic sensory organs respond to stimuli differently from wild types, indicating the chemoreceptive nature of the cephalic organs (Lewis and Hodgkin, 1977).

The phasmids, paired – pore-like sensilla – are located in the lateral field of the tails of many nematode species. Ultrastructural studies on phasmids suggest a chemosensory function although their precise purpose remains unclear (Jones, 2002). Other posterior sense organs include caudal papillae and male bursal rays, which may assist with mating.

Touch cell receptors are found along the length of several nematode species. Touch receptors consist of processes that are embedded in the epidermis and are surrounded by a matrix, termed the mantle. In *C. elegans,* the mantle is needed for the touch receptors to function (Jones, 2002). Deirids, which are considered mechanoreceptive organs, are cuticular sensory organs located at various points along the nematode body. Nematodes also have internal sensory organs, which mostly detect mechanical stimuli. For example, the animal nematode *Strongyloides stercolaris* has inner labial sensilla that do not open to the external environment and probably function as mechanoreceptors (Fine *et al.*, 1997).

Several nematode species, especially marine nematodes, are capable of detecting light cues with internal photoreceptive organs (Jones, 2002). These nematode photoreceptors are in the form of ocelli or pigment spots (McLaren, 1976a; Burr, 1984a).

3.2.1. Chemoreceptors

The amphids, the largest and most complex of the anterior sensory organs, are situated on either side of the nematode mouth. The earliest reports on the chemoreceptive function of the amphids and other sensilla are derived from work on *C. elegans* (Lewis and Hodgkin, 1977). Ultrastructural studies on several species (Trett and Perry, 1985), including *C. elegans* (Bargmann *et al.*, 1990), showed that the amphids are primarily chemosensory structures, although there is evidence indicating that they may serve additional functions.

Neuronal cell bodies innervate the amphids at the anterior end (Bargmann and Horvitz, 1991; Riga *et al.*, 1995a; Bargmann and Mori, 1997). Laser microbeam studies involving ablation of individual amphidial neurones of *C. elegans* have confirmed the chemoreceptive role of the amphids (Bargmann *et al.*, 1990). Immunocytochemistry (Stewart *et al.*, 1993a,b) and electrophysiological studies (Riga *et. al.*, 1995a) also revealed the chemosensory role of these organs. Ultrastructural studies have been performed on several nematode species and, like most chemoreceptive organs, the amphids are exposed to the external environment by a pore in the cuticle (Ashton and Schad, 1995; Perry, 1996). In *C. elegans*, single chemosensory neurones are able to detect different odorants, and some neurones can detect high and low concentrations of a single odorant (Sengupta *et al.*, 1993). For example, the two amphidial neurones of *Syngamus trachea* responded to avian blood and to D-tryptophan (Riga *et al.*, 1995a). Kaplan and Horvitz (1993), using laser ablation techniques on *C. elegans*, reported that the chemosensory neurones of the amphids are multifunctional, with each neurone bearing more than one type of receptor. Ashton and Schad (1995) and Perry (1996) have extensive reviews on nematode chemoreceptive organs, while Jones (2002) provides an overview and detailed diagrams of the amphid structure for multiple nematode species. In addition to their chemosensory role, there is evidence of a chemosecretory role.

The sheath cell of the amphids produces secretions that may function externally to the nematode (Jones, 2002). The amphids of *S. trachea* have a dual role as they also produce a secretory protein (Riga *et al.*, 1995b). In hookworms, the sheath cell produces an anticoagulant (Eiff, 1966). However, the chemosecretory role of the amphids is not well understood and needs further investigation.

Other sensory structures, including the inner labial sensilla and cuticular receptors, probably aid nematodes with chemoreception, but only limited work on these receptors has been completed. The six inner labial sensilla, in most nematodes, are open to the environment and are considered chemosensory structures. Jones (2002) hypothesized that the arrangement of these sensilla around the mouth suggests they are taste organs. In addition, the feeding apparatus of *Xiphinema diversicaudatum* contains neurones that may have either a chemosensory or a mechanosensory function (Robertson, 1975, 1979).

Wang and Chen (1985) reported that the phasmids of *Scutellonema brachyurum* may detect sex pheromones, a finding inconsistent with that of Riga *et al.* (1996a, 1997), who reported that the anterior sense organs detect pheromones. Ablation of *Panagrellus silusiae* spicules inhibited sex pheromone reception, leading to the suggestion that the spicules contain a chemoreceptor (Samoiloff *et al.*, 1973). Additional chemoreceptors are located on the male tail and the sex organs in many nematode species, including the bursa of *C. elegans* with 18 genital papillae (Sulston and Horvitz, 1977).

3.2.2. Mechanoreceptors

More than 16 types of mechanoreceptive organs capable of detecting mechanical stimuli can be found in the nematode anterior end. Kaplan and Horvitz (1993) reported that one of the amphidial neurones of *C. elegans* may be able to detect mechanical stimuli. Although nematode mechanoreceptors are variable among species, a detailed cuticular mechanoreceptor diagram is described by Jones (2002), based on *Pratylenchus penetrans* work by Trett and Perry (1985). Touch receptors have been investigated mostly in *C. elegans* (Chalfie and Sulston, 1981), but studies are needed to understand and compare the function of touch receptors of nematodes living in a wide range of habitats. Hermaphrodites of *C. elegans* have many ciliated neuronal processes embedded within the cuticle that may have a mechanoreceptive function (Chalfie and White, 1988). Mechanoreceptors are also found in the copulatory spicules and bursa. For example, the spicules of *Nippostrongylus brasiliensis* contain nerve axons (Lee, 1973). Males of *C. elegans* have mechanosensory neurones in their tails that function during mating (Sulston and Horvitz, 1977).

McLaren (1976a,b) describes other mechanoreceptors, all poorly characterized, including stretch receptors, deirids, bristles and setae, mostly from marine nematodes. The marine nematode *Bathylaimus tarsioides* has long jointed bristles on its head but their function was not determined (Croll, 1970). The somatic setae and papillae of *Deontostoma californicum* may be capable of perceiving mechanical stimuli (Croll, 1970). Mechanoreceptive organs have also been described in several marine nematodes, for example, *Cyatholaimus*, *Longicyatholaimus* and *Choniolaimus* (Inglis, 1963). Additional studies are needed to understand the sensory role of these structures.

3.2.3. Photoreceptors

Some nematode species have receptors capable of detecting and responding to light, and yet nematode photoreception has received little study (Jones, 2002; Lee, 2002).

With the exception of the insect-parasitic nematode, *Mermis nigrescens*, photore-ceptors have been reported mostly from free-living nematodes from marine or freshwater environments. Nematode eyes or ocelli are easily visible as single or paired dense accumulations of pigment located a short distance behind the anterior tip. Together with head-waving, scanning motions or exploratory locomotion, the various photoreceptive structures provide information about the distribution of light and dark in the nematode environment (Burr, 1979, 1984b; Burr and Babinszki, 1990; Burr *et al.*, 1990). Three types of eye-spot pigmentation have been observed: granular resembling melanin granules, non-granular of unknown com-position (*Araeolaimus elegans*, Croll *et al.*, 1975) and the crystalline oxyhaemoglo-bin of *Mermis* eyes. The latter was identified by absorption spectra *in situ* and in extracts, and has chemical properties and an amino acid sequence characteristic of other nematode haemoglobins (Ellenby, 1964; Burr *et al.*, 1975; Croll *et al.*, 1975; Burr and Harosi, 1985). Pigment composition of freshwater and marine nematodes has been investigated by Croll *et al.* (1972, 1975), Croll (1974) and Bollerup and Burr (1979). With the exception of *A. elegans*, the eye-spots of freshwater and marine species are composed of a granular pigment that resembles melanin granules (melanosomes) (Burr and Webster, 1971; Croll *et al.*, 1972, 1975; Van de Velde and Coomans, 1988). The pigments range from red-brown and purple to bright red or orange. Absorption spectra vary, with the browner pigments resembling eumelanin, which has a broad spectrum in the visible region. Eye-spot size and pigment density increase with age and, at the higher absorbance, the colour becomes browner or blacker (Bollerup and Burr, 1979). The eye-spot pigments vary in stability and chemical properties; in some species they fade after death, while in *Enoplus* they remained stable even in strong HCl. Histochemical tests support the hypothesis that eye-spot pigments in *Enoplus*, *Deontostoma*, *Seuratiella* and *Oncholaimus vesicarius* are either eumelanin or a less stable melanin type (Bollerup and Burr, 1979).

Other eye-spot pigments have been more difficult to identify. The granular eye-spot pigments of *Chromadorina*, *Symplocostoma*, *O. vesicarius* and *O. skawensis* have several properties in common that differ from those indicative of melanins and cannot be identified with any of the main classes of animal pigments. They might be an unusual melanin or a carotenoprotein (Croll *et al.*, 1972; Bollerup and Burr, 1979). Nematodes that do not have ocelli can detect light by having adaptations on their amphids consisting of pigment spots that shade the amphids. Pigmented eye-spots have been found associated with the pharynx and some pigments have occasionally been found in the epidermis (Bollerup and Burr, 1979).

3.2.4. Thermoreceptors

All nematodes are sensitive to temperature as it is important to development. Thermoreceptive neurones have been found in several nematode species (Burman and Pye, 1980). Soil-dwelling nematodes possess sensory neurones with extensive microvillar structures similar to those of *C. elegans*, which can detect thermal gradients of less than 0.1°C (Hedgecock and Russell, 1975). Animal-parasitic and marine nematodes have simpler thermoreceptive neurones (Jones, 2002). An amphid sensory neurone, AFD, has been identified in *C. elegans* as the major thermosensory neurone, and has been studied using mutants and laser ablation (Perkins *et al.*, 1986). Mori and Ohshima (1995) reported that destruction of the AFD neurone prevented *C. elegans* from detecting thermal signals. In addition, two interneurones have been identified with thermoreceptive properties; the amphid interneurone AIY, which is associated with thermophilic movement (preference for higher than cultivation temperature) and the amphid interneurone AIZ, which is associated with cryophilic movement (preference for lower than cultivation temperature) (Mori and Ohshima, 1995). Ashton *et al.* (1999) reported that in *C. elegans* the thermoreceptive cells are located in digitiform amphidial neurones. In *S. stercolaris*, a highly modified amphidial neurone, the lamellar cell, appears to be the main thermoreceptive neurone. Ashton *et al.* (1999) performed a three-dimensional reconstruction of the chemo- and thermosensory sensilla of *Ancylostoma caninum* used in host location and developmental control.

3.2.5. Electric and magnetic field receptors

Nematode receptors capable of detecting electric and magnetic signals remain largely uncharacterized. Sukul *et al.* (1975) suggested that the amphids contain large negatively charged, poorly soluble, organic molecules that attract cations in water. Possibly the amphids sense the potential change resulting under an applied electrical field. However, the above study is an isolated case and further research is needed to find receptors capable of detecting electric and magnetic signals in nematodes.

3.3. Chemo-orientation

Chemo-orientation is the ability to compare the gradient of an attractant or repellent compound to determine if the organism will orient to and move up or down the gradient. Chemo-orientation is essential for detection of host plant exudates, food stimulants, food deterrents and sex pheromones and therefore is vital to survival. Chemical perception is considered to be olfaction or taste and the main chemoreceptive organs are the amphids (Riga *et al.*, 1995a; Bargmann and Mori, 1997).

Nematodes might respond to a chemical gradient using three possible mechanisms (Ward, 1976). In the first mechanism, nematodes use either single or multiple receptors to compare concentrations successively at time intervals separated by forward movement. In the second mechanism, nematodes use two different receptor groups to compare concentrations simultaneously (tropotaxis). In the third mechanism, nematodes use successive comparisons in time by side-to-side displacement of receptors (klinotaxis). Ward (1976) reported that only the third mechanism, klinotaxis, is used by *C. elegans* for chemo-orientation. He rejected the first mechanism based on forward velocity by using mutants who responded to a chemical gradient, although their body muscles had degenerated. The second mechanism, tropotaxis, was rejected because the chemoreceptive organs, the amphids and the inner labial papillae, are closely located and therefore differences in chemical concentration would be too difficult to detect. Orientation using head to tail comparison was also rejected, as mutants with blisters covering the phasmids were still able to orient (Ward, 1973). However, killing neurones either on the left or on the right chemosensilla of *C. elegans* did not perturb nematode chemotaxis, thus lending support to the klinotaxis hypothesis of Bargmann and Horvitz (1991) and Bargmann and Mori (1997). However, it is still unclear if nematode responses to chemical gradients are mediated either with klinotaxis or with tropotaxis (Perry and Aumann, 1998). Further work is needed, preferably in a three-dimensional matrix.

Recent reports show that *C. elegans* detects chemical stimuli with the help of several chemosensory receptor genes. The ODR-10 superfamily is the largest chemoreceptive gene family, consisting of 831 genes and pseudogenes that encode G-protein-coupled, seven-transmembrane domain receptors (Peckol *et al.*, 2001). Additional gene families have been identified with possible chemoreceptive functions based on their sequence and their expression in chemosensory neurones (CSC (*C. elegans* Sequencing Consortium), 1998).

3.3.1. Semiochemicals

Semiochemicals cause inter- and intraspecific interactions between animals and provide signals for nematode orientation. Semiochemicals consist of allelochemicals and pheromones. Allelochemicals cause behavioural or physiological responses between members of different species while pheromones are chemicals that cause responses between members of the same species (Huettel, 1986; Perry and Aumann, 1998). Allelochemicals consist of allomones, which elicit a negative response by the receiver, kairomones, which elicit a positive response by the receiver, synomones, which elicit a positive response both by the receiving and the emitting organism, and apneumones, which are signals produced by non-living matter. Sex pheromones elicit responses between or within the sexes, epidietic pheromones regulate population densities and alarm pheromones are warning or protective signals. Alarm pheromones have not yet been described in nematodes (Huettel, 1986). Most semiochemical work in nematodes has focused on sex and aggregation chemical communication. Sex pheromones have been demonstrated in several plant- (Jonz *et al.*, 2001) and animal-parasitic nematodes (Riga and MacKinnon, 1988). A mate-finding cue is produced by hermaphrodites of the free-living nematode *C. elegans* to attract males (Simon and Sternberg, 2002). Aggregation and attraction have been demonstrated in many nematode species. For example, *Meloidogyne javanica* and *Globodera rostochiensis* juveniles respond to tomato (Prot, 1980) and potato (Rolfe *et al.*, 2000) root diffusates, respectively.

3.3.1.1. Chemoperception

Few compounds involved in chemoperception of plant-parasitic nematodes have been investigated. One of these chemicals is β-myrcene, which is produced by pines and attracts the pine wood nematode, *Bursaphelenchus xylophilus* (Ishikawa *et al.*, 1986). The specificity of nematode responses to semiochemicals has been demonstrated (Riga *et al.*, 1995a, 1996b, 1997; Rolfe *et al.*, 2000). In addition, extracellular electrophysiological assays (EEAs) revealed that *Brugia pahangi* and *Leidynema appendiculata* showed a concentration-dependent response to 100, 10 and 1 mM solutions of acetylcholine (Rolfe *et al.*, 2000, 2001; Perry, 2001).

Chemoperception in nematodes has tended to be evaluated based on assays that count nematodes at different distances from the source. Depending on the biology of the animal, several novel methods have been used to observe nematode chemoperception and orientation to chemical stimuli, including photographing nematode tracks made on agar (Riddle and Bird, 1985; Riga and Webster, 1992), migration chambers filled with liquid medium (Riga and MacKinnon, 1987), narrow agarose tracks that restrict nematode movement to a linear direction (Castro *et al.*, 1989) and recently EEA (Riga *et al.*, 1995a; Rolfe *et al.*, 2000). To analyse and quantify chemoperception and orientation, it is necessary to devise behavioural bioassays that mimic the animal's environment.

Clemens *et al.* (1994) reported that attractants were differentiated from repellents by counting the number of movements towards and away from a root exudate gradient. Infective juveniles of *Heterodera schachtii*, for example, displayed a higher number of forward than away movements from the gradient. Although these behavioural bioassays provide useful chemo-orientation information, they are difficult to quantify. EEA from the cephalic region of nematodes is a more sensitive and quantifiable method than previous behavioural bioassays (Jones *et al.*, 1991; Riga *et al.*, 1995a; Rolfe *et al.*, 2000). Functional information of nematode sense organs derived from electrophysiology can be greatly enhanced if combined with conventional behavioural assays (Riga *et al.*, 1996a), laser ablation (Kaplan and Horvitz, 1993) and immunocytochemistry (Davis *et al.*, 1992; Stewart *et al.*, 1993a,b).

In addition to behavioural assays, changes in nematode physiological responses can be used as means to study responses to chemical stimuli, i.e. behaviours observed before or during nematode copulation can be a useful tool. Physiological responses demonstrated by male *H. glycines* (i.e. coiling) can be used to identify biologically active substances. Jaffe *et al.* (1989) used male coiling behaviour to identify vanillic acid as a putative sex pheromone of *H. glycines*. Males of *H. schachtii* exhibited stylet-thrusting behaviour upon exposure to pheromones produced by females (Jonz *et al.*, 2001). Hatching stimuli are also related to chemoperception in nematodes and evoke behavioural responses and physiological changes. Doncaster and Shepherd (1967) reported that *G. rostochiensis* juveniles responded to potato root diffusates by moving inside the egg and by using the stylet to emerge. They also investigated differences in behaviour patterns of *G. rostochiensis* juveniles in response to different hatching agents. The formation of developmentally arrested dauer juveniles in *C. elegans* is a physiological response to starvation or overcrowding stimuli perceived by the nematodes (Cassada and Russell, 1975). Golden and Riddle (1982) showed that a dauer pheromone produced by *C. elegans* induces dauer formation even in the presence of adequate food and lack of crowding. The *C. elegans* dauer exits arrested development in response to a reduction in pheromone concentration and an increase in food (Riddle, 1988).

Orientation of predatory nematodes to prey attractants was examined by Bilgrami and his co-workers. Diplogasterid predators, *Mononchoides longicaudatus* and *M. fortidens* (Bilgrami and Jairajpuri, 1988), and dorylaim, *Mesodorylaimus bastiani,* and nygolaim predators, *Aquatides thornei* (Bilgrami and Pervez, 2000a), showed attraction towards plant-parasitic and free-living nematodes as prey. The diplogasterid predators were also attracted to bacterial colonies in the agar plates (Bilgrami and Jairajpuri, 1988). Depending upon the attraction responses, Bilgrami and Pervez (2000a) divided prey-searching behaviour of predatory nematodes into random movement, exploration and probing, perception of attractants, switched response, coordinated (stimulated) locomotion and chemotactic response. While moving randomly, predators make chance contacts with prey. Predators may also establish contacts by wilful movements if they perceive prey-emitted attractants. Upon perceiving prey attractants, the predator response changes from uncoordinated (random) to coordinated (stimulated) locomotion showing a more directional approach towards prey nematodes. Upon establishing contact, predators probe the prey body by lip rubbing. Prey attractants appear to play a role in changing predator locomotion from uncoordinated to coordinated. During coordinated locomotion predators show exploratory behaviour to find prey.

Nematodes are also attracted to nematode-parasitic fungi (Jansson and Nordbring-Hertz, 1979, 1980; Jansson, 1982a; Zuckerman and Jansson, 1984), including conidia (Jansson, 1982a,b). Nematodes showed a stronger attraction to obligatory endoparasitic fungi than to saprophytic fungi (Jansson and Nordbring-Hertz, 1979). Balan *et al.* (1976) showed that the nematode-trapping fungus *Monacrosporium rutgeiensis* produces three nematode attractants.

3.3.1.2. Aggregation

Steiner (1925) was among the first to report aggregation behaviour in noting *Meloidogyne* juveniles gathering at root tips. Lucerne roots infected with *Fusarium oxysporum* produced more CO_2 than healthy roots, which resulted in increased attraction of the lesion nematode *P. penetrans* to the infected roots (Edmunds and Mai, 1967). Pitcher (1967) studied the aggregation behaviour of seven species of *Trichodorus* and reported that they responded to root exudates from long distances. The marine nematode *Metoncholaimus* was attracted to cultures of marine fungi (Meyers and Hopper, 1966). Entomopathogenic nematodes use their sensory perception to locate insects in response to host carbon dioxide (Gaugler *et al.*, 1980), faeces (Kondo, 1989; Lewis *et al.*, 1992; Grewal *et al.*, 1994) and plasma (Khlibsuwan *et al.*, 1992).

EEA recordings from the cephalic region of adult males of the potato cyst nematodes, *G. rostochiensis* and *G. pallida*, along with behavioural assays in Petri dishes showed that these nematodes perceive various chemical stimuli (Riga *et al.*, 1997). Exposure to acetylcholine and L-glutamic acid elicited strong electrophysiological responses and positive movement towards the sources. Males of *G. rostochiensis* responded positively towards D-tryptophan and α-aminobutyric acid, whereas males of *G. pallida* responded to γ-aminobutyric acid (Riga *et al.*, 1997). Prot (1977) reported long-distance attraction of *M. javanica* to tomato roots and differential migration of *M. javanica* towards resistant and susceptible tomato roots. Host and non-host trap crops attract plant-parasitic nematodes. For example, Franco *et al.* (1999) reported that a number of plants, including *Oxalis tuberosa*, *Ullucus tuberosus*, *Tropaeolum tuberosum*, *Lupinus mutabilis* and *Chenopodium quinoa*, have potential for use as trap crops against potato cyst nematodes. *Asparagus officinalis* and *Tagetes erecta* release root exudates attractive to a wide range of nematodes (Bilgrami *et al.*, 2001a), yet invading nematodes are killed due to nematicidal plant chemicals such as glycosides (*A. officinalis*) and thiophene (*T. erecta*) (Bilgrami, 1997). Identifying additional compounds and understanding their mode of action could provide a tool for plant nematode management.

Bargmann *et al.* (1993) positioned a volatile organic molecule, isoamyl alcohol, on the lid of the Petri dish, avoiding any contact with the agar, and *C. elegans* responded rapidly. *Panagrellus redivivus* was attracted to four bacteria species (Jansson and Nordbring-Hertz, 1980) and to cell-free filtrates of certain yeasts and fungi (Balanova and Balan, 1991). Other free-living nematodes, *Acrobeloides* and *Mesodiplogaster lheritieri*, are attracted to chemicals produced by their bacterial food (Anderson and Coleman, 1981). The fungal feeder *Neotylenchus linfordi* aggregates at fungal mycelia secretions (Klink *et al.*, 1970). Similarly, four species of marine bacterial-feeding nematodes were attracted to chemicals emitted from decaying marsh grass, *Spartina anglica* (Moens *et al.*, 1999).

Little information is available on the pre- and post-feeding aggregation behaviour of predatory nematodes. Bilgrami and Pervez (2000b) reported that the predatory nematodes *Discolaimus major* and *Laimydorus baldus* showed prolonged post-feeding aggregation at feeding sites which were formed earlier. Post-feeding aggregation behaviour may be due to the lingering effects of attractants left by the consumed prey (Bilgrami and Jairajpuri, 1989a) or to predator oesophageal secretions released into the prey for extracorporeal digestion (Bilgrami and Jairajpuri, 1989b). Bilgrami and Pervez (2000b) hypothesized that aggregation at the feeding site increases food sharing and opportunities for mating.

Entomopathogenic nematodes also display orientation and attraction to their insect hosts (Gaugler *et al.*, 1980). *Steinernema glaseri* displayed orientation and chemo-attraction to wax worms, *Galleria mellonella*, by cruising along the path of the chemical gradient formed by kairomones emitted by this highly susceptible insect host; however, there was no attraction to the non-host insect species *Spodoptera litura*, *Blattella germanica* and *Locusta migratoria* (Bilgrami *et al.*, 2000, 2001c). *Steinernema glaseri* also oriented towards five concentrations of root and leaf homogenates of *A. officinalis* and *T. erecta* (Bilgrami *et al.*, 2001a) and to nitrogenous insect products, including 0.2–1.0% urea, 1% uric acid, 0.4–0.8% allantoin and 0.6–1.0% arginine (Bilgrami *et al.*, 2001b).

Plant-parasitic nematodes are attracted to CO_2 produced by roots and bacteria in the rhizosphere (Prot, 1980; Dusenbery, 1987; Robinson, 1995). *Meloidogyne incognita* responds and orients to *in vitro* CO_2 gradients as low as 1% change per cm (Pline and Dusenbery, 1987). Similarly, *M. incognita, Rotylenchulus reniformis* and *S. glaseri* were attracted to CO_2 when placed on a linear gradient of 0.2% per cm at a CO_2 concentration of 1.2% (Robinson, 1995). Dusenbery (1987) suggested that, although CO_2 is a non-specific stimulus, it might lead nematodes to a certain soil depth. The reniform nematode, *R. reniformis* is attracted to inorganic salts, including Na^+, Mg^+, Cl^-, OAc^- and adenosine monophosphate (AMP) and cyclic AMP, in a concentration-dependent manner (Riddle and Bird, 1985).

Chemo-orientation of nematodes to plant roots is well established (Bird, 1959; Prot, 1980; Moltmann, 1990; Grundler *et al.*, 1991). Infective juveniles of *Meloidogyne naasi* were attracted to excised root tips of susceptible cultivars of barley and wheat, whereas a resistant grass, *Aegilop variabilis*, was less attractive (Balhadère and Evans, 1994). Castro *et al.* (1989) reported that *M. incognita* is attracted to volatile fractions obtained from cucumber roots. Several volatile molecules evoked chemo-orientation in *M. incognita* and *C. elegans* (McCallum and Dusenbery, 1992; Bargmann *et al.*, 1993). Rolfe *et al.* (2000) reported that infective juveniles of *G. rostochiensis* responded to the presence of potato root diffusates, but no response was noted when the juveniles were exposed to diffusates from sugar beets, a non-host.

3.3.1.3. Sex pheromones

Sex pheromones facilitate communication behaviour between males and females of the same species (Huettel, 1986; Bone, 1987). Greet *et al.* (1968) noted that mating in nematodes would be rare if they were to rely on random contact, especially when population densities were low. The first sex attraction was observed in the free-living nematode *Panagrolaimus rigidus* (Greet, 1964), followed by *Cylindrocorpus longistoma, C. cruzii* (Chin and Taylor, 1969) and *Pelodera strongyloides* (Stringfellow, 1974). To date, the chemical communication behaviour of approximately 40 species of free-living and plant-, invertebrate- and animal-parasitic nematodes has been studied. Male *C. elegans* move towards hermaphrodites but it is unclear if this response is due to chemotaxis or kinesis (Emmons and Sternberg, 1997). Chemoattraction of *Steinernema carpocapsae* males to pheromones produced by homospecific females was reported by Lewis *et al.* (2002).

Sex pheromones can be produced either by the females or by both sexes. For example, in *C. longistoma, C. cruzii* and *A. caninum* the sex pheromone is produced by females only (Roche, 1966; Chin and Taylor, 1969), whereas in *Heligmosomoides polygyrus* production is by both males and females (Riga and MacKinnon, 1987). Some nematode species have sedentary females, indicating that the sex pheromone is female-produced (Huettel and Rebois, 1986; Riga *et al.*, 1996a; Jonz *et al.*, 2001).

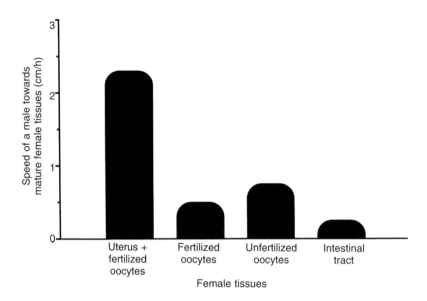

Fig. 3.1. Average speed of a male *Heligmosomoides polygyrus* responding to homogenized tissues of a mature female.

The site of sex pheromone production is known for only a few species. Female cyst nematodes, *H. schachtii* and *G. rostochiensis*, secrete sex pheromones over their entire body surface (Green and Greet, 1972). Cheng and Samoiloff (1972) reported that the gonads of the free-living nematode *P. silusiae* produced the pheromone. Anya (1976) suggested that the site of pheromone production in the female pinworm, *Aspiculuris tetraptera*, might be the pulvillar cells, which are secretory cells in the reproductive system, while male pheromones may be secreted by the caudal glands. Pheromone release by *P. redivivus* females is related to the presence of unfertilized oocytes in the oviducts (Duggal, 1978). The source of pheromones by *H. polygyrus* females is the uterine epithelium (Fig. 3.1; Riga and MacKinnon, 1988).

Pheromone production is influenced by nematode age. Production in *H. glycines* peaked 3 days post-emergence from the host plant and then declined with age (Rende *et al.*, 1981). Several studies have compared the attractiveness of mated and virgin female nematodes. Virgin females of *Rhabditis pellio* (Somers *et al.*, 1977), *P. silusiae* (Cheng and Samoiloff, 1971), *P. redivivus* (Duggal, 1978), *H. schachtii* and *G. rostochiensis* (Green *et al.*, 1970) were more attractive to males than to mated females. Similarly, *S. carpocapsae* males were attracted to virgin but not to mated females (Lewis *et al.*, 2002). Riga and MacKinnon (1988) reported that mated females of *H. polygyrus* were more attractive to males than virgin females; however, the physiological state and age of the females had no effect on the attractiveness of *N. brasiliensis* males to them (Bone and Shorey, 1978).

Characterization of sex pheromones from several nematode species has been reported, although a chemical structure has not been obtained. One of the main reasons is that assays used to analyse nematode behaviour tend to be insufficiently sensitive and are difficult to quantify. EEA provides the advantage of determining whether the difference in recognition of a stimulus is due to differences in sensory responses or to other factors, such as differences in nematode mobility. These factors can complicate the interpretation of Petri dish behavioural bioassays (Riga *et al.*, 1995b). EEA provided detailed analysis of nematode responses and was used to verify the chemosensory function of individual amphids of intact *S. trachea* (Riga *et al.*, 1995a). The response to male sex pheromones of *G. rostochiensis* and *G. pallida* has also been demonstrated using EEA. Males of *G. rostochiensis* only responded to intraspecific sex pheromones from *G. rostochiensis* females; in contrast, *G. pallida* males responded to the sex pheromones from females of both *G. pallida* and *G. rostochiensis*. However, the response of *G. pallida* males was significantly reduced in the heterospecific tests (Riga *et al.*, 1996a).

A few attempts have been made to isolate and analyse nematode pheromones and their chemical and physical properties. For example, the sex pheromone of *Pelodera teres* females was water soluble (Jones, 1966), the pheromone of *N. brasiliensis* was a phenylalanine-based compound consisted of two fractions of 500 and 375 relative molecular weight (Bone *et al.*, 1979), and the crude pheromone of *B. xylophilus* was a highly polar and hydrophilic compound with a relative molecular weight of less than 3000 (Riga, 1992). Vanillic acid has been structurally identified as a putative female sex pheromone of *H. glycines* (Jaffe *et al.*, 1989). Partial isolation of two volatile components of *H. schachtii* pheromone was reported by Aumann *et al.* (1998a), along with a water-soluble component from monoxenic cultures of *H. schachtii* (Aumann *et al.*, 1998b). Jonz *et al.* (2001) partially isolated the sex pheromone of *H. schachtii*, using high-performance liquid chromatography. The putative pheromone was water-soluble, was heat-stable, did not attract second-stage juveniles and was not stored within the female gelatinous matrix. This putative pheromone stimulated adult male chemotaxis in a concentration-dependent manner and caused chemoattractive behaviour after partial isolation.

3.3.1.4. Inhibition, avoidance and repellency

Chemicals can also interfere with or inhibit orientation. The reduced attraction exhibited by males of *N. brasiliensis* (Alphey, 1971) and the fish parasite *Camallanus* (Salm and Fried, 1973) towards their homospecific females when several males were in close proximity was attributed to physical contact interfering with male response. Bone and Shorey (1977a) reported a similar phenomenon in *N. brasiliensis,* but inhibition occurred when males were not allowed physical contact, suggesting the presence of an inhibitory chemical. Likewise, behavioural bioassays using more than one responding male indicated that males inhibit the movement of other males towards target females of *H. polygyrus* (Fig. 3.2; Riga and MacKinnon, 1987), providing further evidence of an inhibitory chemical secreted by competing males. The chemical nature of this inhibitory substance is not known.

Among the survival strategies developed by entomopathogenic nematodes, *Steinernema* suppress invasion of a host previously invaded by conspecific or

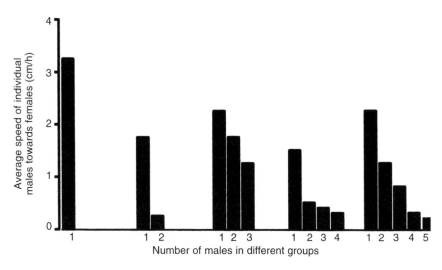

Fig. 3.2. Individual responses of male *Heligmosomoides polygyrus* in groups of one to five males to ten females.

heterospecific nematodes. Infective juveniles of *S. carpocapsae* were repelled from insect hosts infected for at least 4 h with heterospecific nematodes (Grewal *et al.*, 1997). Selvan *et al.* (1993) speculated that an inhibitory substance diffuses from the host and prevents subsequent invasion.

Nematodes show avoidance behaviour as well. Wild-type *C. elegans* display osmotic avoidance behaviour when exposed to high concentrations of a range of chemicals, e.g. NaCl, fructose and sorbitol (Ward, 1973; Culloti and Russell, 1978). However, lower concentrations of the same chemicals evoked a chemo-attractive response. Avoidance responses in *C. elegans* were also observed in the presence of cadmium, zinc and copper ions (Sambongi *et al.*, 1999), D-tryptophan, garlic extracts, camphor, menthol and α-terthienyl (Ward, 1973; Dusenbery, 1974; Bargmann and Mori, 1997) and in the presence of quinine, quinidine and other antimalarial drugs (Bazzicalupo *et al.*, 1994).

Nematodes can also be repelled by chemicals. Males of *G. pallida* moved away from citric acid (Riga *et. al.*, 1997); *C. elegans* is repelled by volatile long-chain ketones and alcohols (Bargmann *et al.*, 1993); *M. incognita* is repelled by a non-volatile fraction obtained from cucumber roots (Castro *et al.*, 1989); and salts of specific ions repelled *M. incognita* sufficiently to protect tomato roots (Castro *et al.*, 1991). Alginate pellets containing hyphae of the nematophagous fungi *Monacrosporium cionopagum*, *Monacrosporium ellipsosporum* and *Hirsutella rhossiliensis* repelled juveniles of *M. incognita* (Robinson and Jaffee, 1996). Although *S. glaseri* is attracted to a range of nitrogenous insect products, the nematodes are deterred by ammonia at all tested concentrations ranging from 0.2 to 1.0% (Bilgrami *et al.*, 2001b).

3.4. Mechanoreception

Nematode responses during hatching, feeding, penetration, moulting and copulation are partly due to mechanical stimulation. For example, *H. schachtii* penetrates hydrophobic nitrocellulose membranes in response to physical contact (Dickinson, 1959). Similarly, juveniles of plant-parasitic nematodes, e.g. *G. rostochiensis*, orient to a slit they make during egg hatching, which is associated with mechanostimulation (Doncaster and Seymour, 1973). Nematodes also respond to tactile stimuli, including *N. brasiliensis*, which displays thigmokinetic behaviour to remain among the host intestinal villi in response to pressure (Lee, 2002).

A number of touch stimuli evoke responses in *C. elegans*, which recoil when they bump into an object (Croll, 1970) and reverse direction when touched near the head with a fine hair (Chalfie and Sulston, 1981). Individuals that do not respond to mild touch stimuli will move away from stronger touch stimuli (Chalfie and Sulston, 1981; Chalfie *et al.*, 1985). In addition, touching *C. elegans* often causes pharyngeal pumping to stop (Chalfie *et al.*, 1985). The neural circuit associated with a gentle touch to the body of *C. elegans* consists of six neurones (Duggan *et al.*, 1996), which are associated with 13 genes (Gu *et al.*, 1996).

Angle sense is an additional nematode behaviour due to mechanical stimulus. Croll (1970) summarized reports on nematode angle sense but there has been little recent research progress. Hookworm juveniles orient at right angles to a solid flat surface which presumably assists in host penetration (Croll, 1970). Wallace (1968) reported that infective juveniles of *M. javanica* orient at right angles to the host plant roots.

3.5. Photo-orientation

Nematode species that either have eyes or respond to light are found sporadically in several orders. Chitwood and Murphy (1964) provided the first clear demonstration of nematode photo-orientation with *Diplolaimella schneideri*. These marine nematodes swam away from horizontal light and reoriented about 5 min after the vessel was rotated 180°. In light projected from below they directed their heads towards the water surface. Burr (1979) investigated photoreception in the marine nematodes *O. vesicarius* and *Enoplus anisospiculus* using heat-blocked white light to provide equal illumination in the arena without a temperature gradient. A significantly negative photo-orientation was observed.

The positive photo-orientation of *M. nigrescens* gravid females has been measured in a variety of different ways. On horizontal or vertical felt surfaces and in the grass, the female's body and elevated anterior are oriented towards the light source as she crawls toward the source (Burr *et al.*, 1989). Phototaxis was maximal over a wide range of wavelengths, 350–540 nm, the approximate range of skylight. The half-maximum response occurred at intensity equivalent in effectiveness to pre-dawn twilight (Burr, 1979). Thus positive photo-orientation would be useful to guide the egg-laying female to upper vegetation, where her insect herbivore host is likely to feed. Immature adult females or pre-adult females of *M. nigrescens,* which lack the shadowing pigment, are not phototactic or are slightly negatively photo-tactic (Burr *et al.,* 2000).

Robinson *et al.* (1990) discovered a transverse photo-orientation in the infec-tive juveniles of *Agamermis catadecaudata* and *Hexamermis.* The infective juveniles of these Mermithidae hatch in the soil and ascend vegetation, where they discover and invade their grasshopper host. They do not have a distinguishable pigment spot or eye structure; however, they display an unusual photo-orientation, in which the juveniles on agar orient precisely perpendicular to the light direction and crawl towards that direction. Several observations indicate that the nematode phototactic responses are due to light direction and not temperature changes or gradients: (i) passing the light through 10 cm of water does not affect the photo-orientation, although they respond to radiant heat with a withdrawal response; (ii) orientation to light occurs within 2–15 s of turning on the light and, on rotation of the agar plate, they quickly reorient; (iii) the nematodes orient perpendicular to light for more than 8 h; and (iv) orientation occurred regardless of angle of incidence until the source was nearly perpendicular. The wavelength dependence, tested between 400 and 1000 nm, peaked at about 450–475 nm and decreased to zero at about 550–600 nm. The authors speculated that transverse photo-orientation might help to guide the juveniles up or down grass stems in the presence of low-angle sunlight. However, the ability of these nematode species to detect light direction without a shadowing pigment is not understood.

Infective juveniles of *Haemonchus contortus* migrate from dung, ascend grass and wait for a warm-blooded ruminant host. In a study of vertical migration, Rees (1950) spread infective juveniles on the soil surface of experimental pots at mid-night and sampled grass at 2 h intervals over 24 h while recording incident light intensity and the maximum and minimum temperature and humidity in the grass. The number of nematodes recovered in grass consistently reached a noticeable maximum when the incident light was low, at dawn or 2 h after dawn and at sunset or 2 h before sunset. Changes in temperature or humidity did not seem to trigger migration. Thus, vertical migration of this species is activated by dim light. As light sensitivity might be especially advantageous to species that climb vegetation, further studies would be desirable in which the light stimulus is sufficiently filtered and temperature sufficiently regulated.

The mechanism of photo-orientation in nematodes is poorly understood. The female *M. nigrescens* elevates the anterior 2 mm of her body periodically, allowing scanning motions and orientational bending of the next 13 mm of the body, the 'neck'. These activities swing the anterior tip and eye in a sampling motion that causes the cylindrical shadowing pigmentation to modulate the illumination of the photoreceptors inside. During photo-orientation, deviations of the base of the head from the direction to the light source are corrected by bending the neck, and loco-motion follows the orientation of the neck.

3.6. Thermo-orientation

Thermoreception has been frequently characterized in nematodes. Response to temperature is valuable to nematodes as it influences their dispersion and assists with food search (Robinson, 1989). Dusenbery (1988) reported that in the thermal environment of the soil and in the absence of other stimuli, nematodes in the top 20 cm of the soil move towards a level of 5 cm, considered optimal for host search. Similarly, nematode parasites of warm-blooded animals use chemical and thermal signals in host finding (Ashton *et al.*, 1999). The dog hookworm, *A. caninum*, and the threadworm *S. stercolaris* use both thermal and chemical signals to locate their hosts (Ashton *et al.*, 1999).

One of the earliest reports on nematode thermotaxis showed that several nematode species moved towards higher temperatures in very shallow gradients, e.g. 0.03°C per cm (El-Sherif and Mai, 1969). Two decades later, Pline *et al.* (1988) reported similar results from *M. incognita* juveniles. When eggs were acclimated to 23°C, the hatching juveniles migrated towards higher temperatures even at a threshold gradient below 0.001°C per cm. Sensitive thermotaxis of this nature is invaluable for nematodes as they are able to detect even the smallest thermal gradients in the soil. El-Sherif and Mai (1969) hypothesized that plant-parasitic nematodes may be able to locate plant roots via thermal gradients created by heat production from the roots during metabolism and from rhizosphere bacteria. The extraordinary sensitivity of *M. incognita* juveniles to thermal gradients bolsters El-Sherif and Mai's premise. Dusenbery (1987), however, suggested that thermal gradients may not lead the nematodes directly to the roots, but rather permit straight path orientation until they detect chemical signals produced by roots.

Entomopathogenic nematodes move towards their host in response to the insect body temperature (Byers and Poinar, 1982). Infective juveniles of *S. carpocapsae* grown at 15, 20 and 25°C migrated towards a gradient corresponding to their respective growth temperatures when assayed immediately after collection from the culture medium (Burman and Pye, 1980).

Robinson (1994) reported that both temperature and gravity can differentially influence vertical dispersal of ecologically different nematodes. He subjected nematodes to fluctuating temperature for 20 to 36 h in tubes filled with moist sand. When heat waves were propagated horizontally, *R. reniformis* moved away from and *M. incognita* moved towards the thermal surface; the movement of *Ditylenchus phyllobius*, *S. glaseri* and *Heterorhabditis bacteriophora* was random. Robinson (1989) studied the thermotactic behaviour of two foliar and two root-parasitic nematodes using water agar horizontal assays. The foliar nematodes *D. phyllobius* and *Ditylenchus dipsaci* were attracted to much lower temperatures than the root nematodes *R. reniformis* and *Tylenchulus semipenetrans*. Wallace (1961) reported negative thermotaxis by *D. dipsaci*, while Croll (1967) reported that the negative thermotaxis displayed by *D. dipsaci* was adaptive as nematodes cultured at 10, 20 and 30°C aggregated on gradients near their culture temperatures. Both juveniles and adults of *C. elegans* when placed on a thermal gradient migrated to the temperature at which they have been cultured (Hedgecock and Russell, 1975). This behaviour can be modified by acclimation of *C. elegans* at a new temperature; however, starved adults and dauer juveniles grown at 20°C will migrate away from the culture temperature.

Thermo-avoidance behaviour, mediated by sensory neurones, has been reported in *C. elegans*. When *C. elegans* was exposed to a deleterious temperature, i.e. 33°C, using an electronically controlled heated platinum wire, the nematodes responded by a withdrawal reflex (Wittenburg and Baumeister, 1999).

3.7. Orientation to Magnetic Fields and Electric Currents

Several nematode species respond to electric currents and magnetic fields. Bird (1959) described galvanotaxis and electrotaxis in plant-parasitic nematodes, and suggested that these nematodes orient along a gradient produced by a lower redox potential at the root surface. In many plants, the root surface has a negative electrical potential. *Turbatrix aceti, H. schachtii* and *D. dipsaci* displayed directional movement towards magnetic poles at 20–30 mV (Croll, 1970). *Tylenchus, Pratylenchus, Aphelenchoides, Dorylaimus, Rhabditis, Noecephalobus* and *Chiloplacus* showed a galvanotactic response towards the cathode when placed in an electric field (Caveness and Panzer, 1959), while juveniles of *P. strongyloides* moved towards the anode (Whittaker, 1969). *Anguina tritici* juveniles migrated towards the cathode when introduced between platinum electrodes at 0.06 mÅ of current in sand saturated with water or salt solution (Sukul *et al.*, 1975). Sukul and Croll (1978) used nematode-tracking techniques to verify galvano-orientation of *C. elegans*. Galvanotaxis in *C. elegans* was dependent on the current, potential difference and ionic concentration of KCl (Sukul and Croll, 1978). Additional reports verified galvanoreception as a genuine nematode response that is sensory-mediated and not due to passive movement of the nematode or to the effect of the current on the muscular physiology of the nematode (Sukul *et al.*, 1975; Sukul and Croll, 1978; Robertson and Forrest, 1989).

3.8. Future Prospects

Disrupting orientation during certain phases of the nematode life cycle, for example during the search for food or mates, could lead to novel nematode control methodologies (Perry, 1994). Such disruption might be achievable using artificial sex pheromones or host cues. Bone and Shorey (1977b) hypothesized that, if the environment is saturated with sex pheromones, nematode sensory organs will also become saturated and unable to respond to signals from the opposite sex. Bone (1987) and Perry (1994) subsequently found that mate-finding abilities of male nematodes could be interrupted with the disruption of their sensory abilities.

The use of chemosensory disruptive chemicals is another possible option to achieve management. Treatment of *P. penetrans* with the nematocide aldicarb changes the structure of the amphids and associated dendritic processes (Trett and Perry, 1985). Exposure of *P. redivivus* to sialic acid-specific lectin reduces attraction to nematophagous fungi (Jansson and Nordbring-Hertz, 1984). The attraction of *C. elegans* and *P. redivivus* towards filtrates of *Escherichia coli* is inhibited following treatment with α-mannosidase and sialidase (Zuckerman and Jansson, 1984).

Exploiting nematode orientation behaviour for practical purposes will require greater development and use of sensitive and quantitative techniques such as electrophysiological assays. This would expand our understanding of the role of chemoreceptors, leading in turn to the isolation and identification of active pheromone and allelochemical components. Further studies are needed to compare various nematode receptors, including amphids, phasmids, deirids, bursal rays and caudal and cephalic papillae with sensory functions and orientation. Finally, the use of chemo-orientation between sexes might one day be applied as a diagnostic method to separate amphimictic sibling species (Riga and Webster, 1992).

Acknowledgements

Thanks to A.H. Burr for assistance with the photoreception section of the chapter and M. Jonz for valuable advice.

References

Alphey, T.J.W. (1971) Studies on the aggregation behaviour of *Nippostrongylus brasiliensis*. *Parasitology* 63, 109–117.

Anderson, R.V. and Coleman, D.C. (1981) Population development and interactions between two species of bacteriophagic nematodes. *Nematologica* 27, 6–19.

Anya, A.O. (1976) Studies on the reproductive physiology of nematodes: the phenomenon of sexual attraction and the origin of the attractants in *Aspiculuris tetraptera*. *International Journal of Parasitology* 6, 173–177.

Ashton, F.T. and Schad, G.A. (1995) Amphids in *Strongyloides stercoralis* and other parasitic nematodes. *Parasitology Today* 12, 187–194.

Ashton, F.T., Li, J. and Schad, G.A. (1999) Chemo- and thermosensory neurons: structure and function in animal parasitic nematodes. *Veterinary Parasitology* 84, 297–316.

Aumann, J., Dietsche, E., Rutencrantz, S. and Ladehoff, H. (1998a) Physiochemical properties of the female sex pheromone of *Heterodera schachtii* (Nematoda: Heteroderidae). *International Journal for Parasitology* 28, 1691–1694.

Aumann, J., Ladehoff, H. and Rutencrantz, S. (1998b) Gas chromatographic characterization of the female sex pheromone of *Heterodera schachtii* (Nematoda: Heteroderidae). *Fundamental and Applied Nematology* 21, 119–122.

Balan, J., Krizkova, L., Nemec, P. and Kolozsvary, A. (1976) A qualitative method for detection of nematode attracting substances and proof of production of three different attractants by the fungus *Monacrosporium rutgeriensis*. *Nematologica* 22, 306–311.

Balanova, J. and Balan, J. (1991) Chemotaxis-controlled search for food by the nematode *Panagrellus redivivus*. *Biologia* 46, 257–263.

Balhadère, P. and Evans, A.A.F. (1994) Characterization of attractiveness of excised root tips of resistant and susceptible plants for *Meloidogyne naasi*. *Fundamental and Applied Nematology* 17, 527–536.

Bargmann, C.I. and Horvitz, H.R. (1991) Control of larval development by chemosensory neurons on *Caenorhabditis elegans*. *Science* 251, 1243–1246.

Bargmann, C.I. and Mori, I. (1997) Chemotaxis and thermotaxis. In: Riddle, D.L., Blumenthal, T., Meyer, B.J. and Preiss, J.R. (eds) *C. elegans II*. Cold Spring Harbor Laboratory Press, Cold Spring Harbor, New York, pp. 717–737.

Bargmann, C.I., Thomas, J.H. and Horvitz, H.R. (1990) Chemosensory cell function in the behaviour and development of *Caenorhabditis elegans*. *Cold Spring Harbor Symposia on Quantitative Biology* 55, 529–538.

Bargmann, C.I., Hartwieg, E. and Horvitz, H.R. (1993) Odorant-selective genes and neurons mediate olfaction in *C. elegans*. *Cell* 74, 515–527.

Bazzicalupo, P., De Roso, L., Maimone, F., Ristoratore, F. and Sebastiano, M. (1994) Chemoreception in Nematodes. In: *Advances in Molecular Plant Nematology*. NATO ASI Series, Series A, Life Sciences, No. 268, Plenum Press, New York, pp. 251–260.

Bilgrami, A.L. (1997) *Nematode Biopesticides*. Aligarh Muslim University, Aligarh, India, 262 pp.

Bilgrami, A.L. and Jairajpuri, M.S. (1988) Attraction of *Mononchoides longicaudatus* and *M. fortidens* (Nematoda: Diplogasterida) towards prey and factors influencing attraction. *Revue de Nématologie* 11, 195–202.

Bilgrami, A.L. and Jairajpuri, M.S. (1989a) Resistance of prey to predation and strike rate of the predators *Mononchoides longicaudatus* and *M. fortidens* (Nematoda: Diplogasterida). *Revue de Nématologie* 12, 45–49.

Bilgrami, A.L. and Jairajpuri, M.S. (1989b) Aggregation of *Mononchoides longicaudatus* and *M. fortidens* (Nematoda: Diplogasterida) at feeding site. *Nematologica* 34, 119–121.

Bilgrami, A.L. and Pervez, R. (2000a) Prey searching and attraction behaviour of *Mesodorylaimus bastiani* and *Aquatides thornei* (Nematoda: Dorylaimida). *International Journal of Nematology* 10, 199–206.

Bilgrami, A.L. and Pervez, R. (2000b) Numerical analysis of the aggregation behaviour of *Discolaimus major* and *Laimydorus baldus* (Nematoda: Dorylaimida) at feeding sites. *Acta Zoological Taiwanica* 11, 83–93.

Bilgrami, A.L., Kondo, E. and Yoshiga, T. (2000) Experimental models for testing attraction and preferential behaviour of *Steinernema glaseri* to several insects. *Japanese Journal of Nematology* 30, 35–46.

Bilgrami, A.L., Kondo, E. and Yoshiga, T. (2001a) Attraction response of *Steinernema glaseri* in presence of plant roots and leaves. *Annals of Plant Protection Science* 9, 258–263.

Bilgrami, A.L., Kondo, E. and Yoshiga, T. (2001b) Absolute and relative preferential attraction of *Steinernema glaseri* to excretory substances of insects. *International Journal of Nematology* 11, 27–34.

Bilgrami, A.L., Kondo, E. and Yoshiga, T. (2001c) Host searching and attraction behaviour of *Steinernema glaseri* using *Galleria mellonella* as its host. *International Journal of Nematology* 11, 168–176.

Bird, A.F. (1959) The attractiveness of roots to the plant-parasitic nematodes *Meloidogyne javanica* and *M. hapla*. *Nematologica* 4, 322–335.

Bollerup, G. and Burr, A.H. (1979) Eyespot and other pigments in nematode esophageal muscle cells. *Canadian Journal of Zoology* 57, 1057–1069.

Bone, L.W. (1987) Pheromone communication in nematodes. In: Veech, J.A. and Dickson, D.W. (eds) *Vistas in Nematology*. Society of Nematologists, Hyattsville, Maryland, pp. 147–152.

Bone, L.W. and Shorey, H.H. (1977a) Interactive influences of male- and female-produced pheromones on male attraction to female *Nippostrongylus brasiliensis*. *Journal of Parasitology* 63, 845–848.

Bone, L.W. and Shorey, H.H. (1977b) Disruption of sex pheromone communication in a nematode. *Science* 197, 694–695.

Bone, L.W. and Shorey, H.H. (1978) The influence of mating and a protein inhibitor on the response of *Nippostrongylus brasiliensis* to sex pheromone. *Proceedings of the Helminthological Society of Washington* 45, 264–266.

Bone, L.W., Gaston, L.K., Hammock, B.D. and Shorey, H.H. (1979) Chromatographic fractionation of the aggregation and sex pheromones of *Nippstrongylus brasiliensis*. *Journal of Experimental Zoology* 208, 311–318.

Burman, M. and Pye, A.E. (1980) *Neoaplectana carpocapsae*: movements of nematode population on the thermal gradient. *Experimental Parasitology* 49, 258–265.

Burr, A.H. (1979) Analysis of phototaxis in nematodes using directional statistics. *Journal of Comparative Physiology A* 134, 85–93.

Burr, A.H. (1984a) Evolution of eyes and photoreceptor organelles in the lower phyla. In: Ali, M.A. (ed.) *Photoreception and Vision in Invertebrates*. Plenum Press, New York, pp. 131–178.

Burr, A.H. (1984b) Photomovement behaviour in simple invertebrates. In: Ali, M.A. (ed.) *Photoreception and Vision in Invertebrates*. Plenum Press, New York, pp. 179–215.

Burr, A.H. and Harosi, F. (1985) Naturally crystalline hemoglobin of the nematode *Mermis nigrescens*. *Biophysical Journal* 47, 527–536.

Burr, A.H. and Webster, J.M. (1971) Morphology of the eyespot and description of two pigment granules in the esophageal muscle of a marine nematode, *Oncholaimus vesicarius*. *Journal of Ultrastructural Research* 36, 621–632.

Burr, A.H., Schiefke, R. and Bollerup, G. (1975) Properties of a hemoglobin from the chromotrope of the nematode *Mermis nigrescens*. *Biochimica Biophysica Acta* 405, 404–411.

Burr, A.H., Eggleton, D.K., Patterson, R. and Leutscher-Hazelhoff, J.T. (1989) The role of hemoglobin in the phototaxis of the nematode *Mermis nigrescens*. *Photochemistry and Photobiology* 49, 89–95.

Burr, A.H., Wagar, D. and Sidhu, P. (2000) Ocellar pigmentation and phototaxis in the nematode *Mermis nigrescens*: changes during development. *Journal of Experimental Biology* 203, 1341–1350.

Burr, A.H.J. and Babinszki, C.P.F. (1990) Scanning motion, ocellar morphology and orientation mechanisms in the phototaxis of the nematode *Mermis nigrescens*. *Journal of Comparative Physiology A* 167, 257–268.

Burr, A.H.J., Babinszki, C.P.F. and Ward, A.J. (1990) Components of phototaxis of the nematode *Mermis nigrescens*. *Journal of Comparative Physiology A* 167, 245–255.

Byers, J.A. and Poinar, G.O. (1982) Location of insect host by the nematode *Neoplectana carpocapsae* in response to temperature. *Behaviour* 79, 1–10.

Cassada, R.C. and Russell, R.L. (1975) The dauer larva, a post-embryonic developmental variant of the nematode *Caenorhabditis elegans*. *Developmental Biology* 46, 326–342.

Castro, C.E., Besler, N.O., McKinney, H.E. and Thomason, I.J. (1989) Quantitative bioassay for chemotaxis with plant-parasitic nematodes: attractant and repellent fractions for *Meloidogyne incognita* from cucumber roots. *Journal of Chemical Ecology* 15, 1297–1309.

Castro, C.E., McKinney, H.E. and Lux, S. (1991) Plant protection with inorganic ions. *Journal of Nematology* 23, 409–413.

Caveness, F.E. and Panzer, J.D. (1959) Nemic galvanotaxis. *Proceedings of the Helminthological Society of Washington* 27, 73–74.

Chalfie, M. and Sulston J.E. (1981) Developmental genetics of the mechanosensory neurons of *Caenorhabditis elegans. Developmental Biology* 82, 358–370.

Chalfie, M. and White, J. (1988) The nervous system. In: Wood, W.B. (ed.) *The Nematode* Caenorhabditis elegans. Cold Spring Harbor Press, Cold Spring Harbor, New York, pp. 337–391.

Chalfie, M., Sulston, J.E., White, J.G., Southgate, E., Thomson, J.N. and Brenner, S. (1985) The neural circuit for touch sensitivity in *Caenorhabditis elegans. Journal of Neuroscience* 5, 956–964.

Cheng, R. and Samoiloff, M.R. (1971) Sexual attraction in the free-living nematode *Panagrellus silusiae* (Cephalobidae). *Canadian Journal of Zoology* 49, 1443–1448.

Cheng, R. and Samoiloff, M.R. (1972) Effects of cyclohexamine and hydroxyurea on mating behaviour and its development in the free-living nematode *Panagrellus silusiae* (de Man 1913) Goodey, 1945. *Canadian Journal of Zoology* 50, 333–336.

Chin, D.A. and Taylor, D.P. (1969) Sexual attraction and mating patterns in *Cylindrocorpus longistoma* and *C. curzii* (Nematoda: Cylindrocorporidae). *Journal of Nematology* 1, 313–317.

Chitwood, B.G. and Murphy, D.G. (1964) Observations on two marine monhysterids – their classification, cultivation and behaviour. *Transactions of the American Microscopical Society* 83, 311–329.

Clemens, C.D., Aumann, J., Spiegel, Y. and Wyss, U. (1994) Attractant-mediated behaviour of mobile stages of *Heterodera schachtii. Fundamental and Applied Nematology* 17, 569–574.

Croll, N.A. (1967) Acclimatization in the eccritic thermal response of *Ditylenchus dipsaci. Nematologica* 13, 385–389.

Croll, N.A. (1970) Responses to chemicals. In: Croll, N.A. (ed.) *The Behaviour of Nematodes.* Edward Arnold, London, pp. 53–67.

Croll, N.A. (1974) *Necator americanus:* activity patterns in the egg and the mechanisms of hatching. *Experimental Parasitology* 35, 80–85.

Croll, N.A., Riding, I.L. and Smith, J.M. (1972) A nematode photoreceptor. *Comparative Biochemistry and Physiology* 42A, 999–1009.

Croll, N.A., Evans, A.A.F. and Smith, J.M. (1975) Comparative nematode photoreceptors. *Comparative Biochemistry and Physiology* 51A, 139–143.

CSC (*C. elegans* Sequencing Consortium) (1998) Genome sequence of the nematode *C. elegans*: a platform for investigating biology. *Science* 282, 2012–2018.

Culloti, J.G. and Russell, R.L. (1978) Osmotic avoidance defective mutants of the nematode *Caenorhabditis elegans. Genetics* 90, 243–256.

Davis, E.L., Aron, L.M., Pratt, L.H. and Hussey, R.S. (1992) Novel immunization procedures used to develop antibodies that bind to specific structures in *Meloidogyne* spp. *Phytopathology* 82, 1244–1250.

Dickinson, S. (1959) The behaviour of larvae of *Heterodera schachtii* on nitrocellulose membranes. *Nematologica* 4, 60–66.

Doncaster, C.C. and Seymour, M.K. (1973) Exploration and selection of penetration site by Tylenchida. *Nematologica* 19, 137–145.

Doncaster, C.C. and Shepherd, A.M. (1967) The behaviour of second stage *Heterodera rostochiensis* larvae leading to their emergence from the egg. *Nematologica* 13, 476–478.

Duggal, C.L. (1978) Sex attraction in the free-living nematode *Panagrellus redivivus. Nematologica* 24, 213–221.

Duggan, A., Ma, C. and Chalfie, M. (1996) Regulation of touch receptor differentiation by the *Caenorhabditis elegans* mec-3 and unc-86 genes. *Development* 125, 4107–4119.

Dusenbery, D.B. (1974) Analysis of chemotaxis in the nematode *Caenorhabditis elegans* by countercurrent separation. *Journal of Experimental Zoology* 188, 41–48.

Dusenbery, D.B. (1987) Prospects for exploiting sensory stimuli in nematode control. In: Veech, J.A. and Dickson, D.W. (eds) *Vistas in Nematology*. E.O. Painter Printing, DeLeon Springs, Florida, pp. 131–135.

Dusenbery, D.B. (1988) Behavioural responses of *Meloidogyne incognita* to small temperature changes. *Journal of Nematology* 20, 351–355.

Edmunds, J.E. and Mai, W.F. (1967) Effects of *Fusarium oxysporum* on movement of *Pratylenchus penetrans* towards alfalfa roots. *Phytopathology* 57, 468–471.

Eiff, J.A. (1966) The effect of extracts of the amphidial glands, excretory glands, and esophagus of adults of *Ancylostoma caninum* on the coagulation of dog's blood. *Journal of Parasitology* 52, 833–843.

Ellenby, C. (1964) Haemoglobin in the chromotrope of an insect parasitic nematode. *Nature* 202, 615–616.

El-Sherif, M. and Mai, W.F. (1969) Thermotactic response of some plant-parasitic nematodes. *Journal of Nematology* 1, 43–48.

Emmons, S.W. and Sternberg, P.W. (1997) Male development and mating behaviour. In: Riddle, D.L., Blumenthal, T., Meyer, B.J. and Priess, J.R. (eds) *C. elegans II*. Cold Spring Harbor Laboratory Press, Cold Spring Harbor, New York, pp. 295–334.

Fine, A.E., Ashton, F.T., Bhopale, V.M. and Schad, G.A. (1997) Sensory neuroanatomy of a skin-penetrating nematode parasite *Strongyloides stercolaris*, II. Labial and cephalic neurones. *Journal of Comparative Neurology* 389, 212–223.

Franco, J., Main, G. and Oros, R. (1999) Trap crops as a component for the integrated management of *Globodera* spp. (potato cyst nematodes) in Bolivia. *Nematropica* 29, 51–60.

Gaugler, R., LeBeck, L., Nakagaki, B. and Boush, G.M. (1980) Orientation of the entomopathogenic nematode *Neoaplectana carpocapsae* to carbon dioxide. *Environmental Entomology* 9, 649–652.

Golden, J.W. and Riddle, D.L. (1982) A pheromone influences larval development in the nematode *Caenorhabditis elegans*. *Science* 218, 578–580.

Green, C.D. and Greet, D.N. (1972) The location of the secretions that attract male *Heterodera schachtii* and *H. rostochiensis* to their females. *Nematologica* 18, 347–352.

Green, C.D., Greet, D.N. and Jones, F.G.W. (1970) The influence of multiple mating on the reproduction and genetics of *Heterodera rostochiensis* and *H. schachtii*. *Nematologica* 16, 309–326.

Greet, D.N. (1964) Observations on sexual attraction and copulation in the nematode *Panagrolaimus rigidus* Schneider. *Nature* 204, 96–97.

Greet, D.N., Green, C.D. and Poulton, M.E. (1968) Extraction, standardization and assessment of the volatility of the sex attractants of *Heterodera rostochiensis* Woll and *H. schachtii* Schm. *Annals of Applied Biology* 61, 511–519.

Grewal, P.S., Gaugler, R. and Selvan, S. (1994) Host recognition by entomopathogenic nematode: behavioural response to contact with host feces. *Journal of Chemical Ecology* 19, 1219–1230.

Grewal, P.S., Lewis, E.E. and Gaugler, R. (1997) Response of infective stage parasites (Nematoda: Steinernematidae) to volatile cues from infected hosts. *Journal of Chemical Ecology* 23, 503–515.

Grundler, F., Schnibbe, L. and Wyss, U. (1991) *In vitro* studies on the behaviour of 2nd-stage juveniles of *Heterodera schachtii* (Nematoda, Heteroderidae) in response to host plant exudates. *Parasitology* 103, 149–155.

Gu, G.O., Caldwell, G.A. and Chalfie, M. (1996) Genetic interactions affecting touch sensitivity in *Caenorhabditis elegans*. *Proceedings of the National Academy of Sciences, USA* 93, 6577–6582.

Hedgecock, E.M. and Russell, R.L. (1975) Normal and mutant thermotaxis in the nematode *Caenorhabditis elegans*. *Proceedings of the National Academy of Sciences, USA* 72, 4061–4065.

Huettel, R.N. (1986) Chemical communication in nematodes. *Journal of Nematology* 18, 3–8.

Huettel, R.N. and Rebois, R.V. (1986) Bioassay comparisons for pheromone detection in *Heterodera glycines*, the soybean cyst nematode. *Proceedings of the Helminthological Society of Washington* 53, 63–68.

Inglis, W.G. (1963) 'Campaniform type' organs in nematodes. *Nature* 197, 618.

Ishikawa, M., Shuto, Y. and Watanabe, H. (1986) beta-Myrcene a potent attractant component of pine wood for the pine wood nematode *Bursaphelenchus xylophilus*. *Agricultural and Biological Chemistry* 50, 1863–1866.

Jaffe, H., Huettel, R.N., DeMilo, A.B., Hayes, D.K. and Rebois, R.V. (1989) Isolation and identification of a compound from soybean cyst nematode *Heterodera glycines*, with sex pheromone activity. *Journal of Chemical Ecology* 15, 2021–2043.

Jansson, H.-B. (1982a) Attraction of nematodes to endoparasitic nematophagous fungi. *Transactions of the British Mycological Society* 79, 25–29.

Jansson, H.-B. (1982b) Predacity by nematophagous fungi and its relation to the attraction of nematodes. *Microbial Ecology* 8, 237–240.

Jansson, H.-B. and Nordbring-Hertz, B. (1979) Attraction of nematodes to living mycelium of nematophagous fungi. *Journal of General Microbiology* 112, 89–93.

Jansson, H.-B. and Nordbring-Hertz, B. (1980) Interactions between nematophagous fungi and plant-parasitic nematodes: attraction, induction of trap formation and capture. *Nematologica* 26, 383–389.

Jansson, H.-B. and Nordbring-Hertz, B. (1984) Involvement of sialic acid in nematode chemotaxis and infection by an endoparasitic nematophagous fungus. *Journal of Genetic Microbiology* 130, 39–43.

Jones, J.T. (2002) Nematode sense organs. In: Lee, D.L. (ed.) *The Biology of Nematodes*. Taylor and Francis, London, pp. 353–368.

Jones, J.T., Perry, R.N. and Johnston, M.R.L. (1991) Electrophysiological recordings of electrical activity and responses to stimulants from *Globodera rostochiensis* and *Syngamus trachea*. *Revue de Nématologie* 14, 467–473.

Jones, T.P. (1966) Sexual attraction and copulation in *Pelodera teres*. *Nematologica* 12, 518–522.

Jonz, M.G., Riga, E., Potter, J.W. and Mercier, A.J. (2001) Partial isolation of a water soluble pheromone from the sugar beet cyst nematode, *Heterodera schachtii*, using novel and conventional bioassays. *Nematology* 3, 55–64.

Kaplan, J.M. and Horvitz, H.R. (1993) A dual mechanosensory and chemosensory neuron in *Caenorhabditis elegans*. *Proceedings of the National Academy of Sciences, USA* 90, 2227–2231.

Khlibsuwan, W., Ishibashi, N. and Kondo, E. (1992) Responses of *Steinernema carpocapsae* infective juveniles to the plasma of three insect species. *Journal of Nematology* 24, 156–159.

Klink, J.W., Dropkin, V.E. and Mitchell, J.E. (1970) Studies on the host-finding mechanism of *Neotylenchus linfordi*. *Journal of Nematology* 2, 106–117.

Kondo, E. (1989) Studies on the infectivity and propagation of entomogenous nematodes,

Steinernema spp. (Rhabditida: Steinernematidae), in the common cutworm, *Spodoptera litura* (Lepidoptera: Noctuidae). *Bulletin of Faculty of Agriculture, Saga University* 67, 1–88.

Lee, D.L. (1973) Evidence for a sensory function for the copulatory spicules of nematodes. *Journal of Zoology* 169, 281–285.

Lee, D.L. (2002) Behaviour. In: Lee, D.L. (ed.) *The Biology of Nematodes*. Taylor and Francis, London, pp. 369–387.

Lewis, E.E., Gaugler, R. and Harrison, R. (1992) Entomopathogenic nematode host finding: response to contact cues by cruise and ambush foragers. *Parasitology* 105, 309–315.

Lewis, E.E., Barbarosa, B. and Gaugler, R. (2002) Mating and sexual communication by *Steinernema carpocapsae*. *Journal of Nematology* 34, 328–331.

Lewis, J.A. and Hodgkin, J.A. (1977) Specific neuroanatomical changes in chemosensory mutants of the nematode *Caenorhabditis elegans*. *Journal of Comparative Neurology* 712, 489–510.

McCallum, M.E. and Dusenbery, D.B. (1992) Computer tracking as a behavioural GC detector – nematode responses to vapor of host roots. *Journal of Chemical Ecology* 18, 585–592.

McLaren, D.J. (1976a) Nematode sense organs. *Advances in Parasitology* 14, 195–265.

McLaren, D.J. (1976b) Sense organs and their secretions. In: Croll, N.A. (ed.) *The Organization of Nematodes*. Academic Press, London, pp. 143–161.

Meyers, S.P. and Hopper, E. (1966) Attraction of the marine nematode *Metoncholaimus* sp. to fungal substrates. *Bulletin of Marine Science* 16, 142–150.

Moens, T., Verbeeck, L., deMaeyer, A., Swings, J. and Vincx, M. (1999) Selective attraction of marine bacterivorous nematodes to their bacterial food. *Marine Ecology Progress Series* 176, 165–178.

Moltmann, E. (1990) Kairomones in root exudates of cereals – their importance in host finding of juveniles of the cereal cyst nematode, *Heterodera avenae* (Woll.) and their characterization. *Zeitschrift für Pflanzerkrankheiten und Pflanzenschutz* 97, 458–469.

Mori, I. and Oshima, Y. (1995) Neural regulation of thermotaxis in *Caenorhabditis elegans*. *Nature* 376, 344–348.

Peckol, E.L., Troemel, E.R. and Bargmann, C.I. (2001) Sensory experience and sensory activity regulate chemosensory receptor gene expression in *Caenorhabditis elegans*. *Proceedings of the National Academy of Sciences, USA* 98, 11032–11038.

Perkins, L.A., Hedgecock, E.M., Thomson, J.N. and Culotti, J.G. (1986) Mutant sensory cilia in the nematode *Caenorhabditis elegans*. *Developmental Biology* 117, 456–487.

Perry, R.N. (1994) Studies on nematode sensory perception as a basis for novel control strategies. *Fundamental and Applied Nematology* 17, 199–202.

Perry, R.N. (1996) Chemoreception in plant-parasitic nematodes. *Annual Review of Phytopathology* 34, 181–199.

Perry, R.N. (2001) Analysis of the sensory responses of parasitic nematodes using electrophysiology. *International Journal of Parasitology* 31, 909–918.

Perry, R.N. and Aumann, J. (1998) Behaviour and sensory responses. In: Perry, R.N. and Wright, D.J. (eds) *The Physiology and Biochemistry of Free-living and Plant-parasitic Nematodes*. CAB International, Wallingford, UK, pp. 75–102.

Pitcher, R.S. (1967) The host–parasite relations and ecology of *Trichodorus viruliferus* on apple roots, as observed from an underground laboratory. *Nematologica* 13, 547–557.

Pline, M. and Dusenbery, D.B. (1987) Responses of the plant-parasitic nematode *Meloidogyne incognita* to carbon dioxide determined by video camera–computer tracking. *Journal of Chemical Ecology* 13, 873–888.

Pline, M., Diez, J.A. and Dusenbery, D.B. (1988) Extremely sensitive thermotaxis of the nematode *Meloidogyne incognita. Journal of Nematology* 20, 605–608.

Prot, J.C. (1977) Amplitude et cinétique des migrations de nématode *Meloidogyne javanica* sous l'nfluence d'un plant de tomate. *Cahiers Office Recherche Scientifique et Technique Outre-Mer Physiologie des Plantes Tropicales Cultivées Série Biologie* 11, 157–166.

Prot, J.C. (1980) Migration of plant-parasitic nematodes toward plant roots. *Revue de Nématologie* 3, 305–318.

Rees, G. (1950) Observations on the vertical migrations of the third-stage larva of *Haemonchus contortus* (Rud.) on experimental plots of *Lolium perenne* S24, in relation to meteorological and micrometeorological factors. *Parasitology* 40, 127–143.

Rende, J.F., Tefft, P.M. and Bone, L.W. (1981) Pheromone attraction in the soybean cyst nematode *Heterodera glycines* race 3. *Journal of Chemical Ecology* 8, 981–991.

Riddle, D.L. (1988) The dauer larva. In: Wood, W.B. (ed.) *The Nematode* Caenorhabditis elegans. Cold Spring Harbor Laboratory, Cold Spring Harbor, New York, pp. 393–412.

Riddle, D.L. and Bird, A.F. (1985) Responses of the plant-parasitic nematode *Rotylenchulus reniformis, Anguina agrostis* and *Meloidogyne javanica* to chemical attractants. *Journal of Nematology* 91, 185–195.

Riga, E. (1992) Multifaceted approach for differentiating isolates of *Bursaphelenchus xylophilus* and *B. mucronatus* (Nematoda), parasites of pine trees. PhD thesis, Simon Fraser University, British Columbia, Canada.

Riga, E. and MacKinnon, B.M. (1987) Sex and aggregation attractants of *Heligmosomoides polygyrus* (Nematoda: Trichostrongylidae). *Canadian Journal of Zoology* 65, 1842–1846.

Riga, E. and MacKinnon, B.M. (1988) Chemical communication in *Heligmosomoides polygyrus* (Nematoda: Trichostrongylidae): the effect of age and sexual status of attracting and responding worms and localization of the sites of pheromone production in the female. *Canadian Journal of Zoology* 66, 1943–1947.

Riga, E. and Webster, J.M. (1992) Use of sex pheromones in the taxonomic differentiation of *Bursaphelenchus* spp. (Nematoda), pathogens of pine trees. *Nematologica* 28, 133–145.

Riga, E., Perry, R.N., Barrett, J. and Johnston, M.R.L. (1995a) Investigation of the chemosensory function of amphids of *Syngamus trachea* using electrophysiological techniques. *Parasitology* 111, 347–351.

Riga, E., Perry, R.N., Barrett, J. and Johnston, M.R.L. (1995b) Biochemical analysis on single amphidial glands, excretory–secretory gland cells, pharyngeal glands and their secretions from the avian nematode *Syngamus trachea. International Journal of Parasitology* 25, 1151–1158.

Riga, E., Perry, R.N., Barrett, J. and Johnston, M.R.L. (1996a) Electrophysiological responses of males of the potato cyst nematodes, *Globodera rostochiensis* and *G. pallida,* to their sex pheromones. *Parasitology* 112, 239–246.

Riga, E., Perry, R. and Barrett, J. (1996b) Electrophysiological analysis of the response of males for *Globodera rostochiensis* and *G. pallida* to their female sex pheromones and to potato root diffusate. *Nematologica* 42, 493–498.

Riga, E., Perry, R.N., Barrett, J. and Johnston, M.R.L. (1997) Electrophysiological responses of *Globodera rostochiensis* and *G. pallida,* to some chemicals. *Journal of Chemical Ecology* 23, 417–428.

Robertson, W.M. (1975) A possible gustatory organ associated with the odontophore in *Longidorus leptocephalus* and *Xiphinema diversicaudatum. Nematologica* 21, 443–448.

Robertson, W.M. (1979) Observations on the oesophageal nerve system of *Longidorus leptocephalus. Nematologica* 25, 245–254.

Robertson, W.M. and Forrest, J.M.S. (1989) Factors involved in host recognition by plant-parasitic nematodes. *Aspects of Applied Biology* 22, 129–133.

Robinson, A.F. (1989) Thermotactic adaptation in two foliar and two root parasitic nematodes. *Revue de Nématologie* 12, 125–131.

Robinson, A.F. (1994) Movement of five nematode species through sand subjected to natural temperature gradient fluctuations. *Journal of Nematology* 26, 46–58.

Robinson, A.F. (1995) Optimal release rates for attracting *Meloidogyne incognita*, *Rotylenchulus reniformis*, and other nematodes to carbon dioxide in sand. *Journal of Nematology* 27, 42–50.

Robinson, A.F. and Jaffee, B.A. (1996) Repulsion of *Meloidogyne incognita* by alginate pellets containing hyphae of *Monacrosporium cionopagum*, *M. ellipsosporum*, or *Hirsutella rhossiliensis*. *Journal of Nematology* 28, 133–147.

Robinson, A.F., Baker, G.L. and Heald, C.M. (1990) Transverse phototaxis by infective juveniles of *Agamermis* sp. and *Hexamermis* sp. *Journal of Parasitology* 76, 147–152.

Roche, M. (1966) Influence of male and female *Ancylostoma caninum* on each other's distribution in the intestine of the dog. *Experimental Parasitology* 19, 327–331.

Rolfe, R.N., Barrett, J. and Perry, R.N. (2000) Analysis of chemosensory responses of second stage juveniles of *Globodera rostochiensis* using electrophysiological techniques. *Nematology* 2, 523–533.

Rolfe, R.N., Barrett, J. and Perry, R.N. (2001) Electrophysiological analysis of responses of adult females of *Brugia pahangi* to some chemicals. *Parasitology* 122, 347–357.

Salm, R.W. and Fried, B. (1973) Heterosexual chemical attraction in *Camallanus* sp. (Nematoda) in the absence of worm-mediated tactile behaviour. *Journal of Parasitology* 59, 434–436.

Sambongi, Y., Nagae, T., Liu, Y., Yoshimizu, T., Takeda, K., Wada, Y. and Futai, M. (1999) Sensing of cadmium and copper ions by externally exposed ADL, ASE, and ASH neurons elicits avoidance response in *Caenorhabditis elegans*. *Neuroreport* 10, 753–757.

Samoiloff, M.R., McNicholl, P., Chen, P. and Balakanich, S. (1973) Regulation of nematode behaviour by physical means. *Experimental Parasitology* 33, 253–262.

Selvan, S., Campbell, J.F. and Gaugler, R. (1993) Density-dependent effects on entomopathogenic nematodes (Heterorhabditidae and Steinernematidae) within an insect host. *Journal of Invertebrate Pathology* 62, 278–284.

Sengupta, P., Colbert, H.A., Kimmel, B.E., Dwyer, N. and Bargmann, C.I. (1993) The cellular and genetic basis of olfactory responses in *Caenorhabditis elegans*. *CIBA Foundation Symposia* 179, 235–250.

Simon, J.M. and Sternberg, P.W. (2002) Evidence of a mate-finding cue in the hermaphrodite nematode *Caenorhabditis elegans*. *Proceedings of the National Academy of Sciences, USA* 99, 1598–1606.

Somers, J.A., Shorey, H.H. and Gaston, L.K. (1977) Sex pheromone communication in the nematode *Rhabditis pellio*. *Journal of Chemical Ecology* 3, 467–474.

Steiner, G. (1925) The problem of host selection and host specialization of certain plant-infecting nemas and its application in the study of nemic pests. *Phytopathology* 15, 499–534.

Stewart, G.R., Perry, R.N., Alexander, J. and Wright, D.J. (1993a) A glycoprotein specific to the amphids of *Meloidogyne* species. *Parasitology* 106, 405–412.

Stewart, G.R., Perry, R.N. and Wright, D.J. (1993b) Studies on the amphid specific glycoprotein gp32 in different life cycle stages of *Meloidogyne* species. *Parasitology* 107, 573–578.

Stringfellow, F. (1974) Hydroxyl ion, and attractant to the male of *Pelodera strongyloides*. *Proceedings of the Helminthological Society of Washington* 41, 4–10.

Sukul, N.C. and Croll, N.A. (1978) Influence of potential difference and current on the electrotaxis of *Caenorhabditis elegans*. *Journal of Nematology* 10, 314–317.

Sukul, N.C., Das, P.K. and Ghosh, S.K. (1975) Cation-mediated orientation of nematodes under electrical field. *Nematologica* 21, 145–150.

Sulston, J.E. and Horvitz, H.R. (1977) Post-embryonic cell lineages of the nematode *Caenorhabditis elegans*. *Developmental Biology* 56, 110–156.

Trett, M.W. and Perry, R.N. (1985) Functional and evolutionary implications of the anterior sensory anatomy of species of root lesion nematodes (genus *Pratylenchus*). *Revue de Nématologie* 8, 341–355.

Van de Velde, M.C. and Coomans, A. (1988) Ultrastructure of the photoreceptor of *Diplolaimella* sp. (Nematoda). *Tissue and Cell* 20, 421–429.

Wallace, H.R. (1961) The orientation of *Ditylenchus dipsaci* to physical stimuli. *Nematologica* 6, 222–236.

Wallace, H.R. (1968) Undulatory locomotion of the plant-parasitic nematode *Meloidogyne javanica*. *Parasitology* 58, 377–391.

Wang, K.C. and Chen, T.A. (1985) Ultrastructure of the phasmids of *Scutellonema brachyurum*. *Journal of Nematology* 17, 175–186.

Ward, S. (1973) Chemotaxis by the nematode *Caenorhabditis elegans*: identification and analysis of the response by use of mutants. *Proceedings of the National Academy of Sciences, USA* 70, 817–821.

Ward, S. (1976) Nervous system of nematodes. In: Croll, N.A. (ed.) *The Organization of Nematodes*. Academic Press, London, pp. 365–382.

Ward, S., Thomson, N., White, J.G. and Brenner, S. (1975) Electron microscopical reconstruction of the anterior sensory anatomy of the nematode *Caenorhabditis elegans*. *Journal of Comparative Neurology* 160, 313–337.

Whittaker, F.W. (1969) Galvanotaxis of *Pelodera strongyloides* (Nematoda, Rhabditidae). *Proceedings of the Helminthological Society of Washington* 36, 40.

Wittenburg, N. and Baumeister, R. (1999) Thermal avoidance in *Caenorhabditis elegans*: an approach to the study of noniception. *Neurobiology* 96, 10477–10482.

Zuckerman, B.M. and Jansson, H.-B. (1984) Nematode chemotaxis and possible mechanism of host/prey recognition. *Annual Review of Phytopathology* 22, 95–113.

4 Feeding Behaviour

ANWAR L. BILGRAMI AND RANDY GAUGLER

Department of Entomology, Rutgers University, New Brunswick, NJ 08901–8524, USA

4.1. Introduction

Nematodes have adapted themselves as parasites, predators and free-living entities to a wide range of habitats. Despite many structural and functional similarities, they show great diversity in food and feeding habits. Nematodes may be monophagous or polyphagous. They obtain nutrients from bacteria, algae, diatoms, protozoa, fungi, nematodes or plant and animal tissues. Some show dual feeding habits (Yeates *et al.*, 1993), alternating in feeding upon different food resources at different stages of the life cycle. This diversity has made nematode characterization into different feeding groups difficult. The separate disciplines working with plant, animal, insect and free-living nematodes have also complicated the characterization of nematode feeding habits. These disparate disciplines have often generated different terminologies for the same feeding structures or behaviours, making comparisons far more difficult. Such intricacies have restricted understanding of nematode feeding behaviour and its role in various soil processes, including food webs, food conversion, biodiversity and biological control.

There is a dearth of general reviews comparing feeding behaviour in plant, animal, insect and free-living nematodes. Nevertheless, many non-comparative studies have been completed on the feeding and related activities of predatory (Jairajpuri and Bilgrami, 1990; Bilgrami, 1997), plant (Hussey and Mims, 1991; Wyss, 2002), insect (Gaugler *et al.*, 1989; Grewal *et al.*, 1993), animal (Jasmer, 1995), bacterial (Young *et al.*, 1996; Avery and Thomas, 1997) and fungal-feeding nematodes (Doncaster and Seymour, 1975; Shepherd and Clark, 1976). This chapter presents a comparative account of feeding behaviour of all groups of nematodes, focusing on feeding types and various aspects of food search and feeding mechanisms.

4.2. Structure and Function of the Feeding Apparatus

Feeding apparatus has been a widely used term for the stoma or mouth in describing nematode feeding mechanisms (Grootaert and Wyss, 1979; Bilgrami, 1997; Wyss, 2002). For the same reason, in this chapter feeding apparatus is used to refer to the main feeding organs (e.g. stylet, mural tooth, buccal cavity, tooth) of nematodes to maintain uniformity in describing feeding mechanisms. Based on the structural and functional similarities of nematodes, a feeding apparatus may be classified into four types: (i) engulfing (e.g. predatory mononchs); (ii) piercing (e.g. plant-parasitic, predatory and fungivores); (iii) cutting (predatory diplogasterids, some marine and animal-parasitic nematodes); and (iv) sucking (bacterial-feeding and some animal-parasitic nematodes).

The structural organization and function of the feeding apparatus form the basis of trophic categorization of nematodes and predict their food and feeding habits. Cheilostom, prostom, mesostom, metastom and telostom (Bird and Bird, 1991) are not only of taxonomic importance but also predict nematode feeding preferences. The main feeding organ of the feeding apparatus is different in its structural composition, shape and size and is referred to by different names by different nematode groups. In many nematodes the main feeding organ is the stylet, a pointed needle-like structure typically having a lumen. Similar to tylenchids, fungal-feeding nematodes possess a stylet, albeit a smaller and less robust one than plant feeders, to puncture hyphal cells and ingest nutrients. Plant-parasitic Triplonchida and predatory Nygolaimina possess a solid but pointed tooth, used for puncturing plant cells and prey cuticle, respectively. The feeding apparatus of predatory nematodes is a buccal cavity that contains specialized structures, e.g. dorsal tooth, ventral teeth and denticles. The subventral lancets, cutting plates, dorsal gutter and dorsal cone are structures associated with animal-parasitic nematode feeding. These specialized structures help nematodes during food capture, puncture, extracorporeal digestion and ingestion.

4.2.1. Engulfing

Only mononchid predators possess this type of feeding apparatus. Mononchs engulf and swallow their prey whole, although they may also feed by first shredding prey (Bilgrami *et al.*, 1986). The engulfing type consists of two sets of three plates each located in the buccal cavity (Fig. 4.1A). The anterior set is more developed, large and vertically placed as compared with the posterior (Jairajpuri and Bilgrami, 1990). The dorsal plate bears a sharply pointed tooth for puncturing prey cuticle,

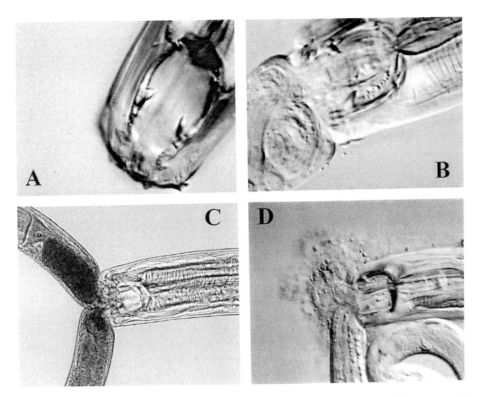

Fig 4.1. A. Feeding apparatus of *Anatonchus tridentatus*. B. *A. tridentatus* engulfing *Panagrellus redivivus*. C. *Prionchulus punctatus* engulfing an entire prey. D. *A. tridentatus* ingesting body contents of *P. redivivus*. (A, B, D, courtesy of U. Wyss; C, from Eisenback and Zunke, 1997.)

whereas the subventral walls bear a smaller tooth, teeth or denticles for grinding or sieving. The buccal armature is supported by a set of subdorsal, subventral and sub-lateral labial muscles that control lip movements during feeding, prey capturing, puncturing or swallowing. The labial muscles contract and pull the lips outwards and backwards to open the feeding apparatus wide enough to allow predators to engulf or swallow the intact prey (Coomans and Lima, 1965; Fig. 4.1B–D). Oesophageal suction also helps in holding the prey firmly (Roggen, 1973).

4.2.2. Piercing

Species with this type of feeding apparatus are known as stylet-bearing nematodes because the main feature of the feeding apparatus is a protrusible stylet or spear, which is pointed and needle-like with a narrow lumen connected with the oesoph-agus (Fig. 4.2A, B). This modification of the buccal cavity is used to pierce plant or fungal cells and suck fluids, but is also found in some predators, as well as insect and animal parasites. The stylet is termed a stomatostyle in tylenchid nematodes, and consists of an anterior conical portion with subterminal ventral opening, shaft and knobs. The stylet of fungal-feeding and predatory dorylaim nematodes is a cuticular extension consisting of an anterior odontostyle and posterior odon-tophore.

4.2.3. Cutting

This type of feeding apparatus is present in diplogasterid, actinolaim and enoplid predators and some animal parasites (e.g. *Haemonchus*). Their feeding apparatus

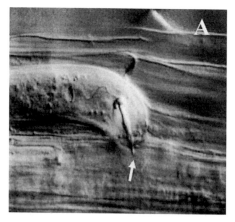 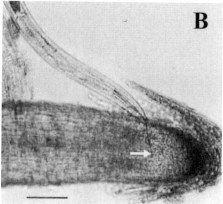

Fig. 4.2. A. Second-stage juvenile of *Heterodera schachtii* feeding from the initial syncytial cell in the root of *Brassica napus*. B. *Xiphinema diversicaudatum* penetrating meristematic cells at the root tip of *Ficus carica* seedlings. Arrows show feeding site. (Courtesy of U. Wyss.)

consists of a buccal cavity armed with a tooth or teeth of variable sizes located at different positions on the dorsal and ventral walls. The mouth opening is terminal, and may be displaced dorsally (e.g. *Pterygodermatites*) or ventrally (e.g. *Ichthyocephaloides*) (Hunt and Sutherland, 1984). Heavily sclerotized dorsal metarhabdions provide a claw-like dorsal movable (*Mononchoides*) or immovable tooth (*Actinolaimus*) with a lumen, the main killing organ for these nematodes. The immovable tooth is called an odontia when located at the cheilostom (e.g. *Ancylostoma*) and an onchia when positioned at the metarhabdion (e.g. *Haemonchus*) (Bird, 1971). The former is used for cutting and the latter for grinding food particles. Cutting and shredding food into pieces or ripping off tissues depends upon the size and number of teeth. There may be one (e.g. *Haemonchus contortus*), three (e.g. *Ternidens deminutus*) or six teeth (e.g. *Streptopharagus pigmentatus*) with sharply pointed (e.g. *Triodontophorus*) or rounded tips (e.g. *Strongylus vulgaris*) or with elaborate ridges (*Strongylus asini*). The teeth of *Ancylostoma* are modified to sublateral lancets that sieve food concomitantly with small denticles. *Haemonchus* uses teeth to tear open tissues, whereas *Ancylostoma* and *Mylonchulus* use denticles for grinding and sieving food particles (Van der Heiden, 1974; Jairajpuri and Bilgrami, 1990).

The cutting type of feeding apparatus may be symmetrical or asymmetrical, rudimentary, narrow, well developed, cylindrical or globular. *Ancylostoma* and *Grammocephalus* have large semilunar and small dorsolateral cutting plates, respectively, located at the outer margin of the feeding apparatus. During feeding, forward and backward movements of the buccal armature are regulated by dialator buccae and protractor and retractor muscles. *Parascaris*, a gut-dwelling nematode, has dentigerous ridges or a prebuccal groove at the outer margin of each lip (Ansel *et al.*, 1974) that possesses sensory interlabia or spirally arranged fibrils. Contractions of these fibrils assist the parasites in securing their position within the host gut.

4.2.4. Sucking

The sucking type of feeding is characteristic of bacterial and carrion feeders (e.g. early juvenile stages of *Mononchus* and *Mylonchulus*). Included in this category are *Thelastoma*, *Mermis*, *Steinernema* and *Heterorhabditis* (insect-parasitic nematodes) and *Chromodora* (marine nematode). This simple, primitive type of feeding apparatus has a wide distal opening of the same width at both ends. The ingestion of nutrients depends upon the concentration and flow of nutrients, turgor pressure at the food resource and suction generated by the median oesophageal bulb. Suction is provided by the oesophageal muscle cells and coordinated by the associated neurones (Albertson and Thomson, 1976). Movement of nutrients through the stoma is regulated by rapid opening and closing of the procorpus and metacorpus simultaneously with peristaltic movements in the oesophagus. Food is crushed by fine finger-like projections located in the metacorpus. In *Ascaris*, the oesophagus functions as a two-stage pump during feeding (Mapes, 1966). Contraction of anterior muscles allows food to pass into the lumen of the oesophagus; the oesophago-intestinal valve then releases food into the intestine. Rate of suction depends upon oesophageal pumping, which occurs at regular (continuous) or irregular (intermittent) intervals.

4.3. Feeding Types

Nematode feeding has evolved from primitive (bacterial feeder) to advanced types (parasitism). Nematodes show food selection and preferential feeding habits permitting discrimination between diverse bacteria, plant and animal tissues, feeding sites and prey animals. Little effort has been made to establish functional correlations between structural components of the feeding apparatus in different nematode groups. Therefore, systematic studies on the entire feeding process by any single nematode species are still lacking. Even for *Caenorhabditis elegans*, the most intensely studied nematode at all organismal, cellular and molecular levels, the food and feeding habits, structure and function of feeding apparatus and digestion are not fully benchmarked. To establish such correlations, a scheme of nematode classification based on similarities in their food and feeding habits and the structure and function of their feeding apparatus is required. Ideally, feeding habits of each nematode species should be determined in their particular ecological setting, a task that is difficult to achieve due to inadequate information, dual feeding habits and different structures performing similar acts of feeding. A classification based on food and feeding structures and the habits of nematodes is required to explain diversity in nematode feeding behaviour. We adopt a scheme of nematode classification similar to that of Yeates *et al.* (1993), based on nematode food and feeding habits.

4.3.1. Plant

Nematodes feeding on plants are stylet-bearing, obligate parasites obtaining nutrients from living cells of roots, stems or leaves. Variations in nematode feeding habits occur for their adaptive needs. As a result, evolutionary adaptations for plant feeding led to the development of a stylet and morphological and physiological modifications of the oesophagus. Nutrients from living cells and tissues are sucked into the oesophagus through the stylet lumen; trichodorids (e.g. *Paratrichodorus*) are an exception, however, in that this stylet is a tooth that lacks a lumen and is used only to puncture the plant cells (Hunt, 1993) (Fig. 4.3A, B).

   Yeates *et al.* (1993) and Hussey and Grundler (1998) divided plant feeders into: (i) ectoparasitic; (ii) migratory ectoparasitic; (iii) migratory endoparasitic; (iv) sedentary ectoparasitic; (v) sedentary endoparasitic; and (vi) semi-endoparasitic nematode categories. The sedentary ectoparasite *Criconemella xenoplax* feeds at a single feeding site on epidermal cells for long durations, causing little tissue damage, although terminal galls are induced when feeding occurs at the root tip. Sedentary

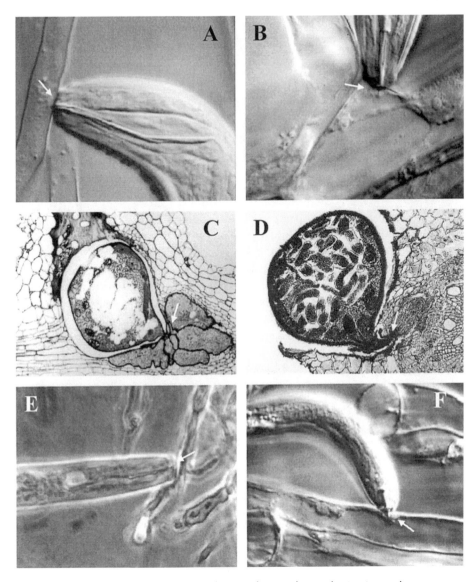

Fig. 4.3. A. *Paratrichodorus anemones* feeding on the root hairs of *Nicotiana tabacum.*
B. *P. anemones* feeding on epidermal cell of *N. tabacum.* C. Cross-section of root showing
feeding by *Meloidogyne incognita* on a giant cell. D. Cross-section of tomato root showing
gravid *M. incognita* female feeding on a giant cell. E. Pinewood nematode, *Bursaphelenchus
xylophilus,* feeding on fungal hyphae. F. Lesion nematode, *Pratylenchus penetrans,*
ectoparasitic feeding behaviour on the root hair of tobacco. Arrows show site of feeding.
(A, B, C, courtesy of U. Wyss; D, E, F, from Eisenback and Zunke, 1997.)

endoparasites, such as *Heterodera, Globodera, Meloidogyne* or *Sphaeronema,* establish
complex feeding relationships with their host by modifying host endodermal cells
into specialized feeding structures, such as giant cells, syncytia or nurse cells

(Fig. 4.3C, D). Ectoparasites, such as the root lesion nematode *Pratylenchus*, feeds on epidermal cells or on root hairs (Fig. 4.3F). *Pratylenchus, Anguina, Radopholus* and other migratory endoparasitic species obtain nutrition from cortical cells through periodical intercellular migration within the root. *Hoplolaimus* and *Telotylenchus* are semi-endoparasites feeding internally and externally on plant root tissues. Ectoparasitic plant feeders obtain nutrition from epidermal (e.g. *Tylenchorhynchus*) or endodermal cells (e.g. *Trichodorus, Xiphinema*); *Psilenchus, Tylenchus* and *Atylenchus* feed only on root hairs (Fig. 4.3F).

4.3.2. Fungal

Stylet penetration into fungal hyphae (Fig 4.3E) is restricted to some genera of Tylenchida (e.g. *Ditylenchus*), Dorylaimida (e.g. *Leptonchus, Tylencholaimellus*) and Aphelenchida (e.g. *Aphelenchus, Aphelenchoides*) (Yeates *et al.*, 1993; Yeates, 1998). Fungal-feeding by nematodes may be obligate (e.g. *Aphelenchoides*) or facultative (e.g. *Beddingia*) with a narrow host range (Ruess *et al.*, 2001). The narrow host range has been attributed to toxic metabolites released by the fungi and morphological differences in conidiophores. Fungal feeders switch food preference to avoid the toxic effects of fungal metabolites. Suitability of food does not depend upon attractiveness of hyphae as even a poor fungal host can emit strong attractants eliciting nematode response.

4.3.3. Bacterial

Although many species feed exclusively on bacteria (Yeates, 1998), *C. elegans* is the foremost example of a bacterial-feeding nematode. *C. elegans* feeds on mixed bacterial populations under natural conditions and is laboratory cultured extensively using *Escherichia coli*. Diplogasterid predators will consume bacteria in the absence of prey (Jairajpuri and Bilgrami, 1990). Entomopathogenic nematodes are primitive nematodes that have evolved the ability to attack insects and yet have not severed their nutritional relationship with bacteria. The first two juvenile stages of the vertebrate parasites *Haemonchus* and *Ancylostoma* feed on bacteria (Munn and Munn, 2002). *Enterobius, Heterakis* and *Syphacia* can be included as bacterial feeders because of the probability that they feed on microorganisms in the digestive tract of the host animal.

Eudiplogaster aphodii, a parasite of dung beetles shows a biphasic life cycle with dual feeding. The adults are bacterial-feeding but the third-stage juveniles become parasitic when ingested by dung beetle larvae (Poinar *et al.*, 1976). The juveniles enter the haemocoel by penetrating the midgut wall, where they develop by absorbing nutrients through the cuticle. The adults exit the host during beetle oviposition and return to feeding on bacteria in dung.

4.3.4. Predacious

Predacious nematodes feed indiscriminately on other nematodes as well as bacteria, rotifers and other microorganisms (Bilgrami *et al.*, 1986). Most nematode predators live in soil but others can be found in fresh (*Mononchus, Tripyla*) or marine waters (*Halicoanolaimus, Sphaerolaimus*). Predators are best known among the Mononchida (Bilgrami, 1992), Dorylaimida (Bilgrami, 1993), Diplogasterida (Bilgrami and Jairajpuri, 1989b) and Aphelenchida (Hechler, 1963), but are limited to a few genera of Enoplida (*Ironus*), Actinolaimoidea (*Paractinolaimus*) and Thalassogeneridae (*Thalassogenus*) (Bilgrami, 1997).

Predatory feeding may be classified into three types depending upon mode of feeding and type of feeding apparatus. One type occurs in mononchid predators, which tend to engulf and swallow their prey whole, although they occasionally feed by cutting larger prey animals into pieces (Bilgrami *et al.*, 1986). The second type involves piercing and sucking as seen in stylet-bearing predators (e.g. dorylaim, nygolaim, aphelenchid). Cuticle perforation and penetration are here achieved by the stylet (*Seinura, Dorylaimus, Aporcelaimellus*) or mural tooth (*Nygolaimus, Aquatides*) (Bilgrami, 1997). The third type of feeding is seen in diplogasterid predators (e.g. *Mononchoides, Butlerius*) (Small and Grootaert, 1983; Bilgrami and Jairajpuri, 1989b), which cut the cuticle and suck out the prey contents.

The aphelenchid predator *Seinura* injects toxic oesophageal secretions to paralyse the prey (Hechler, 1963). Stylet-bearing predators paralyse prey by disrupting internal body organs with the stylet. Mononchid and diplogasterid predators immobilize prey by holding it firmly with high oesophageal suction.

4.3.5. Unicellular eukaryote

Unicellular eukaryotes are a diverse group that includes diatoms, algae and various protozoa. *Achromadora, Enchodelus* (Yeates *et al.*, 1993), *Chromadorita* (Jensen, 1981), *Monhystera microphthalma* and *Diplolaimelloides bruciei* (Romeyn and Bouwman, 1983), *Desmodora* (Nicholas *et al.*, 1988), *Fictor* (Pillai and Taylor, 1968), *Ironus* (Hunt, 1977), *Tobrilus* (Small, 1987) and other related nematode genera have been reported to feed on unicellular eukaryotic food. Otherwise this area of nematode feeding has received minimal effort, in large part due to the difficulty of recognizing structures within the nematode intestine.

4.3.6. Insect

The infective or dauer juvenile of entomopathogenic nematodes is a non-feeding, free-living stage that searches for hosts within the soil. Entomopathogenic nematodes carry symbiotic bacteria within their gut (Bird and Akhurst, 1983), which are released into the insect haemocoel shortly after host penetration. The bacteria have no means of host search other than passive transportation within the nematode. Once in the haemocoel the bacteria proliferate, producing an array of toxins and exoenzymes that kill the host in 24–48 h and convert the host tissues into a nutrient-rich growth medium. The infective juvenile soon exits the dauer stage and begins feeding on the bacteria, producing one or more generations before emerging from the cadaver and dispersing in search of new insect hosts.

The mermithid nematode *Romanomermis culicivorax* uses its stylet only to penetrate the host body wall and not to ingest nutrients. Once inside the mosquito host, the stylet becomes rudimentary. The intestine separates from the oesophagus, transforms into a blind sac and becomes a food storage organ (Poinar, 1975). Absorption of nutrients occurs through the nematode cuticle, which is thin and contains pores (7–11 nm in diameter) (Poinar and Hess, 1977).

The bumblebee nematode *Sphaerularia bombi* is another species that does not use the gut for feeding (Poinar and van der Laan, 1972). The growth of young females is restricted but the uterus grows immensely once it has pushed out through the vulva and into the host body cavity. As a result, a uterine sac several times longer than the female nematode is formed, through which absorption of nutrients from the insect haemolymph takes place.

4.3.7. Omnivorous

Many nematodes are difficult to classify into any particular feeding category because of indiscriminate feeding habits. Omnivorous feeding habits are most pronounced in predatory nematodes, and are probably a strategy for survival in the absence of their primary food source. For example, omnivorous feeding occurs in dorylaim (Russell, 1986) and diplogasterid predators (Bilgrami, 1997), which at low prey densities switch to bacteria, fungi and protozoa. Similarly, *Mononchoides* and *Butlerius* (Bilgrami, 1997) devour bacteria as secondary food but revert back to predation as soon as prey animals are available (Jairajpuri and Bilgrami, 1990).

4.3.8. Animal

The feeding habits of larval and adult animal-parasitic nematodes are diverse, exploiting different food resources at different stages of their life histories. These nematodes ingest blood (e.g. *Haemonchus*) (Murray and Smith, 1994), digesta (e.g. *Ascaris*) (Brownlee and Walker, 1999), secretions (e.g. *Dictyocaulus*) (Munn and Munn, 2002), lymph (e.g. *Wuchereria*) (Munn and Munn, 2002), mucosa (e.g. *Nippostrongylus*) (Croll and Smith, 1977) and hepatic tissues (e.g. *Capillaria*) (Wright, 1974). Moreover, the first two juvenile stages of *Haemonchus* are indiscriminate bacterial feeders, whereas the later stages are parasitic and release anticoagulants to feed on blood (Knox *et al.*, 1993). Similarly, the free-living juvenile stages of *Necator* and *Ancylostoma* feed on bacteria, whereas adults use proteolytic enzymes to gain host access and feed on blood. The whipworms, *Trichuris*, induce syncytia, similar to those of some sedentary plant-parasitic nematodes, and rely on predigestion of host tissues (Artis *et al.*, 1999). In contrast, *Mermis* and *Daubaylia*

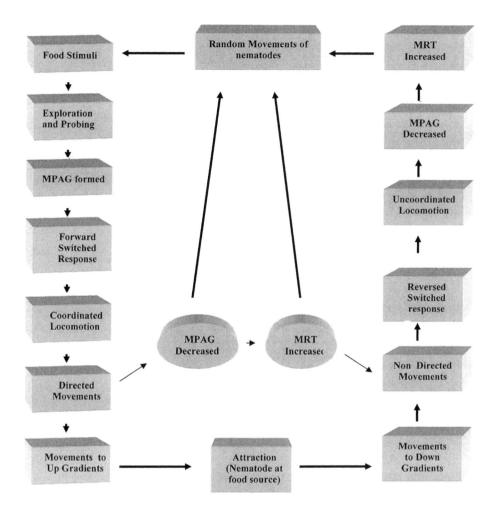

Fig. 4.4. Schematic presentation of the food searching mechanisms in nematodes MPAG, minimum perceptible attraction gradient; MRT, mimimum response threshold. (From Bilgrami, et al., 2001.)

obtain nutrients by feeding on insect and snail haemolymph, respectively (Poinar, 1983). *Acugutturus parasiticus*, an obligate nematode parasite of terrestrial cockroaches (Hunt, 1980), feeds as an ectoparasite on haemolymph, using a feeding apparatus equal to approximately 37% of its body length (Marti *et al.*, 2002).

4.4. Behaviours Leading to Feeding

Behavioural sequences leading to feeding have been described for various nematode groups. Feeding in plant-parasitic nematodes has been classified into widespread and local exploration, stylet thrusting, ingestion and departure from the feeding site (Wyss, 1971). Bilgrami *et al.* (2001a) identified random movements, exploration and probing, perception of attractants, switched response, coordinated movements and chemotactic response during food searching and feeding behaviours of the entomopathogenic species *Steinernema glaseri*. Bacterial-feeding and animal-parasitic nematodes show similar feeding patterns but have not yet been system-atized like those of predacious, plant or entomopathogenic nematodes. The feeding process can be divided into two major behavioural events: (i) food search; and (ii) feeding mechanisms.

4.4.1. Food search

Food search activities of entomopathogenic (Bilgrami *et al.*, 2001a), predatory (Bilgrami *et al.*, 2001b) and plant-parasitic nematodes (Doncaster and Seymour, 1973) depend upon the intensity, concentration and composition of food attrac-tants (Huettel, 1986) as well as the neurosensory abilities of the nematodes. Aggregation of nematodes at feeding sites is a common feature of food search behaviour (Taylor and Pillai, 1967). Large, robust predators, such as mononchs, which can overpower individual prey, however, do not show aggregation (Bilgrami, 1997).

Nematodes display random movements until they perceive a food attraction gradient. Random movement is characterized by short wavelength, high amplitude, increased displacement (i.e. greater wave frequency), high friction index and lack of head movement. Random movement before approaching food has been demon-strated for plant (Wyss, 1971; Bilgrami *et al.*, 1985), insect (Bilgrami *et al.*, 2001a) and predacious (Bilgrami and Jairajpuri, 1988; Bilgrami *et al.*, 2000) nematode species.

4.4.1.1. Exploration and probing

Nematode search after perceiving food stimuli is termed exploration (Fig. 4.4). The ability to perceive food attractants and changes occurring in search parameters provide demarcation between searching (directional) and random movements (non-directional). *Steinernema glaseri* (Bilgrami *et al.*, 2001a), *Mesodorylaimus bas-tiani* (Bilgrami *et al.*, 2000), *Aquatides thornei* (Bilgrami *et al.*, 2001b), *Hirschmanniella oryzae* (Bilgrami *et al.*, 1985) and bacterial feeders (Young *et al.*, 1996) all respond identically to food stimuli. Such an orientation towards food is determined by head probing movements (Ward, 1976; Seymour *et al.*, 1983). Once in the food vicinity, the nematodes use mechanoreceptors, such as labial papillae, to identify the food.

Food search in entomopathogenic nematodes depends upon foraging strategies (Gaugler *et al.*, 1997). Ambushers do not respond to host-released cues as strongly as cruisers (Lewis *et al.*, 1992; Lewis, 2002). The rationale behind this differential behaviour is that ambusher nematodes wait to make contact with mobile hosts whereas cruisers move in search of sedentary hosts. Ambushers nictate (Grewal *et al.*, 1993) or jump towards the food resource to make contact (Campbell and Kaya, 2000). The strong response of *S. glaseri* to host chemical cues (Lewis *et al.*, 1993) and their ability to parasitize relatively sedentary hosts such as *Popillia japonica* indicate a cruising mode. Campbell and Gaugler (1993) grouped *S. glaseri* and *Heterorhabditis bacteriophora* as cruisers on the basis of their foraging strategies. In contrast, the low response of *Steinernema carpocapsae* and *Steinernema scapterisci* to host kairomones and their ability to nictate are good examples of an ambushing strategy. An intermediate type of foraging behaviour (e.g. *S. feltiae*), between cruising and ambushing, has also been demonstrated (Campbell and Gaugler, 1993).

Mononchus and *Monhystera* show vertical undulatory movements similar to nictation in an aquatic environment, possibly to make contact with food. The animal-parasitic nematodes depend upon various modes of dispersal to make contact with food resources, e.g. invertebrate and vertebrate hosts or carriers and vectors. After entering hosts they could perceive food stimuli from blood (e.g. *Bunostomum*), lymph (e.g. *Wuchereria*), haemolymph (e.g. *Beddingia*), gut contents (e.g. *Brumptaemilius*), secretions (e.g. *Litomosoides*), lacrymal secretions (e.g. *Nippostrongylus*) or digesta (e.g. *Ascaris*).

Time of exploration and probing may be short or long and depends upon surface texture, quality, quantity and composition of food. Increased head probing, forward and backward body movements and lateral head movements characterize exploratory and probing activities.

4.4.1.2. Coordinated locomotion

Once nematodes perceive attractants, they move up the gradient. This point, when locomotion changes from uncoordinated (random) to coordinated (oriented) in response to attractants, is called a switched response by Bilgrami *et al.* (2001a). When nematodes move directionally towards food (Fig. 4.4), the locomotion is said to be in coordination. During this phase, wavelength, forward and backward body movements and lateral head movements increase. Amplitude and friction index decrease, head oscillations begin and nematodes move up attraction gradients.

4.4.1.3. Chemotactic response

Nematodes reach food resources by showing chemotactic responses (Fig. 4.4). Upon establishing contact, they identify food with mechanoreceptors (e.g. lip papillae, cephalic setae). After feeding, nematodes move down gradients (away from the feeding site) due to the decrease in attractant concentration. During this phase, locomotion changes from coordinated to uncoordinated and wavelength frequency, forward and backward body movements and lateral head movements decrease, but rates of displacement, wave formation, friction index and degree of wave amplitude increase (Bilgrami *et al.*, 2001a). Head probing decreases due to the decrease in the concentration of perceptible attraction gradient. Reversal behaviour may occur due

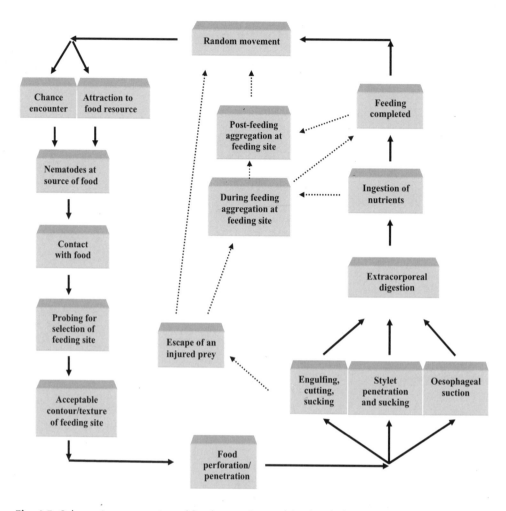

Fig. 4.5. Schematic presentation of food capturing and feeding behaviour of nematodes.

to decreased attraction concentration or an increase in the nematode response threshold.

4.4.2. Feeding mechanisms

Nematode feeding mechanisms vary with feeding apparatus and habitat. Using information from plant-parasitic (Doncaster and Seymour, 1973; Bilgrami, 1997); predatory (Bilgrami, 1992, 1993), entomopathogenic (Grewal *et al.*, 1993; Bilgrami *et al.*, 2001a) and bacterial nematode species (Avery and Horvitz, 1990), a common model of food capturing and feeding mechanisms is presented (Fig. 4.5). Food capturing and feeding mechanisms are divided into five phases: (i) contact with the food resource; (ii) selection of feeding site; (iii) perforation or penetration; (iv) extracorporeal digestion; and (v) ingestion (Fig. 4.5).

4.4.2.1. Contact with food resource

Nematodes contact food during random movements either by chance (*Mononchus, Iotonchus*) (Nelmes, 1974; Grootaert and Maertens, 1976) or, more commonly, by orienting to chemoattractants (*Mononchoides, Caenorhabditis, Meloidogyne, Haemonchus*). Both mechanoreceptors (labial papillae, setae, probolae) and chemoreceptors (amphids) play vital roles in food recognition. Labial contact is necessary to initiate feeding as non-labial contact frequently fails to elicit a feeding response (Bilgrami *et al.*, 1984).

The role of food kairomones in establishing contact with food is well documented in parasitic, predatory and bacterial nematodes. Stylet-bearing predators use food attractants to establish labial contact with prey at right angles (Khan *et al.*, 1991), helping the predators to achieve maximum stylet thrust against cells, tissues or cuticle (Fig. 4.6A–C). Such contacts are particularly important for nygolaim and trichodorid nematodes, whose stylet or mural tooth lacks a lumen. After perforating the food surface, these nematodes ingest nutrients with the help of oesophageal suction through the stoma. Firm right-angle contact maintains vacuum and oesophageal suction within the predator's body by sealing gaps between nematode lips and the food surface during ingestion. Glancing contacts may fail to establish firm contact and disrupt prey capture and ingestion. Contact angle is not important for bacterial-feeding nematodes because ingestion depends upon a flow of nutrients sustained by high oesophageal suction.

4.4.2.2. Selection of feeding site

Labial contact with the food results in probing, which enables nematodes to identify a suitable spot to puncture or penetrate the food resource surface. Probing may or may not be localized, depending upon the contour and texture of the food surface (Bilgrami *et al.*, 1985). Probing begins with side-to-side lip rubbing. At this stage the strengthening rods within the cuticular lining of the oesophageal lumen remain well behind the oral aperture. During probing these rods are drawn forward, pushing the lips outwards to make contact with and assess the food. If the food is recognized as suitable, contact is followed by attempts at perforation (Wyss *et al.*, 1979). The nematodes probe other areas of the food surface if they are unable to penetrate at an earlier location. Each perforation bout lasts several minutes.

Root parasites prefer soft-textured cell surfaces to initiate feeding (Bilgrami, 1994). Predators prefer soft-bodied, less active prey individuals (Bilgrami and Jairajpuri, 1989a). Duration of probing varies both inter- and intraspecifically, being short (e.g. *Mononchoides fortidens*) or long (*Mononchoides longicaudatus*), vigorous (e.g. *Caenorhabditis*) or slow (e.g. *Trichodorus*), intermittent (e.g. *Hirschmanniella*) or continuous (e.g. *Steinernema* or *Xiphinema*). Selection of a suitable feeding site leads to perforation and penetration (Fig. 4.5).

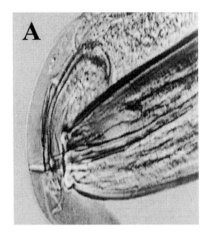

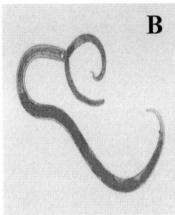
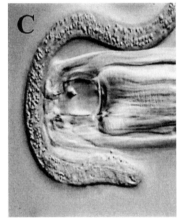

Fig. 4.6. A. *Labronema vulvapapillatum* feeding at the posterior region of *Panagrellus redivivus*. B. *Iotonchus* catching an individual of prey. C. *Anatonchus tridentatus* initiating an attack on *P. redivivus*. (A, C, courtesy of U. Wyss; B, from Eisenback and Zunke (1997.))

4.4.2.3. Perforation and penetration

Parasitic and predacious nematodes use their main feeding organs, e.g. stylet (dorylaim and tylenchid parasites and predators), mural tooth (nygolaim predators), dorsal tooth (mononchid predators), onchia (actinolaim predators, animal parasites), odontia (animal parasites), teeth (enoplid predators, animal-parasitic), cutting plates, dentigerous ridges and sublateral lancets (animal parasites), or combined actions of a movable dorsal tooth and high oesophageal suction (diplogasterid predators) to achieve cuticle perforation.

The labial muscles of predators (e.g. *Mononchus*, *Mylonchulus*) contract to widen the mouth (Jairajpuri and Bilgrami, 1990), bringing the dorsal tooth forward to make contact with the prey cuticle (Grootaert and Wyss, 1979). *Labronema* punctures prey with quick stylet thrusting (Wyss and Grootaert, 1977), whereas *Aquatides* and *Dorylaimus* achieve perforation by gradual and intermittent thrusting against prey cuticle (Shafqat *et al.*, 1987). Diplogasterid predators, *Diplenteron* (Yeates, 1969), *Butlerius* (Grootaert *et al.*, 1977) and *Mononchoides* (Bilgrami and Jairajpuri, 1989b), use their movable dorsal tooth and oesophageal suction to slit open the prey cuticle. *Labronema vulvapapillatum* and *A. thornei* require five to six stylet thrusts (Wyss and Grootaert, 1977; Bilgrami *et al.*, 1985) and *Dorylaimus stagnalis* six to eight thrusts to puncture prey cuticle (Shafqat *et al.*, 1987).

Thick cuticle (e.g. *Hoplolaimus*), coarse body annulations (e.g. *Hemicriconemoides*), gelatinous matrix (e.g. *Meloidogyne* eggs), toxic body secretions (e.g. *Helicotylenchus*) and rapid undulatory movements (e.g. *Rhabditis*) provide resistance for prey animals against predator attack (Small and Grootaert, 1983; Bilgrami and Jairajpuri, 1989a). The most common reasons for prey nematodes escaping predators include an inability of the predator to puncture the prey cuticle or evasive behaviours by the prey animals.

Plant-parasitic nematodes penetrate epidermal or endodermal root cells to obtain nutrients or they feed externally by puncturing the cells and root hairs. Cell wall perforation by endoparasitic nematodes (e.g. second-stage juveniles of *Meloidogyne incognita* and *Heterodera schachtii*) is intracellular (Fig. 4.7A–D), whereas it is extracellular for many ectoparasitic species (e.g. *Xiphinema index*). *Sphaeronema* (Fig. 4.8A) and *Rotylenchulus* (Fig. 4.8B–D) puncture root epidermal cells to ingest nutrients. During cell wall perforation, the base of the odontophore rotates due to the action of a narrow longitudinal slit that runs the entire length of the odontophore (Hunt and Towle, 1979). Such rotational movements help nematodes achieve penetration. Longidorid nematodes use their long needle-like stylet to puncture and penetrate roots (Wyss, 2002; Fig. 4.9A–C). Stylet movements are supported by protractor muscles attached to the rear of the odontophore.

Cell wall perforation is unique in *Trichodorus*. Feeding apparatus movements are slow in the beginning but gain momentum at later stages, depending upon the texture, contour and structure of the feeding surface. During cell wall perforation these nematodes use approximately 50–60% of their feeding apparatus. Once punctured, the feeding apparatus penetrates deep into the cell to initiate ingestion of nutrients.

Animal-parasitic nematodes use three penetration modes to enter hosts:

1. Passive: access to host depends on ingestion of nematode eggs by the host (e.g. *Ascaris*, *Necator* and *Ancylostoma*). The nematodes hatch and remain in the host gut or penetrate the gut wall using their 'boring tooth' (e.g. *Ascaris*) so as to migrate to the jejunum as the fourth-stage juvenile moves through the blood, liver, lungs, trachea and oesophagus (Munn and Munn, 2002).
2. Delivered: the nematodes (e.g. filarial worms) are injected into the final host (e.g. human) by the intermediate host (e.g. mosquitoes). Microfilariae are introduced directly into the host bloodstream, from where they migrate to the lymphatic vessels to feed and complete their life cycle (Flegg, 1985).
3. Active: occurs in nematodes, e.g. entomopathogenic species, which enter hosts through external orifices (e.g. oral, anal or respiratory) or pierce the host cuticle (Renn, 1988).

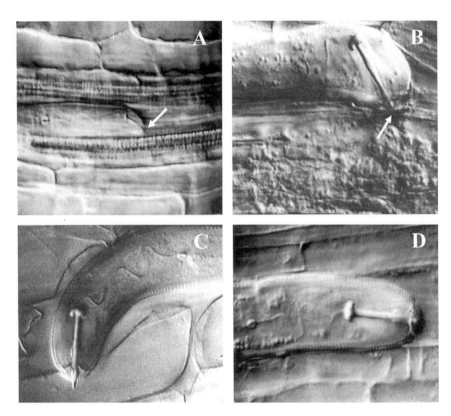

Fig. 4.7. A. Forward migration of second-stage juvenile of *Meloidogyne incognita* in the young vascular cylinder of *Arabidopsis thaliana* root. B. Second-stage juvenile of *Heterodera schachtii* with disintegrated syncytium a few days after the juvenile has died at the beginning of the moulting from J2 to J3. C. Second-stage juvenile of *H. schachtii* penetrating through the cortical cell by cutting the cell wall with the stylet. D. Intercellular migration of a second-stage juvenile of *H. schachtii* inside *B. napus* root. Arrows show feeding site. (A, D, courtesy of U. Wyss; B, C, from Eisenback and Zunke, 1997.)

The guinea worm, *Dracunculus medinensis*, is an excellent example of passive penetration. This parasite gains access to human hosts when water infested with copepods (the intermediate host) is ingested. The juveniles migrate to the subcutaneous connective tissues and mature to males and females. Adult females feed on the host blood and migrate to the skin to produce juveniles. Hookworm juveniles of *Ancylostoma* and *Necator* use cutaneous penetration to reach blood capillaries in mammalian hosts. If they fail to locate a food supply, they wander through the skin, causing cutaneous larval migrans, a condition resulting from an inflammatory host reaction. Adult hookworms attach themselves to the intestinal wall with the buccal capsule and penetrate blood-vessels, using cutting plates and dorsal or ventral teeth. These species also penetrate pulmonary capillaries and intestinal mucosa to feed.

Various mechanical and chemical activities and stage-specific expressions are involved in invasion and penetration, including the action of hyaluronidase and

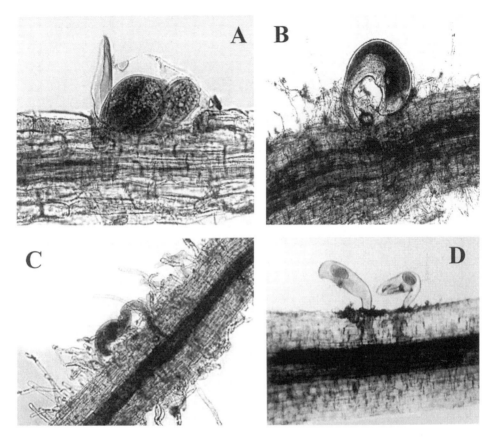

Fig. 4.8. A. *Sphaeronema* feeding on the root of sugar cane. B. *Rotylenchulus reniformis* female feeding on tomato root tissues. C. Female of *R. reniformis* feeding on epidermal cells of tomato roots. D. Females of *R. reniformis* penetrating and feeding on soybean root. (From Eisenback and Zunke, 1997.)

proteolytic enzymes on cells and tissues (Hotez *et al.*, 1992). Metalloproteases are released by *Strongyloides stercoralis*, to dissolve the extracellular matrix of the dermis (McKerrow *et al.*, 1990). Proteolytic enzymes are released by *Necator* to dissolve cells and tissues once they are established at the intestinal mucosa.

4.4.2.4. Extracorporeal digestion

The feeding apparatus of parasitic, predatory and fungal-feeding nematodes has a narrow lumen through which food passes into the intestine. The lumen is, however, too narrow to carry large food molecules without their being digested prior to ingestion. Similarly, for penetration of plant and animal tissues and intercellular migration, nematodes require a means of disrupting tissues. Nematodes achieve these goals through extracorporeal digestion.

Extracorporeal digestion is a physiological condition required to partially digest

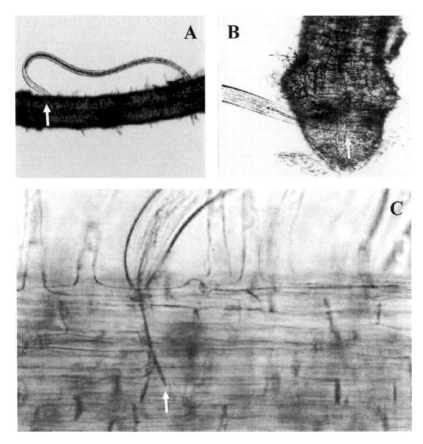

Fig. 4.9. A. *Xiphinema index* feeding from cortex cells deep inside the roots of *Ficus*. B. *Xiphinema diversicaudatum* feeding on a differentiated root region of *Lycopersicum esculentum*. C. *X. diversicaudatum* feeding at the root tip of *Ficus carica*. Arrows point at stylet tip. (A, courtesy of Eisenback and Zunke, (1997); B, C, courtesy of U. Wyss.)

food outside the corpus (Fig. 4.5). It is analogous to salivation in the sense that digestive enzymes are involved. Extracorporeal digestion occurs in many plant-parasitic (Wyss, 1971) and predatory nematodes (Bilgrami and Jairajpuri, 1989b). Animal-parasitic nematodes differ in the sense that they inject anticoagulants (Capello *et al.*, 1995) as well as digestive oesophageal secretions (Knox *et al.*, 1993) to dissolve host tissues prior to penetration or food ingestion. Some oesophageal secretions (e.g. acetylcholinesterase) (Griffiths and Pritchard, 1994) may not have digestive functions.

Extracorporeal digestion does not occur in predacious mononchs (Jairajpuri and Bilgrami, 1990). Presumably they do not need to predigest food since they have a wide buccal cavity capable of engulfing or swallowing prey whole or in pieces. Extracorporeal digestion does occur in diplogasterid predators (Bilgrami and Jairajpuri, 1989b). Nutrients are partially digested within the prey body by oesophageal gland secretions prior to ingestion. Complex food globules are broken down to simple and smaller particles, which are ingested into the intestine through the oesophagus.

The role of digestive enzymes in extracorporeal digestion is most evident in sedentary endoparasitic nematodes e.g. *Meloidogyne*, *Heterodera* and *Globodera*. Their digestive secretions induce and modify plant cells and tissues into giant cells, syncytia or nurse cells for feeding throughout their life. *Ditylenchus*, *Hirschmanniella*, *Tylenchorhynchus* and *Pratylenchus* also digest food contents prior to ingestion. Many plant-parasitic nematodes secrete globules at the outlet of the dorsal gland ampulla, which form homogeneous non-granular saliva for extracorporeal digestion. Although digestive gland secretions were absent in *Paratylenchus projectus*, a dome of granular matter was observed over the stylet tip within the cell (Rhoades and Linford, 1961). The fungal-feeding nematodes *Aphelenchus avenae*, *Aphelenchoides bicaudatus* (Siddiqi and Taylor, 1969) and *Aphelenchoides hamatus* (Zunke *et al.*, 1986) do not show extracorporeal digestion, but *Ditylenchus myceliophagus* predigest hyphal cell contents before ingestion.

Ultrastructural studies of oesophageal gland ampulla showed that its cells modify into an 'end apparatus' surrounded by a membrane-delineated 'end-sac'. The end apparatus acts like a valve, controlling the flow of digestive enzymes from gland ampulla to oesophageal lumen. It was impossible to prove that dorsal oesophageal gland secretions were injected into the food until high-resolution video-enhanced contrast light microscopy became available (Wyss and Zunke, 1986). This technique permitted observation of cell responses several layers below the root surface and confirmed that dorsal oesophageal gland secretions are injected into the cells for extracorporeal digestion (Wyss *et al.*, 1988). During this phase the feeding apparatus is pushed deep into the perforated cells, remaining immobile while oesophageal gland secretions are released into the cytoplasm and immediately dissolving cytoplasm adjacent to the stylet tip (Wyss, 2002). Oesophageal glands generate sufficient pressure to inject their secretions into the host (Doncaster and Seymour, 1975).

Animal-parasitic nematodes accumulate host tissues in the stoma to form a plug. Digestive gland secretions containing proteases and hyaluronidases are then secreted to break down tissues before ingestion or penetration begins (Capello *et al.*, 1995). These enzymes are released through the 'duct tooth' in *H. contortus* (McLaren *et al.*, 1974) and *Bunostomum trigonocephalum* (Wilfred and Lee, 1981), similarly to diplogasterid predators (Bilgrami and Jairajpuri, 1989b).

4.4.2.5. Ingestion

Ingestion of nutrients begins after extracorporeal digestion (Fig. 4.5). Most species of predatory mononchs engulf prey whole (Fig. 4.10A) or shred them into pieces prior to ingestion (e.g. *Anatonchus*) (Fig. 4.10B). Swallowing of prey or its pieces is supported by oesophageal muscle contractions, which pull prey or its pieces into the buccal cavity through vertically positioned oblique plates (Grootaert and Wyss, 1979). During swallowing the oesophageal lumen remains closed by the marginal oesophageal muscles. The oesophageal muscles retract after feeding to restore the

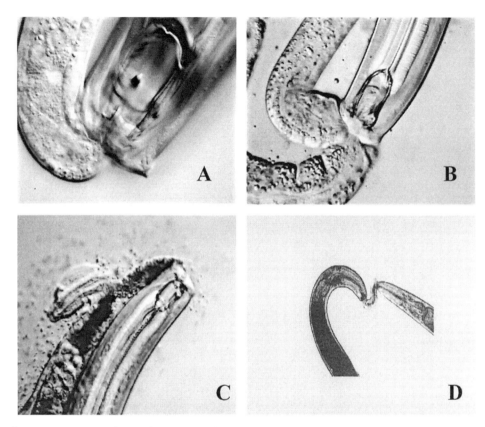

Fig. 4.10. A. *Anatonchus tridentatus* engulfing *Panagrellus redivivus*. B. *Prionchulus punctatus* ingesting nutrients by cutting an individual of *P. redivivus*. C. *P. punctatus* detaching its lip from the prey after feeding. D. Two individuals of *Diplenteron colobocercus* feeding together on single prey. (A, B, C, courtesy of U. Wyss; Fig. D, from Yeates, 1969.)

head and feeding apparatus to their original positions. Some individuals show periods of inactivity after devouring an entire prey, while others initiate further attacks (Nelmes, 1974).

Stylet-bearing predators cannot swallow or engulf or shred prey since their feeding apparatus is of a piercing type. These predators rupture internal prey structures with their stylet. Each protrusion of the stylet is accompanied by oesophageal suction. Suction is generated by stretching of the oesophageal bulb and dilatation of the lumen (Wyss and Grootaert, 1977). Once sufficient suction is generated, nutrients are ingested and forced into the intestine through the oesophago-intestinal valve by rhythmic contraction and relaxation of the oesophageal bulb lumen. Once all the contents are ingested, predators detach their lips from prey cuticle, retract the feeding apparatus (Fig. 4.10C) and move in search of fresh prey.

Diplogasterids can devour intact first- or second-stage juveniles of small prey nematodes (e.g. *Acrobeloides* or *Cephalobus*) but must cut larger prey to pieces. Ingestion is supported by the dorsal tooth and oesophageal suction.

Ingestion by predaceous nematodes is often a group activity. An injured prey attracts other predators which aggregate around and feed together at the feeding site (Fig. 4.10D). This group behaviour is especially common in dorylaim (Bilgrami *et al.*, 2000), nygolaim (Bilgrami *et al.*, 2001b) and diplogasterid predators (Bilgrami and Jairajpuri, 1988). Feeding in groups allows predators to finish the prey quickly and continue hunting. Aggregation of predators at feeding sites becomes most pronounced at low prey densities.

With the exception of *Hexatylus viviparus*, ingestion in other fungal-feeding nematodes depends upon oesophageal pumping. *H. viviparus* feeds on turgid hyphae and ingests nutrients passively (Shepherd and Clark, 1976). Passive ingestion depends upon turgor pressure of fungal hyphae coupled with the pumping of nematode rectal muscles (Seymour, 1975). In the absence of oesophageal pumping, nutrients flow from high (inside the hyphal cell) to low food pressure (inside the nematode).

Plant-parasitic nematodes have a wide range of feeding strategies. The most highly evolved are those of sedentary endoparasitic nematodes, which induce cells to form permanently modified feeding structures: unicellular giant cells, multicellular giant cells or syncytia (Wyss, 2002).

Uninucleate giant cells are the evolutionary precursors of multinucleate giant cells. These are induced in the pericycle and found in non-cyst-forming heteroderids such as *Cryphodera* (Mundo-Ocampo and Baldwin, 1984). *Rotylenchulus macrodoratus* transforms cortical cells into uninucleate giant cells (Cohn and Mordechai, 1977). Stylet thrusting induces cell cytoplasm to accumulate into a mass at the site of the wall perforation, leading to ingestion. During ingestion, the radial metacorporeal bulb muscles contract several times per second, causing the oesophageal pump chamber to dilate and releasing nutrients into the intestine via the oesophago-intestinal junction.

The root-knot nematodes *Meloidogyne* induce multinucleate giant cells for feeding. Giant cells are developed by the expansion of about half a dozen cambial cells. The cells become multinucleated by repeated mitosis without cytokinesis (Gheysen *et al.*, 1997). The cyst-forming nematodes *Heterodera* and *Globodera* induce multinucleate syncytia from expanding cambial cells, whose protoplasts fuse after the cell wall dissolves, initially due to the gradual widening of the plasmodesmata and finally due to the enzymatic actions (Grundler *et al.*, 1998). For further information on the structure and function of giant cells and syncytia the reviews of Bleve-Zacheo and Melillo (1997) and Wyss (1981, 2002) may be consulted.

Feeding tubes and plugs are specialized feeding structures that occur in sedentary plant nematodes. These structures, formed at the distal end of the stylet, play a significant role during prolonged root feeding by *Rotylenchulus*, *Heterodera*, *Meloidogyne* and *Trichodorus*. They begin to develop 12–24 h after syncytia are formed (Sobczak *et al.*, 1999; Wyss, 1992). The feeding tube facilitates withdrawal of cytoplasm and functions as a filter, which allows only molecules less than approximately 40 kDa pass; larger cell particulates would otherwise block the stylet lumen (Rebois, 1980), with fatal consequences. Bockenhoff and Grundler (1994) have demonstrated selective intake of nutrients in *H. schachtii* juveniles, whose stylet orifice diameter is about 100 nm (Wyss, 1988). Upon completion of feeding the tubes are dissolved in the syncytial cytoplasm by disassociating from the stylet orifice. Feeding tubes differ greatly in their shapes between straight and helical but all possess a lumen for cytoplasm transport. An electron-dense feeding plug, extruded from the amphids, is formed around the stylet at the point of penetration to seal the perforated cell wall when the stylet is withdrawn. The plug can be used for subsequent stylet re-entry (Wyss, 1992).

The mechanism for capturing and ingesting bacteria is filter feeding. During ingestion, the oesophageal corpus and anterior isthmus muscles contract and retract concurrently. As a result, ingested food is crushed and pushed posteriorly into the intestine by muscular contractions of the terminal oesophageal bulb. During this process, the contractions pull the oesophageal lumen open into a triangular shape from its resting and closed Y-shaped position, allowing crushed food to pass into the intestine. Food remains suspended in the oesophageal bulb when the oesophagus is in the resting position. Concurrent contractions and retractions of muscles allow the grinder (thick-ridged cuticle on the oesophageal lumen surface) to return to its resting position and the oesophageal bulb lumen to remain closed. As a result the liquid of the bacterial suspension is forced out of the oesophageal bulb and the bacteria are retained.

Precise mechanisms for trapping bacteria remain unclear since contractions and relaxations are so fast (< 17 ms: Avery, 1993) that fluid movements cannot be recorded. However, the speed and rate of relaxation should play a significant role in trapping bacteria. Avery (1993) suggested a role of an oesophageal motor neurone 'M3' in muscle relaxations and in bacterial trapping. Metastomal flaps play some role during bacterial ingestion, a role which may be revealed if laser ablation of muscles is conducted. High-speed video recordings of muscular movements and motions of individual particles within the oesophageal lumen could be useful for understanding trapping mechanisms.

The basic patterns of food ingestion in animal-parasitic nematodes resemble those of parasitic, predatory and bacterial-feeding nematodes. Nutrients are drawn through the feeding apparatus and pass into the oesophageal lumen. In *Ascaris*, the anterior muscles are relaxed, the posterior radial muscles are contracted and the mouth is closed once food is ingested. The radial muscle contractions force open the triangular oesophageal lumen into a circular-shaped opening, allowing food to move towards the oesophago-intestinal junction. The oesophagus begins to refill before the food contents are fully emptied into the intestine through the oesophago-intestinal valve. During ingestion the oesophageal lumen remains half-filled with food at any given time. Movement of nutrients depends upon the rate and intensity of oesophageal pumping, which vary from species to species. Oesophageal pumping may be continuous or intermittent, e.g. short bursts of bulb contractions and retractions at high frequency. The intensity of oesophageal suction is difficult to measure but according to one estimate the oesophagus could generate suction at 101 kPa in *Ascaris* (Munn and Munn, 2002).

Ingestion by tooth-bearing animal-parasitic nematodes such as *H. contortus* is different from those feeding on digesta (digested food). These nematodes use their tooth to cut and puncture host tissues to feed. Third-stage juveniles penetrate hosts to reach the abomasal mucosa, the fourth compartment of the stomach of ruminants, via the rumen and omasum. Feeding on host blood begins soon after moulting. *H. contortus* exposes blood from host tissues by lacerating mucosa with the dorsal tooth. Once blood starts to flow, a capillary-like structure forms around the wound, leading to blood ingestion. The free-living stages of *Necator* also feed on bacteria and their infective stages penetrate host tissues using strategies similar to *H. contortus*.

Feeding and ingestion by the whipworm *Trichuris trichuris* resemble those of *Meloidogyne*, with predigestion and the induction of cell modifications. Infective first-stage juveniles induce cells of striated muscles to form nurse cells, which permit prolonged feeding. This cellular modification results from overproduction of interferon-gamma at the feeding site (Artis *et al.*, 1999). Similarly, the cattle eye worm, *Thelazia*, forms giant cells resembling those of *Meloidogyne* to obtain nutrition (Stoffolano and Yin, 1985).

Ingestion by animal-parasitic nematodes increases with increased temperature but decreases as pH and solute concentrations decrease (Bottjer and Bone, 1985). Short-term food deprivation inhibited but oxygenation enhanced feeding activities (Bone *et al.*, 1985). Filarial nematodes, e.g. *Wuchereria bancrofti* and *Brugia pahangi*, inhabiting lymphatic vessels demonstrate yet another mode of feeding: they have a small mouth but a long and narrow oesophagus that provides increased surface area for nutrient absorption.

The fungal feeder *Tylopharynx foetida* possesses mouth-parts similar to those of *Mononchoides* but feeds on fungal hyphae by ripping apart cell walls with their stegostomatal teeth. Movement of teeth in and out of the buccal cavity is regulated by elongation of the oesophagus during each pumping cycle (von Liveven, 2002).

4.5. Post-feeding Activities

4.5.1. Digestion and absorption

Digestion and absorption of nutrients are post-feeding activities that as important parts of the feeding process are briefly discussed here. Most post-feeding activities, including digestion, absorption, synthesis and storage of nutrients, occur in the intestine. Digestive enzymes secreted by the oesophageal glands regulate digestion of the ingested food. Digested food is absorbed through the intestinal microvilli, which constitute the absorptive surface. The absorptive surface of the intestine together with the rate of nutrient flow determines the rate of absorption. For example, in *C. elegans* the intestine is emptied every 45 s, compared with 3–4 min intervals in *Ascaris* (Munn and Munn, 2002). Absorption of nutrients may also depend on the rate of digested food flow through the intestine, which depends upon the rates of oesophageal pumping and defecation. The intestine plays multiple roles, including storage of proteins, lipids and glycogen, absorption of blood for gut-dwelling parasites, protein synthesis, oxygen uptake and transport of symbiotic bacteria.

4.5.2. Defecation

Three distinct motor steps are involved in defecation: (i) posterior body muscle contraction; (ii) anterior body muscle contraction; and (iii) expulsion muscle contraction. Together these steps constitute the defecation motor programme (Avery and Thomas, 1997). Coordinated contractions of the intestinal rectal muscles and posterior and anterior body wall muscles control defecation in nematodes (Croll and Smith, 1978). Contraction and relaxation of posterior body wall muscles push intestinal contents forward. Cephalic muscles then relax, causing the oesophagus to retract, thereby forcing intestinal, sphincter and rectal muscles to open the intestinal–rectal valve and the proctodeum, through which intestinal contents are expelled as excreta (Avery and Thomas, 1997).

4.6. Feeding Preferences

Some nematode species feed on more than one type of primary food resource, whereas others are host-specific and rely on one food resource. Preference or no preference, both categories of feeding behaviours have their own advantages and benefits for nematode growth and survival. In general, host-specific species tend to be compensated with greater longevity (e.g. *Ascaris*), environmentally resistant stages (e.g. cyst, ova, dauer juveniles, gelatinous egg sacs), low metabolic rates (e.g. anaerobic nematodes) and enhanced persistence.

Prey preferences have been observed by direct observations in laboratory experiments (Bilgrami, 1992), gut content (Bilgrami *et al.*, 1986) and faecal matter analysis (Small, 1987). Feeding by mononchs is specific only in the sense that they feed exclusively on nematodes. In prey choice experiments, mononchs and dorylaims preferred species of *Meloidogyne*, *Pratylenchus* and *Paratylenchus* (Esser, 1963). *Prionchulus* preferred to prey upon *Anguina* and *Aphelenchoides* (Nelmes, 1974). *Mylonchulus* and *Dorylaimus* preferred second-stage juveniles of *Meloidogyne*, *Anguina* and *Heterodera* and adults of *Hirschmanniella* and *Tylenchorhynchus*. *Labronema* preferred *Aphelenchus*, *Panagrellus* and *Anguina* over either *Pratylenchus* or *Xiphinema*. *Butlerius* and *Mononchoides* preferred second-stage juveniles of *Heterodera*, *Meloidogyne* and *Anguina*, but did not attack *Hoplolaimus* or *Hemicriconemoides*. Prey species belonging to *Rotylenchus*, *Tylenchus*, *Meloidogyne* and *Pratylenchus* were attacked by *Butlerius* only when artificially wounded (Grootaert *et al.*, 1977). *Diplenteron* fed equally on *Mesorhabditis*, *Panagrolaimus*, *Acrobeloides* and *Zeldia* (Yeates, 1969). Prey selection depends upon the activity, size and behaviour of prey organisms (Bilgrami *et al.*, 1983).

Analysis by Bilgrami *et al.* (1986) of the intestinal contents of predacious mononchs revealed more free-living microbivore (75%) nematodes than parasitic tylenchs (45%), dorylaims (42%) or predators (27%). The heterogeneity of prey indicates that mononchid predators are polyphagous. In general, mononchs preferred *Pratylenchus*, *Tylenchorhynchus*, *Tylencholaimus*, *Aporcelaimus*, *Thornenema*, *Mylonchulus*, *Rhabditis*, *Acrobeloides*, *Chiloplacus*, *Hirschmanniella* and second-stage juveniles of *Meloidogyne, Heterodera* and *Anguina*. Arpin (1979) found a significant correlation between mononchs and free-living nematodes, but Nelmes and McCulloch (1975) did not find this correlation during their studies with mononchid predators.

The predacious estuarine nematode *Enoploides longispiculosus* exhibited a distinct preference for *Pellioditis marina* over *Chromadora nudicapitata* and *Diplolaimella dievengatensis* (Moens *et al.*, 2000). On the other hand, *Adoncholaimus fuscus* showed no preference when given a choice between *Diplolaimelloides meyli* and *Monhystera*. These predators may have facultative predatory feeding since they also feed on organically enriched matters (Lopez *et al.*, 1979; Moens *et al.*, 1999).

Most plant-feeding nematode species have specific feeding sites (Fig. 4.11). Trichodorid nematodes prefer mitochondrial cytoplasm and plastids while feeding on epidermal cells in the elongation region of growing roots (Wyss, 1982). *Xiphinema, Paralongidorus* and *Longidorus* prefer to feed at the root tips and deeper tissues by transforming feeding sites into terminal galls (Wyss, 2002). The galls are composed of necrotic and modified cells that provide food for these nematodes. Feeding by *X. index* is restricted to grape and fig roots (Pitcher *et al.*, 1974) where

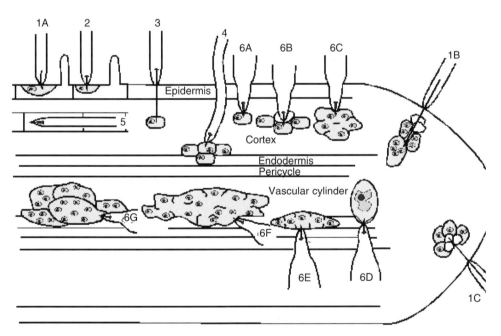

Fig. 4.11. Schematic representation of feeding sites of selected root-parasitic nematodes. 1. Dorylaimid migratory ectoparasites: 1A, *Trichodorus*; 1B, *Xiphinema index*; 1C, *Longidorus elongatus*. 2–6. Tylenchid nematodes: 2, migratory ectoparasite: *Tylenchorhynchus dubius*; 3: sedentary ectoparasite: *Criconemella xenoplax*. 4: migratory ectoparasites: *Helicotylenchus*; 5, Migratory endoparasites: *Pratylenchus*; 6, sedentary endoparasites: 6A, *Trophotylenchulus obscurus*; 6B, *Tylenchulus semipenetrans*; 6C, *Verutus volvingentis*; 6D, *Cryphodera utahensis*; 6E, *Rotylenchulus reniformis*; 6F, *Heterodera*; 6G, *Meloidogyne*. (From Wyss, 1997.)

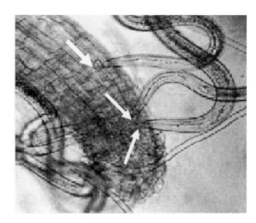

Fig. 4.12. The sting nematode, *Belonolaimus longicaudatus*, showing aggregation, penetration and feeding on maize roots. Arrows show feeding sites. (Courtesy of J.O. Becker.)

meristematic cells transform into hypertrophied multinucleated cells within root tips. *Longidorus* similarly modifies root tips into terminal galls, which can resemble the galls characteristic of root-knot nematodes. *Tylenchorhynchus claytoni* aggregates around root tips and feeds on epidermal cells at the elongation region. *Hemicycliophora* obtain food from the root cortex; *Tetylenchus joctus* draws nutrients from the region of differentiation; and *Tetylenchus dubius* feeds from the elongation zone or root hairs. The sting nematode, *Belonolaimus longicaudatus*, tends to prefer epidermal cells near the root tip, but also penetrates to the root cortex of maize (Huang and Becker, 1997; Fig. 4.12). The cyst nematode *Heterodera cruciferae* prefers root hairs for transient feeding but endodermis for permanent feeding sites. Root-knot nematodes have a host range that encompasses more than 2500 plant species, but their feeding sites are restricted to epidermal and cortical cells of growing host roots. *Rotylenchus robustus* and *Naccobus serendipiticus* prefer to feed on root hairs and epidermal and cortical cells in succession. The long stylets of criconematid nematodes enable feeding on tissues within the root cortex.

Although *Ditylenchus dipsaci*, *D. triformis* and *D. destructor* all feed on fungal hyphae, the former two species also feed on stem and root epidermal cells. *Neotylenchus linfordi*, a fungal-feeding nematode, obtains nutrients from septate hyphae, whereas *A. avenae* draws nutrients from root hairs of higher plants besides fungal cells. Foliar nematodes, e.g. *Aphelenchoides ritzemabosi*, feed on leaf epidermis and obtain nutrients from mesophyll.

Few studies have addressed preferential host selection (Lewis *et al.*, 1996; Gaugler *et al.*, 1997; Bilgrami *et al.*, 2001a) of entomopathogenic nematodes. Some entomopathogenic species, including *S. carpocapsae* and *H. bacteriophora*, possess a host range that encompasses several insect orders. Other species show more host discrimination, including *S. feltiae*, which prefers dipteran hosts, *S. glaseri*, which prefers coleopteran hosts, and *S. scapterisci*, which prefers orthopteran hosts, particularly crickets. Symbiotic associations of nematodes with their bacteria partners are, however, highly specific, e.g. *S. carpocapsae* and *Xenorhabdus nematophila*.

4.7. Conclusions and Future Directions

Feeding is one of the most important and challenging aspects of nematode behaviour. No systematic study has been made of the feeding behaviour of any single nematode that conveys feeling for the depth and breadth of the details required to understand food and feeding habits, digestion and ingestion of nutrients and food conversion. The full complement of feeding molecules is unknown even in *C. elegans*. Most studies on food and host search behaviours have been made on plant and insect nematodes and relatively few on predatory and animal-parasitic nematodes. Extracorporeal digestion and the role of the feeding tubes and plugs need to be thoroughly understood, since they highlight modes of parasitism and enlighten the relationships between the host and parasites and predator and prey organisms. Digestion, absorption, food conversion efficiency, metabolism, food storage, nutritional requirements, food preferences, feeding mechanisms and food utilization by infective or pre-infective juvenile stages are other aspects of feeding that deserve special attention. If food preferences and choices are known, feeding cycles of parasitic nematodes may potentially be disrupted to reduce parasite damage or feeding habits of predatory nematodes altered to stimulate their biological control potential. Future studies on nematode feeding should lead to an understanding of the relationships between nematode food and feeding patterns and the role of such relationships in manipulating host–parasite or predator–prey interactions for management purposes.

References

Alberston, D.G. and Thomson, J.N. (1976) The pharynx of *Caenorhabditis elegans*. *Philosophical Transactions of the Royal Society, London* 275, 299–325.

Ansel, M., Thibaut, M. and Saez, H. (1974) Scanning electron microscopy on *Parascaris equorum* (Goeze, 1782) Yorke and Maplestone, 1926. *International Journal for Parasitology* 4, 17–23.

Arpin, P. (1979) Ecologie et systématique des nématodes Mononchidae des zones forestiéres et herbacées sous climat, tempéré humide: 5-Types de sol et groupements spécifiques. *Revue de Nématologie* 2, 211–221.

Artis, D., Potten, C.S., Else, K.J., Finkelman, F.D. and Grencis, R.K. (1999) *Trichuris muri*: host intestinal epithelial cell hyper proliferation during chronic infection is regulated by interferon-gamma. *Experimental Parasitology* 92, 144–153.

Avery, L. (1993) The genetics of feeding in *Caenorhabditis elegans*. *Genetics* 133, 897–917.

Avery, L. and Horvitz, H.R. (1990) Effects of starvation and neuroactive drugs on feeding in *Caenorhabditis elegans*. *Journal of Experimental Zoology* 253, 263–270.

Avery, L. and Thomas, J.H. (1997) Feeding and defecation. In: Riddle, D.L. and Blumenthal, T. (eds) *C. elegans II*. Cold Spring Harbor Laboratory Press, Cold Spring Harbor, New York, pp. 679–716.

Bilgrami, A.L. (1992) Resistance and susceptibility of prey nematodes to predation and strike rate of the predators *Mononchus aquaticus*, *Dorylaimus stagnalis* and *Aquatides thornei*. *Fundamental and Applied Nematology* 15, 265–270.

Bilgrami, A.L. (1993) Analyses of relationships between predation by *Aporcelaimellus nivalis* and different prey trophic categories. *Nematologica* 39, 356–365.

Bilgrami, A.L. (1994) Attraction and feeding behaviour of *Tylenchorhynchus mashhoodi*. *Annals of Plant Protection Sciences* 2, 33–38.

Bilgrami, A.L. (1997) *Nematode Biopesticides*. Aligarh Muslim University Press, Aligarh, 262 pp.

Bilgrami, A.L. and Jairajpuri, M.S. (1988) Attraction of *Mononchoides longicaudatus* and *M. fortidens* (Nematoda: Diplogasterida) towards prey and factors influencing attraction. *Revue de Nématologie* 11, 195–202.

Bilgrami, A.L. and Jairajpuri, M.S. (1989a) Resistance of prey to predation and strike rate of the predators *Mononchoides longicaudatus* and *M. fortidens* (Nematoda: Diplogasterida). *Revue de Nématologie* 12, 45–49.

Bilgrami, A.L. and Jairajpuri, M.S. (1989b) Predatory abilities of *Mononchoides longicaudatus* and *M. fortidens* (Nematoda: Diplogasterida) and factors influencing predation. *Nematologica* 35, 475–488.

Bilgrami, A.L., Ahmad, I. and Jairajpuri, M.S. (1983) Some factors influencing predation by *Mononchus aquaticus*. *Revue de Nématologie* 6, 325–326.

Bilgrami, A.L., Ahmad, I. and Jairajpuri, M.S. (1984) Predatory behaviour of *Aquatides thornei* (Nygolaimina: Nematoda). *Nematologica* 30, 457–462.

Bilgrami, A.L., Ahmad, I. and Jairajpuri, M.S. (1985) Responses of adult *Hirschmanniella oryzae* towards different plant roots. *Revue de Nématologie* 8, 265–272.

Bilgrami, A.L., Ahmad, I. and Jairajpuri, M.S. (1986) A study on the intestinal contents of some mononchs. *Revue de Nématologie* 9, 191–194.

Bilgrami, A.L., Pervez, R., Yoshiga, T. and Kondo, E. (2000) Feeding, attraction and aggregation behaviour of *Mesodorylaimus bastiani* and *Aquatides thornei* at the feeding site using *Hirschmanniella oryzae* as prey. *International Journal of Nematology* 10, 207–214.

Bilgrami, A.L., Kondo, E. and Yoshiga, T. (2001a) Host searching and attraction behaviour of *Steinernema glaseri* using *Gallaria mellonella* as its host. *International Journal of Nematology* 11, 168–176.

Bilgrami, A.L., Pervez, R., Kondo, E. and Yoshiga, T. (2001b) Attraction and aggregation behaviour of *Mesodorylaimus bastiani* and *Aquatides thornei* (Nematoda: Dorylaimida). *Applied Entomology and Zoology* 36, 243–249.

Bird, A.F. (1971) Digestive system of *Trichodorus porosus*. *Journal of Nematology* 3, 50–57.

Bird, A.F. and Akhurst, R.J. (1983) The nature of the intestine vesicle in nematodes of the family Steinernematidae. *International Journal for Parasitology* 13, 599–606.

Bird, A.F. and Bird, J. (1991) *The Structure of Nematodes*. Academic Press, London, 316 pp.

Bleve-Zacheo, T. and Melillo, M.T. (1997) The biology of the giant cells. In: Fenoll, C., Grundler, F.M.W. and Ohl, S.A. (eds) *Cellular and Molecular Aspects of Plant Nematode Interactions*. Kluwer Academic Publishers, Dordrecht, The Netherlands, pp. 65–79.

Bockenhoff, A. and Grundler, F.M.W. (1994) Studies on the nutrient uptake by the beet cyst nematode *Heterodera schachtii* by *in situ* microinjection of fluorescent probes into the feeding structures in *Arabidopsis thaliana*. *Parasitology* 109, 249–254.

Bone, L.W., Markiw, A. and Bottjer, K.P. (1985) Feeding of *Nippostrongylus brasiliensis* in oxygen and serum. *Journal of Parasitology* 71, 126–128.

Bottjer, K.P. and Bone, L.W. (1985) *Nippostrongylus brasiliensis*: feeding activity in the mouse. *International Journal for Parasitology* 15, 9–14.

Brownlee, D.J.A. and Walker, R.J. (1999) Actions of nematode FMRFamide-related peptides on the pharyngeal muscle of the parasitic nematode, *Ascaris suum*. *Annals of the New York Academy of Sciences* 897, 228–238.

Campbell, J.F. and Gaugler, R. (1993) Nictation behaviour and its ecological implications in the host search strategies of entomopathogenic nematodes (Heterorhabditidae and Steinernematidae). *Behaviour* 126, 155–169.

Campbell, J.F. and Kaya, H.K. (2000) Influence of insect associated cues on the jumping behaviour of entomopathogenic nematodes (*Steinernema* spp.). *Behaviour* 137, 591–609.

Capello, M., Vlasuk, G.P., Bergum, P.W., Huang, S. and Hotez, P.J. (1995) *Ancylostoma caninum* anticoagulant peptide: a hookworm-derived inhibitor of human coagulation factor Xa. *Proceedings of the National Academy of Sciences, USA* 92, 6152–6156.

Cohn, E. and Mordechai, M. (1977) Uninucleate giant cell induced in soybean by the nematode *Rotylenchulus macrodoratus*. *Phytoparasitica* 5, 85–93.

Coomans, A. and Lima, M.B. (1965) Description of *Anatonchus amiciae* n. sp. (Nematoda: Mononchida) with observations on its juvenile stages and anatomy. *Nematologica* 11, 413–431.

Croll, N.A. and Smith, J.M. (1977) The location of parasites within their hosts: the behaviour of *Nippostrongylus brasiliensis* in the anaesthetized rat. *International Journal of Parasitology* 7, 195–200.

Croll, N.A. and Smith, J.M. (1978) Integrated behaviour in the feeding phase of *Caenorhabditis elegans* (Nematoda). *Journal of Zoology* 184, 507–517.

Doncaster, C.C. and Seymour, M.K. (1973) Exploration and selection of penetration site by Tylenchida. *Nematologica* 19, 137–145.

Doncaster, C.C. and Seymour, M.K. (1975). Passive ingestion in a plant nematode, *Hexatylus viviparus* (Neotylenchidae: Tylenchida). *Nematologica* 20, 297–307.

Eisenback, J. and Zunke, U. (1997) *NemaPix*, Vol. I, *A Journal of Nematological Images*. Mactode Blacksburg, VA.

Esser, R.P. (1963) Nematode interactions in plates of non-sterile water agar. *Proceedings of Soil and Crop Science Society of Florida.* 23, 121–138.

Flegg, J. (1985) *Parasitic Worms*. Shire Publications, Princes Risborough, UK, 24 pp.

Gaugler, R., Campbell, J.F. and McGuire, T.R. (1989) Selection for enhanced host-finding in *Steinernema feltiae*. *Journal of Invertebrate Pathology* 54, 363–372.

Gaugler, R., Lewis, E. and Stuart, R. (1997) Ecology in the service of biological control. *Oecologica* 109, 483–489.

Gheysen, G., de Almeida Engler, J. and Van Montagu, M. (1997) Cell cycle regulation in nematode feeding sites. In: Fenoll, C., Grundler, F.M.W. and Ohl, S.A. (eds) *Cellular and Molecular Aspects of Plant Nematode Interactions*. Kluwer Academic Publishers, Dordrecht, Netherlands, pp. 120–132.

Grewal, P., Gaugler, R. and Lewis, E.E. (1993) Host recognition behaviour by entomopathogenic nematodes during contact with insect gut contents. *Journal of Parasitology* 79, 495–503.

Griffiths, G. and Pritchard, D.I. (1994) Purification and biochemical characterization of acetylcholinesterase (AchE) from the excretory/secretory products of *Trichostrongylus colubriformis*. *Parasitology* 108, 576–586.

Grootaert, P. and Maertens, D. (1976) Cultivation and life cycle of *Mononchus aquaticus*. *Nematologica* 22, 173–181.

Grootaert, P. and Wyss, U. (1979) Ultrastructure and function of the anterior feeding apparatus in *Mononchus aquaticus*. *Nematologica* 25, 163–173.

Grootaert, P., Jaques, A. and Small, R.W. (1977) Prey selection in *Butlerius* sp. (Rhabditidae: Diplogasteridae). *Medelingen Faculteit Landbouwwetenschappen Rijksuuniversiteit Gent* 24, 1559–1563.

Grundler, F.M.W., Sobczak, M. and Golinowski, W. (1998) Formation of wall openings in root cells of *Arbidopsis thaliana* following infection by the plant-parasitic nematode *Heterodera schachtii*. *European Journal of Plant Pathology* 104, 545–551.

Hechler, H.C. (1963) Description, developmental biology and feeding habits of *Seinura tenuicaudata* (de Man) J.B. Goodey, 1960 (Nematoda: Aphelenchida), a nematode predator. *Proceedings of the Helminthological Society of Washington* 30, 182–195.

Hotez, P.J., Narsimhan, S., Haggerty, J., Milstone, L.B., Bhopale, V., Schad, G.A. and Richards, F.F. (1992) Hyaluronidase from infective *Ancylostoma* hookworm larvae and its possible function as a virulence factor in tissue invasion and in cutaneous larva migrans. *Infection and Immunity* 60, 1018–1023.

Huang, X. and Becker, J.O. (1997) *In vitro* culture and feeding behaviour of *Belonolaimus longicaudatus* on excised *Zea mays* roots. *Journal of Nematology* 29, 411–415.

Huettel, R.N. (1986) Chemical communicators in nematodes. *Journal of Nematology* 18, 3–8.

Hunt, D.J. (1977) Observations on the feeding of *Ironus longicaudatus* (Enoplida: Ironidae). *Nematologica* 23, 478–479.

Hunt, D.J. (1980) *Acugutturus parasiticus* n.g., n.sp., a remarkable ectoparasitic aphelenchoid nematode from *Periplaneta americana* (L.), with proposal of *Acugutturinae* n. subf. *Systematic Parasitology* 1, 167–170.

Hunt, D.J. (1993) *Aphelenchidae, Longidoridae and Trichodoridae: the Systematics and Bionomics.* CAB International, Wallingford, UK, 352 pp.

Hunt, D.J. and Sutherland, J.A. (1984) *Ichthyocephaloides dasyacanthus* n.g., n.sp. (Nematoda: Rhigonematoidea) from millipede from Papua New Guinea. *Systematic Parasitology* 6, 141–146.

Hunt, D.J. and Towle, A. (1979) Feeding studies on *Xiphinema vulgare* Tarjan, 1964 (Nematoda: Longidoridae). *Revue de Nématologie* 2, 37–40.

Hussey, R.S. and Grundler, F.M.W. (1998) Nematode parasitism of plants. In: Perry, R.N. and Wright, D.J. (eds) *The Physiology and Biochemistry of Free-living and Plant-parasitic Nematodes.* CAB International, Wallingford, UK, pp. 213–243.

Hussey, R.S. and Mims, C.W. (1991) Ultra-structure of feeding tubes formed in giant-cells induced in plants by the root-knot nematode *Meloidogyne incognita*. *Protoplasm* 162, 99–107.

Jairajpuri, M.S. and Bilgrami, A.L. (1990) Predatory nematodes. In: Jairajpuri, M.S., Alam, M.M. and Ahmad, I. (eds) *Nematode Bio-control: Aspects and Prospects.* CBS Publishers and Distributors, New Delhi, pp. 95–125.

Jasmer, D.P. (1995) *Trichinella spiralis*: subversion of differentiated mammalian skeletal muscle cells. *Parasitology Today* 11, 185–188.

Jensen, P. (1981) Phyto-chemical sensitivity and swimming behaviour of the free-living marine nematode *Chromadorita tenuis*. *Marine Ecology – Progress Series* 4, 203–206.

Khan, Z., Bilgrami, A.L. and Jairajpuri, M.S. (1991) Some studies on the predation abilities of *Aporcelaimellus nivalis* (Nematoda: Dorylaimida). *Nematologica* 37, 333–342.

Knox, D.P., Redmond, D.L. and Jones, D.G. (1993) Characterization of proteinases in extracts of adult *Haemonchus contortus*, the ovine abomasal nematode. *Parasitology* 106, 395–404.

Lee, E.E. (1996) Why do some nematode parasites of the alimentary tract secrete acetylcholinesterase? *International Journal of Parasitology* 26, 499–508.

Lewis, E.E. (2002) Behavioural ecology. In: Gaugler, R. (ed.) *Entomopathogenic Nematology.* CAB International, Wallingford, UK, pp. 190–205.

Lewis, E.E., Gaugler, R. and Harrison, R. (1992) Entomopathogenic nematode host finding: response to host contact cues by cruise and ambush foragers. *Parasitology* 105, 309–319.

Lewis, E.E., Gaugler, R. and Harrison, R. (1993) Response of cruiser and ambusher ento-mopathogenic nematodes (Steinernematidae) to host volatile cues. *Canadian Journal of Zoology* 71, 765–769.

Lewis, E.E., Ricci, M. and Gaugler, R. (1996) Host recognition behaviour reflects host suit-ability for the entomopathogenic nematode, *Steinernema carpocapsae*. *Parasitology* 113, 573–579.

von Lieven, A.F. (2002) Functional morphology, origin and phylogenetic implications of the feeding mechanism of *Tyloesophagus foetida* (Nematoda: Diplogasterina). *Russian Journal of Nematology* 10, 11–23.

Lopez, G., Riemann, F. and Schrage, M. (1979) Feeding biology of the brackish-water oncholaimid nematode *Adoncholaimus thalassophygas*. *Marine Biology* 54, 311–318.

McKerrow, J.H., Brindley, P., Brown, M., Gam, A.A., Staunton, C. and Neva, F.A. (1990) *Strongyloides stercoralis*: identification of a protease that facilitates penetration of skin by the infective larvae. *Experimental Parasitology* 70, 134–143.

McLaren, D.J., Burt, J.S. and Ogilvie, B.M. (1974) The anterior glands of the adult *Necator americanus* (Nematoda: Strongyloidea): II. Cytochemical and functional studies. *International Journal of Parasitology* 4, 39–46.

Mapes, C.J. (1966) Structure and function in the nematode pharynx. III. The pharyngeal pump of *Ascaris lumbricoides*. *Parasitology* 56, 137–149.

Marti, O.G., Jr, Silvain, J.F. and Adams, B.J. (2002) Speciation in the Acugutturidae (Nematoda: Aphelenchida). *Nematology* 4, 489–504.

Moens, T., Verbeeck, L. and Vincx, M. (1999) The feeding biology of a predatory and a facultative predatory marine nematode *Enoploides longispiculosus* and *Adoncholaimus fuscus*. *Marine Biology* 134, 585–593.

Moens, T., Herman, P., Verbeeck, L., Teyaert, M. and Vincx, M. (2000) Predation rates and prey selectivity in two predaceous estuarine nematode species. *Marine Ecology Progress Series* 205, 185–193.

Mundo-Ocampo, M. and Baldwin, J.G. (1984) Comparison of host response of *Cryphodera utahensis* with other heteroderidae and a discussion of phylogeny. *Proceedings of the Helminthological Society of Washington* 51, 25–31.

Munn, E.A. and Munn, P.D. (2002) Feeding and digestion. In: Lee, D.L. (ed.) *The Biology of Nematodes*. Taylor and Francis, New York, pp. 233–260.

Murray, J. and Smith, W.D. (1994) Ingestion of host immunoglobulin by three non-blood-feeding nematode parasites of ruminants. *Research in Veterinary Science* 57, 387–389.

Nelmes, A.J. (1974) Evaluation of the feeding behaviour of *Prionchulus punctatus* (Cobb), a nematode predator. *Journal of Animal Ecology* 43, 553–565.

Nelmes, A.J. and McCulloch, J.S. (1975) Number of mononchid nematodes in soil sown to cereals and grasses. *Annals of Applied Biology* 79, 231–242.

Nicholas, W.L, Stewart, A.C. and Marples, T.G. (1988) Field and laboratory studies of *Desmodora gerlach*, 1956 (Desmodoridae: Nematoda) from mangrove mud flats. *Nematologica* 34, 331–349.

Pillai, J.K. and Taylor, D.P. (1968) Biology of *Paroigolaimella bernensis* and *Fictor anchico-prophaga* (Diplogasterina) in laboratory culture. *Nematologica* 14, 159–170.

Pitcher, R.S., Siddiqi, M.R. and Brown, D.J.F. (1974) *Xiphinema diversicaudatum*. CIH Descriptions of Plant-Parasitic Nematodes, Set 4, No. 60, 4 pp.

Poinar, G.O., Jr (1975) *Entomogenous Nematodes: a Manual and Host List of Insect-Nematode Association*. E.J. Brill, Leiden, 317 pp.

Poinar, G.O., Jr (1983) *The Natural History of Nematodes*. Prentice Hall, Englewood Cliffs, New Jersey, 323 pp.

Poinar, G.O., Jr and Hess, R. (1977) *Romanomermis culicivorex*: morphological evidence of transcuticular uptake. *Experimental Parasitology* 42, 27–33.

Poinar, G.O., Jr and van der Laan, P.A. (1972) Morphology and life history of *Sphaerularia bombi*. *Nematologica* 18, 239–252.

Poinar, G.O., Jr, Triggiani, O. and Merritt, R.W. (1976) Life history of *Eudiplogaster aphodii* (Rhabditida: Diplogasteridae), a facultative parasite of *Aphodius fimetarius* (Coleoptera: Scarabaeidae). *Nematologica* 22, 79–86.

Rebois, R.V. (1980) Ultrastructure of a feeding peg and tube associated with *Rotylenchulus reniformis* in cotton. *Nematologica* 26, 396–405.

Renn, N. (1988) Routes of penetration of the entomopathogenic nematode *Steinernema feltiae* attacking larval and adult houseflies. *Journal of Invertebrate Pathology* 72, 281–287.

Rhoades, H.I. and Linford, M.B. (1961) A study of the parasitic habit of *Paratylenchus projectus* and *P. dianthus*. *Proceedings of the Helminthological Society of Washington* 28, 185–190.

Roggen, D.R. (1973) Functional morphology of the nematode pharynx I. Theory of the soft walled cylindrical pharynx. *Nematologica* 19, 349–365.

Romeyn, K. and Bouwman, L. (1983) Food selection and consumption by estuarine nematodes. *Hydrobiological Bulletin* 17, 103–109.

Ruess, L., Garcia Zapata, E.J. and Dighton, J. (2001) Food preferences of a fungal-feeding *Aphelenchoides* species. *Nematology* 2, 223–230.

Russell, C.C. (1986) The feeding habits of a species of *Mesodorylaimus*. *Journal of Nematology* 18, 641.

Seymour, M.K. (1975) Defecation in a passively-feeding plant nematode, *Hexatylus viviparus* (Neotylenchidae: Tylenchida). *Nematologica* 20, 355–360.

Seymour, M.K., Wright, K.A. and Doncaster, C.C. (1983) The action of the anterior feeding apparatus of *Caenorhabditis elegans* (Nematoda: Rhabditida). *Journal of Zoology* 201, 527–539.

Shafqat, S., Bilgrami, A.L. and Jairajpuri, M.S. (1987) Evaluation of the predation abilities of *Dorylaimus stagnalis* Dujardin, 1845 (Nematoda: Dorylaimida). *Revue de Nématologie* 10, 455–461.

Shepherd, A.M. and Clark, S.A. (1976) Structure of the anterior alimentary tract of the passively feeding nematode *Hexatylus viviparus* (Neotylenchidae: Tylenchida). *Nematologica* 22, 332–342.

Siddiqi, I.A. and Taylor, D.P. (1969) Feeding mechanisms of *Aphelenchoides bicaudatus* on three fungi and algae. *Nematologica* 15, 503–509.

Small, R.W. (1987) A review of the prey of predatory soil nematodes. *Pedobiologia* 30, 179–206.

Small, R.W. and Grootaert, P. (1983) Observations on the predation abilities of some soil dwelling predatory nematodes. *Nematologica* 29, 109–118.

Sobczak, M., Golinowski, W. and Grundler, F.M.W. (1999) Ultrastructure of feeding plugs and feeding tubes formed by *Heterodera schachtii*. *Nematology* 1, 363–374.

Stoffolano, J.G. and Yin, L.R.S. (1985) The ultrastructure of the giant cell surrounding larvae of *Thelazia* spp. (Nematoda: Thelazioidea) in the face fly, *Musca autumnalis* De Geer. *Canadian Journal of Zoology* 63, 2352–2363.

Taylor, D.P. and Pillai, J.K. (1967) *Paraphelenchus acontioides* n. sp. (Nematoda: Paraphelelenchidae), a mycophagous nematode from Illinois, with observations on its feeding habits and a key to the species of *Paraphelenchus*. *Proceedings of the Helminthological Society of Washington* 34, 51–59.

Van der Heiden, A. (1974) The structure of the anterior feeding apparatus in members of the Ironidae (Nematoda: Enoplida). *Nematologica* 20, 419–436.

Ward, S. (1976) Chemotaxis by the nematode *Caenorhabditis elegans*: identification of attractants and analysis of the response by use of mutants. *Proceedings of the National Academy of Sciences, USA* 70, 817–821.

Wilfred, M. and Lee, D.L. (1981) Observations on the buccal capsule and associated glands of adult *Bunostomum trigonocephalum* (Nematoda). *International Journal for Parasitology* 11, 485–492.

Wright, K.A. (1974) The feeding site and probable feeding mechanism of the parasitic nematode *Capillaria hepatica* (Bancroft 1893). *Canadian Journal of Zoology* 52, 1215–1220.

Wyss, U. (1971) Der Mechanismus der Nahringsaufname bei *Trichodorus similes*. *Nematologica* 17, 508–518.

Wyss, U. (1981) Ectoparasitic root nematodes: feeding behaviour and plant cell responses. In: Zuckerman, B.M. and Rhode, R.A. (eds) *Plant Parasitic Nematodes*, Vol. III. Academic Press, New York, pp. 325–351.

Wyss, U. (1982) Virus-transmitting nematodes: feeding behavior and effect on root cells. *Plant Disease* 66, 639–644.

Wyss, U. (1987) Video assessment of root cell responses to dorylaimid and tylenchid nematodes. In: Veech, J.A. and Dickson, D.W. (eds) *Vistas on Nematology*. Society of Nematologists, Hyattsville, pp. 325–351.

Wyss, U. (1988) Pathogenesis and host–parasite specificity in nematodes. In: Singh, R.S., Singh, U.S., Hess, W.M. and Webster, D.J. (eds) *Experimental and Conceptual Plant Pathology*. Vol. II. Gordon and Breach, New York, pp. 417–432.

Wyss, U. (1992) Observations on the feeding behaviour of *Heterodera schachtii* throughout development, including events during moulting. *Fundamental and Applied Nematology* 15, 75–89.

Wyss, U. (1997) Root parasitic nematodes: an overview. In: Fenoll, C., Grundler, F.M.W. and Ohl, S.A. (eds) *Cellular and Molecular Aspects of Plant-Nematode Interactions*. Kluwer Academic Publishers, Dordrecht, pp. 5–22.

Wyss, U. (2002) Feeding behaviour of plant-parasitic nematodes. In: Lee, D.L. (ed.) *The Biology of Nematodes*. Taylor and Francis, New York, pp. 233–259.

Wyss, U. and Grootaert, P. (1977) Feeding mechanism of *Labronema* sp. *Mededelingen van de Faculteit. Landbouwwetenschappen Rijksuniversitet Gent* 42, 1521–1527.

Wyss, U. and Zunke, U. (1986) Observations on the behaviour of second-stage juveniles of *Heterodera schachtii* inside host roots. *Revue de Nématologie* 9, 153–165.

Wyss, U., Jank-Ladwig, R. and Lehmann, H. (1979) On the formation and ultrastructure of feeding tubes produced by trichodorid nematodes. *Nematologica* 25, 385–390.

Wyss, U., Robertson, W.M. and Trudgill, D.L. (1988) Esophageal bulb function of *Xiphinema index* and associated root cell responses, assessed by video-enhanced light microscopy. *Revue de Nématologie* 11, 253–261.

Yeates, G.W. (1969) Predation by *Mononchoides potohikus* (Nematoda: Diplogasterida) in laboratory cultures. *Nematologica* 15, 1–9.

Yeates, G.W. (1998) Feeding in free-living soil nematodes. In: Perry, R.N. and Wright, D.J. (eds) *The Physiology and Biochemistry of Free-living and Plant-parasitic Nematodes*. CAB International, Wallingford, UK, pp. 244–269.

Yeates, G.W., Bongers, T., de Goede, R.G.M., Freckman, D.W. and Georgieva, S.S. (1993) Feeding habits in soil nematode families and genera – an outline for soil ecologists. *Journal of Nematology* 25, 315–331.

Young, I.M., Griffiths, B.G. and Robertson, W.M. (1996) Continuous foraging by bacterial-feeding nematodes. *Nematologica* 42, 378–382.

Zunke, U., Rossner, J. and Wyss, U. (1986) Parasitierungsverhalten von *Aphelenchoides hamatus* an funf verschiedenen phytopathogenen Pilzen und an *Agricus campestris*. *Nematologica* 32, 194–201.

5 Reproductive Behaviour

ROBIN N. HUETTEL

Department of Entomology and Plant Pathology, Auburn University, AL 36849, USA

5.1. Introduction

The phylum Nematoda is very diverse and occupies many different ecological habitats, from free-living bacterial feeders in the soil to obligate parasites in plants and animals. This diversity has allowed for the evolution of many different reproductive schemes and reproductive behavioural characteristics. Generally, discussions on reproductive behaviour would only focus on cases where genetic exchange takes place between sexes. However, nematodes have developed very interesting and unusual ways of maintaining populations through both sexual and asexual reproduction. This chapter will consider all types of reproductive mechanisms and the traditional and non-traditional reproductive behavioural characteristics that accompany them.

Historically, nematodes were used as models to study reproduction as well as other biological functions. Some early studies on cell structure and cell functions used *Ascaris* because of the ease in observing cell division, fertilization and other related cytological activities in its egg and various stages of its life cycle. In the late 19th century, observations using these nematodes resulted in the major scientific discovery of meiosis. The first observation of chromosome recombination was demonstrated in the animal-parasitic nematode *Ascaris megalocephala* by Van Beneden in 1883. In his classic study, Van Beneden demonstrated that the chromosome number in the egg was reduced by half in the polar body and doubled again when the egg and sperm fused. This came from his observations that this nematode had four chromosomes in the somatic cells but the egg and sperm only contained two chromosomes each. This was the basis for understanding the role of sexual reproduction in chromosome recombination and exchange of genetic material. In the early 1900s, Boveri and others conducted studies using *Ascaris* and other nematodes that helped biological scientists to understand sex determination, chromosome exchange and chromosome diminution (Boveri, 1909, 1910). These important discoveries included mechanisms such as the X-O and multiple X chromosomes in nematodes.

With the advent of research in molecular biology and advances in microscopy and microtechnologies, again a nematode has become the basic biological model. *Caenorhabditis elegans*, a free-living nematode, has been used extensively to better understand biology in general and to serve as a model for genetics (Wood, 1988; Riddle *et al.*, 1997). Some of the areas of interest related to this chapter are studies that have advanced the understanding of reproductive mechanisms and behaviours of this nematode at the molecular level, including sex determination and signalling, olfaction in mate finding, sperm morphology and function and chromosome counting in determination of sexes. These studies have helped elucidate what mechanisms are involved in the role of reproduction, as well as providing methodologies that have been utilized in research on other nematodes, such as animal and plant-parasitic nematodes and other free-living nematodes. Due to the large volume of literature available on *C. elegans*, many subject areas will only be covered briefly through highlights of relevant information that might help to better understand nematode reproductive behaviour in general. In-depth references are listed for those interested in obtaining more details on many of the areas covered.

5.2. Modes of Reproduction, Gamete Formation and Sex Determination

5.2.1. Reproduction and gamete production

Every known mode of reproduction occurs in nematodes and often both sexual and asexual reproduction occurs within a genus. In species where males and females occur at roughly equal numbers, reproduction is generally through mating and sexual exchange of genetic material. Sexual reproduction is called amphimixis, which is the term that will be used in this chapter. In amphimixis, gametes are produced by the male and female, fuse and undergo meiosis to form the zygote. During meiosis, the diploid chromosomes (2n) are reduced by half to haploid (n) through reduction divisions in the spermatocytes in the testis and the oocytes in the ovary. The chromosomes are recombined when the nuclei of the sperm and ovum unite, back to the diploid (2n) chromosome number. The progeny therefore receive half of their chromosomes from each parent. The combining of the two sets of chromosomes allows for reciprocal gene exchange, which increases variability in the offspring. Amphimictic reproduction is common in many animal, plant and free-living nematode species.

Other nematode species reproduce asexually and this mode is called parthenogenesis. In parthenogenic nematodes, males do not occur or may rarely occur. There are three types of parthenogenesis: apomixis, facultative apomixis and automixis. Apomixis is the most common type of asexual reproduction and gamete production is through mitosis. In mitotic reproduction, the diploid chromosomes replicate themselves in the oocytes in the ovaries of the females and maintain 2n through duplication of the chromosomes. The nucleus splits into two identical daughter nuclei so there is no exchange of genetic material. The second type of asexual reproduction, facultative apomixis, is also mitotic. Males can be produced but generally only in response to environmental stress. These males are generally non-functional but in rare instances may be capable of providing genetic exchange. In some species of plant-parasitic nematodes, including *Meloidogyne*, adverse conditions, including overcrowding, reduction in food sources through plant stress, temperature extremes and host susceptibility, have been considered to play a role in change in male development (Triantaphyllou, 1973). Hormonal signals from either the nematodes or the plant may control the genetic switch to males in species in which males occur only under adverse conditions (Papadopoulou and Triantaphyllou, 1982). The third type of parthenogenesis is called automixis. In this mode, females produce oocytes that divide in half and each daughter oocyte has a haploid (n) number of chromosomes. Through meiotic recombination, one oocyte, called the polar body (n), and another oocyte, called a daughter cell (n), fuse to form the zygote (2n). This type of reproduction has been described for many plant-parasitic nematodes where males are absent, such as *Pratylenchus scribneri*, *Meloidogyne hapla* and *Aphelenchus avenae* (Triantaphyllou, 1971).

Another type of reproduction found in nematodes is pseudogamy, also referred to as hermaphroditism, where both egg and sperm are produced in the same organism. In some cases, the gonad produces eggs first, followed by sperm. In other hermaphroditic nematodes, such as *C. elegans*, the gonad produces the sperm first and then the oocytes. In hermaphroditic species, males tend to be rare, occurring at a frequency of around 1:1000 in *C. elegans*. However, these males are completely functional and reproduce sexually through meiosis. This type of reproduction is most common in the Rhabditidae but other types of reproduction also occur with this family of nematodes.

5.2.2. Genetics of sex determination

Sex determination in nematodes is different from that in other organisms, based on studies with *C. elegans*. In mammalian systems, the chromosome Y, if present, determines whether the offspring is a male. In *C. elegans*, however, there is a chromosome counting mechanism that controls the fate of the sex. The sex of the offspring is determined by the expression of the X-chromosome gene through dosage compensation (Meyer, 1997). Even though understanding the molecular basis for sex determination in *C. elegans* is of great importance in understanding sex genetics in general, few other nematodes have been studied at this level. Due to the large volume of research published on this subject, it is impractical to cover this subject in depth here. For a complete review of sex determination and dosage compensation, see Meyer (1997). Based on the complexities of the genetic mechanisms that control sex in nematodes, the presence or absence of males therefore does not necessarily indicate the mode of reproduction.

5.3. Nematode Sperm Behaviour

5.3.1. Sperm morphology and mobility

The behaviour of nematode sperm has intrigued scientists since the turn of the century as the sperm have no flagella and thus lack a perceived mechanism for movement. In 1979, Abbas and Cain conducted studies on the spermatozoa of *Ascaris lumbricoides* var. *suum* that renewed interest in sperm behaviour. This study, as well as subsequent observations on *C. elegans* by Ward *et al.* (1981), demonstrated that nematode spermatozoa had a pseudopod that contained material different from the cell body. This pseudopod was determined to be responsible for sperm locomotion. A scanning electron micrograph shows an isolated spermatozoon from *C. elegans* extending the pseudopod (Fig. 5.1). The shape and presence of a pseudopod at first led investigators to assume that it moved in an amoeboid motion; however, it was determined in the early 1980s that there was little actin and no myosin as would be considered necessary for such movement in other organisms (Nelson *et al.*, 1982). By use of mutants that were fertilization- and muscle-defective, it was determined that the translocation of sperm in *C. elegans* was through projections on the pseudopods that were responsible for movement. The sperm actually changes its conformation to control its movement. Roberts and Ward (1982) demonstrated that the movement of the spermatozoa was through centripetal flow, which was responsible for the movement of the pseudopod and which propelled the sperm. Sperm lack vesicles, filaments and tubules and are filled with amorphous, granular cytoplasm, which leads to a novel mechanism for movement (Roberts and Ward, 1981; Nelson *et al.*, 1981). This involves the transporting of membranes components through the cytoplasm for assembly at the tip of the pseudopod.

A unique group of proteins, called major sperm proteins (MSPs), were identi-

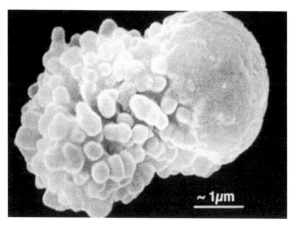

Fig. 5.1. Scanning electron micrograph of a spermatozoon from *Caenorhabditis elegans*. (Courtesy of P. Muhlrad and S. Ward.)

fied that comprised 15% of the protein in *C. elegans* sperm (Klass *et al.*, 1982; Burke and Ward, 1983). The MSPs appear to be unique to the Nematoda. These proteins were shown to be synthesized in the primary spermatocytes and assembled and reassembled into filaments in the pseudopod. This allows the sperm to move in a treadmill-type action. Ward *et al.* (1988) identified more than 50 MSP genes in the testis of *C. elegans*, whereas there was only one MSP gene in *A. lumbricoides* var. *suum* (Nelson and Ward, 1981; Bennett and Ward, 1986). There is no actin in the sperm of either species.

Scott *et al.* (1989) compared the genes for MSPs and actin in other species of nematodes. Since MSPs appear to be highly conserved in nematodes and also appear to be a mechanism for the biosynthesis required by sperm during their rapid development, their presence in other nematodes and the number of MSP genes were of interest. In this study, Scott *et al.* (1989) examined: *A. lumbricoides, Ancylostoma caninum, Dirofilaria immitis, Burgia pahangi* and *Strongyloides ratti* (animal-parasitic nematodes); *Meloidogyne incognita* and *Heterodera glycines* (plant-parasitic nematodes); *Steinernema carpocapsae* (insect-parasitic nematode); *Onchocerca volvulus, Burgia malayi* and *Litomosoides carinii* (human-parasitic nematodes); and *Acanthocheilonema viteae, Panagrellus redivivus* and *C. elegans* (free-living nematodes). In comparisons of DNA samples, there were more similarities among the filarial nematodes, which are both human and animal-parasitic nematodes, than the other groups. There were other interesting similarities between unlikely related nematodes, such as *H. glycines* and *C. elegans*. Even though a large range of numbers of MSP genes exists in the various nematodes species, the general conclusion was that nematode sperm need biosynthesis of MSPs for mobility. It was suggested that in short-lived nematodes, such as *C. elegans* (3-day life cycle), where the spermatozoa live about 90 min, multiple genes increase the rate of MSP production. In long-living nematodes, such as *Ascaris* and the filarial worms, less MSP production is needed since sperm lifespan is much longer and thus fewer genes are required.

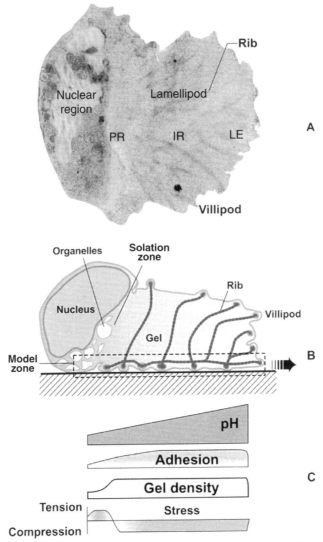

Fig. 5.2. A. Top view of a crawling *Ascaris* sperm. The lamellipod can be divided into three major regions: the leading edge (LE) where polymerization and gel condensation into macrofibres and ribs take place; in *Ascaris,* but not in other nematode sperm, hypercomplexed branched major sperm protein (MSP) 'ribs' are prominent and originate in protuberances called villipodia; the perinuclear region (PR), where the MSP gel solates and generates a contractile stress; and the intermediate region (IR) between the LE and PR, where the gel density is nearly constant. The proximal-to-distal pH gradient affects the polymerization and depolymerization rates at the LE and in the PR. B. Schematic diagram showing the *Ascaris* sperm lamellipod in cross-section. The ventral-fibre complexes branch dorsally. The MSP gel forms at the leading edge and is connected mechanically to the substratum through the membrane. As the cell moves forwards, the gel remains stationary with respect to the substratum, eventually entering the perinuclear region, where it solates and contracts. C. The pH, adhesion, gel density and elastic stress vary with position in the lamellipod. (From Bottino *et al.,* 2002 with permission from *Journal of Cell Science* 115, 367–384.)

The locomotion of nematode sperm without actin has been of interest at both the mechanical and the molecular levels (Bullock *et al.*, 1998). *Ascaris* sperm can crawl at up to 50 μm/min and the cytoskeleton filaments that are extended in the pseudopod can be observed by light microscopy (Roberts and Stewart, 1995, 1997; Stewart *et al.*, 1998). Bottino *et al.* (2002) described the mechanisms that allow for the crawling motion of nematode sperm. The cytoskeleton that contains MSPs in *Ascaris* is organized into a meshwork called fibre complexes. These occur across the lamellipod from the edge of the base to the cell body (Fig. 5.2). These filaments form a thixotropic (shear-thinning) gel that has bonds or cross-links that are labile and can reform. The cell crawls forward in a treadmilling of the cytoskeleton that moves rearward continuously through the lamellipod (Roberts *et al.*, 1998; Italiano *et al.*, 1999). An electron micrograph of the sperm shows that the lamellipod can be divided into three regions in *Ascaris* (Fig. 5.2A). In Fig. 5.2B, a schematic diagram of the *Ascaris* sperm lamellipod in cross-section shows that the MSP gel forms the leading edge and is connected mechanically to the substratum. Figure 5.2C shows the various positions of tension and compression in the lamellipod as affected by pH, adhesion, gel density and elastic stress. A computational model was developed by Bottino *et al.* (2002) as a quantitative biophysical formulation of the 'push-pull' hypothesis as described by Roberts and Stewart (2000). The computation model Bottino *et al.* (2002) developed involved the various components that affect movement, such as the leading of the lamellipod, the solation regions and the forces that affect the physiochemical basis for the model, such as protrusion, adhesion, retraction and depolymerization.

5.3.2. Sperm competition in mating

LaMunyon and Ward (1989) indicated that larger sperm were able to physically displace smaller sperm in *C. elegans*. Sperm competition is of evolutionary significance in this hermaphrodite since the sperm from male nematodes, which are larger, outcompete the smaller sperm produced by the hermaphrodite. Since males only make up about 0.1% of the population, the more competitive the male sperm the better chance of outcrossing, thus adding increased genetic variability into the species. In the hermaphrodite, spermatogenesis takes place first, with the spermatozoa appearing in the ovotestis. The female subsequently produces oocytes, which push the sperm from the ovotestis into the uterus. For fertilization to occur, the sperm must crawl back to the spermatheca to fertilize the other oocytes. Not all oocytes are self-fertilized, which allows some oocytes to be available for male-produced sperm. When the males copulate with a hermaphrodite, their sperm crawl up the uterus, displace the self-produced sperm and fertilize the remaining unfertilized oocytes (Ward and Carrel, 1979).

To determine whether sperm size was also of importance in amphimictic species, LaMunyon and Ward (1989) compared 19 species of Rhabditidae, of which six were amphimictic and 13 were hermaphrodites. In their observations, sperm from hermaphrodites were always smaller than sperm from the males of their species, and sperm from amphimictic species were always larger. Size may have a cost; the larger the sperm, the more energy needed for their production and thus a reduction in the rate at which they can be produced.

5.3.3. The role of major sperm proteins in oocyte maturation

Miller *et al.* (2001) demonstrated another role for MSPs in signalling oocyte maturation in *C. elegans*. During oocyte development, meiosis is arrested until the oocyte is ready for maturation. Ovulation is promoted through the sperm and gonadal sheath cell contraction. When the oocyte enters into the spermatheca, fertilization occurs. By use of deletion mutants of *C. elegans* (one lacking the COOH-terminal region that promotes maturation and the second able to prevent sheath cell contraction), the MSPs were shown to have extracellular signalling and intercellular functions for reproduction.

5.4. Mate Location and Selection Behaviour

Nematodes use chemotaxis to locate members of their own species, members of the opposite sex, food and other resources that aid in their survival (Zuckerman and Janssen, 1984; Huettel, 1986). Chemosensory interactions are received through a group of semiochemical signals called pheromones (Norlund and Lewis, 1981). In nematode species that are amphimictic, the male must be able to locate the female in the host or in the environment where they live. Chance encounters would be too few to allow for successful mating to occur, especially with the low vagility of nematodes (White, 1978). The cues used in nematode mate finding are referred to as sex pheromones (Karlson and Luscher, 1959).

Green (1966) conducted early studies on the orientation of male nematodes to females of the plant-parasitic nematode, *Heterodera schachtii*. This study as well as subsequent studies by other researchers used various media for bioassays in which actual movement of males to an attraction source were observed with microscopy (Green, 1980; Huettel *et al.*, 1986; Rende *et al.*, 1982). In most of these studies, males were readily attracted to females of their own species or to partial or pure extracts from these females (Bone *et al.*, 1979; Jaffe *et al.*, 1989). Males were observed to move randomly on a bioassay until they came in contact with a gradient source from an attractant placed on the bioassay. The males then moved up the gradient source and towards the attractant even though their paths were sometimes in irregular arcs and in spirals around the source. This type of movement continued until contact was made with the point where the attractant had been applied (Green, 1971).

Jaffe *et al.* (1989) isolated a sex pheromone compound that attracted males to females in *H. glycines*. Bioassays were used to determine if males were attracted to the females and/or to purified extracts from females. In *H. glycines*, vanillic acid was isolated from females and was highly attractive to males. Males moved to a female, a spot on agar plates where females had been placed or where vanillic acid had been placed. Once the male reached the spot where the females and/or purified vanillic acid had been placed, a coiling behaviour was observed. This behaviour was similar to that observed during mating in *in vitro* cultures of *H. glycines* on soybean roots. However, when a male *H. glycines* approached a female, the male would probe the female body while coiling and before copulation. This indicated that a mechanosensory cue was received by the male before copulation occurred.

In entomopathogenic nematodes that mate within a host cadaver, a close proximity attractant may be utilized. Lewis *et al.* (2002) observed that *S. carpocapsae* males responded strongly to virgin females but only in their immediate proximity. An ethogram was developed using observational events, which showed that 80% of the males crawled to the females, with 37.5% approaching the female head, 6.25% approaching the female tail and 56.25% approaching the female middle section where the vulva is located. This response occurred only within 2.5 mm of the females, indicating that close-range attractants may be more important than long-range or mate location attractants in this species.

Molecular studies using *C. elegans* are beginning to help elucidate how olfactory stimuli result in various behavioural characteristics. Signal transduction mechanisms that are critical components for sensory response in *C. elegans* are being unravelled. By use of *C. elegans* mutants, the role of different classes of genes, such as nicotinic acetylcholine receptor genes, excitatory and inhibitory glutamate receptor genes and gap junction genes, and different G-protein signalling pathways is being understood (Bargmann and Kaplan, 1998). Ongoing research is beginning to link some of these functions to the genes that interact with sensory neurones involved in recognition of chemoattractants (Bargmann and Mori, 1997).

5.5. Sexual Dimorphism and Male Copulatory Structures

In adult male nematodes, sexual characteristics differentiate them from female nematodes. In some species, the males are morphologically similar to the females except for male reproductive structures. An example of sexual dimorphism can be observed in *Syphacia obvelata*, the mouse pinworm (Fig. 5.3). The males of this species are extremely small compared with the females and differ in their morphology. In *Trichosomoides* the male is so small it lives in the uterus of the female nematode (Little and Orihel, 1972). In some early classifications of nematodes, the males were so different they were identified as a different species from the female, such as in *Radopholus*. The males were named *Tylenchus similis* (Cobb, 1893) and the females *Tylenchus biformis* (Cobb, 1917) until it was discovered that they were the same species and renamed *Radopholus similis* by Thorne (1949). Biologically, there is not a good explanation for sexual dimorphism in nematodes. Since location of the opposite sex is through chemosensory perception and there are no visual abilities in nematodes, the completely different morphology and diverse copulatory structures appear to have no behavioural significance as related to mate attraction or successful copulation.

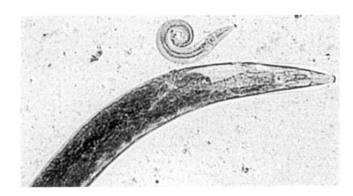

Fig. 5.3. *Syphacia obvelata,* a mouse pinworm that illustrates sexual dimorphism in nematodes. The large nematode is an egg-laden female and the small nematode is the male. (Courtesy of the Missouri Research Animal Diagnostic Laboratory.)

Male nematodes have spicules, which are the copulatory structures that play a role in sperm transfer. In some animal-parasitic and marine nematodes, the copulatory structures are complex and vary greatly in size and shape (Fig. 5.4). The spicules themselves do not have a lumen through which the sperm can pass but aid in physically opening the vulval area and provide channels directing the sperm into the vagina. Bird (1976) showed that spicules in some species can lock to form a tube that may direct the sperm into the vagina. Most male nematodes have two spicules but some species have only one spicule or the spicules may be fused. Spicules can be extended by erector muscles or retracted by protractor muscles, which are attached to their proximal end. During the last moult into adult males, muscles are added to the tail area and other muscles, such as the sphincter muscle, are enlarged. The sphincter muscle closes off the gut area during ejaculation (Sulston *et al.,* 1980).

Some male nematodes have copulatory accessories, such as a bursa, bursal rays and/or ray sensilla, which are found on their posterior region (Fig. 5.4). These structures are moved along the female body before mating occurs, indicating that they might have a mechanosensory function (Chalfie and White, 1988). The flap-like tail modifications may be used to stabilize the female as copulation occurs. The behaviour of males has been altered through the removal of the accessory copulatory apparatuses by the use of lasers (Chalfie and White, 1988). In *C. elegans,* the removal of the hook sensilla prevent the males from locating the vulva, whereas removal of the ray sensilla prevent male tail movements (Chalfie and White, 1988; White, 1988).

Extensive studies have been conducted on *C. elegans* related to mating behaviour as it is regulated by sensilla on the tail of the males (Emmons and Sternberg, 1997). The ability to produce mutant males of this species that are chemosensory-defective, motor system-defective and/or mechanosensory-defective has helped to pinpoint distinct behavioural responses to hermaphrodites. Male *C. elegans* copulatory accessories include nine pairs of sensory rays, ventral sensilla, hooks, fans and

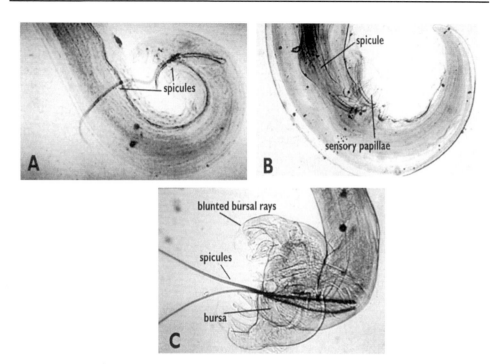

Fig. 5.4. A. The non-bursate tail of a male *Cylicospirura* from a bobcat stomach, showing caudal alae and uneven spicule length, as often seen in the superfamily *Spiruroidea*. B. A non-bursate tail with caudal alae of a male *Physaloptera rara*, showing stalked sensory papillae within the alae and on the body, which is also seen in the Spiruroidea. C. Bursate tail of a male *Metastrongylus apri*, showing the blunted rays characteristic of the Metastrongyloidae. (Courtesy of J. Fox.)

postcloacal sensilla. Each of the accessories has specific functions that initiate specific behaviours in the male.

Liu and Sternberg (1995) developed a detailed model for mating behaviour in *C. elegans* based on specific cues elicited from hermaphrodites and the reciprocal response from males. Males responded to hermaphrodites by backing up and apposing the ventral side of the hermaphrodite's body. The rays and ventral sensilla mediate this response. The sensory rays mediated the cues from the hermaphrodite that cause the male to continue backing and turning around the hermaphrodite. As the male moves down the body of the hermaphrodite, the vulva is located by a distinct subset of the rays. A cue is then sensed by the hook structures, which cause the males to stop near the vulva. The males remain close to the vulva and another cue is sensed by the postcloacal sensilla, which cause the male to protrude his spicules. Once the spicules are inserted into the vulva there are two cues that involve specific neurones. The SPD sensory neurone appears to sense protrusions into the vulva and the SPV neurone appears to inhibit sperm transfer until there is a uterine signal. After the sperm are released, the spicules are retracted and the male moves away. No other genera of nematodes have been studied at this level. Since many male nematode species lack copulatory accessories, their response to females may not be the same as that of *C. elegans.*

5.6. Pre-mating and Copulatory Behaviour

Mating has been reported by many nematode researchers. One of the first detailed reports of copulation was in 1964 by Anderson and Darling. They observed mating in the plant-parasitic nematode, *Ditylenchus destructor.* Their observation included pre-mating behaviour, in which the male passed the female several times before penetration by the spicules occurred. The males were only attracted to certain females and the males would move away if copulation did not occur within 5–10 min. Females were receptive to males for only 1 week, while the males were capable of mating for up to 3 weeks. This behaviour indicated that chemosensory production might be age-dependent in the female. Females were observed to mate with more than one male but the total number of matings was not stated. In the case of *D. destructor,* the bursa appeared to play a role in holding the female in place while mating occurred. After the spicules were inserted, about six to 20 sperm were injected, after which the male quickly moved away.

In the migratory plant-parasitic nematodes *R. similis* and *R. citrophilus,* females were approached by males, which began to probe the female body before copulation occurred. The males were observed to occasionally flare their bursa and extend the spicules prior to mating. Copulation took place with the head of the male and the female in the same direction and lasted up to 5 min (Fig. 5.5; Huettel *et al.,* 1982). During mating, the bursa of the male was extended away from the vulva and was not used to secure the female (Fig. 5.5). This nematode is an endoparasitic nematode that feeds within the roots of its host. The observations of mating took

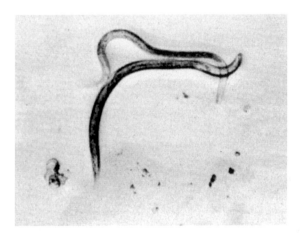

Fig. 5.5. Males of *Radopholus* align with their heads in the same direction as the females. The male tail and bursa is held up away from the female when the spicules enter the vulva, which is located in the middle of the female.

place under culture conditions where the nematodes were placed on water agar plates. It is not known whether mating takes place within the root itself or in the soil surrounding the root.

In the ectoparasitic nematode *Belonolaimus longicaudatus*, mating occurred with the male and female heads oriented in the same direction, as in *Radopholus* (Huang and Becker, 1999). This nematode has a female to male sex ratio of approximately 3:2 (Todd, 1989; Huang and Becker, 1999). Interestingly, even though there are more females than males, several males gathered around a single female. These females may produce a strong sex pheromone signal to ensure that males find them to mate since these nematodes mate in the soil environment under *in vivo* conditions. Males begin a pre-mating process by rubbing the sides of the females with the lateral side of their lip region. Once the bursa contacts the female, the rubbing continues until the bursa reaches the vulva region. The female then twists her body until the spicules penetrate the vulva. The nematodes became quiescent while sperm transfer presumably took place, lasting from 6 to 10 min.

In a sedentary plant-parasitic nematode, *H. glycines*, the males do not undergo their final moult into migratory adults until 1 day after the females have moulted to adults and produced a gelatinous matrix (Lauritis *et al.*, 1983). This may allow the female, which is attached to the root, time to produce a sex pheromone to help the male locate her (Jaffe *et al.*, 1989). The male approached the female and a coiling behaviour was observed (Huettel and Rebois, 1982). The male probed the female several times with his head before copulation occurred. Copulation lasted several minutes before the male moved away. There are some 'super' females of *H. glycines* that are often surrounded by multiple males (Fig. 5.6). However, observa-

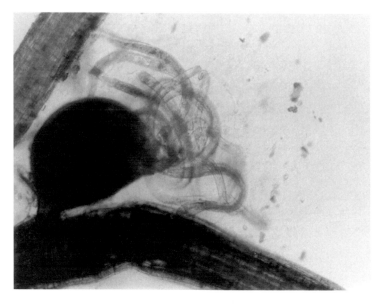

Fig. 5.6. Multiple male nematodes of the soybean cyst nematode, *Heterodera glycines*, coiling around a female.

Fig. 5.7. A *Steinernema carpocapsae* male nematode completely coiled around the middle of the female while copulating. (Courtesy of D. Gouge.)

tion of these females occurred under root explant culture conditions, so it is not known if this behaviour occurs *in vivo*. Under *in vitro* conditions, 'super' females are often not mated in these circumstances since the males tend to push each other away while undergoing coiling behaviour (Huettel and Rebois, 1982).

The mating behaviours of two entomopathogenic nematodes, *S. carpocapsae* and *Heterorhabditis bacteriophora*, are very different from each other even though these nematodes share many morphology and life history traits. Observations of *S. carpocapsae* on culture plates indicated that the males were attracted to females after 1–5 min of random movement (Lewis *et al.*, 2002). Males probed the females in all regions of the body but more than 50% of the time in the middle region around the vulva. When mating occurred, a male coiled where the vulva is located and copulation took place (Fig. 5.7). In *H. bacteriophora* the position for mating is different. The male nematode aligns approximately parallel to the female body and does not coil around her as in *S. carpocapsae*. The male head is pointed in the opposite direction from that of the female, with much of its body away from the female (Fig. 5.8). There is no morphological reason for the two mating positions to be so different.

Observations of copulatory behaviour in *C. elegans* indicated that mating is similar to that in *H. bacteriophora*, where the males back into position before copulation. In another free-living nematode, *Rhabditis pellio*, male and female copulatory behaviour involved the male moving its entire body along the female before copulation occurred (Somers *et al.*, 1976). The males also extended and retracted their spicules, which was the same as the behaviour observed in *Radopholus* (Huettel *et al.*, 1982). During both pre-mating and copulation both males and females continued to feed. This continuing feeding behaviour during mating has also been

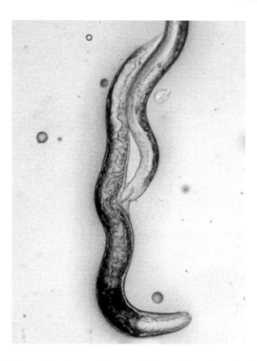

Fig. 5.8. *Heterorhabditis bacteriophora* male copulating with a female. The heads of the nema-todes are in opposite directions. (Courtesy of R.U. Ehlers.)

reported by Chin and Taylor (1969) in *Cylindrocorpus longistoma* and *Cylindrocorpus curzii*. In *Cruznema lambdiensis* only the females fed while copulating (Ahmad and Jairajpuri, 1981).

There have been fewer observations of copulatory behaviour in animal-parasitic nematodes than in free-living or plant-parasitic nematodes since mating tends to take place within a specific organ of the host animal. Observation of copulatory behaviour in the stronglyoid nematode *Nematospiroides dubius* (=*Heligmosomoides polygyrus*) was reported by Sommerville and Weinstein (1964). The nematodes mate with their anterior ends in opposite directions. The male used its lateral and ventral bursal rays to clasp the female. The spicules, once inserted in the vulva guided sperm into the uterus. In the sheep-parasitic nematode, *Haemonchus* spp., interspecific mating is known to occur between closely related species. In studies on *H. contortus* and *H. placei*, in both natural infections and experimental tests, the hybrids can be distinguished by spicule length (Lichtenfels *et al.*, 1988). Hybridization would indicate that a weak chemosensory signal is produced within these closely related species, which are not behaviourally specific enough to prevent cross-mating.

There is some unusual behaviour of other animal-parasitic nematodes reported in the literature. One of these is in *Anatrichomosa buccalis*, a nematode parasite of opossums that inhabits the host buccal mucosa. The male lacks copulatory structures and moves its posterior end into the female up to the uterus to deposit its sperm (Little and Orihel, 1972).

In a marine nematode, *Oncholaimus oxyuris*, mating was described by Chabaud *et al.* (1983) as a traumatic event. The female nematode of this species lacks a vulva. To get sperm into the female, the male nematode punctures the female body with his spicules. To accomplish this, the male uses its genital setae and ventral tail papillae to attach to the female. The male caudal gland then secretes a material that secures the female and allows the puncture to take place. After the male punctures the body, the spicules move separately to enlarge the hole and form a funnel-like subcuticular cavity. Once the sperm enters the female, it moves to the uterus through a sperm-transfer organ known as the Demanian system. Mating lasts up to 20 min, followed by the withdrawal of the spicules (Maertens and Coomans, 1978).

5.7. Post-mating Behaviour

Chitwood (1929) described a substance produced by male *Rhabditis* (=*Pelodera*) *strongyloides* after copulation that adhered to the vulva. He referred to this excreted material as a copulatory sac. Since this early description, similar copulatory sacs have been reported by other researchers in many nematodes, including the plant-parasitic nematodes *Scutellonema cavenessi* (Demeure *et al.*, 1980), *Neodolichodorus rostrulatus* (Sarr *et al.*, 1987) and *H. glycines* (personal observation, unpublished), the free-living nematodes *Acrobeloides* (Jairajpuri and Azmi, 1977) and *C. elegans* (Barker, 1994; Liu and Sternberg, 1995) and some marine nematodes. The copulatory sac can be secreted by either the male or the female, depending on the species. The secreted material is also referred to as a copulatory plug as it appears to be similar to copulatory plugs in other organisms, such as insects and snakes (Parker, 1970; Shine *et al.*, 2000). In snakes, the copulatory plug has been described as a chastity belt as it serves as a physical barrier to mating by other males after copulation has occurred.

In the marine nematode *Calomicrolaimus compridus*, the female secretes a copulatory plug in two parts. The dorsal gland secretes a hyaline plug that covers the vulva and the ventral gland produces a granular secretion that encloses the invaginated region of the cuticle, where the vulva is located (Gourbault and Vincx, 1988). It was speculated that the plug was produced either to help grip the female during copulation or to prevent copulation of the female by other males. In observations of *H. glycines* in *in vitro* root explant cultures by video monitoring, a copulatory plug was seen to be secreted by the female after insemination had occurred. There was often a delay of up to 10 min before the plug was extruded from the vulva. Even though another mating could have taken place during that time-lag, this was never observed (personal observations, unpublished).

In other species, males produce the copulatory plugs. In *Neodolichodorus rostrulatus*, the secretion may come from the male reproductive system in the vas deferens, especially in the ventral and lateral parts of its wall (Sarr *et al.*, 1987). Large, refractive granules were observed in the vas deferens, which moved into the cloaca and were excreted, presumably to hold the female in place. The secretion was described as a large, yellowish mass that remained on the female and sealed off the vulva after copulation occurred. They proposed, as did Chitwood (1929), that the male may release another substance that dissolves enough of this material to free him from the female but not remove the copulatory plug. More recent studies by Barker (1994), however, suggest that this copulatory plug is more likely to provide paternal assurance than to hold the nematodes together. Two male strains of *C. elegans*, one that produced a plug and the other that did not, were compared to determine if the plug interfered with secondary matings. Observations were made on the duration of mating and number of encounters between males and plugged and non-plugged females. The presence of the plug decreased the number of second matings when compared with non-plugged females and the males were less attracted to females that had plugs.

5.8. Males as the Colonizing Sex

In the insect-parasitic nematodes, *Steinernema* spp., male infective juveniles have been shown to locate and parasitize hosts before females arrive. The steinernematid juveniles emerge from a host cadaver as infective third-stage juveniles, and then penetrate the host, where maturation to adult males and females as well as subsequent mating occurs. In a study by Grewal *et al.* (1993) comparing *S. anomali*, *S. carpocapsae*, *S. feltiae*, *S. glaseri* and *S. scapterisci*, all species but *S. feltiae* had a high percentage of male juveniles (58–100%) within the first 24 h after penetrating the host from a distance of 10 cm. In another test in this study, all species had significantly more infective juvenile males than females that dispersed more than 9 cm from the inoculation point. This might indicate that the males have developed more sensitivity to host cues, which help them to disperse first. In other comparisons in this study, infective juvenile females of *S. glaseri* were more attracted to hosts that had been exposed to infective juvenile males. By entering the host first, the males assume all the dangers of the host immune response as well as increasing the bacterial food source for females. This behaviour may help to ensure that the insect host is more readily attractive for the infective juvenile females and also that they are inseminated, since the males are already present.

5.9. Evolutionary Considerations on Sexual Reproduction

In considering the many types of reproduction found in nematodes, one must ponder the evolutionary significance of so many different mating strategies. No one type of mating scheme is necessarily consistent within a genus nor is there consistency in mating types across habitats (i.e. parasitic or free-living species). Amphimixis seems to be preferred for overcoming selection pressures, such as new host varieties, especially by the obligate parasitic nematodes, which must have a host to complete their life cycle. Increased genetic variability is most readily accomplished through amphimixis because of the segregation and recombination of genes. There are several current hypotheses on the advantages of amphimixis (Kondrashov, 1993). One is the 'immediate benefit' hypothesis, which proposes that amphimictic reproduction is advantageous in that it directly increases progeny fitness, reduces deleterious mutations and benefits in selection, regardless of reciprocal gene exchange. Another is the 'variation and selection' hypothesis, which proposes that reciprocal gene exchange is what alters variability and response to selection. Regardless of the mechanisms involved, arguments point to the advantage of amphimixis over apomixis.

Another hypothesis which suggests that host–parasite relationships are best maintained through amphimixis is the Red Queen principle (Van Valen, 1973). The name of the principle refers to the scene from *Alice in Wonderland* in which Alice and the Queen remain in place even though they are running (Carroll, 1872). For host–parasite relationships, changes in the host would require a similar change or adaptation from the parasite; if one species changes, then the other would have to do the same (run) just to keep even. Other examples of the advantages of amphimixis using the Red Queen hypothesis are that sexual reproduction is advantageous in biotic interactions (Bell, 1982) and that sexual reproduction in parasites may help to evade the host immune response (Gemmill *et al.*, 1997). Read and Viney (1996) suggest that if a parasite were able to generate new genotypes through sexual reproduction then they could more quickly evade the genotype-specific immunity.

The Red Queen hypothesis was proposed for animal-parasitic nematodes that produce males only under certain environmental stress conditions. For example, Gemmill *et al.* (1997) showed that juveniles of *S. ratti* were more likely to develop into sexual adults from rats that had acquired immunity. This study suggested that the switch from asexual to sexual males in animal-parasitic nematodes that produce mates only under certain environmental stress conditions, such as under host immunity may increase the likelihood of progeny with new allelic combinations (Gemmill *et al.*, 1997; Vine *et al.*, 1983). West *et al.* (2001) conducted a study to determine whether the Red Queen hypothesis was the explanation for this behaviour. This experiment compared the rat-parasitic nematodes *S. ratti*, *Strongyloides venezuelensis* and *Nippostrongylus brasiliensis*. Both *S. ratti* and the congenetic *S. venezuelensis* are apomixis nematodes that produce eggs through mitosis. After moulting to the third stage, juveniles can develop either into females asexually or into sexually reproducing males and females. In *S. ratti* and *S. venezuelensis*, the proportion of juveniles that become males is dependent on both genetic and environmental controls. *Nippostrongylus brasiliensis* has an obligate and conventional sexual reproduction mode. Sheep red blood cells were used as a mammalian antigenic source in the study. Rats were challenged with various combinations of *S. ratti*, *S. venezuelensis*, *N. brasiliensis* and sheep red blood. There was a positive correlation to the strength of the protective immunization and the proportion of offspring that developed sexually. However, the increase in the number of males occurred in all immunizations, demonstrating that the increase was due to immune responses to all antigenic sources, not just *S. ratti*. Thus, these results did not support the Red

Queen hypothesis since the increase in males was from the genotype-specific response both to the parasite, as would be expected, and to the antigenic source of sheep red blood cells.

5.10 Conclusions

The Nematoda have developed many strategies for maintaining populations through varied reproductive mechanisms. Accompanying these mechanisms are unusual and interesting behaviours involving complex molecular signals. Continuing research on *C. elegans* and the application of these results to other nematodes will help to unravel why some of these behaviours have evolved. Understanding nematode behaviour is not just important in the biological arena. Since many nematode species are harmful parasites of both animals and plants, manipulation of behaviours may provide management tools for these pests. Disruptions in mate finding through interference of chemosensory mechanisms in the environments where these parasitic nematodes occur may reduce infestations in the soil or infections in animal hosts. A new era of nematode management that is species-specific and does not affect the host immune systems may be made available in the near future through altering chemosensory mechanisms such as sex pheromones or through genetic manipulations.

Acknowledgements

I thank S.F. Meyer and the Office of Communications, Alabama Agricultural Experiment Station, for their review of the manuscript.

References

Abbas, M. and Cain, G.D. (1979) *In vitro* activation and behaviour of the amoeboid sperm of *Ascaris suum* (Nematoda). *Cell and Tissue Research* 200, 273–284.

Ahmad, I. and Jairajpuri, M.S. (1981) The copulatory behaviour of *Cruznema lambdiensis* (Nematoda: Rhabditidae). *Revue de Nématologie* 4, 151–156.

Anderson, R.V. and Darling, H.M. (1964) Embryology and reproduction of *Ditylenchus destructor* Thorne, with emphasis on gonad development. *Proceedings of the Helminthological Society of Washington* 31, 240–256.

Bargmann, C.I. and Kaplan, J.M. (1998) Signal transduction in the *Caenorhabditis elegans* nervous system. *Annual Review of Neuroscience* 21, 279–308.

Bargmann, C.I. and Mori, I. (1997) Chemotaxis and thermotaxis. In: Riddle, D.L., Blumenthal, T., Meyer, B.J. and Priess, J.R. (eds) *C. Elegans II*. Cold Spring Harbor Laboratory Press, Cold Spring Harbor, New York, pp. 717–737.

Barker, D.M. (1994) Copulatory plugs and paternity assurance in the nematode, *Caenorhabditis elegans*. *Animal Behaviour* 48, 147–156.

Bell, G. (1982) *The Masterpiece of Nature. The Evolution and Genetics of Sexuality*. University of California Press, Berkeley, 635 pp.

Bennett, K.L. and Ward, S. (1986) Neither a germ line-specific nor several somatically expressed genes are lost or rearranged during embryonic chromatin diminution in the nematode, *Ascaris lumbricoides* var. *suum*. *Developmental Biology* 118, 141–147.

Bird, A. (1976) The development and organization of skeletal structures in nematodes. In: Croll, N.A. (ed.) *The Organisation of Nematodes*. Academic Press, New York, pp. 107–137.

Bone, L.W., Gaston, L.K., Hammock, B.D. and Shorey, H.H. (1979) Chromatographic fractionation of aggregation and sex pheromones of *Nippostrongylus brasiliensis* (Nematoda). *Journal of Experimental Zoology* 208, 311–318.

Bottino, D., Mogilner, A., Roberts, T., Stewart, M. and Oster, G. (2002) How nematode sperm crawls. *Journal of Cell Science* 115, 367–384.

Boveri, T. (1909) Die Blastomerenkern von *Ascaris megalocephala* und die Theorie der Chromom-somenindividualität. *Archiv für Zellforschung* 3, 181–268.

Boveri, T. (1910) Die Protenzen der *Ascaris* Blastomeren bei abgeanderter Furchung. *Festschrift zum für Richard Hertwig* 3, 131–214.

Bullock, T.L., McCoy, A.J., Kent, H.M., Roberts, T.M. and Stewart, M. (1998) Structural basis for amoeboid motility in nematode sperm. *Nature of Structural Biology* 5, 184–189.

Burke, D.J. and Ward, S. (1983) Identification of a large multigene family encoding the major sperm protein of *Caenorhabditis elegans*. *Journal of Molecular Biology* 171, 1–29.

Carroll, L. (1872) *Through the Looking Glass and What Alice Found There*. Macmillan, London.

Chabaud, A., Bain, O., Hugot, J., Rausch, R.L. and Rausch, V.R. (1983) Organe de Monsieur de Man et insémination traumatique. *Revue de Nématologie* 6, 127–131.

Chalfie, M. and White, J. (1988) The nervous system. In: Wood, W.B. (ed.) *The Nematode Caenorhabditis elegans*. Cold Spring Harbor Laboratory Press, Woodbury, New York, pp. 383–384.

Chin, D.A. and Taylor, D.P. (1969) Sexual attraction and mating patterns in *Cylindrocorpus longistoma* and *C. curzii* (Nematoda: Cylindrocorporidae). *Journal of Nematology* 1, 313–317.

Chitwood, B.G. (1929) Note on the copulatory sac of *Rhabditis strongyloides* Schneider. *Journal of Parasitology* 15, 282–283.

Cobb, N.A. (1893) *Nematodes, mostly Australian and Fijian*. Miscellaneous Publication No. 13, Department of Agriculture, New South Wales, 1–50 pp.

Cobb, N.A. (1917) *Tylenchus biformis*, the cause of a root disease of sugar cane and banana. *Journal of Agricultural Research: United States Department of Agriculture* 4, 5–6.

Demeure, Y., Netscher, C. and Queneherve, P. (1980) Biology of the plant-parasitic nematode, *Scutellonema cavenessi* Sher, 1964: reproduction, development and life cycle. *Revue de Nématologie* 3, 213–225.

Emmons, S.W. and Sternberg, P.W. (1997) Male development and mating behaviour. In: Riddle, D.L., Blumenthal, T., Meyer, B.J. and Priess, J.R. (eds) *C. Elegans II*. Cold Spring Harbor Laboratory Press. Cold Spring Harbor, New York, pp. 295–334.

Gemmill, A., Viney, M.E. and Read, A.F. (1997) Host immune status determines sexuality in a parasitic nematode. *Evolution* 51, 393–401.

Gourbault, N. and Vincx, M. (1988) *Calomicrolaimus compridus* (Gerlach, 1956) n. comb., a marine nematode with a female producing a copulatory plug. *Revue de Nématologie* 11, 39–43.

Green, C.D. (1966) Orientation of male *Heterodera rostochiensis* Woll. and *H. schachtii* Schm. to their females. *Annals of Applied Biology* 58, 327–339.

Green, C.D. (1971) Mating and host finding behaviour of plant nematodes. In: Zuckerman,

B.M., Mai, W.F., and Rohde, R.A. (eds) *Plant Parasitic Nematodes*. Academic Press, New York, pp. 247–266.

Green, C.D. (1980) Nematode sex attractants. *Helminthological Abstracts Series B, Plant Nematology* 49, 81–94.

Grewal, P.S., Selvan, S., Lewis, E.E. and Gaugler, R. (1993) Male insect-parasitic nematodes: a colonizing sex. *Experientia* 49, 605–608.

Huang, X. and Becker, J. (1999) Life cycle and mating behaviour of *Belonolaimus longicaudatus* in gnotobiotic culture. *Journal of Nematology* 31, 70–74.

Huettel, R.N. (1986) Chemical communicators in nematodes. *Journal of Nematology* 18, 3–8.

Huettel, R.N. and Rebois, R.V. (1986) Bioassay comparisons for pheromone detection in *Heterodera glycines*, the soybean cyst nematode. *Proceedings of the Helminthological Society of Washington* 53, 63–68.

Huettel, R.N., Dickson, D.W. and Kaplan, D.T. (1982) Sex attractants and behaviour in the two races of *Radopholus similis*. *Nematologica* 28, 360–369.

Italiano, J.E., Jr, Steward, M. and Roberts, T.M. (1999) Localized depolymerization of the major sperm protein cytoskeleton correlates with forward movement of the cell body in the amoeboid movement of nematode sperm. *Journal of Cell Biology* 146, 1087–1095.

Jaffe, H., Huettel, R.N., DeMilo, A.B., Hayes, D.K. and Rebois, R.V. (1989) Isolation and identification of a compound from soybean cyst nematode, *Heterodera glycines*, with a sex pheromone activity. *Journal of Chemical Ecology* 15, 2031–2043.

Jairajpuri, M.S. and Azmi, M.J. (1977) Reproductive behaviour of *Acrobeloides* sp. *Nematologica* 23, 202–212.

Karlson, P. and Luscher, M. (1959) 'Pheromone,' a new term for a class of biologically active substances. *Nature* 183, 155–176.

Klass, M., Dow, B. and Herndon, M. (1982) Cell-specific transcriptional regulation of the major sperm protein in *Caenorhabditis elegans*. *Developmental Biology* 93, 152–164.

Kondrashov, A.S. (1993) Classification of hypotheses on the advantage of amphimixis. *Journal of Heredity* 84, 372–387.

LaMunyon, C.W. and Ward, S. (1998) Larger sperm outcompete smaller sperm in the nematode, *Caenorhabditis elegans*. *Proceeding of the Royal Society of London B* 265, 1997–2002.

LaMunyon, C.W. and Ward, S. (1999) Evolution of sperm size in nematodes: sperm competition favours larger sperm. *Proceedings of the Royal Society of London B* 266, 263–267.

Lauritis, J.A., Rebois, R.V. and Graney, L.S. (1983) Development of *Heterodera glycines* Ichinohe on soybean, *Glycine max* (L) Merr. under gnotobiotic conditions. *Journal of Nematology* 15, 272–281.

Lewis, E.E., Barbarosa, B. and Gaugler, R. (2002) Mating and sexual communication by *Steinernema carpocapsae*. *Journal of Nematology* 34, 328–331.

Lichtenfels, J.R., Pilitt, P.A. and Le Jambre, L.F. (1988) Spicule lengths of the ruminant stomach nematodes, *Haemonchus contortus*, *Haemonchus placei* and their hybrids. *Proceedings of the Helminthological Society of Washington* 55, 97–100.

Little, M.D. and Orihel, T.C. (1972) The mating behaviour of *Anatrichosoma* (Nematoda) Trichuroidea. *Journal of Parasitology* 58, 1019–1020.

Liu, K.S. and Sternberg, P.A. (1995) Sensory regulation of male mating behaviour in *Caenorhabditis elegans*. *Neuron* 14, 79–89.

Maertens, D. and Coomans, A. (1978) The function of the demainian system and an atypical copulatory behaviour in *Oncholaimus oxyuris*. *Annals of the Royal Society of Zoology of Belgium* 108, 83–87.

Meyer, B.J. (1997) Sex determination and X chromosome dosage. In: Riddle, D.L,
    Blumenthal, T., Meyer, B.J., and Priess, J.R. (eds) C. Elegans *II*. Cold Spring Harbor
    Laboratory Press. Cold Spring Harbor, New York, pp. 209–240.

Miller, M., Nguyen, V.Q., Lee, M., Kosinski, M., Schedl, T., Caprioli, R. and Greenstein,
    D. (2001) The sperm cytoskeletal protein that signals oocyte meiotic maturation and
    ovulation. *Science* 291, 2144–2147.

Nelson, G.A. and Ward, S. (1982) Amoeboid motility and actin in *Ascaris lumbricoides*
    sperm. *Experimental Cell Research* 131, 149–160.

Nelson, G.A., Roberts, T.M. and Ward, S. (1982) *Caenorhabditis elegans* spermatozoan
    locomotion: amoeboid movement with almost no actin. *Journal of Cell Biology* 92,
    121–131.

Norland, D.A. and Lewis, W.J. (1976) Terminology of chemical releasing stimuli in
    intraspecific and interspecific interactions. *Journal of Chemical Ecology* 2, 211–220.

Papadopoulou, J. and Triantaphyllou, A.C. (1982) Sex differentiation in *Meloidogyne incog-
    nita* and anatomical evidence of sex reversal. *Journal of Nematology* 14, 549–566.

Parker, G. (1970) Sperm competition and its evolutionary consequences in the insects.
    Biological Review 45, 525–567.

Read, A.F. and Viney, M.E. (1996) Helminth immunogenetics: why bother? *Parasitology
    Today* 12, 337–343.

Rende, J.F., Tefft, P.M. and Bone, L.W. (1982) Pheromone attraction in the soybean cyst
    nematode, *Heterodera glycines* race 3. *Journal of Chemical Ecology* 8, 13–20.

Riddle, D.L, Blumenthal, T., Meyer, B.J. and Priess, J.R. (eds) (1997) C. Elegans *II*. Cold
    Spring Harbor Laboratory Press, Cold Spring Harbor, New York, 1222 pp.

Roberts, T.M. and Steward, M. (1995) Nematode sperm locomotion. *Current Opinions in
    Cell Biology* 7, 13–17.

Roberts, T.M. and Steward, M. (1997) Nematode sperm: amoeboid movement without
    actin. *Trends in Cell Biology* 7, 368–373.

Roberts, T.M. and Stewart, M. (2000) Acting like actin: the dynamics of the nematode
    major sperm protein (MSP) cytoskeleton indicate a push–pull mechanism for amoeboid
    cell motility. *Journal of Cell Biology* 149, 7–12.

Roberts, T.M. and Ward, S. (1981) Membrane flow during nematode spermiogenesis. *Cell
    Biology* 92, 113–120.

Roberts, T.M. and Ward, S. (1982) Centripetal flow of pseudopodial surface components
    could propel the amoeboid movement of *Caenorhabditis elegans* spermatozoa. *Journal of
    Cell Biology* 92, 132–138.

Roberts, T.M., Salmon, E.D. and Stewart, M. (1998) Hydrostatic pressure shows that lam-
    melipoidal motility in *Ascaris* sperm requires membrane-associated major sperm protein
    filament nucleation and elongation. *Journal of Cell Biology* 140, 367–376.

Sarr, E., Coomans, A. and Luc, M. (1987) Development and life cycle of *Neodolichodorus
    rostrulatus* (Siddiqi, 1976), with observations on the copulatory plug (Nematoda:
    Tylenchina). *Revue de Nematologie* 10, 87–92.

Scott, A.L., Dinman, J., Sussman, D.J. and Ward, S. (1989) Major sperm protein and actin
    genes in free-living and parasitic nematodes. *Parasitology* 98, 471–478.

Shine, R., Olsson, M.M. and Mason, R. (2000) Chastity belts in gartersnakes: the
    functional significance of mating plugs. *Biological Journal of the Linnean Society* 70,
    377–390.

Somers, J.A., Shorey, H.H. and Gaston, L.K. (1976) Reproductive biology and behaviour
    of *Rhabditis pellio* (Schneider) (Rhabditida: Rhabditidae). *Journal of Nematology* 9,
    143–148.

Sommerville, R.I. and Weinstein, P.P. (1964) Reproductive behaviour of *Nematospiroides dubius in vivo* and *in vitro*. *Journal of Parasitology* 50, 401–409.

Stewart, M., Roberts, T.M., Italiano, J.E., Jr, King, K.L., Hammel, R., Parathasathy, G., Bullock, T.L., McCoy, A.J., Kent, H., Haaf, A. and Neuhaus, D. (1998) Amoeboid motility without actin: insights into the molecular mechanism of locomotion using the major sperm protein (MSP) of nematodes. *Biological Bulletin* 194, 342–344.

Sulston, J.E., Albertson, D.G. and Thomson, J.N. (1980) The *Caenorhabditis elegans* male: postembryonic development of nongonadal structures. *Developmental Biology* 78, 542–576.

Thorne, G. (1949) On the classification of the Tylenchida, new order (Nematoda, Phasmidia). *Proceedings of the Helminthological Society of Washington* 16, 37–73.

Todd, T.C. (1989) Population dynamics and damage potential of *Belonolaimus* sp. on corn. *Journal of Nematology* 21, 697–702.

Triantaphyllou, A.C. (1971) Genetics and cytology. In: Zuckerman, B.M., Mai, W.F. and Rohde, R.A. (eds) *Plant Parasitic Nematodes*. Academic Press, New York, pp. 1–32.

Triantaphyllou, A.C. (1973) Environmental sex differentiation of nematodes in relation to pest management. *Annual Review of Phytopathology* 11, 441–462.

Van Beneden, E. (1883) *Recherches sur la maturation de l'oeuf, la fécondation et la division cellulaire*. Gand et Leipzig, Paris, 265 pp.

Van Valen, L. (1973) A new evolutionary law. *Evolutionary Theory* 1, 1–30.

Vine, M.E., Matthews, B.E. and Walliker, D. (1993) Mating in the nematode parasite *Strongyloides ratti*: proof of genetic exchange. *Proceedings of the Royal Society of London B* 254, 213–219.

Ward, S. and Carrel, J.S. (1979) Fertilization and sperm competition in the nematode, *Caenorhabditis elegans*. *Developmental Biology* 73, 304–321.

Ward, S., Argon, Y. and Nelson, G.A. (1981) Sperm morphogenesis in wild-type and fertilization defective mutants of *Caenorhabditis elegans*. *Journal of Cell Biology* 91, 26–44.

Ward, S., Roberts, T.M., Nelson, G.A. and Agron, Y. (1982) The development and motility of *Caenorhabditis elegans* spermatozoa. *Journal of Nematology* 14, 259–266.

Ward, S., Burke, D.J., Sulston, J.E., Coulson, A.R., Albertson, D.G., Ammons, D., Klass, M. and Hogan, E. (1988) Genomic organization of major sperm protein genes and pseudogenes in the nematode *Caenorhabditis elegans*. *Journal of Molecular Biology* 199, 1–13.

West, S.A., Gemmill, A.W., Graham, A., Viney, M.E. and Read, A.F. (2001) Immune stress and facultative sex in a parasitic nematode. *Journal of Evolutionary Biology* 14, 333–337.

White, J. (1988) The anatomy. In: Wood, W.B. (ed.) *The Nematode* Caenorhabditis elegans. Cold Spring Harbor Laboratory Press, Cold Spring Harbor, New York, pp. 81–122.

White, M.J.D. (1978) *Modes of Speciation*. W.H. Freeman, San Francisco, 455 pp.

Wood, W.B. (ed.) (1988) *The Nematode* Caenorhabditis elegans. Cold Spring Harbor Laboratory Press, Cold Spring Harbor, New York, 677 pp.

Zuckerman, B.M. and Janssen, H.B. (1984) Nematode chemotaxis and possible mechanisms of host/prey recognition. *Annual Review of Phytopathology* 22, 95–113.

6   Ageing and Developmental Behaviour

E.E. Lewis and E.E. Pérez

Department of Entomology, Virginia Tech, Blacksburg, VA 24061–0319, USA

6.1. Introduction

There is great popular and scientific interest in changes that occur in organisms as they age. Efforts to understand, and perhaps delay, human ageing have been the driving force behind much of the work discussed in this chapter. Indeed, research focused on how and why humans age has become a realm of investigation that encompasses many disciplines. Because in this chapter we are interested in how nematode behaviour changes with age, we shall not venture into the vast literature on human ageing. However, gerontological studies, which have human ageing and lifespan as main interests, are relevant to this chapter and so require attention, for two reasons. First, many of the studies to which we refer were conducted because nematodes serve as model organisms for studying the human ageing process. The best known is *Caenorhabditis elegans*, which has been used in ageing studies since the mid-1960s (Gershon and Gershon, 2001). Secondly, much of the vocabulary and theoretical framework of ageing studies were developed for studies of human ageing and have been used in studies of nematode ageing as well.

For a working definition of ageing, we shall adopt that of Arking (1999): the time-independent series of cumulative, progressive, intrinsic and deleterious functional and structural changes that usually begin to manifest themselves at reproductive maturity and eventually culminate in death. This definition must be clear to allow us to determine which behavioural changes are age-related and which are not. Arking (1999) suggests that to determine whether an event is age-related, ask whether it is cumulative, progressive, intrinsic and deleterious. For example, these criteria could be used with nematodes to separate the effects of cuticle degradation, which do fit these conditions, from desiccation, which is not intrinsic but fits the other criteria. A key aspect of this definition is the term 'time-independent'. We usually measure ageing with a calendar, and indeed many of the studies discussed in this chapter have done just that. However, when trying to make generalizations about the impact of ageing, other biological markers are often more suitable. The changes that take place in an individual organism with time are the result of cascades of biological and physical events that manifest in measurable differences between the young and old. Chronological time, then, should be thought of as a dependent variable not the independent variable, when regressing character changes with age. For nematodes, perhaps simple measures like lipid content, muscle tissue degradation or reproductive status could serve well for the independent variable in studies of their age.

Several variables have been measured with age in nematodes. For example, many different physiological markers that change with age have been measured and may serve to reflect ageing rates. Acetyl cholinesterase activity and malic dehydrogenase activity have been measured with time in *Turbatrix aceti* (Gershon and Gershon, 1970) and esterase and phosphatase activity have been measured over time in *Meloidogyne javanica* (van Gundy *et al.*, 1967). Zuckerman *et al.* (1972) measured an increase in specific gravity with age of *Caenorhabditis briggsae*, which could also be a potential non-chronological timekeeper.

Another phrase in Arking's (1999) definition, 'usually manifesting themselves at reproductive maturity', deserves special clarification to make the definition more relevant to nematodes. This phrase separates changes associated with ageing from developmental changes. Many nematode species enter a non-developmental stage called the dauer stage. For some parasites, this stage is analogous to the infective-stage juvenile, which is often the only free-living stage. Changes that occur during the dauer stage are legitimate subjects for this chapter because their behaviours change in a cumulative, progressive, intrinsic and deleterious manner. The searching and infection behaviours of entomopathogenic nematodes becoming less effective over time during the infective stage (Lewis *et al.*, 1995) and *C. elegans* dauers experiencing declining motility through the course of the stage (Klass and Hirsh, 1976) are examples of age-related changes during the dauer stage. The difference between ageing in dauers and ageing in adults is that the changes in dauers with age are reversible (Houthoofd *et al.*, 2002) whereas changes in adults are not. For example, these authors found that mitochondrial function declined with age through the dauer stage, but upon the resumption of development, mitochondrial function returned to the levels of younger worms.

Age has different types of impacts on nematode behaviour which have been studied in detail in both parasitic and free-living species. The more obvious changes are due to the diminished capacity of older animals to function. Many studies of nematode behaviour with respect to ageing illustrate how behaviours gradually slow with time until they stop. There are characters that change with age that do not necessarily signify a loss of vigour; greying hair and balding in humans are examples. An example in nematodes is that egg production is usually low at the beginning of the adult stage, then increases and then decreases again (Beck and Rankin, 1993). Where possible, we address the underlying causes of behavioural changes due to age as well as the changes themselves, to present a more complete picture of behaviour as it relates to age. Hopefully, a mechanistic approach to the problem of ageing in nematodes will facilitate meaningful comparisons among systematic and functional groups.

The overwhelming majority of studies of nematode ageing focus on the nematode *C. elegans*, although *C. briggsae*, *Panagrellus redivivus* and *T. aceti* have been popular models as well (Reznick and Gershon, 1999). That these nematode species have many ideal characters of model organisms for genetic and developmental studies has led to detailed studies of how the ageing process affects them. Studies of how their behaviour is affected by ageing are not as common as those of their genetics and development. We shall summarize these ageing studies later in the chapter. In addition to *C. elegans* serving as a model for human ageing, it is also a model for nematode ageing and behaviour.

Studies of the ageing and lifespan of nematode species of economic importance are conducted because of the need to better understand their specific biology. For nematodes that parasitize animals and humans, development, behaviour and lifespan are studied to enable predictions about their impact on host populations. Data essential for developing epidemiological models that predict the spread of disease include the duration of development inside the host, the timing and duration of reproduction and the duration of infective stage viability (Basáñez and Ricárdez-Esquinca, 2001). Age-related events are also essential to predicting population dynamics of plant-parasitic nematodes in agriculture, with the same characteristics remaining important. Finally, beneficial entomopathogenic nematodes have been studied for ageing and longevity due to questions of how to extend their shelf-life as pest control products (Grewal *et al.*, 1999). In these cases, where practical information about 'how long' a nematode remains functional is the goal, chronological age is a most important variable, despite our preferred definition by Arking (1999).

6.2. Ageing versus Development versus Senescence

These three terms are used extensively in the ageing literature and they have meanings that are specific, at least within the field. Ageing has been defined above. Development is the stage of the life of an individual that leads to 'adulthood', or reproductive maturity. Arking (1999) proposes that development ends when the age-specific mortality of the population is at the minimum. In human populations, the argument can be made that this age is reached immediately before puberty, after which the individual begins to senesce. The end of development is not a fixed point, however. For example, for nematodes, some development is evident after the last moult and this is discussed below. Senescence is the state that occurs between the onset of reproductive maturity and death. So there are positive outcomes to development (i.e. reproductive maturity and improved chances of survival) whereas there are no positive outcomes of senescence. This definition is easier to envision in organisms with distinct developmental stages than it is with humans, where ageing and development are not punctuated by moults. When a nematode achieves the last moult, reproduction and senescence begin soon thereafter.

6.3. General Nematode Development

There are four juvenile stages of postembryonic development in nematodes, a constant number within the phylum. The constant number of stages is evidence of great evolutionary stability. The third and fourth juvenile stages (J3 and J4, respectively) of plant-parasitic and free-living nematodes are considered the critical periods of postembryonic development and are the fastest-growing stages (Paramonov, 1962, as cited in Malakhov, 1994). Rudimentary gonoducts are present in the J3 and most of the reproductive system develops during the third and fourth stages. The reproductive system is completely formed in the J4 but is not yet functional (Malakhov, 1994). Nematodes exhibit two distinct types of growth patterns, which are classified according to the amount of the total growth that occurs after the final moult (Malakhov, 1994). In free-living and many plant-parasitic nematodes, the body mass growth during the adult stage is similar to or less than the body mass growth during any of the other developmental stages. In other words, growth is accompanied by periodic moults. In contrast, animal parasites display a growth pattern where most of the nematode's growth is accomplished after the final moult to adulthood. For relatively unspecialized animal parasites, such as entomopathogenic nematodes, the body mass growth in the adult is 50–80% of the total growth (Malakhov, 1994). For highly specialized animal parasites, such as *Ascaris*, body mass growth in the adult stage is up to 99% of total growth. The transfer of basic growth to the last stage of postembryonic development is a specific adaptation of nematodes to the parasitic form of life. The interaction of behaviour with this type of growth pattern is that specialized parasites such as *Ascaris* are relatively sedentary once adulthood is reached, whereas free-living nematodes remain active and mobile.

6.4. Types of Ageing Studies

We classify studies of ageing into three broad categories. The first and most traditional examines changes that take place in individuals or populations after adulthood is reached. Most studies that use *C. elegans* as a model of human ageing fit this group. The second category includes studies where lifespan is of primary importance. All of the work done with ageing mutants of *C. elegans* would be included. Third are comparative studies that examine how behaviour changes as nematodes moult from stage to stage. For example, one could ask whether or not a J2 feeds differently from an adult of the same species.

6.5. Ageing of Free-living Nematodes

The adult stage of animals is defined by reproductive maturity. This is also the stage at which ageing, in the strict sense, begins. For nematodes there is a distinct adult stage, separated from juvenile stages by a moult. Most studies of *C. elegans* ageing are conducted with worms that have completed their last moult. *Caenorhabditis elegans* develops from the egg, through four juvenile stages and finally to adult in about 3 days at 20°C under laboratory conditions with unlimited food in the form of a lawn of *Escherichia coli* grown on an agar plate. The lifespan of adult wild-type *C. elegans* is in the range of 2 weeks at 20°C; peak egg laying occurs on day 4 and ceases on day 7 or 8 after the last moult (Beck and Rankin, 1993). One of the reasons why *C. elegans* is such a valuable model is that the adult stage, from the beginning of reproductive maturity through senescence and death, occurs in about 2 weeks. Studies of the behaviour of adult stage parasitic nematodes are not as common as for *C. elegans* for two reasons. First, the stage of interest in parasites is usually the infective stage because this is the stage that causes infection, e.g. the one that we want to limit for disease prevention. Secondly, studying adult parasites within the host is a challenging and costly task (Sukhedeo *et al.*, 2002).

The behavioural repertoire of free-living nematode adults, at least those that have been studied, can be represented by *C. elegans*. Several behaviours of *C. elegans* adults have been studied in part to screen for ageing mutants early in adult life (Yang and Wilson, 2000). With more than 50 longevity mutants currently under study (Herndon *et al.*, 2002), the need for quick, easy and repeatable screening methods is obvious. The long-lived mutants can realize a lifespan increase of up to 300% compared with the wild-type N2 strain. The premise behind this screening is that certain behavioural traits can be associated with a long lifespan and may be used in a predictive manner to choose mutants from a population while the worms are still reproductively viable. Below we discuss specific behaviours that have been studied in both adult and arrested juvenile stages. Some of these behaviours are used as screening tools.

6.5.1. Locomotion

Adult *C. elegans* gradually slow in their movement as they age. Hosono *et al.* (1980) measured surprisingly high amounts of inter-individual variation in behavioural changes that relate to ageing *C. elegans*. This amount of variation is surprising because the nematodes tested had been in culture for thousands of generations. They showed that nematodes quickly decreased locomotion and feeding behaviours after the peak of egg laying, but that there was intrinsic variation in relation to when the decrease began. Herndon *et al.* (2002) also studied changes in spontaneous locomotion and response to prodding in age-synchronized, isogenic groups of adult nematodes while on their culture plates. They classed individuals, in a way similar to the classifications of Hosono *et al.* (1980), according to how the nematodes moved: class A nematodes moved spontaneously through the bacterial lawn, class B nematodes moved when prodded but did not move spontaneously, and class C nematodes responded to touch only by twitching the head and tail. All nematodes were graded as class A at the beginning of adulthood, the first class B adults were seen after 6–7 days and class C adults after 9–10 days. Even among these individual nematodes, which were selected purposefully for genetic homogeneity, there was a remarkable level of inter-individual variation in the rate of change in their locomotory behaviour and the concomitant degradation in some tissues. The degradation in the nematode's movement was attributed to deterioration of muscle tissue (among other kinds of tissue) and various stochastic processes, but not to a decrease in neural function. These data support the 'disposable soma' theory of ageing. This theory postulates that deterioration with ageing is not genetically based, but due to the high cost of cell repair. Evolutionarily, it makes more sense to spend metabolic energy on increasing reproduction than on extending lifespan. Thus, the amount of energy spent on maintenance of soma is enough to keep the individual alive long enough for it to reproduce, but not for an extended duration past that time. Herndon *et al.* (2002) also found that movement class was a better predictor of lifespan than was chronological age, which agrees with Arking's (1999) definition of ageing as 'time-independent'. This study also supports Yang and Wilson's (2000) technique of detecting longevity mutants by behavioural screening early in adulthood, suggesting that motility might be a viable character for screening.

Zuckerman *et al.* (1971) recorded that *C. briggsae* adults were less able to osmoregulate as they aged. They attributed this to either an increased solute concentration of pseudocoelomic fluid or to degradation of cuticular components such as collagen. Older nematodes also lose elasticity in their cuticle (Zuckerman *et al.*, 1971). These cuticular degradations could be partially responsible for the loss of mobility recorded in ageing nematodes.

6.5.2. Sensory responses and learning

Nematodes respond to a variety of different stimuli, including mechanosensory, chemical and thermal cues. Many of the genetic underpinnings of these responses have been described in detail, and, at least with *C. elegans*, mutants have been developed that respond to cues in ways different from those of wild-type nematodes. The relationships between the response to stimuli and ageing have been studied with free-living nematodes in two ways: nematodes' responses to the stimuli directly as they age and the effects of age on non-associative learning.

Responses to mechanosensory cues, such as a gentle touch along the body, a touch to the nose, a touch to the side of the head, a tap to the dish in which they are cultured, and response to the texture of the substrate have been used to study the links between anatomy and behaviour and the coordination of locomotory behaviour (reviewed by Ernstrom and Chalfie, 2002). Herndon *et al.* (2002) showed that the changes they measured in both spontaneous locomotion and response to mechanical stimulation were not due to degradation in sensory capability. Similarly, Beck and Rankin (1993) studied the responses of adult worms to a series of taps on the culture dish. They measured the frequency and magnitude of reversals in directional movement in response to the taps. Adult worms responded differently as they aged; the responses to the taps were of smaller magnitude, but not less frequent, as the worms grew older. These authors also concluded that there was little change in sensitivity to mechanical stimulus, but the overall response degraded in magnitude.

Learning is inextricably linked to ageing because it takes time to gain experience; thus a nematode must age if it is to learn. Nematodes are capable of non-associative learning, which involves the modification of behaviour due to repeated exposure to a single cue and includes habituation, dishabituation and sensitization (Bernhard and van der Kooy, 2000). Habituation is a decrement in response to the same repeated stimulus which is not due to adaptation, e.g. receptor fatigue. Dishabituation is the return from a state of habituation to baseline levels of response in response to a novel or noxious cue. Dishabituation is not just recovery from habituation spontaneously. True habituation can always be reversed by dishabituation. Sensitization is the enhancement of a response to a stimulus that results from repeated exposures. Beck and Rankin (1993) studied the effects of ageing on habituation and dishabituation. The test consisted of a series of taps on the culture dish and the nematodes' responses measured were the frequency and magnitude of a reversal of direction. Worms of all ages tested (4, 7 and 12 days after the final moult) showed significant degrees of habituation, and there was no difference in the frequency of response decrement among the ages. This experiment showed that even the oldest worms tested were capable of learning. However, older worms showed a greater degree of habituation than did the 4-day-old worms, and the 12-day-old worms recovered from habituation more slowly than did the younger worms. The slower and smaller magnitude of responses supports the data of Herndon *et al.* (2002) showing that ageing does not have an impact on the sensory capabilities of adult nematodes, but that other types of tissue degradation are responsible for an overall reduction of response to stimuli.

Other studies of non-associative learning of nematodes have examined habituation and dishabituation to olfactory cues. Bernhard and van der Kooy (2000) showed that *C. elegans* became habituated to diacetyl when exposed to low concentrations, but at high concentrations they became adapted, probably due to receptor or sensory fatigue. Several studies have examined habituation to olfactory cues in *C. elegans*, but how habituation changes with ageing has not been addressed. We shall not delve further into this area, and include olfactory habituation in our chapter only because non-associative learning has a time component involved and, with the passage of time, ageing occurs.

6.5.3. Reproduction

The relationship between ageing and reproduction is not complicated for *C. elegans*, as long as only the wild type is considered. An entire chapter in this volume is dedicated to nematode reproductive behaviour (see Huettel, Chapter 5, this volume), so we shall not provide a detailed description here. However, specific examples of how mating and reproductive success change with age are discussed. Also, the impact of *C. elegans* mutations that increase lifespan on fecundity and other rhythmic adult activities has been studied in detail and provides opportunities to test relationships between lifespan and fecundity. For example, mutations in any of the genes *clk*-2, *clk*-3 or *gro*-1 lead to alteration of developmental and behavioural timing, and *clk*-1 mutants have a general lengthening or slowing of processes such as defecation, development, pharyngeal pumping and others (Ewbank *et al.*, 1997). Chen *et al.* (2001) explored the differences in demographic traits between wild-type (Bristol N2 strain) and *clk*-1 and *age*-1 longevity mutants of *C. elegans*. They found that there was a positive relationship between lifespan and number of eggs produced only in worms that died before the end of their reproductive phase, which ended about 8 days after it started. In other words, the more days that worms laid eggs, the more eggs they produced. This was the case for both wild-type and mutant *C. elegans*. Interestingly, unlike other adult traits, such as feeding or defecation, the duration of the egg-laying phase and the days on which it began and ended were essentially the same for the wild type and both of the mutants. There was no correlation between egg production and post-reproductive lifespan among individuals with the same mutation or the wild type, but the longer-lived mutants produced fewer eggs over their entire lifespan than did the wild type. There is, apparently, some cost to a long life. The authors of this study caution that, since hermaphrodite egg production is sperm-limited, if males were added to the picture their results could change.

Gender-specific differences in lifespan have been recorded and are sometimes associated with whether or not the nematodes mate. Females of *T. aceti* lived 14 days longer than males, and both virgin males and females had greater longevity than did their mated counterparts (Kisiel and Zuckerman, 1974). Males of *C. elegans* have not been studied extensively as to their lifespan or their behaviour. Interestingly, mating increases the average male lifespan whereas mating decreases the average hermaphrodite lifespan (Gems and Riddle, 1996). The lifespans of unmated males and hermaphrodites are similar (Gems and Riddle, 1996) and reproduction by self-fertilization has no measurable effect on hermaphrodite lifespan (Kenyon *et al.*, 1993). In mixed populations, males have a different mortality pattern from that of hermaphrodites and males tend to live longer (Gems and Riddle, 2000). However, in populations of males only, high population density has a negative effect on male lifespan whereas there is no density-dependent effect on hermaphrodites (Gems and Riddle, 2000). This is an unusual scenario, since in many species females tend to outlive males (Comfort, 1979).

Other examples of mating behaviour either influencing lifespan or being affected by age are spotty. The timing of mating and related behavioural events has been studied. *Panagrellus redivivus* females are capable of copulating immediately after their final moult and their ability to copulate does not decline with age. As adult females age, their attractiveness to young males declines. Both sexes are attracted by adults of the opposite sex, but pre-adult males and females did not respond to adults of the opposite sex (Duggal, 1978). The age at which females mate can have an impact on fecundity. *Rhabditis pellio* is a facultative parasite of earthworms. According to Somers *et al.* (1977) females mated at 3 days of age produced an average of 253 juveniles over their lifetime and had the maximum egg production on their fourth day. However, when females were mated later, their juvenile production dropped; if they were mated on day 5, they produced 71 juveniles and, if they were mated on day 11, they produced only four juveniles. *Turbatrix aceti* adults have a similar pattern with age at mating. Kisiel and Zuckerman (1974) showed that egg production drops when progressively older females are mated with young males and when progressively older males are mated with young females. These authors suggest that the effects of age on female and male gonads are similar. Thus, the greater attraction to younger potential mates may have a great fitness benefit to choosy adults.

6.5.4. Ageing in the dauer stage of free-living nematodes

Ageing in the dauer stage needs to be treated somewhat differently from the ageing of adults. The dauer (enduring) stage of nematodes is a resistant life stage that occurs in response to adverse environmental conditions. The relevance to ageing in the dauer is that, when a nematode enters this stage, its lifespan can be increased tremendously. The diversity among nematode species in their lifespan is significant, partly due to the durations of these enduring stages (Fig. 6.1). Dormancy has been classified and defined in a number of ways. We summarize these classifications below for background information. We group parasitic and free-living nematodes to avoid repetition later.

6.5.5. Classification of dormancy

6.5.5.1. Diapause and quiescence

One reason why nematodes have been the focus of ageing research is because they can improve their chances of survival by entering a dormant state that significantly

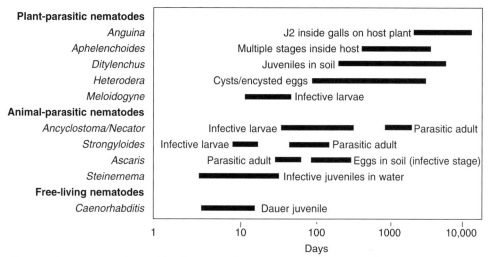

Fig. 6.1. The range, in days, of the duration of various nematode species' lifespan as adults and their respective dauer or infective stages. This figure is meant to provide an appreciation for the diversity in nematode lifespan.

extends their lifespan. During this dormant state, ageing is delayed by slowing or ceasing development, thereby conferring an increased lifespan (Evans and Womersley, 1980). During the dormant state, nematodes alter their metabolic rate and may or may not stop ontogenetic development, which enables them to withstand, or in some cases avoid, adverse environmental conditions (Cooper and van Gundy, 1971; Evans, 1982; Riddle *et al.*, 1987). Diapause is a dormant state associated with a type of morphogenesis that involves the cessation of ontogentic development mediated by endogenous factors (van Gundy, 1965; Evans, 1982). Diapause may be a facultative state that is initiated by unfavourable environmental stimuli, after which normal development is resumed when favourable conditions are restored (Evans, 1982). It can also be an obligate state where development does not resume unless or until certain physiological processes have been completed, even when the environment is favourable (van Gundy, 1965). Previous to diapause some nematodes go into prediapause or oligopause. During this state, ontogenetic development is either slowed or stopped while metabolism remains largely unaffected (Evans, 1982).

Diapause is contrasted with quiescence, which refers to the dormant state caused by unfavourable environmental conditions and ends when favourable conditions are restored. Quiescence is not associated with a type of morphogenesis and does not affect ontogentic development. During quiescence, somatic development is affected only as a result of slowed metabolic rate (van Gundy, 1965; Evans, 1982). Dormant parasitic nematodes are infective dispersal forms with characters that enhance tolerance to environmental extremes in periods between host infections (Evans, 1987). Dormant free-living nematodes are also dispersal forms with enhanced fitness and some species have been shown to be non-ageing with a longer than normal lifespan (Riddle *et al.*, 1987).

Dormancy in eggs of plant-parasitic nematodes synchronizes hatch with the host growing cycle. Most evidence of diapause in eggs of plant-parasitic nematodes comes from the family Heteroderidae, which includes cyst and root-knot nematodes. Because host plant factors during cyst formation seem to trigger diapause in juveniles inside eggs of *Globodera rostochiensis*, it can be considered a facultative response (Hominick *et al.*, 1986; Evans, 1987). Photoperiod and temperature possibly provide cues for the duration of diapause (Evans, 1982). Diapause in *Meloidogyne* unembryonated eggs is not linked to an environmental stimulus, and Evans (1987) proposed that this form of obligate diapause is genetically programmed. The persistence of *Meloidogyne* infestations in soil devoid of host plants seems to be enabled by the ability of females to lay dormant eggs (Ishibashi, 1969; Evans, 1987).

6.5.5.2. Cryptobiosis

Cryptobiosis can be defined as a dormant state associated with an undetectable metabolic rate where structural integrity is maintained even after exposure to ionizing radiation, chemicals or temperature extremes (van Gundy, 1965; Cooper and van Gundy, 1971). Several species of nematodes are capable of cryptobiosis. This state is similar to dormancy in that it permits survival in unfavourable environmental conditions and delays the ageing process. Qualitative and quantitative differences between quiescence and cryptobiosis are sometimes difficult to recognize, but both are reversible states of metabolic depression that may be caused by deprivation of water or oxygen or exposure to cold temperatures (Bhatt and Rohde, 1970). The cryptobiotic state may be named according to the conditions that stimulated cryptobiosis. Anhydrobiosis is caused by loss of water; cryobiosis is caused by lowering of temperature; anoxybiosis is caused by low levels of oxygen; and osmobiosis is caused by removal of water by an external solution with high osmotic potential. Osmobiosis and cryobiosis are similar to anhydrobiosis since all three are induced by the lack of water needed for metabolic reactions (Demeure and Freckman, 1981). Cryptobiosis is considered more resistant than quiescence, with no measurable metabolism, whereas quiescent nematodes have a diminished but detectable metabolism (Cooper and van Gundy, 1971). Metabolic changes leading to the accumulation of trehalose were shown to occur during dehydration of nematodes (Behm, 1997; Solomon *et al.*, 1999; Gal *et al.*, 2001). During the dehydration and rehydration process, trehalose possibly protects the cell membrane from damage by conserving the structure and functional integrity (Crowe *et al.*, 1992).

Cryptobiosis can extend the lifespan of the nematode from several months to several years. Because of the importance of cryptobiosis to nematode ageing, we list a few of the many examples of lifespan extensions, usually associated with plant-parasitic nematodes, due to cryptobiosis. Second-stage juveniles of the wheat seed gall nematode, *Anguina tritici*, and fourth-stage juveniles of the stem and bulb nematode, *Ditylenchus dipsaci*, were revived after 23 years of cryptobiosis (Fielding, 1951) and *Tylenchus polyhypnus* after 39 years (Steiner and Albin, 1946). Studies of some plant-parasitic nematodes during the induction and exit of anhydrobiosis revealed behavioural and metabolic changes (Crowe and Madin, 1975; Wharton *et al.*, 1985, 1988, 2000). Nematodes in anhydrobiosis lose 95–99% of their water, coil their bodies and have decreased body volume (Crowe and Madin, 1975; Wharton *et al.*, 1985). The *D. dipsaci* J4 is a non-feeding infective stage with large quantities of lipid droplets and protein bodies stored in its intestine (Wharton and Barrett, 1985). After desiccated J4 *D. dipsaci* are immersed in water, their metabolic rate and water content return rapidly to near-normal levels but movement does not start until after a 2–3 h lag phase (Barrett, 1982; Wharton *et al.*, 1985). During this lag phase physiological functions of the nerves and muscles are restored (Wharton *et al.*, 2000). For more in-depth information on this topic, see Wharton, Chapter 13 this volume.

6.5.6. Dauer larvae of *C. elegans*

Most nematode species have a single juvenile stage that is able to enter a dormant state. For *C. elegans*, temperature, food supply and population density during the first two juvenile stages determine whether the juvenile develops into a dauer or continues directly to adulthood. Briefly, a constitutively expressed *C. elegans*-specific pheromone accumulates in a culture and is an indication of population density. Food also produces chemical cues. When a high ratio of pheromone to food signal occurs, dauer formation is favoured. When the temperature is warm, the likelihood of dauer formation is further increased. The stimuli leading to dauer formation for other nematode species are less completely understood. For most parasitic species, the infective-stage juvenile is analogous to a dauer and is able to endure the relatively harsh and unpredictable conditions outside the host. We are interested in changes in dauer behaviour as they 'age'.

In *C. elegans*, dauer juveniles do not change morphologically as time passes. By definition, this is a non-developmental stage and there is no feeding or defecation. The dauer stage ends with the nematode's recovery, which is a term used to indicate a resumption of development. The first extensive work on dauer biology was conducted by Klass and Hirsh (1976). *Caenorhabditis elegans* dauers have mobility characteristics that are similar to non-dauers for the first 2–3 weeks of the stage. Their motility gradually declines until the end of life, a maximum of about 70 days for wild-type strains. As the dauer stage progresses, fewer nematodes recover when presented with appropriate cues, usually food. If the dauer stage lasts 10 days, virtually all nematodes recover. Remember that 10 days represents nearly a 50% extension of the normal lifespan, from egg through the adult stage, for *C. elegans*. After 60 days in the dauer stage, approximately 32% of the nematodes recovered.

What is the fate of the nematodes that recover? Curiously, nematodes that develop into hermaphrodites from dauer juveniles suffer no ill effects on fecundity, regardless of how long the dauer stage lasted. If the individual recovers, there is no impact on subsequent development or reproductive success in the conditions of the experiments conducted.

6.5.7. Limitations of *C. elegans* as a model for ageing

It is appropriate to note the limitations of work conducted on *C. elegans*. Almost all *C. elegans* research has been conducted using strains of nematodes isolated from mushroom compost over 40 years ago (Dougherty *et al.*, 1959). The N2 strain, which is used as a 'wild-type' strain against which mutant strains are compared, was first used in the mid-1970s (Brenner, 1974). This strain has undergone thousands of generations in laboratory cultures and their behaviour may bear little resemblance to a *C. elegans* strain that would be isolated from soil today. Another difficulty is that due to differences in culture methods among different laboratories, there are differences in lifespan among N2 strains (Gems and Riddle, 2000). However, standard N2 cultures can be obtained from the *Caenorhabditis* Genetics Center at the University of Minnesota. Reznick and Gershon (1999) contend that the genetic make-up of laboratory strains of *C. elegans* is very different from that of truly wild strains. The ability of laboratory strains to cope with stress is likely to be reduced compared with strains recently isolated from soil. These authors suggest that one of the reasons single gene mutations result in life extension is because these mutations cause the laboratory strains to become more like naturally occurring soil nematodes, and that the 'extended' lifespan is actually more like the 'real' lifespan of wild worms in the absence of natural stressors. Reznick and Gershon (1999) also suggest that the hermaphroditic *C. elegans* may not be an ideal model for bisexual eukaryotes.

6.6. Ageing of Parasitic Nematodes

Nematodes that are parasites must be adapted to at least two very different habitats. Part of their lives is spent within a host and part is spent in a transmission stage outside the host. Because of the disparity between these two habitats, the behaviour of parasitic nematodes is in many cases more stage-specific than that of free-living nematode species. In other words, infective stages behave very differently from parasitic (non-infective) stages. Thus, infective stages are affected by age in very different ways from parasitic stages.

6.6.1. Heterogonous and homogonous development

Earlier we discussed 'general nematode development'. Here, we focus on the development of parasitic nematodes because their development is intimately linked to the biology of their host and therefore more specialized than that of free-living nematodes. Many species of parasitic nematode moult only after they parasitize their hosts. Entomopathogenic and some plant-parasitic nematodes are examples of this type of life cycle. On the other hand, there are some parasitic nematode species that alternate between free-living and parasitic generations. Free-living generations complete their life cycle, including reproduction, without a host (the heterogonic cycle), whereas parasitic generations require a host to complete their life cycles (the homogonic cycle). The regulation of the alternating life cycles has been studied in several nematode species.

One of the most studied examples of alternating life cycles is that of *Strongyloides ratti*, an intestinal parasite of the rat. Briefly, parasitic female adults are anchored in the host intestinal mucosa and produce eggs that hatch as they move throughout the gut. The J1 nematodes exit with the faeces and remain there to either develop into a developmentally arrested J3 infective stage or continue development into free-living adults. Free-living adults mate and may give rise to free-living and parasitic individuals in the next generation. The infective stages penetrate a new host through the skin and via tissue migration find the intestinal mucosa. Nematode behaviour plays an integral role in the regulation of this life cycle because, in response to a combination of environmental cues, the nematodes at some point develop into either parasitic or free-living forms. Kimura *et al.* (1999) reported that parasitic *S. ratti* females had two peaks of egg production: the first one with maximum egg production at 7–8 days after infection and a second, less conspicuous peak, at around day 25. The second peak of egg production was associated with adults that dislodged from the small intestine and resettled in the caecum and the colon of the host. Most eggs from the second peak developed into free-living adults, whereas most from the first peak became infective juveniles that developed into parasitic adults.

In *S. ratti*, there is a complicated interaction between environmental conditions experienced by the parental parasitic female while inside the host intestine and environmental conditions outside the host experienced by the J1 nematodes (Viney, 2002). The environmental conditions inside the host that have an impact on the adult parasites are related to the host immune system. As the infection progresses, the adult worms become older and two things happen to their progeny: the sex ratio of the progeny becomes more male-biased and the proportion of female progeny that go through subsequent heterogonous development increases (Harvey *et al.*, 2000). These changes in the biology of progeny are not the result of parasitic worms ageing, but of the increasingly strong anti-*S. ratti* immune response of the host, which will eventually eliminate the infection. Once the eggs and J1s are expelled with faeces, the external temperature exerts influence on whether female juveniles will go through homogonous or heterogonous development (Viney, 1996). At lower temperatures, females develop directly into developmentally arrested infective J3 juveniles, which will infect a new host, whereas warmer temperatures favour indirect development into free-living adults. The developmental pathway remains uncertain until the end of the J2 stage (Viney, 1996). The interplay between the conditions inside and outside the host determines the female's developmental pathway. As the infection progresses and the host immune response strengthens, the female progeny become increasingly sensitive to temperature changes outside the host. The result is that the proportion of female juveniles that develop into free-living adults increases as the infection progresses if outside temperatures remain constant, and also increases as the external temperature warms. One remarkable aspect of this system is that this interaction between the two environments suggests some sort of 'transgenerational memory' from the parasitic adult to its progeny outside the host (Viney, 2002).

6.6.2. Feeding

All stages of free-living nematodes feed in more or less the same way, with the exception of dauer stages which do not feed. Parasitic nematodes feed in a variety of ways on their hosts. Juveniles, adult males and females of the same species may differ radically in their feeding habits. The most obvious difference in feeding habit occurs between free-living and parasitic stages within the same life cycle. Entomopathogenic nematodes and some species of plant-parasitic nematodes are examples of species where all feeding and most development occurs within the host. However, there are other parasitic species where feeding, development and mating may occur both outside and inside the host. Additionally, some nematode species use more than a single host, which requires further feeding specialization. An interesting aspect of parasitic nematodes is that some species have several stages in the life cycle where no feeding occurs and yet moulting does occur. Though feeding may change with age within a life stage, most marked changes in feeding habits are among different life stages.

6.6.2.1. Plant-parasitic nematode feeding

Root-feeding plant-parasitic nematodes can be classified into three main groups according to their feeding behaviour on roots. Ectoparasites feed externally and rarely enter the roots. Endoparasites have at least one feeding life stage that entirely penetrates the root tissue. Some endoparasites migrate inside root tissues and are called migratory endoparasites, whereas sedentary endoparasites establish a feeding site where they remain for most of their life cycle. Semi-endoparasites feed with the anterior end of the body embedded in root tissue and adult females do not move from the feeding site. All juvenile stages of ectoparasites and migratory endoparasites are infective, whereas only the J2 is infective in the sedentary endoparasites.

Within feeding groups, different life stages of the nematode often have different preferences of feeding location. Why these preferences occur is uncertain. For example, young juveniles of the dagger nematode, *Xiphinema index*, are ectoparasites feeding primarily on the root hair region of *Vitis vinifera*, whereas older juveniles and adults feed only on apical meristem tissue (Fisher and Raski, 1967). The reniform nematode *Rotylenchus robustus* (syn. *Rotylenchus reniformis*) is a semi-endoparasite; the J2 hatches from the egg and undergoes three moults prior to feeding. The infective stage is the immature adult female while still in the vermiform stage. Females penetrate the root cortex with the anterior third of the body and establish a permanent feeding site. Feeding females are sedentary and the posterior portion of the body swells and protrudes from the root surface and has a characteristic kidney shape. Males have a degenerative oesophagus and apparently do not feed (Birchfield, 1962). In the citrus nematode, *Tylenchulus semipenetrans*, J2 nematodes hatch from eggs. Males pass through three moults and become mature adults without feeding. Second-, third- and fourth-stage females feed on the root hypodermal and epidermal cells. Young adult females that are still vermiform penetrate deeply into the root and initiate nurse cells to serve as feeding sites in the cortical parenchyma. When the females mature the posterior end becomes globose and protrudes from the root surface (van Gundy, 1958; Siddiqi, 1974).

Root-knot nematodes are classified as sedentary endoparasites because they lose their mobility after establishing a feeding site. Second-stage juveniles are the only infective stage and penetrate the host near the root tip, where they establish permanent feeding sites in the protophloem and protoxylem (Endo and Wergin, 1973). Development of the nematode is dependent on its being able to stimulate the host cells to differentiate into specialized multinucleate giant cells (Starr, 1998). Development of the J3 to adult happens in approximately 6 days and the nematode does not feed during this period. After the last moult, only the adult females resume feeding and grow rapidly, achieving a pear-shaped body within a few days (Bird, 1972). After the final moult, males become vermiform and motile and leave the root (Starr and Veech, 1986). Sex ratio changes due to food-related stimuli were reported to occur in root-knot nematodes (Triantaphyllou, 1960; Triantaphyllou and Hirschmann, 1973). Unfavourable conditions, such as overcrowding, cause a higher than normal proportion of males to develop in these parthenogenetic populations (Bird, 1970; Triantaphyllou and Hirschmann, 1964). Developing females that have two gonads change sex and become males with two testes, rather than normal males with a single testis. Because adult males do not feed, this sex reversal is thought to occur as a survival mechanism. Only nematodes that continue developing as females use the available food source (Hansen *et al.*, 1973). Also adult males may migrate and inseminate females infesting other plants.

6.6.2.2. Animal-parasitic nematode feeding

Different juvenile stages of animal parasites show dramatic changes and adaptations in their feeding habits as they age and progress through their life cycles. *Rhabdias bufonis* and *Rhabdias ranae*, common parasites in the lungs of toads and frogs, have heterogonic development (Schmidt and Roberts, 1981). The parasitic adult is a protandrous hermaphrodite that produces eggs that hatch in the intestine of the frog. First-stage juveniles are excreted with the frog faeces. The J1 has a rhabditiform oesophagus (the posterior end of the oesophagus has a bulb separated from the anterior portion by an isthmus). These juveniles undergo four moults and produce a generation of free-living males and females. This non-parasitic generation feeds on bacteria and other inhabitants of the soil during all life stages. The progeny of this generation hatch inside the mother and consume her internal organs, destroying her (*endotokia matricida*). They exit the spent female body as J3 infective juveniles, which comprise the filariform stage because the oesophagus has no terminal bulb or isthmus. The filariform juveniles are developmentally arrested until they infect a new host. After penetration those juveniles that reach the lungs mature into hermaphroditic adults, whereas those that fail to reach the lungs are absorbed by the host.

Members of the family Mermithidae, parasites of invertebrates, also have different feeding habits at different stages of their life cycles. The J2 hatches from the egg inside the host gut and gains access to the body cavity by means of a piercing tooth. In the body cavity the mermithid absorbs nourishment through the cuticle. Developing female nematodes remain longer in the host and reach a greater body length and diameter than do developing males. The pre-adult emerges from the host to mature in the soil to a free-living non-feeding adult. Females live up to 2 years whereas males die soon after mating (Webster and Thong, 1984). *Sphaerularia bombi* (a mermithid parasite of bumble-bees) undergo two moults inside the eggs and hatch as J3s (Poinar and van der Laan, 1972). Third-stage juveniles leave the host and become adults after two moults in the soil. After leaving the host, juveniles do not feed. Males and females mate and the males die, but female nematodes may survive for many months before entering a host.

6.6.3. Reproduction

There is limited information concerning how reproductive behaviour changes with age for parasitic nematodes. Most literature is concerned with monitoring changes in the production of progeny (or infective stages) as adults age or, more commonly, how progeny production changes as the infection progresses. This paucity of information may be due to the difficulty of studying the behaviour of parasitic nematodes within their hosts.

In parasitic nematodes females generally have a longer lifespan than males. Males of *Trichinella spiralis* die shortly after copulation while the female gives birth to several hundred juveniles over a period of 4–16 weeks (Schmidt and Roberts, 1981). Females of *Dracunculus medinensis* are fertilized inside the host by the third month after infection. Males die between the third and seventh months, become encysted and degenerate. Females mature, migrate to the host's skin and can produce eggs for up to a year thereafter.

In some animal parasites, hatching occurs spontaneously when nematodes are free-living before becoming parasitic. This may be the result of lipid-hydrolysing enzymes synthesized in the subventral oesophageal glands in the terminal phases of embryogenesis (Bird and Bird, 1991). Eggs of other parasitic species hatch only when stimulated by conditions that are typical of the guts of a potential host. Dormant infective *Ascaris* eggs hatch only under certain combinations of temperature, oxidation–reduction potential, CO_2 concentration and pH that are found inside the guts of warm-blooded vertebrates (Schmidt and Roberts, 1981). Previous to hatching, the egg lipid layer becomes permeable to chitinase secreted by the juvenile. The juvenile emerges from the egg through a hole produced on the eggshell by the action of sterases and proteinases (Fairbairn, 1960). Similar cues to egg hatching of *Ascaris* are required for completion of ecdysis of the J2 cuticle in the J3 infective stage of some vertebrate parasites of the genera *Haemonchus* and *Trichostrongylus* (Schmidt and Roberts, 1981).

6.6.4. Behavioural changes in the infective stage with age

The infective-stage juvenile of a parasitic nematode is analogous to the dauer juvenile of *C. elegans*. The main difference is that, in addition to 'recovering' from the infective stage, the parasite must retain the ability to infect a host, an ability that declines over time as energy reserves are depleted. For example, as the lipid reserves of entomopathogenic nematode infective juveniles decline with age, so does their ability to establish an infection in a susceptible host (Fig. 6.2A, B). In contrast with the *C. elegans* dauer stage that suffers no ill effects due to a prolonged dauer stage, infective stage parasites may be alive but unable to infect a host successfully.

Lewis *et al.* (1995) recorded changes in several behaviours of infective juvenile

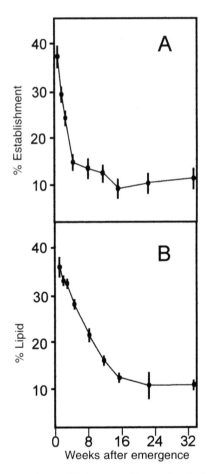

Fig. 6.2. A. Changes in nematode establishment with age for *Steinernema glaseri* within *Galleria mellonella* last instars as hosts. Hosts were dissected 2 days after exposure to infective juveniles. Nematodes were considered established if they had moulted from the infective stage. B. Changes in the percentage of infective juvenile dry weight as lipid. (All data from Lewis *et al.*, 1995.)

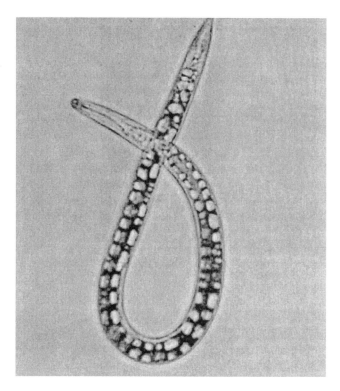

Fig. 6.3. Infective juvenile of *Steinernema glaseri* after several weeks of aqueous storage. The transparent areas within the nematode are water and the opaque areas are the remaining lipid reserves.

entomopathogenic nematodes with age. Entomopathogenic nematodes have a relatively simple life cycle, with a single free-living infective stage that occurs as a J3 that does not feed or develop. The nematode carries in its gut symbiotic bacteria that are released into the host after the nematode reaches the host haemocoel. The host dies within a few days, the bacteria proliferate and the nematodes feed on the bacteria and degrading host tissue. Entomopathogenic nematodes reproduce for up to three generations inside a single host cadaver. As the nutritional quality of the host declines, infective juveniles are produced that exit the cadaver to search for new hosts. The time scale for the infection is 10–15 days for most species.

Locomotion, infectivity and lipid reserves (the infective juveniles' primary energy source) were shown to decrease with age for *Heterorhabditis bacteriophora*, *Steinernema carpocapsae* and *Steinernema glaseri* (Lewis *et al.*, 1995). Figures 6.2 and 6.3 show the effects of ageing on *S. glaseri* infective juveniles. An aspect of entomopathogenic nematodes that does not occur with other parasitic nematodes is their reliance on symbiotic bacteria, which also decline in numbers with age. These three nematode species experienced a linear decline in all behaviours measured, including infectivity. Their metabolic rates were associated with the way they search for hosts. Entomopathogenic nematode foraging behaviour has been studied for several different species (Lewis, 2002) and is a continuum, with some species classed as cruisers (with a high metabolic rate), which move through the soil searching for a host, and others as ambushers (with a low metabolic rate), which stand on their tails and wave their bodies awaiting attachment to a passing host. *Heterorhabditis bacteriophora* (a cruiser) had the highest basal metabolic rate and survived for about 6 weeks. Cohen *et al.* (2002) found that the decline in *H. bacteriophora* behaviour was linear, but also that the time spent in different types of behaviours altered with age. *Steinernema carpocapsae* (an ambusher), had the slowest metabolic rate and lipid decline, and survived for about 12 weeks. *Steinernema carpocapsae* became less effective ambushers with age and lost their ability to stand on their tails and wave, a behaviour linked to their success at infecting hosts (Lewis *et al.*, 1997). *Steinernema glaseri* had a metabolic rate intermediate between the other two species, but due to their larger size (approximately eight times the volume of the other species) they were hypothesized to contain more energy stores and survived for up to 32 weeks.

A phenomenon recorded in some species of entomopathogenic nematodes called 'phased activation' illustrates another time-based change in behaviour during the infective stage. Fan and Hominick (1991) suggested that only a proportion (less than 40%) of *Steinernema feltiae* infective juveniles that emerged from a host cadaver were infectious at any time, regardless of host availability. Campbell *et al.* (1999) found that for three *Steinernema* species (including *S. feltiae*) the proportion of infective juveniles infecting a host was related to host density and that the distribution of nematodes within hosts was more clumped than predicted for random infection. However, Campbell *et al.* (1999) and Perez *et al.* (2003) found that phased activation may occur in *H. bacteriophora*. This suggests that there is significant variability among individuals emerging from the same host cadaver in the timing of their maximum infectivity. The basis of the variability remains unknown.

The duration of the dauer stage for *C. elegans* has been shown to have no effect on the fitness of an individual once they recover. Since *H. bacteriophora* is in the same order as *C. elegans*, we would expect similar results with this parasite. However, this has not been tested. *Trichostrongylus colubriformis*, an intestinal parasite found in lambs, shows a similar trend to *C. elegans*. When infective juveniles were stored for up to 22 weeks before exposure to a host, no decrease in infectivity of the juveniles or their fecundity after they had developed into adults was recorded (Mallet and Kerboeuf, 1985).

6.7. Future Prospects

Research on the biology of ageing mostly focuses on how to prolong life. The concentration on creating the many ageing or longevity mutants of *C. elegans* is evidence of this interest. These mutants allow glimpses into the reasons why behaviour changes with age, but behavioural data collected in most cases are not central to the work described. The overwhelming majority of the research on any aspect of nematode ageing has been conducted with the free-living *C. elegans*. At present, because of the distinct advantages in the realm of genetics offered by *C. elegans*, the diversity of the Nematoda is not seen as an opportunity. This has in some cases resulted in the work on *C. elegans* remaining somewhat isolated from the rest of nematology. We suggest that opportunities may arise for meaningful work on the ageing of economically important nematode species based upon the remarkable accomplishments of *C. elegans* researchers. This may lead to the development of new management practices for nematode-related maladies of economic importance or to fine-tuning the use of nematodes as indicator organisms for ecological studies of soil health.

One of the limitations of the laboratory-based research on *C. elegans* ageing is that it is difficult to extrapolate these results to natural conditions. One avenue of research could be to compare ageing data between *C. elegans* isolates that are wild (i.e. recently isolated from soil, not the N2 'wild type' used in genetic studies) and laboratory strains of mutants and wild types. This would provide an opportunity to test hypotheses in a broader context. Further, as these studies become broader in scope, comparative studies based on the phylogenetic relatedness of the species included may offer clues to the evolutionary origins of genes associated with ageing.

Several studies have shown that individual *C. elegans* that enter the dauer stage extend their lifespan by up to 70 days, with no apparent loss of fecundity or any other measurable decrease in fitness after recovery. These experiments were conducted in conditions where resources were unlimited for the individuals who recovered from the dauer stage and the nematodes faced no competition. One could pose the question of whether nematodes that recover from the dauer stage after 50 days would be at a competitive disadvantage compared with individuals that recover after 10 days in the dauer stage, or with nematodes that had not been induced to enter the dauer stage at all. These types of questions may also be of interest to parasitologists. One study, referenced above, showed that an intestinal parasite of lambs showed no negative impacts of prolonged storage as infective-stage juveniles. However, this study was also carried out in the absence of competition. What impact would measures of nematode fitness after they have infected a host and moulted from the infective stage have on epidemiological models when competition for limited resources is considered?

Measuring age is a challenging undertaking in that there are so many ways to do it. The difficulty lies in choosing a metric that is comparable among diverse taxa with a wide range of behavioural, physiological and morphological characteristics. The calendar is not, in most cases, the proper tool for this job. Perhaps a measure of optical density or light transmittance could be an easy, non-destructive and repeatable way to measure lipid content indirectly. A tool such as this would probably yield measurements that may be comparable among species. This type of measure would also predict characters like lifespan or, for parasites, infectivity better than would chronological age (Patel *et al.*, 1997). The changes in optical density as nematodes age are also relatively obvious (Fig. 6.3), and would not require specialized training to measure.

Choosing behavioural characters that may be used to predict lifespan has become a well-developed technique for choosing longevity mutants from *C. elegans* populations. This tactic is not, to our knowledge, used with other nematode species. This could be a fruitful area, especially for studies focused on parasite population dynamics. For example, if infective stage parasites are collected from the field, behavioural characters could be identified that would indicate how old an individual infective stage parasite is, whether or not it will be able to infect a host successfully and how long it will continue to be infective in the field. This could be especially valuable when studying the variation of natural populations of parasites in regard to their impact on host populations.

References

Arking, R. (1999) *Biology of Aging*, 2nd edn. Sinauer Associates, Sunderland, Massachusetts, 570 pp.

Barrett, J. (1982) Metabolic responses to anabiosis in the fourth stage juveniles of *Ditylenchus dipsaci* (Nematoda). *Proceedings of the Royal Society of London B* 216, 159–177.

Basáñez, M.G. and Ricárdez-Esquinca, J. (2001) Models for the population biology and control of human onchocerciasis. *Trends in Parasitology* 17, 430–438.

Beck, C.D.O. and Rankin, C.H. (1993) Effects of aging on habituation in the nematode *Caenorhabditis elegans*. *Behavioural Processes* 28, 145–164.

Behm, C.A. (1997) The role of trehalose in the physiology of nematodes. *International Journal for Parasitology* 27, 215–229.

Bernhard, N. and van der Kooy, D. (2000) A behavioural and genetic dissection of two forms of olfactory plasticity in *Caenorhabditis elegans*: adaptation and habituation. *Learning and Memory* 7, 199–212.

Bhatt, B.D. and Rohde, R.A. (1970) The influence of environmental factors on the respiration of plant-parasitic nematodes. *Journal of Nematology* 2, 277–285.

Birchfield, W. (1962) Host–parasite relations of *Rotylenchulus reniformis* on *Gossypium hirsutum*. *Phytopathology* 52, 862–865.

Bird, A.F. (1970) The effect of nitrogen deficiency on the growth of *Meloidogyne javanica* at different population levels. *Nematologica* 16, 13–21.

Bird, A.F. (1972) Quantitative studies on the growth of syncytia induced in plants by rootknot nematodes. *International Journal for Parasitology* 2, 157–170.

Bird, A.F. and Bird, J. (1991) *The Structure of Nematodes*, 2nd edn. Academic Press, New York, 316 pp.

Brenner, S. (1974) The genetics of *Caenorhabditis elegans*. *Genetics* 77, 71–94.

Campbell, J.F., Koppenhofer, A.M., Kaya, H.K. and Chinnasri, B. (1999) Are there temporarily non-infectious dauer stages in entomopathogenic nematode populations? A test of the phased infectivity hypothesis. *Parasitology* 118, 499–508.

Chen, J., Carey, J.R. and Ferris, H. (2001) Comparative demography of isogenic populations of *Caenorhabditis elegans*. *Experimental Gerontology* 36, 431–440.

Cohen, N.E., Brown, I.M. and Gaugler, R. (2002) Physiological ageing and behavioural plasticity of *Heterorhabditis bacteriophora* infective juveniles. *Nematology* 4, 81–87.

Comfort, A. (1979) *The Biology of Senescence*, 3rd edn. Elsevier, New York, 414 pp.

Cooper, A.F. and van Gundy, S.D. (1971) Senescence, quiescence and cryptobiosis. In: Zuckerman, B.M., Mai, W.F. and Rhode, R.A. (eds) *Plant Parasitic Nematodes*, Vol. II. Academic Press, New York, pp. 297–318.

Crowe, J.H. and Madin, K.A.C. (1975) Anhydrobiosis in nematodes: evaporative water loss and survival. *Journal of Experimental Zoology* 193, 323–334.

Crowe, J.H., Hoekstra, F.A. and Crowe, L.M. (1992) Anhydrobiosis. *Annual Review of Physiology* 54, 579–599.

Demeure, Y. and Freckman, D.W. (1981) Recent advances in the study of anhydrobiotic nematodes. In: Zuckerman, B.M. and Rohde, R.A. (eds) *Plant Parasitic Nematodes*, Vol. III. Academic Press, New York, pp. 205–226.

Dougherty, E.C., Hansen, E.L., Nicholas, W.L., Mollett, J.A. and Yarwood, E.A. (1959) Axenic cultivation of *Caenorhabditis briggsae* with unsupplemented and supplemented chemically defined media. *Annals of the New York Academy of Science* 77, 176.

Duggal, C.L. (1978) Initiation of copulation and its effect on oocyte production and life span of adult female *Panagrellus redivivus*. *Nematologica* 24, 213–221.

Endo, B.Y. and Wergin, W.P. (1973) Ultrastructural investigations of clover roots during early stages of infection by the root-knot nematode *Meloidogyne incognita*. *Protoplasma* 78, 365–379.

Ernstrom, G.G. and Chalfie, M. (2002) Genetics of sensory mechanotransduction. *Annual Review of Genetics* 36, 411–453.

Evans, A.A.F. (1987) Diapause in nematodes as a survival strategy. In: Veech, J.A. and Dickson, D.W. (eds) *Vistas on Nematology: a Commemoration of the Twenty-fifth Anniversary of the Society of Nematologists*. Society of Nematologists, Hyattsville, Maryland, pp. 180–187.

Evans, A.A.F. and Womersley, C. (1980) Longevity and survival in nematodes: models and mechanisms. In: Zuckerman, B.M. (ed.) *Nematodes as Biological Models*, Vol. II, *Aging and Other Model Systems*. Academic Press, New York, pp. 193–211.

Evans, K. (1982) Effects of host variety, photoperiod and chemical treatments on hatching of *Globodera rostochiensis*. *Journal of Nematology* 14, 203–207.

Ewbank, J.J., Barnes, T.M., Lakowski, B., Lussier, M., Bussey, H. and Hekimi, S. (1997) Structural and functional conservation of the *Caenorhabditis elegans* timing gene clk-1. *Science* 275, 980–983.

Fairbairn, D. (1960) The physiology and biochemistry of nematodes. In: Sasser, J.N. and Jenkins, W.R. (eds) *Nematology Fundamentals and Recent Advances with Emphasis on Plant Parasitic and Soil Forms*. University of North Carolina Press, Chapel Hill, pp. 267–296.

Fan, X. and Hominick, W.M. (1991) Effects of low storage temperature on survival and infectivity of two *Steinernema* species (Nematoda: Steinernematidae). *Revue de Nématologie* 14, 407–412.

Fielding, M.J. (1951) Observations on the length of dormancy in certain plant infecting nematodes. *Proceedings of the Helminthological Society of Washington* 18, 110–112.

Fisher, J.M. and Raski, D.J. (1967) Feeding of *Xiphinema index* and *X. diversicaudatum*. *Proceeding of the Helminthological Society of Washington* 34, 68–72.

Gal, T.Z., Solomon, A., Glazer, I. and Koltai, H. (2001) Alterations in the levels of glycogen and glycogen synthase transcripts during desiccation in the insect-killing nematode *Steinernema feltiae* IS-6. *Journal of Parasitology* 87, 725–732.

Gems, D. and Riddle, D.L. (1996) Longevity in *Caenorhabditis elegans* reduced by mating but not gamete production. *Nature* 379, 723–725.

Gems, D. and Riddle, D.L. (2000) Genetic, behavioral and environmental determinants of male longevity in *Caenorhabditis elegans*. *Genetics* 154, 1597–1610.

Gershon, H. and Gershon, D. (1970) Detection of inactive enzyme molecules in ageing organisms. *Nature* 227, 1214–1217.

Gershon, H. and Gershon, D. (2001) *Caenorhabditis elegans* – a paradigm for aging research: advantages and limitations. *Mechanisms of Ageing and Development* 123, 261–274.

Grewal, P.S., Converse, V. and Georgis, R. (1999) Influence of production and bioassay methods on virulence of two ambush foragers (Nematoda: Steinernematidae). *Journal of Invertebrate Pathology* 73, 40–44.

Hansen, E.L., Buecher, E.J. and Yarwood, E.A. (1973) Alteration of sex of *Aphelenchus avenae* in culture. *Nematologica* 19, 112–116.

Harvey, S.C., Gemmill, A.W., Read, A.F. and Viney, M.E. (2000) The control of morph development in the parasitic nematode *Strongyloides ratti*. *Proceedings of the Royal Society of London Series B* 267, 2057–2063.

Herndon, L.A., Schmeissner, P.J., Dudaronek, J.M., Brown, P.A., Listner, K.M., Sanano, Y., Paupard, M.C., Hall, D.H. and Driscoll, M. (2002) Stochastic and genetic factors influence tissue-specific decline in ageing *C. elegans*. *Nature* 419, 808–814.

Hominick, W.M., Forrest, J.M. and Evans, A.A.F. (1985) Diapause in *Globodera rostochiensis* and variability in hatching trials. *Nematologica* 31, 159–170.

Hosono, R., Sato, Y., Aizawa, S.I. and Mitsui, Y. (1980) Age-dependent changes in mobility and separation of the nematode *Caenorhabditis elegans*. *Experimental Gerontology* 15, 285–289.

Houthoofd, K., Braeckman, B.P., Lenaerts, I., Brys, K., De Vreese, A., Van Eygen, S. and Vanfleteren, J.R. (2002) Ageing is reversed, and metabolism is reset to young levels in recovering dauer larvae of *C. elegans*. *Experimental Gerontology* 37, 1015–1021.

Ishibashi, N. (1969) Studies on the propagation of the root-knot nematode *Meloidogyne incognita* (Kofoid and White) Chitwood, 1949. *Review of Plant Protection Research* 2, 125–128.

Kenyon, C., Chang, J., Grensch, E., Rudener, A. and Tabtiang, R. (1993) A *C. elegans* mutant that lives twice as long as wild type. *Nature* 366, 461–464.

Kimura, E., Shintoku, Y., Kadosaka, T., Fujiwara, M., Kondo, S. and Itoh, M. (1999) A second peak of egg excretion in *Strongyloides ratti*-infected rats: its origins and biological meaning. *Parasitology* 119, 221–226.

Kisiel, M.J. and Zuckerman, B.M. (1974) Studies on aging of *Turbatrix aceti*. *Nematologica* 20, 277–282.

Klass, M. and Hirsh, D. (1976) Non-ageing developmental variant of *Caenorhabditis elegans*. *Nature* 260, 523–525.

Lewis, E.E. (2002) Behavioural ecology. In: Gaugler, R. (ed.) *Entomopathogenic Nematology*, CAB International, Wallingford, UK, pp. 205–224.

Lewis, E.E., Selvan, S., Campbell, J.F. and Gaugler, R. (1995) Changes in foraging behaviour during the infective stage of entomopathogenic nematodes. *Parasitology* 110, 583–590.

Lewis, E.E., Campbell, J.F. and Gaugler, R. (1997) The effects of aging on the foraging behavior of *Steinernema carpocapsae* (Rhabdita: Steinernematidae). *Nematologica* 43, 355–362.

Malakhov, V.V. (1994) *Nematodes: Structure, Development, Classification, and Phylogeny*. Smithsonian Institution Press, Washington, DC, 286 pp.

Mallet, S. and Kerboeuf, D. (1985) *Trichostrongylus colubriformis*, relationship between ageing of infective larvae, infectivity and egg production by adult female worms. *Annales de Recherches Vétérinaires* 16, 99–104.

Patel, M.N., Stolinski, M. and Wright, D.J. (1997) Neutral lipids and the assessment of infectivity in entomopathogenic nematodes: observations on four *Steinernema* species. *Parasitology* 114, 489–496.

Perez, E.E., Lewis, E.E. and Shapiro-Ilan, D.I. (2003) Impact of the host cadaver on survival and infectivity of entomopathogenic nematodes (Rhabditida: Steinernematidae and Heterorhabditidae) under desiccating conditions. *Journal of Invertebrate Pathology* 82, 111–118.

Poinar, G. and van der Laan, P. (1972) Morphology and life history of *Sphaerularia bombi*. *Nematologica* 18, 239–252.

Reznick, A.Z. and Gershon, D. (1999) Experimentation with nematodes. In: Yu, B.P. (ed.) *Methods in Aging Research*, 2nd edn. CRC Press, Boca Raton, Florida, pp. 167–190.

Riddle, D.L., Golden, J.W. and Albert, P.S. (1987) Role of dauer larvae in survival of *Caenorhabditis elegans*. In: Veech, J.A. and Dickson, D.W. (eds) *Vistas on Nematology: A Commemoration of the Twenty-fifth Anniversary of the Society of Nematologists*. Society of Nematologists, Hyattsville, Maryland, pp. 174–179.

Schmidt, G.D. and Roberts, L.S. (1981) *Foundations of Parasitology*, 2nd edn. C.V. Mosby, St Louis, Missouri, 795 pp.

Siddiqi, M.R. (1974) *Tylenchulus semipenetrans*. CIH Descriptions of Plant-parasitic Nematodes, Set 3, No. 34. CAB International, Wallingford, UK, 4 pp.

Solomon, A., Paperna, I. and Glazer, I. (1999) Desiccation survival of the entomopathogenic nematode *Steinernema feltiae*: induction of anhydrobiosis. *Nematology* 1, 61–68.

Somers, J.A., Shorey, H.H. and Gaston, L.K. (1977) Reproductive biology and behaviour of *Rhabditis pellio* (Schneider) (Rhabditida: Rhabditidae). *Journal of Nematology* 9, 143–148.

Starr, J.L. (1998) Cotton. In: Barker, K.R., Pederson, G.A. and Windham, G.L. (eds) *Plant and Nematode Interactions*. ASA, CSSA, SSSA, Madison, Wisconsin, pp. 359–379.

Starr, J.L and Veech, J.A. (1986) Comparison of development, reproduction and aggressiveness of *Meloidogyne incognita* race 3 and 4 on cotton. *Journal of Nematology* 18, 413–415.

Steiner, G. and Albin, F.E. (1946) Resuscitation of the nematode *Tylenchus polyhypnus* n. sp. after almost 39 years' dormancy. *Journal of the Washington Academy of Sciences* 36, 97–99.

Sukhdeo, M.V.K., Sukhdeo, S.C. and Bansemir, A.D. (2002) Interactions between intestinal nematodes and vertebrate hosts. In: Lewis, E.E., Campbell, J.F. and Sukhdeo, M.V.K. (eds) *Behavioural Ecology of Parasites*. CAB International, Wallingford, UK, pp. 223–242.

Triantaphyllou, A.C. (1960) Sex determination in *Meloidogyne incognita* Chitwood, 1949, and intersexuality in *M. javanica* (Treub, 1885) Chitwood, 1949. *Annals of Institute of Phytopathology, Benaki* 3, 12–31.

Triantaphyllou, A.C. and Hirschmann, H.H. (1964) Reproduction of plant and soil nematodes. *Annual Review of Phytopathology* 2, 57–80.

Triantaphyllou, A.C. and Hirschmann, H.H. (1973) Environmentally controlled expression in *Meloidodera floridensis*. *Journal of Nematology* 5, 181–185.

van Gundy, S.D. (1958) The life history of the citrus nematode *Tylenchulus semipenetrans* Cobb. *Nematologica* 3, 283–294.

van Gundy, S.D. (1965) Factors in survival of nematodes. *Annual Review of Phytopathology* 3, 43–68.

van Gundy, S.D., Bird, A.F. and Wallace, H.R. (1967) Ageing and starvation of larvae in *Meloidogyne javanica* and *Tylenchulus semipenetrans*. *Phytopathology* 57, 559–571.

Viney, M.E. (1996) Developmental switching in the parasitic nematode *Strongyloides ratti*. *Proceedings of the Royal Society of London Series B* 263, 201–208.

Viney, M.E. (2002) Environmental control of nematode life cycles. In: Lewis, E.E.,

Campbell, J.F. and Sukhdeo, M.K.V. (eds) *Behavioural Ecology of Parasites.* CAB International, Wallingford, UK, pp. 111–128.

Webster, J.M. and Thong, C.H.S (1984) Nematodes parasites of orthopterans. In: Nickle, W.R. (ed.) *Plant and Insect Nematodes.* Marcel Dekker, New York, pp. 697–726.

Wharton, D.A. and Barrett, J. (1985) Ultrastructural changes during recovery from anabiosis in the plant parasitic nematode, *Ditylenchus. Tissue Cell* 17, 79–96.

Wharton, D.A., Barrett, J.B. and Perry, R.N. (1985) Water uptake and morphological changes during recovery from anabiosis in the plant-parasitic nematode, *Ditylenchus dipsaci. Journal of Zoology* 206, 391–402.

Wharton, D.A., Preston, C.M., Barrett, J. and Perry, R.N. (1988) Changes in cuticular permeability associated with recovery from anhydrobiosis in the plant-parasitic nematode, *Ditylenchus dipsaci. Parasitology* 97, 317–330

Wharton, D.A., Rolfe, R.N. and Perry, R.N. (2000) Electrophysiological activity during recovery from anhydrobiosis in fourth stage juveniles of *Ditylenchus dipsaci. Nematology* 2, 881–886.

Yang, Y. and Wilson, D.L. (2000) Isolating aging mutants: a novel method yields three strains of the nematode *Caenorhabditis elegans* with extended life spans. *Mechanisms of Ageing and Development* 113, 101–116.

Zuckerman, B.M., Himmelhoch, S., Nelson, B., Epstein, J. and Kisiel, M. (1971) Aging in *Caenorhabditis briggsae. Nematologica* 17, 478–487.

Zuckerman, B.M., Nelson, B. and Kisiel, M. (1972) Specific gravity increase of *Caenorhabditis briggsae* with age. *Journal of Nematology* 4, 261–262.

7

Osmoregulatory and Excretory Behaviour

DENIS J. WRIGHT

Department of Biological Sciences, Imperial College London, Silwood Park Campus, Ascot, Berkshire SL5 7PY, UK

7.1. Introduction

Thompson and Geary (2002) reviewed excretion, secretion, ionic and osmotic regulation in nematodes, with particular emphasis on the molecular cell biology of animal-parasitic species. In the present chapter, the main emphasis is on the ecological and behavioural physiology of osmoregulation and the end products of nitrogen catabolism in this very widespread, numerous and species-rich group of meiofauna. A comprehensive account of the excretion of CO_2, alcohols and organic acids, the end-products of energy metabolism in nematodes, is given by Thompson and Geary (2002).

The habitats in which nematodes occur are extremely diverse but the niches that nematodes occupy are all essentially aquatic and they must deal with the opposing problems of osmosis and diffusion. Such problems can be relatively minor in the case of marine species, which may live in a near iso-osmotic environment. For species living under stable but hypo-osmotic freshwater conditions, the challenge is greater. It is greater still for species that inhabit fluctuating, brackish-water environments (0.5–30 parts per thousand total dissolved solids), such as estuaries and some soils, or as parasites having a free-living stage as well as stages in one or more host environments during their life cycle. Finally, some free-living and parasitic species have stages that can lose water due to desiccation or freezing. The strategies used by nematodes to survive such conditions have been considered elsewhere in this volume (see Wharton, Chapter 13).

In nematodes, as in other groups of animals, there is a close relationship between the environment, the ability to regulate water and ionic content, and the excretion of waste products, particularly nitrogenous compounds. Animals, unlike plants, cannot store the end-products of nitrogen catabolism in appreciable amounts so excess nitrogen must be excreted (Wright, 1995; Randall *et al.*, 2001). Where water can be freely exchanged with the environment, the most energetically efficient nitrogenous excretory product is ammonia. Where water exchange is more limited, ammonia, which is very toxic, is generally converted in animals to either urea or a purine (usually uric acid or guanine) prior to excretion.

7.2. Osmotic and Ionic Regulation

Animals faced with changing osmotic and ionic conditions in the environment can either 'conform' or use various mechanisms to maintain homoeostasis (Randall *et al.*, 2001). While at least some nematode species exposed to a temporarily unfavourable environment might be expected to be osmoconformers (organisms in which the osmotic pressure of internal fluids varies with the osmotic pressure of the external environment), species that live in hypo- or hyperosmotic conditions for any extended period should be able to regulate their water content to some extent (Wright, 1998).

Nematodes that are unable to regulate their water content and thus body volume can suffer loss of locomotory function due to the anisometric nature of the nematode cuticle, which results in an almost linear change in body length with change in water content/body volume (Fig. 7.1; Wright and Newall, 1980). Nematode locomotion (see Burr and Robinson, Chapter 2, this volume) depends upon contractions of longitudinal body wall muscles acting against a hydrostatic skeleton and changes in body length of more than ±15% will impair muscle function and thus normal movement. Osmotic factors are also important in feeding (see Bilgrami and Gaugler, Chapter 4, this volume), reproduction (see Huettel, Chapter 5, this volume) and hatching (see Perry and Maule, Chapter 8, this volume).

7.2.1. Physiological ecology

7.2.1.1. Marine and brackish water species

In an early study on osmoregulation in a sublittoral marine species, *Deontostoma californicum*, Croll and Viglierchio (1969) observed a marked ability to regulate

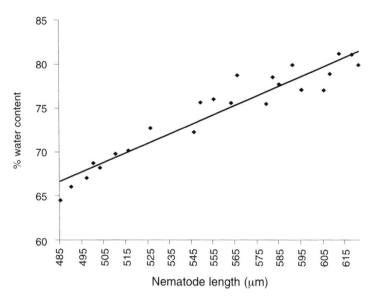

Fig. 7.1. Linear regression analysis of water content vs. length of infective juveniles of *Steinernema* sp. (SSL 85); $R^2 = 0.91$, $P < 0.05$). (Adapted from Piggott *et al.*, 2002.)

hypo-osmotically but no ability to regulate hyperosmotically. Similar findings were reported for two congeneric species by Viglierchio (1974) and for another sub-littoral species, *Sabatieria punctata* (Forster, 1998). Other species have been found to have very different osmoregulatory abilities, which can be correlated to some extent with habitat (Table 7.1). For example, the sublittoral marine species, *Enoplus brevis*, had no osmoregulatory ability, whereas the brackish-water estuarine species, *E. brevis*, showed a marked hyperosmotic regulation and some hypo-osmotic regulation (Wright and Newall, 1976). An upper littoral zone species, *Axonolaimus paraspinosus*, showed a particularly marked ability for hyper- as well as hypo-osmotic regulation (Table 7.1), suggesting that an ability to overcome fluctuations in salinity is a factor determining the horizontal distribution of nematode assemblages in littoral habitats (Forster, 1998).

Most marine nematodes (Table 7.1) can regulate their body volume hypo-osmotically but this may be misleading. A survey of nematodes in East African mangrove swamps (Olafsson, 1995) showed that nematode diversity (and perhaps the ability to hypo-osmotically regulate) was reduced in hypersaline areas (>100 parts per thousand total dissolved solids) where the salinity exceeded that of sea water. A survey of free-living nematodes from littoral benthic sediments of four Ethiopian lakes found the lowest number of species was in the most saline lake (Abebe *et al.*, 2001). However, Warwick *et al.* (2002) reported 13 species or putative species of free-living nematodes from the hypersaline Salton Sea, the largest inland lake in California, thereby doubling the number of invertebrate species known from this site. The nematode taxa identified were representative of those found in marine coastal environments and it was suggested that a wide range of marine nematodes are capable of adapting to hypersaline habitats.

Studies on the meiobenthic fauna of five European estuaries (Ems, Westerschelde, Somme, Gironde and Tagus) showed that salinity was more important in explaining nematode community structure than latitudinal differences (Soetaert et al., 1995). Nematode diversity was almost entirely determined at the genus level and was positively related to salinity. Such observations suggest that the ability of many estuarine nematode species to hyperosmoregulate in low-saline regions may be relatively poor.

7.2.1.2. Freshwater and soil species and stages

These species or life-cycle stages usually exist in a dilute environment, which has an osmotic and ionic concentration considerably less (perhaps sixfold) than their body fluids (Wright and Newall, 1976) although this may fluctuate and in some cases become hyperosmotic to nematodes (Wright, 1998).

As might be expected, the species and stages of nematodes examined (including free-living stages of plant and animal parasites) have tended to be good hyperosmotic regulators (Table 7.1). They are generally poorer hypo-osmotic regulators. An exception is the free-living rhabditid *Panagrellus silusiae*, where a marked ability to regulate hypo-osmotically (Table 7.1) can be correlated with its presence in decaying and fermenting material in soils, where relatively high osmotic and ionic conditions can occur. Marked changes in the solute concentration and thus osmotic pressure of the soil solution can occur due to climatic factors and the use of agricultural amendments (Bednarek and Gaugler, 1997). This is particularly common in upper soil layers where the changes in the net evaporative rate have the greatest impact, most notably during a drought. In some agricultural areas, irrigation systems can result in very high soil salinities.

The relative importance of the solute concentration versus the soil matric potential on the water balance of nematodes has yet to be resolved (Wright, 1998). Robinson et al. (1984) have pointed out that the matric potential of soil water potentials can be tenfold greater than the solute component and that the former is recognized as the more important for plant–water relations. However, there have been no experiments to identify the effects of the different components of the soil water potential on nematode osmoregulation.

Rhabditid species (*Steinernema* and *Heterorhabditis*) pathogenic to insects are examples of parasitic nematodes in which an infective stage is exposed to a marked variation in its osmotic environment. In these entomopathogenic species, a modified third stage, 'dauer' juvenile, emerges from the cadaver of the insect host into the soil, where it may remain in a mobile (and infective) state for a number of weeks or months. The ability of these dauer juveniles to remain active in distilled water suggests they can all regulate body volume in hypo-osmotic conditions (Thurston et al., 1994; Patel et al., 1997). The ability to regulate in hyperosmotic conditions would also be expected for dauer juveniles and other stages within the insect host.

Studies suggest that interspecific and intraspecific variation in hypo-osmotic regulation by dauer juveniles exists (Table 7.1; Piggott et al., 2000, 2002). Such variation may relate to differences in their normal soil environment. For example,

Table 7.1. Osmoregulatory ability of nematodes in relation to habitat.

Species[a]	Osmoregulatory ability[c]		Reference
	Hyperosmotic regulation	Hypo-osmotic regulation	
Marine: free-living, iso-osmotic with environment			
Sabatieria punctata	+	+++	Forster (1998)
Deontostoma antarcticum	−	+++	Viglierchio (1974)
Deontostoma timmerchioi	−	+++	Viglierchio (1974)
Deontostoma californicum	−	+++	Croll and Viglierchio (1969)
Monohystera disjuncta	−	−	Viglierchio (1974)
Enoplus communis	−	−	Wright and Newall (1976)
Marine intertidal: free-living, hypo- to hyperosmotic with environment			
Daptonema oxycerca	++	++	Forster (1998)
Axonolaimus paraspinosus	+++	+++	Forster (1998)
Cervonema tenuicauda	++	+++	Forster (1998)
Brackish–estuarine: free-living, hyper- to iso-osmotic with environment			
Enoplus communis	+++	++	Wright and Newall (1976)
Freshwater–soil: free-living, hyperosmotic with environment			
Panagrolaimus davidi	ND	−	Viglierchio (1974)
Rhabditis terrestris	+++	++	Stephenson (1942)
Panagrellus silusiae	ND	+++	Prencepe *et al.* (1984)
Panagrellus redivivus	+	++	Myers (1966)
Caenorhabditis elegans	+++	ND	Nelson and Riddle (1984)
Freshwater–soil: infective stage of plant-parasitic species, hyperosmotic with environment			
Globodera rostochiensis J2	+++	++	Wright and Newall (1976) Clarke *et al.* (1978)
Heterodera schachtii J2	+++	++	Perry *et al.* (1980)
Freshwater–soil: infective stage of animal-parasitic species, hyperosmotic with environment			
Haemonchus contortus J3	+++	ND	Atkinson and Onwuliri (1981)
Nippostrongylus brasiliensis J3	+++	ND	Atkinson and Onwuliri (1981)
Trichostrongylus colubriformis J3	+++	−	Wharton *et al.* (1983)
Steinernema carpocapsae J3	ND	+++	Piggott *et al.* (2002)
Steinernema spp. J3[c]	ND	++/+++	Piggott *et al.* (2000)
Heterorhabditis indicus J3	ND	+++	Piggott *et al.* (2000)
Heterorhabditis bacteriophora J3	ND	+	Piggott *et al.* (2002)
Heterorhabditis spp. J3	ND	++	Piggott *et al.* (2002)
Animal-parasitic: parasitic stages, hypo- to hyperosmotic with environment			
Pseudoterranova decipiens J3	+++	+++	Davey (1995)
Anguillicola crassus	−	−	Kirk *et al.* (2002)

[a] Adult unless specified as a juvenile (J) stage.
[b] +++, Good regulation; ++, some regulation; +, very limited regulation; -, no regulation observed; ND = not determined.
[c] Osmoregulatory ability varied with isolate/species.

hypo-osmotic regulation was relatively poor in a *Steinernema* isolated from a heavy, inland soil in the humid tropics compared with two isolates (*Heterorhabditis indicus* and *Steinernema*) from sandy coastal soils (Piggott *et al.*, 2000). In the latter, marked fluctuations in the soil water solute concentration is more likely to occur, due to changes in the net evaporation rate with season and the proximity of the sea, resulting in high soil salinities (Finnegan *et al.*, 1999; Piggott *et al.*, 2000). The distribution of *Heterorhabditis* is often highly correlated with the marine shoreline (e.g.

Hara *et al.*, 1991) and more detailed studies on the ecological physiology of these nematodes are required.

The infective juveniles of several plant-parasitic species (*Heterodera* and *Globodera*) have been shown to be good hyperosmotic regulators (Table 7.1). This would be expected, as an ability to retain normal locomotory activity under the normal range of soil water solute concentrations is critical for plant parasites, which in almost all species actively infect the host. The infective juveniles of tri-chostrongyle parasites of ruminants (e.g. *Haemonchus contortus*) are also good hyperosmotic regulators (Table 7.1). These nematodes passively infect the host but need to move on to and survive on foliage for trophic transmission to occur.

7.2.1.3. Animal- and plant-parasitic stages

The adult stages of animal-parasitic nematodes are generally considered to be less tolerant of hypo-osmotic environments when compared with free-living stages of the same or different species (Wright and Newall, 1976; Thompson and Geary, 2002), although relatively few studies have been conducted (Table 7.1).

Davey (1995) reviewed the capacity of some vertebrate parasites to regulate the solute concentration of their pseudocoelomic fluids, at least in the short term, and considered this important for their survival and development. He looked in partic-ular at *Pseudoterranova* (=*Phocanema*) *decipiens*, which as third-stage juveniles live in a dormant state in the muscles of cod and some other marine fish, and as adults in the intestine of seals. Third-stage juveniles are initially able to maintain the osmotic pressure of their body fluids in either very dilute or hyperosmotic solutions. However, after 48 h exposure to osmotic stress, regulation of internal osmotic pres-sure was less marked and after 10 days the worms are effectively osmoconformers (Davey, 1995).

In contrast, *Anguillicola crassus*, which lives in the liquid/gaseous environment of the swim-bladder lumen and feeds on the capillary network of the European eel, *Anguilla anguilla*, is buffered from changes in the osmotic environment to some extent by the osmoregulatory ability of its host, a euryhaline species adapted to a wide range of salinities. Such a parasite might be expected to be an osmoconformer and studies suggest that > 90% of adult *A. crassus* do conform closely to host osmo-larity (Kirk *et al.*, 2002; Table 7.1).

There have been no studies on osmotic and ionic regulation by plant-parasitic stages of nematodes although presumably they tolerate marked fluctuations in water potentials within plants, particularly at times of drought or, indeed, nematode-induced stress. In the case of sedentary endoparasites, such as *Meloidogyne*, *Heterodera* and *Globodera* species, following root invasion by the infective juvenile, the sedentary juvenile and adult female stages might be predicted to osmoconform. Adult males, where present, may retain their osmoregulatory ability since they leave the root after the final moult.

7.2.2. Behavioural aspects

The infective juveniles of *Meloidogyne javanica* migrate away from areas of high salinity (Prot, 1978). Studies on the orientation of infective juveniles of *Meloidogyne incognita* and *Rotylenchulus reniformis* have shown that they respond to the constitutive cations and anions of salts, respectively; the former species finds ammonium salts to be strongly repellent, while the latter species finds chloride salts to be repellent (Le Saux and Queneherve, 2002). How such differences in chemotactic responses relate to the behaviour and distribution of nematodes in soils of varying ionic content remains to be investigated.

Studies on the behaviour of third-stage (J3) infective juveniles of a rodent parasite, *Strongyloides ratti*, showed that the overall direction of their movement in a sodium chloride concentration gradient depended on the initial point of placement (Tobata-Kudo *et al.*, 2000). They appeared to prefer a concentration of *c.* 80 mM and would move towards that from areas of lower or higher concentrations. Movement was typical of a kinetic behaviour, with a unidirectional avoidance movement from unfavourable conditions and a random dispersal movement in favourable areas.

In tidal marine environments, some nematode species show marked vertical migration in sediments. For example, two estuarine species, the predator *Enoploides longispiculosus* and the deposit-feeder *Daptonema normandicum*, migrate upwards with the incoming tide and downwards as the tidal flat becomes exposed (Steyaert *et al.*, 2001). Whether this migration pattern is associated principally with feeding behaviour or with changes in osmolarity or other environmental factors remains to be determined.

The amphids and phasmids (Bird and Bird, 1991), both of which have a chemosensory function (Perry and Aumann, 1998), are linked to osmotic avoidance behaviour in *Caenorhabditis elegans* and undoubtedly in other nematode species. In *C. elegans*, several mutants (*osm-1*, *osm-3*, *osm-5* and *osm-6*) associated with amphid and phasmid neurones (Shakir *et al.*, 1993) fail to avoid high concentrations of NaCl or fructose (Hodgkin, 1997). A pair of polymodal aphidial neurones (ASH) are known to be involved (Bargmann and Mori, 1997), whose response to mechanical, chemical and osmotic stimuli is mediated by post-synaptic glutamate receptors (Mellem *et al.*, 2002). Osmotic avoidance requires both *N*-methyl-D-aspartate (NMDA) and non-NMDA glutamate receptors, while the response to mechanical stimuli requires only the latter receptors. Prolonged exposure of wild-type *C. elegans* to NaCl, NaAc and NH_4Cl can result in a reversible loss of chemosensory behaviour. The heterotrimeric G-protein gamma subunit, gpc-1, is involved in this adaptation, together with *osm-9* and another locus, *adp-1* (Jansen *et al.*, 2002).

7.3. Nitrogen Excretion

7.3.1. Nitrogen catabolism

7.3.1.1. Excretory products

Most aquatic animals, including nematodes, are 'ammonotelic' (Section 7.1). In nematodes, ammonia has been found to constitute *c.* 40–90% of non-protein nitrogen (NPN) eliminated from the body (Table 7.2). Urea is excreted by some nematodes (up to *c.* 20% NPN) but not by others.

Table 7.2. Nitrogenous excretion by selected nematodes (adapted from Wright, 1975a; Thompson and Geary, 2002).

| Species (stage) | Habitat | Experimental conditions | % total nitrogen eliminated (% NPN) | | | | Total N eliminated (µM/g wet wt/24 h) |
			NH_3^+ NH_4^+	Urea	Amino acids	Protein/ peptides	
Panagrellus redivivus (mixed stages)	Soil	Aerobic	34 (59)	7 (11)	10	42	213
Ditylenchus triformis (mixed stages)	Soil/plant	Aerobic	39 (56)	0	+	28	10
Trichinella spiralis (J3)	Gut	Aerobic	33 (42)	0	28	21	200
Nematodirus spp. (adults)	Gut	Aerobic	41 (63)	14 (22)	–	33	97
		Anaerobic	29 (45)	4 (6)	–	43	123
Ascaridia galli (adults)	Gut	Aerobic	56 (66)	8 (9)	–	15	25
		Anaerobic	62 (70)	9 (9)	–	12	26
Ascaris lumbricoides (adults)	Gut	Aerobic	68 (86)	7 (9)	+	21	28
		Anaerobic	72 (88)	7 (8)	+	18	29

NPN, Non-protein nitrogen (ex. amino acids).

Other nitrogen-containing compounds reported to be eliminated by nematodes are free amino acids, peptides, proteins and amines (Wright and Newall, 1976; Wright, 1998; Thompson and Geary, 2002), although whether such compounds are excretory end-products is doubtful (Wright, 1998).

The high levels of free amino acids eliminated by *Panagrellus redivivus* (Table 7.2) were probably due mainly to defecation of non-assimilated food material (Wright and Newall, 1976). This view is supported by the observed elimination of almost the same amount of free amino acids by this species after 6 h compared with 24 h incubation (Wright, 1975a). Increased amounts of amino acids are released by *P. redivivus* under unfavourable incubation conditions (Wright, 1975a), which suggests that the large amounts of amino acids eliminated by the animal parasite *Trichinella spiralis* (Table 7.2) could be an experimental artefact.

The large amounts of proteinaceous material eliminated by most species (Table 7.2) are probably due to defecation and to secretory processes (Wright and Newall, 1976; Thompson and Geary, 2002). Low levels of short-chain aliphatic amines are also eliminated by several animal parasitic species, including *T. spiralis* and *Ascaris suum*, but their role remains to be determined (Thompson and Geary, 2002).

7.3.1.2. Biochemistry

Amino acids are the principal source of waste nitrogen in animals. Their catabolism in nematodes, as in other animals, is by L- and D-amino acid oxidases and in particular by deaminases (Barrett and Wright, 1998). Glutamate dehydrogenases have been detected in various nematode species, including *P. redivivus* and several animal-parasitic species (Wright, 1975b; Wright and Newall, 1976; Grantham and Barrett, 1986), and transdeamination via glutamic acid is thought to be the most important pathway for ammonia production (Grantham and Barrett, 1986). Another ammonia-producing enzyme, glutaminase has been detected in *P. redivivus* (Wright, 1975b; Grantham and Barrett, 1986) and in *C. elegans*, where it is linked with glutaminergic neurotransmission (Rankin and Wicks, 2000). However, it is not known whether glutaminase is involved in nitrogen catabolism in nematodes. In ureotelic mammals, glutaminase is involved in acid–base balance and ammonia transport (Karim *et al.*, 2002).

Urea may be produced from dietary sources, from purine and pyrimidine degradation, from tissue arginine or from more general amino acid catabolism via the ornithine–arginine (urea) cycle. However, there is no conclusive evidence for the pathways involved in urea formation in nematodes (Barrett and Wright, 1998).

7.3.2. Ecological and behavioural aspects

Nematodes are dominant members of the meiofauna in soils and in the benthos, in terms of both species abundance and biomass (e.g. Bloemers *et al.*, 1997; Olafsson and Elmgren, 1997; Brown *et al.*, 2001), and might be expected to make a significant contribution to matter transformation in these environments.

Studies are limited but Ekschmitt *et al.* (1999) have shown that the nematode fauna can have an important influence on the nitrogen status of mineral grassland soils, both directly by excretion and indirectly by affecting microbial activity. They estimated that nematode nitrogenous excretion could, under the most favourable conditions, contribute up to 27% of soluble soil nitrogen.

It might be expected that high concentrations of ammonia, which is highly toxic to nematodes (Wright, 1975a), would cause avoidance behaviour. Such conditions could occur where nematode populations are severely overcrowded or where diffusion of ammonia is limiting.

Ammonia in insect faeces may initiate avoidance behaviour by infective juveniles of entomopathogenic species (Grewal *et al.*, 1993). Shapiro *et al.* (2000) found infective juveniles of *Heterorhabditis bacteriophora* attracted to low concentrations and repelled by high concentrations of ammonium hydroxide. The latter authors showed that considerable nitrogen was released as ammonia from the insect host (*Galleria mellonella*), and that most ammonia was released during the first 3 days of infection. Thus, nematodes may be attracted only to the low levels of ammonia produced by uninfected hosts.

7.4.   Structures and Mechanisms

The basic nematode body plan comprises an elongate cylinder, enclosing a pseudo-coelomate body cavity, with an anterior mouth and a posterior anus connected by an alimentary canal. The body wall comprises an outer living cuticle, an underlying epidermis (also termed 'hypodermis') and an inner layer of longitudinal muscle cells (Bird and Bird, 1991; 1998). The pseudocoelom is separated by only one cell layer from the external environment, is in contact with almost all body cells (Thompson and Geary, 2002) and has been suggested to act as a primitive circulatory system (Wharton, 1986; McKerrow et al., 1999). In moving nematodes, sinusoidal waves of contraction and accompanying internal pressure changes passing along the body result in some mixing of the pseudocoelomic fluid, although the degree to which mixing is localized is unknown.

There is little direct evidence for the major routes of fluid ('urine') excretion in nematodes (Wright, 1998; Thompson and Geary, 2002). The principal limitation is the inability to isolate and analyse fluid excretion via the alimentary canal or from any other route (Thompson and Geary, 2002).

7.4.1. Osmotic and ionic regulation

Most nematodes examined have the ability to regulate their volume (Table 7.1) and ionic content to some extent but relatively little is known about the precise homeostatic mechanisms. The intestine, the body wall, including various epidermal glands, and the 'secretory–excretory' system have all been suggested as sites of urine production in nematodes capable of volume regulation in hypo-osmotic environments (Wright and Newall, 1980; Wright, 1998; Thompson and Geary, 2002).

Two basic mechanisms are involved in the regulation of body volume and the maintenance of internal ionic stability in animals: iso-osmotic intracellular regulation, where the intracellular osmolarity is adjusted (often by changes in the intracellular pool of amino acids) to conform with the extracellular osmotic pressure; and aniso-osmotic extracellular regulation, where the extracellular fluid is maintained hypo- or hyperosmotic to the external environment. There is no direct evidence for the former mechanism in nematodes, although it might at least be expected to occur among euryhaline, brackish-water species (Wright and Newall, 1980). The latter mechanism, which can involve a non-ionic component, complements iso-osmotic intracellular regulation in some animals and in nematodes it has been demonstrated in *P. decipiens* (Fusé et al., 1993a), *Ascaris lumbricoides* and, more tentatively, *Angusticaecum* sp. (Wright and Newall, 1976, for review).

7.4.1.1. Cuticle and epidermis

The nematode cuticle is an extracellular matrix comprised mainly of cross-linked collagen-like proteins, which protects nematodes against adverse conditions, including host defence reactions in animal-parasitic species. Passive factors associated with osmotic and ionic regulation are likely to involve the permeability characteristics of the nematode cuticle or the underlying epidermal membrane. Where estimates of the permeability of nematodes to water have been made, they show a positive relationship with the osmotic pressure of the normal environment; marine species have permeability values about two orders of magnitude greater than those of freshwater or soil species (Wright and Newall, 1976; Wright, 1998). In the animal parasite *P. decipiens*, which is capable of at least short-term hyper- and hypo-osmotic regulation (Section 7.2.1.3), the principal site for fluid removal is thought to be the body wall (Fusé *et al.*, 1993a).

The outer, epicuticle appears to act as a hydrophobic barrier (Blaxter and Robertson, 1998) and this may be an important permeability barrier to water in some species, particularly those living in very hypo-osmotic environments. For example, a reduction in the ability to withstand osmotic stress with increasing age in *Caenorhabditis briggsae* has been linked with changes in the epicuticle and increased permeability to water (Searcy *et al.*, 1976).

In the human parasite *A. suum* and in some other species, water-filled pores in the cuticle provide access to the outer membrane of the epidermis, and this structure rather than the epicuticle may be the true limiting permeability barrier (Ho *et al.*, 1990; Thompson and Geary, 2002). In *A. suum*, both the outer and inner epidermal membranes contain large conductance channels that transport Cl⁻ and organic anions (Blair *et al.*, 2003). In the plant parasite *Xiphinema index*, invaginations of the outer epidermal membrane, which have continuity with the endoplasmic reticulum, have also suggested a transport function (Roggen *et al.*, 1967).

Electron microscope studies on adenophoren species have identified various structures within the epidermis that may have a transport function. In three animal-parasitic genera within the family Trichuroidea, epidermal gland cells (termed 'bacillar bands') closely resemble cells involved in osmotic and ionic regulation in arthropods and vertebrates (Wright and Chan, 1973). Similar cells have been found in the free-living species *Acanthonchus duplicatus* (Wright and Hope, 1968).

7.4.1.2. Pseudocoelom

The pseudocoelom is the principal extracelluler fluid compartment in nematodes although little is known of its ionic and non-ionic composition apart from in a few large animal-parasitic species (Wright and Newall, 1976; Thompson and Geary, 2002). In *P. decipiens*, uptake studies with [³H]₂O have provided evidence that the pseudocoelomic fluid has two compartments, and that exchange of water between these compartments is limited and slow (Fusé *et al.*, 1993b; Davey, 1995). Such properties within the pseudocoelom may account for the fast and slow compartments reported for *E. brevis* in a kinetic analysis of radioactive Na⁺ release (Wright and Newall, 1976).

7.4.1.3. Intestine

The tubular intestine lies within the pseudocoelom and consists of a single layer of epithelial cells. It connects the muscular pharynx with the anus and is closed by a sphincter between the intestine and the rectum. The intestine has no muscles attached and food is moved primarily as a result of body movements. In most species food is also pumped into the intestine by the muscular pharynx against the higher internal pressure in the pseudocoelom. The anus is kept closed by this pressure. A dilator muscle pulls the anus open, and the rectum contents are expelled under pressure.

Caenorhabditis elegans normally defecates at about 45 s intervals and the control of the defecation rhythm in this species involves a gene encoding a novel ion channel of the degenerin/epithelial Na^+ channel superfamily, which is only expressed in the intestine (Take-uchi *et al.*, 1998). While mutations of the gene encoding this channel affected defecation, no apparent effect on osmoregulation was observed.

Most nematodes retain a functioning gut (Wright and Newall, 1976; Thompson and Geary, 2002) and the regular removal of material by defecation in actively feeding nematodes suggests that the intestine has an important role in fluid excretion. There is evidence in support of this for two free-living species, *Rhabditis terrestris* (Stephenson, 1942) and *E. brevis* (Wright and Newall, 1976), both of which are good hyperosmotic regulators with effective means of removing excess water (Table 7.1). In the non-feeding, survival or infective stages of free-living and parasitic nematodes, or in nematodes without a functional gut, as in some filarial and insect-parasitic species, other routes are likely to be more important. In the 'dauer' J3 stage of *C. elegans* (Section 7.2.1.2), for example, pharyngeal pumping is suppressed and the intestine appears non-functional (Popham and Webster, 1979).

7.4.1.4. 'Secretory–excretory' system

Structures that have been described as 'excretory' are present in most nematodes (Bird and Bird, 1991). There are two basic types of systems, a glandular type characteristic of the Adenophora, and a tubular type characteristic of the Secernentea. 'Secretory–excretory' (Nelson *et al.*, 1983) is the most widely used term for such structures, particularly for the tubular type.

The glandular type consists of a single large cell called the 'ventral gland' or 'renette cell'. This cell usually lies within the pseudocoelom and is connected to an 'excretory pore' by a duct that terminates at its external end in a contractile region called the 'ampulla'. The tubular type varies in its precise form between species, with a longitudinal canal running within, or near, one (tylenchid, anisakid) or both (e.g. *Ascaris*) of the lateral epidermal cords (Wright and Newall, 1976), and a transverse canal that connects the longitudinal canal(s) to an excretory pore via a medial duct with a distal ampulla. In some (rhabditid) species, a pair of ventral glands is connected to the canal system. The walls of the canals are lined with microvilli, and in *A. suum* and *C. elegans* the canals are in direct contact with the pseudocoelomic fluid along most of their length (Thompson and Geary, 2002). A close association

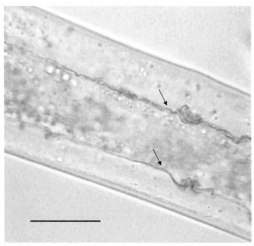

Fig. 7.2. Lateral 'secretory–excretory canals' (arrows) in mid-region of an adult female *Panagrellus redivivus* stained with the redox dye, 1 napthol-2-sodium sulphonate indophenol; bar line = 10 μm. (Adapted from Smith, 1965.)

between the lateral canals and the pseudocoelom has also been found in *P. redivivus* (Smith, 1965; Fig. 7.2).

The tubular type of secretory–excretory system appears to have an important role in the removal of excess fluid in several secernentean species, particularly in non-feeding stages. There are two main types of evidence: (i) the rate of pulsation of the 'excretory ampulla' (Section 7.4.1) is inversely related to the osmotic pressure of the incubation medium in the infective J3 stages of six animal-parasitic species and in the 'dauer' stage of *C. elegans* (Wright, 1998; Thompson and Geary, 2002, for reviews); and (ii) *C. elegans* placed in hypo-osmotic media following laser ablation of their (secretory–excretory) duct, pore or gland cell show an uncontrolled increase in body volume and die within a few days (Nelson and Riddle, 1984).

The 'excretory' canal system in *A. suum* accumulates dyes injected in the pseudocoelomic fluid, and ultrastructural and radio-isotope studies on *A. lumbricoides* suggest that fluid enters the canal system by a secretory process, as in the insect Malpighian tubule (Lee and Atkinson, 1976; Wright and Newall, 1976).

7.4.1.5. Water and ion transport

The ionic composition of the extracellular fluid in several large animal-parasitic species implies that regulation of specific ions must take place (Wright and Newall, 1976). However, in contrast to studies on the ion channels and pumps that maintain electrochemical gradients across nematode muscle and nerve membranes, there has been little work on similar functions in nematode epidermal and intestinal cells (Thompson and Geary, 2002).

The differential permeability of isolated segments of the body wall of *A. suum* to various inorganic and organic ions suggests that active transport is involved in the maintenance of epidermal electrochemical gradients (Pax *et al.*, 1995). This assumption is supported by the presence of large anion conductance channels in the epidermal membranes of this species (Section 7.4.1.1). Metabolic poisons, including cyanide, have been shown to abolish the ability of *P. decipiens* to osmoregulate, providing indirect evidence for the active transport of ions in this species (Fusé *et al.*, 1993a). There is more direct evidence for the active transport of Na^+ and glucose across the intestine of *A. suum*, although it is uncertain whether an Na^+/K^+-adenosine triphosphatase (ATPase) pump is involved (Thompson and Geary, 2002, for review). Na^+/K^+-ATPases are present in *C. elegans*, one of which is involved in pharyngeal pumping (Davis *et al.*, 1995).

The mechanism(s) involved in water transport in nematodes are unknown but in annelids, insects and vertebrates epithelial Na^+ channels actively transport Na^+, with water following passively. The novel degenerin-like epithelial Na^+ channel found to be constitutively expressed in the intestine of *C. elegans* does not appear to be involved in osmoregulation (Section 7.4.1.3). Studies on the expression and function of other genes encoding Na^+ channels that have been identified in *C. elegans* are required (Thompson and Geary, 2002).

Water transport across eukaryote membranes occurs via water channels formed by transmembrane proteins called aquaporins (Zardoya and Villalba, 2001), which belong to the major intrinsic protein (MIP) superfamily. The *C. elegans* genome project has identified eight MIP complementary DNAs (cDNAs); two have been cloned, expressed in *Xenopus* oocytes and confirmed as aquaporins (designated AQP-CE1 and AQP-CE2; Kuwahara *et al.*, 1998, 2000). AQP-CE1 and AQP-CE2 increase the osmotic water permeability of *Xenopus* oocytes by 11- and 10-fold, respectively. AQP-CE1 also increases permeability to urea (4.5-fold) but this channel is only expressed in the embryonic stages of *C. elegans* (Kuwahara *et al.*, 1998). An aquaporin-like gene has also been cloned from the animal-parasitic species *Toxocara canis*, *Brugia pahangi* and *Onchocerca volvulus* (Loukas *et al.*, 1999).

In *P. decipiens* there is indirect evidence that osmoregulation is under the control of diuretic and antidiuretic hormones (Davey *et al.*, 1993). In vertebrates and insects, various neuropeptides are involved in the regulation of ion and water transport. However, to date no candidate peptides have been identified through sequence homology in the *C. elegans* genome database (Thompson and Geary, 2002).

7.4.2. Nitrogen excretion

The relative importance of the intestine, secretory–excretory system and body wall in the removal of ammonia and other nitrogenous end-products will vary from species to species and sometimes from stage to stage of a particular species. This is likely to correspond closely with the relative importance of these structures for the removal of water (Section 7.4.1). A number of ovoid or branched cells (pseudocoelomocytes) have been observed in fixed positions within the pseudocoelom of some nematodes, and excretory/storage, respiratory and scavenging roles have been suggested in different species (Wright, 1998, for review). However, it is unlikely that these cells have a significant role as stores of nitrogenous waste products (Section 7.1).

Most nematodes have a relatively small radius (< 10–20 μm) and a large surface area to volume ratio, with virtually all body cells no more than one cell layer from the external environment and the alimentary canal (Section 7.4). This makes it likely that the body wall and, where functional, the alimentary canal are major routes for the removal of freely diffusible nitrogenous waste products (Wright and Newall, 1976; Thompson and Geary, 2002).

At physiological pH, nearly all ammonia present will be in its charged state (NH_4^+). However, it is likely that most ammonia is excreted by nematodes in its non-ionized form (NH_3), which, unlike NH_4^+, is freely diffusible across membranes, thus driving the conversion of NH_4^+ to NH_3 (Wright and Newall, 1976; Thompson and Geary, 2002). Transport of NH_4^+ across membranes requires a channel or transporter protein and, while a putative ammonium transporter has been identified by sequence homology in the *C. elegans* genome (Marini *et al.*, 1994), nothing is known of its expression or function in nematodes (Thompson and Geary, 2002). The significance of urea excretion in nematodes is uncertain but it is of interest to note that some vertebrate aquaporins (Ishibashi *et al.*, 1994) and at least one nematode homologue (Section 7.4.1.5) can also transport urea.

7.5. Summary

Thompson and Geary (2002) provide an excellent account of the pivotal role of *C. elegans* genomics and proteomics in improving our understanding of homeostatic regulation and secretory–excretory processes in nematodes. Stage- and tissue-specific expression studies combined with reverse genetics (loss of gene product) approaches are particularly useful tools that can help to define the functional role(s) of particular genes. Studies on membrane channels, transporters and pumps in nematodes of medical or veterinary importance may, for example, identify new putative targets for anthelmintics. However, as Thompson and Geary (2002) point out, studies on other nematode species, particularly animal parasites, are not necessarily as straightforward as in *C. elegans*.

Ecological studies are also required. Some species or stages of nematodes, particularly those exposed to wide variations in the osmolarity of their natural environments, are efficient osmoregulators under both hypo- and hyperosmotic conditions. Other species or stages have more limited degrees of volume regulation or, in a few cases, appear to osmoconform under all conditions. Further work on the comparative osmoregulatory abilities of nematodes in assemblages from low- to high-saline environments would be of particular interest and would help improve our understanding of the ecology of this dominant group within the meiobenthic fauna.

References

Abebe, E., Mees, J. and Coomans, A. (2001) Nematode communities of Lake Tana and other inland water bodies of Ethiopia. *Hydrobiologia* 462, 41–73.

Atkinson, H.J. and Onwuliri, C.O.E. (1981) *Nippostrongylus brasiliensis* and *Haemonchus contortus*: function of the excretory ampulla of the third stage larva. *Experimental Parasitology* 52, 191–198.

Bargmann, C.I. and Mori, I. (1997) Chemotaxis and thermotaxis. In: Riddle, D.L., Blumenthal, T., Meyer, B.J. and Priess, J.R. (eds) *C. elegans II*. Cold Spring Harbor Laboratory Press, Cold Spring Harbor, New York, pp. 717–737.

Barrett, J. and Wright, D.J. (1998) Intermediary metabolism. In: Perry, R.N. and Wright, D.J. (eds) *The Physiology and Biochemistry of Free-living and Plant-parasitic Nematodes*. CAB International, Wallingford, UK, pp. 331–353.

Bednarek, A. and Gaugler, R. (1997) Compatibility of soil amendments with entomopathogenic nematodes. *Journal of Nematology* 29, 220–227.

Bird, A.F. and Bird, J. (1991) *The Structure of Nematodes*, 2nd edn. Academic Press, New York, 316 pp.

Bird, A.F. and Bird, J. (1998) Introduction to functional organization. In: Perry, R.N. and Wright, D.J. (eds) *The Physiology and Biochemistry of Free-living and Plant-parasitic Nematodes*. CAB International, Wallingford, UK, pp. 1–24.

Blair, K.L., Geary, T.G., Mensch, S.K., Vidmar, T.J., Li, S.K., Ho, N.F.H. and Thompson, D.P. (2003) Biophysical characterization of a large conductance anion channel in hypodermal membranes of the gastrointestinal nematode, *Ascaris suum*. *Comparative Biochemistry and Physiology A – Molecular and Integrative Physiology* 134, 805–818.

Blaxter, M.L. and Robertson, W.M. (1998) The cuticle. In: Perry, R.N. and Wright, D.J. (eds) *The Physiology and Biochemistry of Free-living and Plant-parasitic Nematodes*. CAB International, Wallingford, UK, pp. 25–48.

Bloemers, G.F., Hodda, M., Lambshead, P.J.D., Lawton, J.H. and Wanless, F.R. (1997) The effects of forest disturbance on diversity of tropical soil nematodes. *Oecologia* 111, 575–582.

Brown, C.J., Lambshead, P.J.D., Smith, C.R., Hawkins, L.E. and Farley, R. (2001) Phytodetritus and the abundance and biomass of abyssal nematodes in the central, equatorial Pacific. *Deep-Sea Research Part 1 – Oceanographic Research Papers* 48, 555–565.

Clarke, A.J., Perry, R.N. and Hennessy, J. (1978) Osmotic stress and the hatching of *Globodera rostochiensis*. *Nematologica* 24, 384–392.

Croll, N.A. and Viglierchio, D.R. (1969) Osmoregulation and the uptake of ions in a marine nematode. *Proceedings of the Helminthological Society of Washington* 36, 1–9.

Davey, K.G. (1995) Water, water compartments and water regulation in some nematodes parasitic in vertebrates. *Journal of Nematology* 27, 433–440.

Davey, K.G., Sommerville, R.I. and Fusé, M. (1993) Stress-induced failure of osmoregulation in the parasitic nematode *Pseudoterranova decipiens*: indirect evidence for hormonal regulation. *Journal of Experimental Biology* 180, 263–272.

Davis, M.W., Somerville, D., Lee, R.Y.N., Lockery, S., Avery, L. and Fambrough, D.M. (1995) Mutations in the *Caenorhabditis elegans* Na,K-ATPase 1–subunit gene, *eat-6*, disrupt excitable cell function. *Neuroscience* 15, 8408–8418.

Ekschmitt, K., Bakonyi, G., Bongers, M., Bongers, T., Bostrom, S., Dogan, H., Harrison, A., Kallimanis, A., Nagy, P., O'Donnell, A.G., Sohlenius, B., Stamou, G.P. and Wolters, V. (1999) Effects of the nematofauna on microbial energy and matter transformation rates in European grassland soils. *Plant and Soil* 212, 45–61.

Finnegan, M.M., Downes, M.J., O'Regan, M. and Griffin, C.T. (1999) Effect of salt and temperature stresses on survival and infectivity of *Heterorhabditis* spp. IJs. *Nematology* 1, 69–78.

Forster, S.J. (1998) Osmotic stress tolerance and osmoregulation of intertidal and subtidal nematodes. *Journal of Experimental Marine Biology and Ecology* 224, 109–125.

Fusé, M., Davey, K.G. and Sommerville, R.I. (1993a) Osmoregulation in the parasitic nematode *Pseudoterranova decipiens. Journal of Experimental Biology* 175, 127–142.

Fusé, M., Davey, K.G. and Sommerville, R.I. (1993b) Water compartments and osmoregulation in the parasitic nematode *Pseudoterranova decipiens. Journal of Experimental Biology* 175, 143–152.

Grantham, B.D. and Barrett, J. (1986) Amino acid catabolism in the nematodes *Heligmosomoides polygyrus* and *Panagrellus redivivus*. I. Removal of the amino group. *Parasitology* 93, 481–493.

Grewal, P.S., Gaugler, R. and Selvan, S. (1993) Host recognition by entomopathogenic nematodes: behavioral response to contact with host feces. *Journal of Chemical Ecology* 19, 1219–1231.

Hara, A.H., Gaugler, R., Kaya, H.K. and LeBeck, L.M. (1991) Natural populations of entomopathogenic nematodes (Rhabditida, Heterorhabditidae, Steinernematidae) from the Hawaiian Islands. *Environmental Entomology* 20, 211–216.

Ho, N.F.H., Geary, T.G., Raub, T.J., Barsuhn, C.L. and Thompson, D.P. (1990) Biophysical transport properties of the cuticle of *Ascaris suum. Molecular and Biochemical Parasitology* 41, 153–166.

Hodgkin, J. (1997) Appendix 1. Genetics. In: Riddle, D.L., Blumenthal, T., Meyer, B.J. and Priess, J.R. (eds) *C. elegans II*. Cold Spring Harbor Laboratory Press, Cold Spring Harbor, New York, pp. 881–1047.

Ishibashi, K., Sasaki, S., Fushimi, K., Uchida, S., Kuwahara, M., Saito, H., Furukawa, T., Nakajima, K., Yamaguchi, Y., Gojobori, T. and Marumo, F. (1994) Molecular cloning and expression of a member of the aquaporin family with permeability to glycerol and urea in addition to water expressed at the basolateral membrane of kidney collecting ducts. *Proceedings of the National Academy of Sciences, USA* 91, 6269–6273.

Jansen, G., Weinkove, D. and Plasterk, R.H.A. (2002) The G-protein gamma subunit gpc-1 of the nematode *C. elegans* is involved in taste adaption. *EMBO Journal* 21, 986–994.

Karim, Z., Attmane-Elakeb, A. and Bichara, M. (2002) Renal handling of NH_4^+ in relation to the control of acid-base balance by the kidney. *Journal of Nephrology* 15, S128–S134.

Kirk, R.S., Morritt, D., Lewis, J.W. and Kennedy, C.R. (2002) The osmotic relationship of the swimbladder nematode *Anguilla crassus* with seawater eels. *Parasitology* 124, 339–347.

Kuwahara, M., Ishibashi, K., Gu, Y., Terada, Y., Kohara, Y., Marumo, F. and Sasaki, S. (1998) A water channel of the nematode *Caenorhabditis elegans* and its implications for channel selectivity of MIP proteins. *American Journal of Physiology – Cell Physiology* 275, C1459–C1464.

Kuwahara, M., Asai, T., Sata, K., Shimbo, I., Terada, Y., Marumo, F. and Sasaki, S. (2000) Functional characterization of a water channel of the nematode *Caenorhabditis elegans. Biochimica et Biophysica Acta – Gene Structure and Expression* 1517, 107–112.

Lee, D.L. and Atkinson, H.J. (1976) *The Physiology of Nematodes*, 2nd edn. Macmillan, London, 215 pp.

Le Saux, R. and Queneherve, P. (2002) Differential responses of two plant-parasitic nematodes, *Meloidogyne incognita* and *Rotylenchulus reniformis*, to some inorganic ions. *Nematology* 4, 99–105.

Loukas, A., Hunt, P. and Maizels, R.M. (1999) Cloning and expression of an aquaporin-like gene from a parasitic nematode. *Molecular and Biochemical Parasitology* 99, 287–293.

McKerrow, J.H., Huima, T. and Lustigman, S. (1999). Do filarid nematodes have a vascular system? *Parasitology Today* 15, 123.

Marini, A.M., Vissers, S., Urrestarazu, A. and Andre, B. (1994) Cloning and expression of the MEP1 gene encoding an ammonium transporter in *Saccharomyces cerevisiae. EMBO Journal* 13, 3456–3463.

Mellem, J.E., Brockie, P.J., Zheng, Y., Madsen, D.M. and Maricq, A.V. (2002) Decoding of polymodal sensory stimuli by postsynaptic glutamate receptors in *C. elegans. Neuron* 36, 933–944.

Myers, R.F. (1966) Osmoregulation in *Panagrellus redivivus* and *Aphelenchus avenae. Nematologica* 12, 579–586.

Nelson, F.K. and Riddle, D.L. (1984) Functional study of the *Caenorhabditis elegans* secretory–excretory system using laser microsurgery. *Journal of Experimental Zoology* 231, 45–56.

Nelson, F.K., Albert, P.S. and Riddle, D.L. (1983) Fine structure of the *Caenorhabditis elegans* secretory–excretory system. *Journal of Ultrastructural Research* 82, 156–171.

Olafsson, E. (1995) Meiobenthos in mangrove areas of East Africa with emphasis on assemblage structure of free-living marine nematodes. *Hydrobiologia* 312, 47–57.

Olafsson, E. and Elmgren, R. (1997) Seasonal dynamics of sublittoral meiobenthos in relation to phytoplankton sedimentation in the Baltic Sea. *Estuarine Coastal and Shelf Science* 45, 149–164.

Patel, M.N., Stolinski, M. and Wright, D.J. (1997) Neutral lipids and the assessment of infectivity in entomopathogenic nematodes: observations on four *Steinernema* species. *Parasitology* 114, 489–496.

Pax, R.A., Geary, T.G., Bennett, J.L. and Thompson, D.P. (1995) *Ascaris suum* – characterization of transmural and hypodermal potentials. *Experimental Parasitology* 80, 85–97.

Perry, R.N. and Aumann, J. (1998) Behaviour and sensory responses. In: Perry, R.N. and Wright, D.J. (eds) *The Physiology and Biochemistry of Free-living and Plant-parasitic Nematodes*. CAB International, Wallingford, UK, pp. 75–102.

Perry, R.N., Clarke, A.J. and Hennessy, J. (1980) The influence of osmotic pressure on the hatching of *Heterodera schachtii. Revue de Nématalogie* 3, 3–9.

Piggott, S.J., Perry, R.N. and Wright, D.J. (2000) Hypo-osmotic regulation in entomopathogenic nematodes: *Steinernema* spp. and *Heterorhabditis* spp. *Nematology* 2, 561–566.

Piggott, S.J., Liu, Q.-Z., Glazer, I. and Wright, D.J. (2002) Does osmoregulatory behaviour in entomopathogenic nematodes predispose desiccation tolerance? *Nematology* 4, 483–487.

Popham, J.D. and Webster, J.M. (1979) Aspects of the fine structure of the dauer larva of *Caenorhabditis elegans. Canadian Journal of Zoology* 57, 794–800.

Prencepe, A., Bianco, M., Viglierchio, D.R. and Scognamiglio, A. (1984) Response of the nematode *Panagrellus silusiae* to hypotonic solutions. *Proceedings of the Helminthological Society of Washington* 51, 36–41.

Prot, J.C. (1978) Influence of concentrations of salts on the movement of second stage juveniles of *Meloidogyne javanica. Revue de Nématologie* 1, 12–26.

Randall, D., Burggren, W. and French, K. (2001) *Eckert Animal Physiology*, 5th edn. W.H. Freeman, New York, 120 pp.

Rankin, C.H. and Wicks, S.R. (2000) Mutations of the *Caenorhabditis elegans* brain-specific inorganic phosphate transporter eat-4 affect habituation of the tap–withdrawal response without affecting the response itself. *Journal of Neuroscience* 20, 4337–4344.

Robinson, A.F., Orr, C.C. and Heintz, C.E. (1984) Effects of the ionic composition and water potential of aqueous solutions on the activity and survival of *Orrina phyllobia*. *Journal of Nematology* 16, 26–30.

Roggen, D.R., Raski, D.J. and Jones, N.O. (1967) Further electron microscope observations on *Xiphinema index*. *Nematologica* 13, 1–16.

Searcy, D.G., Kisiel, M.J. and Zuckerman, B.M. (1976) Age-related increase in cuticle permeability in the nematode *Caenorhabditis briggsae*. *Experimental Aging Research* 2, 293–301.

Shakir, M.A., Miwa, J. and Siddiqui, S.S. (1993) *C. elegans osm-3* gene mediating osmotic avoidance behaviour encodes a kinesin-like protein. *Neuroreport* 4, 891–894.

Shapiro, D.I., Lewis, E.E., Paramasivam, S. and McCoy, C.W. (2000) Nitrogen partitioning in *Heterorhabditis bacteriophora*-infected hosts and the effects of nitrogen on attraction/repulsion. *Journal of Invertebrate Pathology* 76, 43–48.

Smith, L. (1965) The excretory system of *Panagrellus redivivus* (T. Goodey, 1945). *Comparative Biochemistry and Physiology* 15, 89–92.

Soetaert, K., Vincx, M., Wittoeck, J. and Tulkens, M. (1995) Meiobenthic distribution and nematode community structure in 5 European estuaries. *Hydrobiologia* 311, 185–206.

Stephenson, W. (1942) The effect of variations in osmotic pressure upon a free-living soil nematode. *Parasitology* 34, 253–265.

Steyaert, M., Herman, P.M.J., Moens, T., Widdows, J. and Vincx, M. (2001) Tidal migration of nematodes on an estuarine tidal flat (the Molenplaat, Schelde Estaury, SW Netherlands). *Marine Ecology – Progress Series* 224, 299–304.

Take-uchi, M., Kawakami, M., Ishihara, T., Amano, T., Kondo, K. and Katsura, I. (1998) An ion channel of the degenerin/epithelial sodium channel superfamily controls the defecation rhythm in *Caenorhabditis elegans*. *Proceedings of the National Academy of Sciences, USA* 95, 11775–11780.

Thompson, D.P. and Geary, T.G. (2002) Excretion/secretion, ionic and osmotic regulation. In: Lee, D.L. (ed.) *The Biology of Nematodes*. Taylor and Francis, London, pp. 291–320.

Thurston, G.S., Yansong, N. and Kaya, H.K. (1994) Influence of salinity on survival and infectivity of entomopathogenic nematodes. *Journal of Nematology* 26, 345–351.

Tobata-Kudo, H., Higo, H., Koga, M. and Tada, I. (2000) Chemokinetic behaviour of the infective third-stage larvae of *Strongyloides ratti* on a sodium chloride gradient. *Parasitology International* 49, 183–188.

Viglierchio, D.R. (1974) Osmoregulation and electrolyte uptake in Antarctic nematodes. *Transactions of the American Microscopical Society* 93, 325–338.

Warwick, R.M., Dexter, D.M. and Kuperman, B. (2002) Free-living nematodes from the Salton Sea. *Hydrobiologia* 473, 121–128.

Wharton, D.A. (1986) *The Functional Biology of Nematodes*. Johns Hopkins University Press, Baltimore, Maryland, 192 pp.

Wharton, D.A., Perry, R.N. and Beane, J. (1983) The effect of osmotic stress on behaviour and water content of infective larvae of *Trichostrongylus colubriformis*. *International Journal for Parasitology* 13, 185–190.

Wright, D.J. (1975a) Elimination of nitrogenous compounds by *Panagrellus redivivus* Goodey, 1945 (Nematoda: Cephalobidae). *Comparative Biochemistry and Physiology* 52B, 247–253.

Wright, D.J. (1975b) Studies on nitrogen catabolism in *Panagrellus redivivus* Goodey, 1945 (Nematoda: Cephalobidae). *Comparative Biochemistry and Physiology* 52B, 255–260.

Wright, D.J. (1998) Respiratory physiology, nitrogen excretion and osmotic and ionic regulation. In: Perry, R.N. and Wright, D.J. (eds) *The Physiology and Biochemistry of Free-living and Plant-parasitic Nematodes.* CAB International, Wallingford, UK, pp. 103–131.

Wright, D.J. and Newall, D.R. (1976) Nitrogen excretion, osmotic and ionic regulation in nematodes. In: Croll, N.A. (ed.) *The Organization of Nematodes.* Academic Press, London, pp. 163–210.

Wright, D.J. and Newall, D.R. (1980) Osmotic and ionic regulation in nematodes. In: Zuckerman, B.M. (ed.) *Nematodes as Biological Models*, Vol. 2. Academic Press, New York, pp. 143–164.

Wright, K.A. and Chan, J. (1973) Sense receptors in the bacillary band of trichuroid nematodes. *Tissue and Cell* 5, 373–380.

Wright, K.A. and Hope, W.D. (1968) Elaborations of the cuticle of *Acanthonchus duplicatus* Wieser, 1959 (Nematoda: Cyatholaimidae) as revealed by light and electron microscopy. *Canadian Journal of Zoology* 46, 1005–1011.

Wright, P.A. (1995) Nitrogen excretion – three end-products, many physiological roles. *Journal of Experimental Biology* 198, 273–281.

Zardoya, R. and Villalba, S. (2001) A phylogenetic framework for the Aquaporin family in eukaryotes. *Journal of Molecular Evolution* 52, 391–404.

8

Physiological and Biochemical Basis of Behaviour

ROLAND N. PERRY[1] AND AARON G. MAULE[2]

[1]Plant Pathogen Interactions Division, Rothamsted Research, Harpenden, Hertfordshire AL5 2JQ, UK; [2]Parasitology Research Group, School of Biology and Biochemistry, Queen's University Belfast, Medical Biology Centre, 97 Lisburn Road, Belfast BT9 7BL, UK

8.1. Introduction

Behavioural activities are the overt expression of integrated physiological and biochemical processes. The functional subtlety and physiological complexities needed to coordinate and integrate nematode behavioural responses are catered for by a rich milieu of diverse intercellular signalling molecules. As with many other aspects of nematode biology, studies on the behaviour of *Caenorhabditis elegans* are far more detailed and extensive than behavioural studies on other species. A review by Hobert (2003), for example, indicates the information available on behavioural paradigms of *C. elegans* and their underlying neuronal circuits and the genes involved. In this chapter the emphasis will be on a broader review of the physiological and biochemical (including some molecular) information that pertains to behavioural responses of parasitic and free-living nematodes.

8.2. The Nervous System and Sensory Physiology

8.2.1. The nervous system

Although structurally simple, the nematode nervous system provides the central conduit between stimulus reception and behavioural output. The basic nerve element, the neurone, is a single nerve cell with typically one (monopolar), two (bipolar) or more elongate processes. A nerve process may be either an axon or a dendrite. An axon is a long process that usually forms a synapse with another type of cell, such as muscle, or with another neurone (interneurone). A dendrite in nematodes typically terminates within a sensillum and, near the tip of the dendrite, the cytoplasm forms microtubules, resembling sensory 'cilia'. Occasionally, an individual neurone may have multiple synapses combining, for example, interneurone, dendrite and/or axon functions (Chalfie and White, 1988). The majority of nerve processes run longitudinally, as ventral and dorsal nerve cords, or circumferentially, as commissures. The nervous system comprises central, peripheral and sympathetic systems and a complex array of sensilla (Baldwin and Perry, 2004). The circumpharyngeal nerve ring is the main mass of the central nervous system and is the site of most sensory integration. The peripheral nervous system includes neurones and small commissures just beneath the cuticle surface. The sympathetic nervous systems include the nematode pharynx and rectal/cloacal (including ovijector/vulva) region, which are semi-independent of the central nervous system. Chalfie and White (1988), Bird and Bird (1991) and Baldwin and Perry (2004) provide detailed descriptions of the structure of the nematode nervous system.

8.2.2. Sensilla

8.2.2.1. Sensilla structure

The ultrastructure of the many different types of sensilla of various parasitic and free-living species has been examined but, except for work done with *C. elegans*, there is much less experimental evidence of sensilla function or the associated physiological and biochemical processes. There are two basic types of sensilla: internal and cuticular. Whereas the internal sensilla are mechanoreceptors or, less frequently, photoreceptors, the cuticular sensilla detect a much greater variety of stimuli, including chemical, mechanical, temperature and osmotic pressure.

The main concentration of cuticular sensilla is at the anterior. Amphids are anterior paired organs, positioned laterally and with external openings, whose structure is fundamentally the same as that of other sensilla (Jones, 2002). An amphid consists of three basic cell types: a glandular sheath cell, a supporting socket cell and a number of dendritic processes. The socket cell surrounds the amphidial duct, which encloses the distal regions of the receptors. The sheath cell is deeply folded, resulting in a very large surface area, and produces secretions that bathe the dendritic processes.

8.2.2.2. Sensilla secretions

The socket cell and sheath cell both produce secretions, which appear to have multiple roles, some of which may be specialized for parasitism. However, the source of secretions in the amphidial duct needs to be more clearly resolved (Baldwin and Perry, 2004). The receptor cavity of the sheath cell appears to be closed and sheath cell secretions are not in direct contact with the amphidial duct. In contrast, there is continuity of secretory vesicles within the socket cell and with material in the duct. Several functions have been suggested for these secretions, including maintainance of electrical continuity between the bases and tips of the dendritic processes (Trett and Perry, 1985) and protection of the dendritic endings of sensory nerve cells against desiccation and microbial attack (Aumann, 1993). It is probable that the proteins in the secretions have an important role in chemoreception, perhaps similar to that of the odorant binding proteins (OBPs) and odorant degrading enzymes that are present in insect sense organs. The possibility that OBPs are present in nematodes was investigated by Jones *et al.* (1992) but convincing evidence that the secretions of nematode sensilla contain the same suite of molecules as found in insect sense organs is still lacking.

Early work, reviewed by Perry and Aumann (1998), used lectins to demonstrate the presence of carbohydrate residues in amphidial secretions of *Heterodera schachtii* and species of *Meloidogyne* and *Globodera*, and some of these residues were constituents of glycoproteins. However, sialic acids, common glycoproteins of vertebrates and higher invertebrate taxa, have not been found in nematodes. Components of amphidial secretions are thought to include *N*-acetylgalactosamine and fucose in second-stage juveniles (J2s) of *Meloidogyne incognita* and *O*-glycans in J2s of *H. schachtii*, with mannose or glucose, *N*-acetylglucosamine and galactose and/or *N*-acetylgalactosamine forming the oligosaccharide chains.

Several blood-feeding animal-parasitic nematodes secrete anticoagulants from the amphids. Stanssens *et al.* (1996) have described a family of small proteins from the hookworm *Ancylostoma caninum* that inhibit coagulation of host blood. Amphidial secretions from many animal-parasitic nematodes contain acetylcholinesterase and Pritchard (1993) suggested various functions for secretory cholinesterases. Cholinesterases may have a role as a biochemical holdfast, enabling some species to maintain position in the host's gut, and they may alter the permeability of the host membranes to facilitate leakage of nutrients. Cholinesterase has also been detected in plant-parasitic nematodes and *C. elegans*, and in other sensilla in addition to amphids, so this enzyme may have a more general role in chemoreception in addition to specialized functions in some species. Secretions from posterior supplementary sensilla are likely to function as cement in copulation.

In *Haemonchus contortus*, amphids differ between the first and third stages (Li *et al.*, 2001) and it is clear from research on several species of nematodes that at separate stages of the life cycle sensilla, and in particular the amphids, may have different functions or a different combination of functions. Davis *et al.* (1992) found differences in the composition of amphidial secretions between J2s and females of *M. incognita*. Stewart *et al.* (1993a) found a 32 kDa glycoprotein in the amphidial secretions and the sheath cell of J2s of six species of *Meloidogyne* but it was not present in representatives from eight other genera, which included *Globodera* and *Heterodera*, indicating a more specialized function for this protein in *Meloidogyne*. The glycoprotein was found in all active stages of the life cycle of *Meloidogyne javanica* but not in the sedentary adult female, where the amphids appear to be non-functional. Incubation of infective *M. javanica* J2s in the antiserum against the protein significantly retarded chemotaxis (Stewart *et al.*, 1993b). Fioretti *et al.* (2002) found that antibodies recognizing the cuticle surface and the amphids of *Globodera pallida* adversely affected nematode movement and delayed nematode penetration of roots. However, in both these studies sensory perception was not permanently blocked and the period that nematodes remain desensitized to specific stimuli before, presumably, turnover of secretions 'unblocks' the amphids is between 0.5 and 1.5 h.

8.2.3. Chemoreception

Chemicals that cause interactions between organisms are called semiochemicals, a term that includes allelochemicals, mediating interspecific responses, and pheromones, mediating intraspecific responses. For intestinal animal-parasitic nematodes, non-volatiles may be of importance for mate finding and food acquisition. For plant-endoparasitic nematodes, non-volatiles are also central to certain aspects of the life cycle such as movement within the host and selection of the feeding site. However, volatile components are important for long distance orientation of infective stages of plant-parasitic nematodes to their hosts (Robinson, 2002) and have been implicated in the location of insect hosts by entomopathogenic nematodes (Wright and Perry, 2002). Host triggers and chemotactic cues that control the migration and development of plant-parasitic nematodes within the host are little understood. Animal-parasitic nematodes show mate-finding responses involving sex pheromones but such orientated responses have not been demonstrated in habitat selection of intestinal parasites (Sukhdeo *et al.*, 2002). The cues that animal-parasitic and plant-parasitic nematodes use to sense a host immune response (Viney, 2002) and a host resistant response (Atkinson, 2002), respectively, require detailed investigation. Research on learning and memory in nematodes is largely confined to *C. elegans* (reviewed by Hobert, 2003), although there is evidence that a plant-parasitic nematode, *H. schachtii*, was more successful at finding a host after previous contact with host root allelochemicals (Clemens *et al.*, 1994).

The use of electrophysiology to examine the chemosensory perception of live, intact nematodes has provided a detailed analysis of responses (Perry, 2001). Adults of the filarial parasite *Brugia pahangi* did not respond to haemoglobin and it is likely that haemoglobin is not involved as a host cue or feedant for adults. In contrast, exposure of *B. pahangi* to glutathione (a tripeptide found widely in animals and plants) and to heat-inactivated fetal calf serum (a culture medium ingredient) elicited significant increases in spike activity. The increase in spike activity induced by heat-inactivated fetal calf serum could be prevented by a prior 30 min exposure of the nematode to a 1 : 1000 dilution of the nematocide ivermectin (Rolfe *et al.*, 2001a). Such blocking of action potentials indicates that one possible mode of action of ivermectin may directly involve chemoreception at the anterior sensilla, presumably in the amphids. However, the main mode of action of ivermectin appears to be through the modulation of ion channels (Rohrer and Arena, 1995); ivermectin could act to inhibit steps within the odorant signal transduction cascade and affect the ion channels on the surface of the nerve cell.

Disruption of movement and chemotaxis is a feature of most chemical control strategies. However, the nematocide aldicarb and the anthelmintic levamisole impair chemotaxis at concentrations much lower than those required to inhibit locomotion. Exposure of adult females of the plant-parasitic nematode, *Pratylenchus penetrans*, to 5 ppm of aldicarb resulted in ultrastructural changes in the amphids, including hypertrophy of the dendrite terminals within the sheath cell and large granules in the sheath cell cytoplasm (Trett and Perry, 1985). Winter *et al.* (2002) proposed an uptake pathway for aldicarb, levamisole, peptide mimetics and other soluble molecules by retrograde transport along chemosensory dendrites to their site of action at cell bodies and synapses. These authors suggest that this may be a general mechanism for the low-dose effects of some nematocides and anthelmintics.

Analysis of responses of plant-parasitic nematodes to specific plant compounds will provide information that may relate to plant host suitability and the resistant response (Perry, 1996). The electrophysiological technique was used to examine nematode responses to known phagostimulatory compounds for insects. Exposure of J2s of *Globodera rostochiensis* to D-glutamic acid resulted in increased spike activity but there was no response to L-glutamic acid (Rolfe *et al.*, 2000). In insects, the D-isomers of many amino acids usually elicit a phagostimulatory response, while many L-amino acids are feeding deterrents. Glycine is another amino acid that is usually a phagostimulant in insects but the lack of response by J2s of *G. rostochiensis* on exposure to glycine (Rolfe *et al.*, 2000) may correlate with low levels of glycine in root diffusates or within the host plant. Thus, orientation of infective J2s may not be mediated by glycine; alternatively, the threshold concentration, above which J2s respond, may be higher than the 10 mM tested.

Spike activity of J2s of *G. rostochiensis* increased on exposure to potato root diffusate (PRD) (Rolfe *et al.*, 2000; Fig. 8.1) and it is likely that this response is

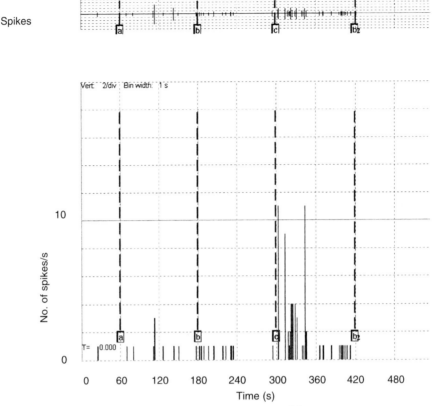

Fig. 8.1. Extracellular recordings from the anterior of an individual J2 of *Globodera rostochiensis* in response to female sex pheromones (a) washed off with artificial tap water (ATW) (b) subsequently exposed to potato root diffusate (c) and washed off with ATW (b₂). (From: Rolfe *et al.*, 2000.)

associated with the role of allelochemicals in PRD in attracting J2s to host roots (Perry, 1997). A monoclonal antibody (mAb) showing immunoreactivity to amphidial secretions of *Globodera* (Fioretti *et al.*, 2002) blocked electrophysiological responses of J2s of *G. rostochiensis* to PRD (Rolfe, 1999). There was no response from J2s to root diffusate from the non-host sugar beet, thus indicating that responses to diffusates may be host-specific. It appears that the allelochemicals in PRD responsible for attracting infective J2s of *G. rostochiensis* and *G. pallida* to host roots are not the same as those known to stimulate hatching (Devine and Jones, 2002). Citric acid stimulated hatching of J2s from cysts of *G. rostochiensis* (Voinilo, 1976) but there is no evidence that citric acid attracts J2s to host roots and J2s and adult males gave no electrophysiological response on exposure to 10 mM citric acid (Riga *et al.*, 1997b; Rolfe *et al.*, 2000).

In contrast to J2s, males of *G. rostochiensis* and *G. pallida* did not respond to PRD (Riga *et al.*, 1996a). Males emerge from the roots and require only sex pheromones to locate the sessile females protruding out of the roots. Of the wide variety of pheromones known to influence behaviour of invertebrates, only sex (Perry and Aumann, 1998) and dauer-inducing pheromones (Riddle and Albert, 1997) have been studied extensively in nematodes. Phasmids, posterior sensilla thought to be chemoreceptors, may be receptors for female sex pheromones (Wang and Chen, 1985). However, this could be a specialized function in only a limited number of species, as phasmids are present on females of many species and, in species where phasmids are absent, other sensilla must assume the pheromone-sensing role. Electrophysiological analysis of the responses of *G. rostochiensis* and *G. pallida* males to sex pheromones from virgin females (Riga *et al.*, 1996b) showed that *G. rostochiensis* males exhibited specific mate recognition, while *G. pallida* males responded to the sex pheromones from females of both species. However, the homospecific response of *G. pallida* was much greater and may be dominant in the soil environment. The sex pheromone of *G. rostochiensis* is composed of several polar compounds that are less polar than PRD and are weakly basic (Riga *et al.*, 1997a); chemical characterization of these compounds and of the two polar compounds in the sex pheromone of *H. schachtii* (Aumann *et al.*, 1998) is required. To date, only one male attractant has been identified for a plant-parasitic nematode: vanillic acid for *Heterodera glycines* (Jaffe *et al.*, 1989). In agar plate behavioural assays, vanillic acid did not attract males of *G. rostochiensis* but possible concentration-dependent responses warrant testing. Greet and Perry (1992) reviewed early work on the identification of sex pheromones of animal-parasitic nematodes.

J2s of *G. rostochiensis* and *G. pallida* showed no change in spike activity in response to sex pheromones (Rolfe *et al.*, 2000; Fig. 8.1). The different responses of J2s and males to PRD and to sex pheromones accord with the priority for J2s to hatch and locate a host, and the priority for adult males to locate and fertilize the females. Amphids of some animal-parasitic nematodes become altered at certain stages of the life cycle in order to fulfil a modified role. The absence of secretory material and the shrunken state of the sheath cell of amphids of unhatched J2s of *G. rostochiensis* (Jones *et al.*, 1994) indicate that the amphids are not functional before hatching and, thus, have no role in detecting allelochemicals involved in the stimulation of hatching (Perry, 2002). After hatching, the ultrastructure of the amphids of *G. rostochiensis* indicated that they were functional but they changed little during subsequent life-cycle stages. Thus, the difference in responses of J2s and males could be mediated by changes in the expression of receptors in the chemosensilla.

The increase in spike activity in *B. pahangi* (Rolfe *et al.*, 2001a), the insect-parasitic nematode *Leidynema appendiculata* (Rolfe *et al.*, 2001b) and *G. rostochiensis* (Rolfe *et al.*, 2000) upon exposure to acetylcholine could indicate a role for acetylcholine in chemosensory neurones, with entry through the cuticle stimulating the nerve ring either directly or via the amphidial pores. Alternatively, it could be the result of a chemotactic response, indicating an acetylcholine receptor in the sensilla, probably the amphids, or a receptor that binds acetylcholine non-specifically. Evidence for the role of acetylcholine in *C. elegans* comes from a variety of experimental approaches (Rand and Nonet, 1997) but, as yet, no neurotransmitter specificity has been assigned to neurones associated with the amphids.

With *B. pahangi*, *L. appendiculata*, *G. rostochiensis* and *G. pallida*, the electrophysiological recordings have been from the cephalic region. With the much larger animal-parasitic nematode *Syngamus trachea*, direct electrophysiological recordings were obtained from the cephalic papillae (Jones *et al.*, 1991) and amphids (Riga *et al.*, 1995) in response to blood, confirming their chemosensory role. Studies on *C. elegans* showed that the amphids are the primary chemosensilla. Mutants of this species with an abnormal response to chemical stimulants had structural defects in the dendritic processes or in the sheath cells (Lewis and Hodgkin, 1977) and laser ablation of individual amphidial neurones altered the nematodes' chemotactic responses (Bargmann *et al.*, 1990). Laser ablation studies and genetic analysis have led to the identification of odorants detected by each of the sensory neurones of *C. elegans* (Bargmann and Mori, 1997). However, it should be remembered that the amphids are not exclusively chemosensilla. The 12 sensory neurones that enter the amphid have been divided into three classes. The eight neurones of the first class, which have dendritic processes that end in the amphidial duct and are exposed directly to the external environment via the amphidial pore, detect water-soluble odorants, including salts, amino acids and the dauer pheromone. The three neurones, called wing cells, of the second class have flattened, branched dendritic processes embedded within the amphidial sheath cell and detect volatile odorants. The third class of neurone, comprising a single neurone called the finger cell, has a sensory ending bearing microvillar projections that are embedded in the sheath cell; this neurone is thought to detect thermal cues (Mori and Ohshima, 1995). Different neurones may detect the same odorant and individual neurones can detect multiple odorants. For example, the ASH neurones (see Section 8.4.3), which have sensory endings exposed to the environment, are polymodal sensory neurones with two chemosensory functions (avoidance of osmotic and volatile repellents) and a mechanosensory function, involving an avoidance response when the nematode touches an object head-on (Bargmann *et al.*, 1990; Kaplan and Horvitz, 1993). Thus, the amphids have chemosensory, mechanosensory and thermosensory roles.

Stimulation of a particular chemosensory neurone of *C. elegans* results in a distinct behavioural response (attraction or avoidance). Thus, the neurones are 'hardwired' and it is the neurone that determines the response, not the nature of the receptor molecule being stimulated (Bargmann and Horvitz, 1991). This was demonstrated by expressing the receptor for diacetyl, an attractant for *C. elegans*, in a sensory neurone that normally detects repellent chemicals; the transgenic nematodes showed a reversal of the wild-type behaviour by avoiding diacetyl (Troemel *et al.*, 1995). Some receptors are expressed in a single sensory neurone while others are present in up to five sensory neurones. The odorant receptors are members of the seven transmembrane domain superfamily of receptors that are linked to G-proteins and referred to as G-protein-coupled receptors (GPCRs; see Section 8.4.4). After binding of signal molecules to their receptors, G-proteins activate enzymes that catalyse the formation of second messengers from inactive precursors; for example, adenylate cyclase catalyses the formation of cyclic adenosine monophosphate (cAMP). These second messengers bind to ion channels, causing them to open, and the resulting ion influx causes voltage-gated ion channels to open, thus generating electrical activity in neurones. However, no data are yet available for the second messenger systems in nematode chemosensilla. In *C. elegans* the G-protein G_o plays a role in different forms of behaviour (Mendel *et al.*, 1995; Ségalat *et al.*, 1995) and the identification of GPCRs as olfactory receptors (Bargmann *et al.*, 1993; Sengupta *et al.*, 1996) implies a role for the second messenger systems in the transduction of olfactory signals.

8.2.4. Mechanoreception

Sensilla specialized for mechanoreception may terminate beneath the cuticle surface without ducts opening to the exterior. However, socket cells of mechanoreceptors may also form a fluid-filled chamber near the surface that might aid in mechanical transmission. Mechanosensory perception is involved in locomotion and responses to external environmental stimuli as well as internal responses associated, for example, with egg-laying and pharyngeal pumping. Sensilla in the tail region are often tactile receptors involved in copulation. In males of *C. elegans*, more than one-fifth of the 381 neurones are sexually specialized, including 18 neurones that each innervate a sensillum of the nine pairs of sensory rays, as well as additional neurones innervating the post-cloacal sensilla, hook sensilla, spicules and spicule musculature as well as interneurones. In addition, each spicule possesses one or more nerves that end at a pore or pores in the tip and are likely to have a chemosensory or mechanosensory function before or during copulation. The sensory rays help in the orientation of the male tail to the hermaphrodite female and the male then locates the approximate position of the vulva using the hook sensillum. The spicules and post-cloacal sensilla are used to locate the vulva precisely (Emmons and Sternberg, 1997). Laser ablation of caudal papillae demonstrated that these sensilla are tactile and essential in mating behaviour and in locating the vulva (White, 1988).

Information on hatching of nematodes (reviewed by Perry, 2002) indicates that mechanosensory perception plays a major role in the process. For example, prior to eclosion of *G. rostochiensis*, the anterior end of the J2 is pressed and/or rubbed against the eggshell, and the stylet starts to probe; subsequent coordinated thrusts cause a regular pattern of perforation of the eggshell, resulting in a slit through which the J2 escapes. The involvement of mechanoreception (as well as chemoreception) may also be inferred from observations on root location and penetration and movement through the roots by plant-parasitic nematodes (Wyss, 2002). However, apart from *C. elegans*, no direct information is available about the functioning of the anterior mechanoreceptors. Mutants of *C. elegans* that are defective in their response to one type of mechanical stimulus do not necessarily respond abnormally to other mechanical stimuli, indicating that different molecular mechanisms are involved in detecting different mechanical stimuli.

Touch receptors, running along the body length, detect gentle mechanical stimuli. Driscoll and Kaplan (1997) proposed a model for the functioning of these receptors in *C. elegans*, which are presynaptic to many other neurones, including motor neurones. Thus, a mechanical stimulus at the anterior end causes backward movement, whereas forward movement is elicited by a stimulus at the posterior end. The model proposes that the cuticle is linked, via proteins, to a movement-sensitive ion channel. Microtubules within the receptor cell may also be linked, via other proteins, to the ion channel. Ion flow could be stimulated by movement of proteins or microtubules. Several genes have been identified that could encode sub-units of a mechanically gated ion channel (Chalfie and Sulston, 1981) and the pro-posed functioning of the touch receptors has been summarized by Jones (2002).

8.2.5. Thermoreception and photoreception

Nematodes can exhibit extremely sensitive thermotaxis (Robinson, 2002). Changes in membrane fluidity may be important as temperature cues, and it is interesting that greatly increased areas of membranes are present on neurones that detect thermal cues. In *C. elegans*, the *tax*-2 and *tax*-4 mutants are defective in their response to several stimuli, including thermal cues, and the disrupted genes encode subunits of cyclic nucleotide-gated ion channels (Komatsu *et al.*, 1996). The cyclic nucleotides appear to be important second messengers in thermoreception. The downstream processing of thermal information is associated with the AIY and AIZ interneurones (Mori and Ohshima, 1995). It is not known whether thermorecep-tive molecules are present in nematodes.

Photoreception is usually mediated either by ocelli, a specialized pair of red/brownish pigmented sensilla that occur on the lateral or dorsolateral sides of the pharynx, or by pigment spots that shade photoreceptive sensory neurones. However, many species of nematodes, including *C. elegans*, that lack both these structures can still respond to visible light (Burr, 1985). Nothing is known of the physiological or molecular mechanisms underlying photoreception.

8.3. Neurotransmitters and Neuromodulators

8.3.1. Mechanism of neurotransmission

A consequence of sensory nerve ending encounters with appropriate stimuli is the release of a neurotransmitter that, following interaction with a specific nerve or muscle-based receptor, triggers a downstream signalling event/cascade that results, either directly or indirectly, in a behavioural change, the response. Neurotransmitters, in their many guises, are the intercellular signalling molecules that shuttle the incoming sensory signal between different neurones and between neurones and muscle cells. They also coordinate normal behavioural activities of the nematode, such as locomotion, which are performed even in the absence of specific external stimuli.

Following the arrival of an electrical signal (nematode neurones do not display all-or-none action potentials but can effectively conduct signals due to high-resistance membranes) and concomitant opening of voltage-gated Ca^{2+} channels, the neurotransmitter is released from membrane-bound secretory vesicles after their fusion with the presynaptic membrane. Once in the synaptic cleft, the neurotransmitter (= ligand) is free to interact with specific membrane-bound receptors on the post-synaptic membrane. In nematodes, as well as normal neuronal synapses, bundles of neurones run in parallel, enabling the formation of multiple synapses that are *en passant* and gap junction connectivity is common (White, 1985; White *et al.*, 1986). Ligand ionotropic receptor interaction leads directly to ion channel opening in the post-synaptic membrane and, if excitatory, the resulting depolarizing potential can be relayed to other cells in a similar manner. Conversely, the ionotropic receptor may regulate anion channels that open to facilitate hyperpolarization and, therefore, a reduced chance of excitation. At any point in this signalling chain, the released neurotransmitter may have selectivity for non-ionotropic (metabotropic) receptors and may initiate an intracellular transduction pathway that allows for signal diversification (may trigger a number of distinct intracellular events) and/or amplification. Individual neurones are highly interconnected locally and display predominantly triangular patterns of connectivity (White, 1985), which will allow for much cross-talk, essential if such a simple nervous system is to control the complex behavioural patterns that are seen in nematodes.

Further plasticity in the behavioural response that is generated is provided for by neuromodulators, intercellular signalling molecules that modify or regulate the activities of the pre- and/or post-synaptic cell. These neuromodulators can up- or down-regulate components in the signalling pathway that ultimately influence the end-point response. The following subsections will examine the neurotransmitters and neuromodulators of nematodes and how they influence behavioural parameters.

8.3.2. Acetylcholine

A considerable body of evidence has amassed to support the role of acetylcholine (ACh) as the main excitatory neurotransmitter in nematodes (Rand and Nonet, 1997; Martin *et al.*, 2002). Ionotrophic nicotinic ACh receptors (nAChRs) have a highly conserved pentameric arrangement of subunits that are organized around the ion channel or pore. In this respect, a series of alpha and beta subunit-encoding genes have been identified and characterized (Ballivet *et al.*, 1996; Wiley *et al.*, 1996; Hoekstra *et al.*, 1997; Walker *et al.*, 2001). At least eight distinct alpha subunit genes are known from *C. elegans* (Mongan *et al.*, 1998) and, assuming that multiple different subunits can be interchanged to generate functional receptor subtypes, indicate a high degree of heterogeneity in nAChR organization. This molecular heterogeneity is somewhat supported by electrophysiological studies on the pig gastrointestinal parasite, *Oesophagostomum dentatum*, which have revealed variation in conductance characteristics consistent with the occurrence of at least four distinct channels on the somatic musculature (Robertson *et al.*, 1999).

Although nAChRs are the dominant cholinergic receptor subtype in nematodes, a variety of results support the occurrence of muscarinic AChRs (mAChRs) in both motor neurones and muscle (Segerberg and Stretton, 1993; Lee *et al.*, 1999). The *C. elegans gar-1* gene was found to code for a novel form of G-protein-coupled AChR that is distinct from, but related to, the muscarinic subtype of AChR (Park *et al.*, 2000). Further evidence of heterogeneity in nematode AChRs was revealed when splice variants of *gar-1* were found to generate three distinct isoforms of the receptor. Even though cholinergic transmission is the most thoroughly studied component of nematode neurotransmission, specific channel arrangements and expression patterns *in vivo* are poorly understood.

Other components of cholinergic signalling pathways have been identified in nematodes. Acetylcholinesterase (AChE) is the enzyme responsible for ACh signal termination and has been of particular interest as the target for organophosphate and carbamate nematocides (Wright, 1981; Opperman and Chang, 1992). AChEs have been characterized from a broad range of free-living as well as animal- and plant-parasitic nematodes (Wright and Perry, 1998). Nematodes appear to be unusual in that they have multiple AChE-encoding (*ace*) genes (Combes *et al.*, 2000, 2001). The expression of nematode AChEs is not confined to the sensory and neuromuscular systems, and some isoforms are believed to play roles in other aspects of nematode biology such as in host–parasite interaction (Hussein *et al.*, 2002a,b).

Genes encoding choline acetyltransferase (ChAT; *cha-1*), the enzyme responsible for the synthesis of ACh, and a vesicular ACh transporter (VAChT; *unc-17*), responsible for loading synaptic vesicles with ACh, have been identified in *C. elegans* and linked to cholinergic neurotransmission (Alfonso *et al.*, 1993, 1994; Zhu *et al.*, 2001).

8.3.3. Amino acids

Exposure of nematodes to gamma-aminobutyric acid (GABA) causes flaccid paralysis following hyperpolarization of somatic body wall muscle that is triggered by the opening of ligand-gated Cl^- channels. GABA has been localized to a wide range of nematodes, including *Goodeyus ulmi*, *G. rostochiensis*, *M. incognita* and *Panagrellus redivivus* (Leach *et al.*, 1987; Stewart *et al.*, 1994), and to inhibitory motor neurones of the ventral nerve cord of *A. suum* (Johnson and Stretton, 1987). GABA has also been identified in excitatory neurones in *C. elegans* that are involved with the control of anal muscles and defecation (McIntire *et al.*, 1993a).

Molecular studies have identified multiple subunits (UNC-49A, UNC-49B and UNC-49C) of a heteromultimeric GABA receptor at the *unc-49* locus in *C. elegans* (Bamber *et al.*, 1999) and suggest that UNC-49B and UNC-49C co-assemble to form the GABA receptor at neuromuscular junctions. The pharmacology of the GABA receptor is distinct from that of the vertebrate GABA receptor (Martin *et al.*, 2002); most notably it is not antagonized by picrotoxin (Bascal *et al.*, 1996). Glutamic acid decarboxylase and GABA deaminase, enzymes responsible for the synthesis and degradation of GABA, have been detected in *C. elegans* homogenates (Schaeffer and Bergstrom, 1988). More recently, the *unc-47* gene in *C. elegans* was found to encode a novel vesicular GABA transporter (VGAT) that loads synaptic vesicles with cytosolic GABA ready for release into the synapse (McIntire *et al.*, 1997).

Glutamate induces both excitatory and inhibitory responses in nematodes through its interaction with a variety of ionotropic glutamate receptor channels that are either gated by *N*-methyl-D-aspartate (NMDA) or are not NMDA-sensitive. At least ten ionotropic glutamate receptor subunits have been identified in *C. elegans*, eight of which belong to the non-NMDA sensitive subtype. The processing of tactile sensory inputs involves the non-NMDA subtype of glutamate receptor subunit, designated GLR-1 and expressed in command interneurones, where it is required for nose-touch sensitivity (Hart *et al.*, 1995; Maricq *et al.*, 1995). Analysis of the expression patterns of the various subunits in *C. elegans* neurones reveals a high level of differential expression (Brockie *et al.*, 2001). Indeed, this has been studied in detail in ASH sensory neurones, which detect a range of environmental stimuli and trigger an avoidance response through glutamate release. This polymodal function for ASH has been linked to the differential release of glutamate and an unidentified neuropeptide at ASH-interneurone synapses (Kaplan, 1996). More recently, ASH functional diversity has been associated with post-synaptic glutamate receptors in that both subtypes of glutamate receptor are needed for osmotic sensitivity, whereas response to mechanical stimuli requires only the non-NMDA subtype (Mellem *et al.*, 2002). Also, there is some evidence that differential activation of post-synaptic glutamate receptor subtypes is central to the polymodal function of at least some *C. elegans* neurones (Mellem *et al.*, 2002). GLR-1 has also been implicated in olfactory associative and non-associative learning, with *glr-1* mutants being deficient in their responses to diacetyl stimuli (Morrison and van der Kooy, 2001).

Glutamate has also been shown to play a significant role in other nematode species, including *Ascaris suum* (Davis, 1998a,b). Glutamate injections into the pseudocoel produced a static 'paralysed' posture and the dorsal excitatory type 2 (DE2) neurones were depolarized; NMDA receptor antagonists blunted spontaneous DE2 activity. Davis (1998a) proposed that a glutamatergic pathway was involved in nematode locomotory behaviour. Further evidence for a glutamatergic signalling system in *A. suum* came from the identification of an electrogenic glutamate transporter in hypodermal tissue and DE2 motor neurones.

Glycine may have a role in the nervous system of nematodes. There is evidence that glycine activates an avermectin-sensitive, glutamate-gated Cl⁻ channel in *C. elegans* (Laughton *et al.*, 1995), and an absence of glycine receptors results in a lack of pressure antagonism of ethanol immobility (Eckenhoff and Yang, 1994).

8.3.4. Amines

Dopamine has been localized to eight cells of *C. elegans* hermaphrodites and three pairs of ray neurones associated with the male tail (Sulston *et al.*, 1975; Sulston and

Horvitz, 1977) and has been detected in *P. redivivus* and *M. incognita* (Stewart *et al.*, 2001). Evidence for the role of dopamine in nematode neurotransmission came when a dopamine GPCR (CeDOP-1) that has high sequence homology with D1-like dopamine receptors was identified in *C. elegans* (Suo *et al.*, 2002). Transient expression of CeDOP-1 in COS-7 cells revealed a high affinity for dopamine and it now appears that dopamine plays a variety of roles in nematode neurotransmission. For example, the *C. elegans* behavioural response comprising slow locomotory activity following bacterial encounter is regulated by a dopaminergic signalling pathway (Sawin *et al.*, 2000).

5-Hydroxytryptamine (5-HT, or serotonin) has been demonstrated in neurones in several nematode species. Two large neurones (NSM cells) of the *C. elegans* pharynx were identified as serotoninergic following formaldehyde-induced fluorescence (Albertson and Thomson, 1976; Horvitz *et al.*, 1982), and a number of other neurones have been reported to contain serotonin, including cells in the head, the vulval region of the ventral nerve cord and the male tail (Chalfie and White, 1988; Desai *et al.*, 1988; McIntire *et al.*, 1992); a similar pattern of immunostaining was reported for *G. ulmi* (Leach *et al.*, 1987). In *A. suum*, serotonin was localized to neurones with immunostaining, being reported in pharyngeal cells (analogous to *C. elegans* NSMs) and associated nerve fibres as well as cells in the ventral nerve cord in the tail region of male worms (Brownlee *et al.*, 1994; Johnson *et al.*, 1996). Its distribution in nematode nervous systems supports a role for serotonin in pharyngeal activity as well as reproductive function in some species (see Section 8.5).

Other components of nematode serotoninergic neurotransmission have been characterized, including *C. elegans* complementary DNAs (cDNAs) encoding both metabotropic and ionotropic serotonin receptors. Several metabotropic receptors have been identified and designated 5HT-Ce (Olde and McCombie, 1997) and 5-HT2Ce (Hamdan *et al.*, 1999). 5HT-Ce had structural and pharmacological features that were similar to the mammalian 5-HT1a receptor subtype; 5-HT2Ce was similar to the mammalian 5-HT2 subtype. The involvement of GPCRs in serotonin-mediated behaviours was supported by the fact that mutations in a *C. elegans* G-protein subunit-encoding gene (*goa-1*) caused behavioural changes similar to those seen in serotonin receptor mutants (Ségalat *et al.*, 1995). In 2000, Ranganathan *et al.* characterized a novel ionotropic serotonin-gated chloride channel (MOD-1) from *C. elegans* that has similarity to known GABA and glycine-gated Cl⁻ channels and which modulates locomotory behaviour. Type 2 serotonin receptor isoforms have also been identified in *A. suum*, with alternative splice variants being expressed in pharynx and body wall muscle (Huang *et al.*, 1999, 2002).

One of the enzymes involved in serotonin biosynthesis, aromatic amino acid decarboxylase (AAAD), has been identified biochemically in *C. elegans* (Sulston *et al.*, 1975), *Phocanema decipiens* (Goh and Davey, 1976) and *Aphelenchus avenae* (Wright and Awan, 1978). A series of behavioural and neuroendocrine defects were reported for *C. elegans* tryptophan hydroxylase (TPH-1; involved in serotonin biosynthesis) mutants, including reduced rates of feeding and egg-laying, increased fat storage, increased reproductive lifespan and, sometimes, developmental arrest at the dauer stage (Sze *et al.*, 2000).

Table 8.1. Predicted *Caenorhabditis elegans* neuropeptide signatures.

Neuropeptide-encoding gene	Peptide signature (amino acids)[a]	
	N terminus	C terminus
flp-1		PNFLRFG
flp-2		EPIRFG
flp-3, 20, 22		MRFG
flp-4, 5, 12		FIRFG
flp-6, 17		KSxxxRFG
flp-7, 9, 11, 16, 19		VRFG
flp-8, 14		KxExxRFG
flp-10		YIRFG
flp-13		PLIRFG
flp-15, 21		PLRFG
flp-18		PGVLRFG
nlp-1, 7, 13		MSFG
nlp-2,		FRPG
nlp-3,		S(I/L/M)G
nlp-4	DYDPRT	GEDRV
nlp-5, 6		MG(L/F)G
nlp-8	(A/S)FDRM	
nlp-9, 21	GG(A/G)RAF	
nlp-10		GGMYG
nlp-11		LxDxG
nlp-12		YRPLQFG
nlp-14		GFGF(D/E)
nlp-15	AFDSLAG(S/Q)GF	
nlp-16		HHGHQ
nlp-17		MRIG
nlp-18, 20		FAFA
nlp-19		GLRLPN(M/F)L
nlp-22, 23		G(F/M)RPG
nlp-26		QFGG
nlp-24, 25, 27-32		YGG(Y/W)G

[a]Single letter notation for amino acids. Only the most characteristic signature sequences are provided. Most peptide signatures are derived from the C-terminal sequence; *nlp-4* peptides possess N- and C-terminal sequence motifs. A C-terminal G residue is likely to undergo post-translational modification to provide a C-terminal amide moiety.

Other amines, including octopamine, phenolamine and histamine, have been detected in nematodes, although compelling evidence to support a role in neurotransmission is lacking. Of these, the strongest candidate is octopamine which has been identified in extracts of *C. elegans* and, in contrast to serotonin, was found to stimulate motor activity and inhibit pharyngeal function and egg-laying (Horvitz *et al.*, 1982; Rogers *et al.*, 2001). SER-2, a putative *C. elegans* octopamine/serotonin receptor, was expressed in COS-7 cells and found to exhibit selectivity for tyramine over octopamine (Rex and Komuniecki, 2002). Also, tyramine reduced forskolin-stimulated cAMP levels following stable heterologous expression, supporting the notion that SER-2 is a tyramine receptor.

8.3.5. Neuropeptides

To date, some 54 neuropeptide-encoding genes, which encode over 150 distinct neuropeptides, have been identified in *C. elegans* (Li *et al.*, 1999b; Nathoo *et al.*, 2001). These have been assigned to distinct families of neuropeptides that have specific amino acid signature sequences (Table 8.1). The largest of these families is the FMRFamide-related peptides (FaRPs) or FMRFamide-like peptides (FLPs) (Table 8.2), which are widely expressed in all neuronal subtypes. A large body of information has accumulated on FaRP physiology and reveals a complex communication system that undoubtedly plays a key role in the coordination of many, if not most, nematode behavioural activities. The non-FaRP neuropeptides (Nathoo *et al.*, 2001) have been designated neuropeptide-like proteins (NLPs), which are encoded on *nlp* genes. In contrast to FaRPs, little is known about the roles played by NLPs in nematodes.

Comparative structural analyses of all known or predicted *C. elegans* FaRPs reveal a level of diversity that has not been recorded for any other peptide family in any other animal species. This begs the question, is the FaRP structural diversity seen in *C. elegans* common to other nematode species? In the absence of the entire expressed protein complement from any other species, this is not known. Nevertheless, a number of facts indicate that neuropeptide complexity is a feature common to nematodes: (i) the abundance of FaRP immunoreactivity is broadly comparable between widely divergent nematode species; (ii) biochemical studies indicate that *A. suum* possesses at least 20 FaRPs and that many more await structural characterization; and (iii) expressed sequence tag (EST) data indicate considerable *flp* gene diversity in a broad range of nematode species.

Although only a limited number of studies examine the neuronal actions of nematode FaRPs, they reveal a complex array of actions commensurate with multifunctional roles (Cowden *et al.*, 1989; Davis and Stretton, 2001). Davis and Stretton (2001) examined the effects of 18 *A. suum* FaRPs on the input resistance and membrane potential of dorsal inhibitory (DI) and DE2 motor neurones, as well as the frequency of DE2 excitatory post-synaptic potentials. Both neuronal types are central to the control of somatic body wall muscle activity and, presumably, locomotion. To complement the study of *A. suum* peptide effects on motor neurones, Davis and Stretton (2001) also examined the effects of these peptides on the behaviour of *A. suum* following injection into the pseudocoel. The effects of the peptides tested were tentatively divided into five response types, each indicative of a distinct receptor interaction or signalling pathway; the relevance of these response types to the normal behaviour of *A. suum* is unknown.

The *flp-6* gene encodes six copies of KSAYMRFamide, a peptide that has also been characterized from *P. redivivus* (designated PF3; Maule *et al.*, 1994a), *A. suum* (designated AF8; Cowden and Stretton, 1995), *H. contortus* (Marks *et al.*, 1999b) and *G. pallida* (Kimber *et al.*, 2001); available EST information suggests that it is common to many other nematode species. This is the only peptide known to induce relaxation of dorsal somatic muscle strips and contraction of ventral somatic muscle strips in *A. suum* (Maule *et al.*, 1995a). Such differential activity could be

Table 8.2. Structurally characterized neuropeptides from nematodes.

Neuropeptide sequence[a]	Code	*C. elegans* gene[b]	Species	References
KNEFIRFa	AF1	*flp-8*	*Ascaris suum*	Cowden *et al.*, 1989
			Caenorhabditis elegans	Nelson *et al.*, 1998
KHEYLRFa	AF2	*flp-14*	*Ascaris suum*	Cowden and Stretton, 1993
			Caenorhabditis elegans	Marks *et al.*, 1996
	HF1		*Haemonchus contortus*	Marks *et al.*, 1999b
			Panagrellus redivivus	Maule *et al.*, 1994b
AVPGVLRFa	AF3	*flp-18*	*Ascaris suum*	Cowden and Stretton, 1995
GDVPGVLRFa	AF4	*flp-18*	*Ascaris suum*	Cowden and Stretton, 1995
SGKPTFIRFa	AF5	*flp-4*	*Ascaris suum*	Cowden and Stretton, 1995
FIRFa	AF6		*Ascaris suum*	Cowden and Stretton, 1995
AGPRFIRFa	*AF7*		*Ascaris suum*	Cowden and Stretton, 1995
KSAYMRFa	AF8	*flp-6*	*Ascaris suum*	Cowden and Stretton, 1995
			Caenorhabditis elegans	Marks *et al.*, 1998
	HF2		*Haemochus contortus*	Marks *et al.*, 1999b
	PF3		*Panagrellus redivivus*	Maule *et al.*, 1994a
GLGPRPLRFa	AF9	*flp-21*	*Ascaris suum*	Cowden and Stretton, 1995
GFGDEMSMPGVLRFa	AF10	*flp-18*	*Ascaris suum*	Cowden and Stretton, 1995
SDIGISEPNFLRFa	AF11	*flp-1*	*Ascaris suum*	Davis and Stretton, 1996
FGDEMSMPGVLRFa	AF12	*flp-18*	*Ascaris suum*	Davis and Stretton, 1996
SDMPGVLRFa	AF13	*flp-18*	*Ascaris suum*	Davis and Stretton, 1996
SMPGVLRFa	AF14	*flp-18*	*Ascaris suum*	Davis and Stretton, 1996
AQTFVRFa	AF15	*flp-16*	*Ascaris suum*	Davis and Stretton, 1996
ILMRFa	AF16		*Ascaris suum*	Davis and Stretton, 1996
FDRDFMHFa	AF17		*Ascaris suum*	Davis and Stretton, 1996
XXXPNFLRFa	AF18	*flp-1*	*Ascaris suum*	Davis and Stretton, 1996
AEGLSSPLIRFa	AF19	*flp-13*	*Ascaris suum*	Davis and Stretton, 1996
GMPGVLRFa	AF20	*flp-18*	*Ascaris suum*	Davis and Stretton, 1996
SDPNFLRFa		*flp-1*	*Caenorhabdiris elegans*	Rosoff *et al.*, 1993
	PF1		*Panagrellus redivivus*	Geary *et al.*, 1992
SADPNFLRFa		*flp-1*	*Caenorhabditis elegans*	Rosoff *et al.*, 1993
	PF2		*Panagrellus redivivus*	Geary *et al.*, 1992
APEASPFIRFa		*flp-13*	*Caenorhabditis elegans*	Marks *et al.*, 1997
KPSFVRFa		*flp-9*	*Caenorhabditis elegans*	Marks *et al.*, 1999a
AADGAPLIRFa		*flp-11*	*Caenorhabditis elegans*	Marks *et al.*, 2001
SVPGVLRFa		*flp-18*	*Caenorhabditis elegans*	Marks *et al.*, 2001
KPNFIRFa	PF4	*flp-1*	*Panarellus redivivus*	Maule *et al.*, 1995b
LQPNFLRFa	HF3	*flp-1*	*Haemonchus contortus*	Geary *et al.*, 1999

[a]Single letter notation for amino acids (a = amide).
[b]*C. elegans* gene that encodes peptide homologues.

involved in locomotory wave-form development and could provide a mechanism whereby opposite effects are induced in adjacent ventral and dorsal muscle fields (see Section 8.5.3). Muscle tension studies in *H. contortus* revealed that worms that were less sensitive to cholinomimetics were also less sensitive to the excitatory actions of PF3, implicating cholinergic neurones in the actions of this peptide (Marks *et al.*, 1999b).

Immunocytochemical studies have localized FaRPs to sensory nerves in a number of nematodes including cephalic papillary nerves, amphids and deirids, and nerves innervating caudal papillae (Brownlee *et al.*, 1993; Wikgren and Fagerholm, 1993; Kimber *et al.*, 2001). The most compelling evidence for their involvement in sensory perception comes from studies with *C. elegans*, in which a range of sensory abilities were perturbed following *flp-1* gene knockout (Nelson *et al.*, 1998). Among other things, *flp-1* knockouts were unresponsive to both nose-touch and osmotic stimuli, which, in wild-type animals, result in rapid backward movements and avoidance behaviours, respectively (Nelson *et al.*, 1998). In the same study, *flp-1* expression was localized to a series of eight neurones in *C. elegans*, including AVA and AIA, which are implicated in nose-touch and osmotic stimulus transduction. A role for neuropeptides in sensory function has also been implicated by the alteration in mechanosensory responses in worms with dysfunctional EGL-3 proprotein convertase, an enzyme that cleaves proproteins at dibasic residues within secretory vesicles (Kass *et al.*, 2001). Little is known about neuropeptide signal termination in nematodes, although it is now known that a family of neprilysin-like enzymes is expressed in *C. elegans* in a tissue-specific manner and these have been implicated in the metabolism of mammalian peptides (Isaac *et al.*, 2000).

8.3.6. Other neuromediators

The volatile gas nitric oxide (NO), which is synthesized by nitric oxide synthase (NOS) from L-arginine and molecular oxygen, has been implicated in the signalling pathway of the neuropeptide PF1 (SDPNFLRFamide), which causes a slow flaccid paralysis of *A. suum* somatic body wall muscle (Bowman *et al.*, 1996). Histochemical studies localized NOS activity to most parts of the nervous system in *A. suum*, including sensory, pharyngeal and motor neurones (Bascal *et al.*, 1995).

8.4. Motor Function and Somatic Musculature

8.4.1. Electrophysiology of somatic muscle

Numerous texts give comprehensive descriptions of the structure and organization of nematode somatic muscle (Rosenbluth, 1965a,b; Waterson, 1988; Moerman and Fire, 1997). It is at the neuromuscular junction that neuronal signals are relayed to the muscle to induce a motor response. The branching fingers of each muscle cell have multiple neuromuscular junctions that provide for excitatory and inhibitory neuronal inputs. Due to the occurrence of multiple tight junctions, the electrical signals can be shared by groups of muscle cells in the same vicinity, thus ensuring similar muscle cell responses that will facilitate coordinated motor activity (Jarman, 1959). Although nematodes are not segmented, the arrangement of their motor nervous system is repeated along the body such that the somatic muscles display segmental responses that allow the propagation of the locomotory waveform. A number of studies examine how the structural relationships of nematode motor neurones, somatic muscle, cuticle and hydrostatic skeleton could interplay to enable locomotion (Lee and Atkinson, 1976; Walrond and Stretton, 1985; Alexander, 2002).

Most electrophysiology studies on nematode somatic muscle have employed *A. suum* or *A. lumbricoides*, primarily due to their large size. The resting membrane potential of *A. suum*, *A. lumbricoides*, *Ascaridia galli*, *Dipetalonema viteae* and *H. contortus* somatic muscle was reported to be around −30 mV (Jarman, 1959; Brading and Caldwell, 1971; Wann, 1987; Rohrer *et al.*, 1988; Atchinson *et al.*, 1992), with extracellular Cl⁻ having the greatest influence and, to a lesser extent, Na⁺ (Caldwell and Ellory, 1968). Ca^{2+}-dependent Cl⁻ channels have been proposed to have an excitatory function and to play a major role in maintenance of the somatic muscle membrane potential (Blair *et al.*, 1998). Also, nematode somatic muscles have been reported to display spontaneous depolarizing and hyperpolarizing currents as well as large action potentials that originate in the muscle cells (Wann, 1987).

The consistent 'segmented' arrangement of the motor neurones in most nematodes, coupled with the reported tonic release of excitatory and inhibitory neurotransmitters, provided a foundation on which to build a model of locomotory function in nematodes. One of these was based on structural and electrophysiological studies on *A. suum* and resulted in the reciprocal oscillator model (Walrond and Stretton, 1985).

8.4.2. Classical transmitters and somatic muscle

Voltage clamp methods have identified ACh receptors (AChRs) both synaptically and extrasynaptically on *A. suum* muscle (Del Castillo *et al.*, 1963; Martin, 1982; Pennington and Martin, 1990). Basically, ACh triggers the opening of non-selective cation channels in nematode muscle and these channels facilitate the influx of Na⁺ and K⁺ to induce depolarization and excitation. Pharmacological studies have catalogued cholinergic agonist and antagonist profiles on *A. suum* somatic muscle (Colquhoun *et al.*, 1991). These and other data indicate that the dominant *A. suum* AChR is activated by cholinergic agonists (ACh and nicotine) and that the actions of ACh are suppressed by the nicotinic antagonist, D-tubocurarine (Natoff, 1969; Rozhova *et al.*, 1980; Segerberg and Stretton, 1993), consistent with its designation as an nAChR.

Numerous studies report the hyperpolarizing action of GABA on somatic muscle and concomitant increase in input conductance caused by the direct gating of Cl⁻ channels (Martin *et al.*, 2002; Section 8.4.3). GABA-activated single-channel currents have been detected and characterized from the bag region of somatic muscle cells in *A. suum* (Martin, 1985). Available pharmacological evidence indicates that, although GABA agonist profiles are similar to those seen for the mammalian GABA-A receptor, the antagonist profiles are very different (Holden-Dye *et al.*, 1988, 1989; Bascal *et al.*, 1996).

The opposite actions of ACh and GABA on somatic muscle and their tonic release, coupled to the reciprocal inhibitory arrangement of the motor neurones provide a compelling model for their coordinated activity (Walrond and Stretton, 1985; Section 8.5.3). Consistent with their proposed action in the control of nematode locomotory activity, numerous *C. elegans* cholinergic and GABAergic mutants display uncoordinated (*unc* phenotypes) movement and/or changes in locomotory behaviour (McIntire *et al.*, 1993b; Jin *et al.*, 1994; Richmond and Jorgensen, 1999; Shingai, 2000). However, it is well recognized that this model represents only a simplified overview of the *in vivo* situation, which involves multiple other neurotransmitters and neuropeptides that modulate somatic muscle activity.

Although serotonin does not appear to have direct actions on somatic body wall muscles in nematodes, several studies indicate that it alters muscle responses to ACh (Walker *et al.*, 2000; Trim *et al.*, 2001). Receptor binding studies identified a serotonin receptor on muscle membranes from *A. suum* (Chaudhuri and Donahue, 1989). Also, serotonin raised cAMP levels in *A. suum* somatic muscle and its muscle-based receptor had a pharmacological profile that was similar to but distinct from mammalian serotonin type 2 receptors (Williams *et al.*, 1992; Walker *et al.*, 2000; Trim *et al.*, 2001). Injections of serotonin into the pseudocoel of *A. suum* inhibited locomotory wave-forms and increased body length, indicating the relaxation of somatic muscle (Reinitz and Stretton, 1996); the actions were believed to be via a pre-synaptic mechanism that altered the activities of selected motor-neurones.

8.4.2.1. Classical transmitters and signalling heterogeneity

The structural simplicity of the nervous system demands a significant level of signalling sophistication to facilitate the complex behavioural repertoires of nematodes. One theme that is becoming ever more apparent in classical transmitter action on nematode muscle is the inherent heterogeneity in their cognate receptors. For example, numerous AChR subtypes, which have the ability to form multiple distinct receptors, have been identified in *A. suum* and *C. elegans* and physiological evidence indicates a similar situation in *O. dendatum* (Treinen and Chalfie, 1995; Ballivet *et al.*, 1996; Martin *et al.*, 1997). Such receptor heterogeneity is also seen in serotonin and glutamate receptors and can come about through distinct combinations of subunit proteins that are transcribed by different genes or that are differentially spliced (Hamdan *et al.*, 1999; Huang *et al.*, 1999, 2002). One significant deficiency in the understanding of these receptors in nematodes is the absence of definitive information on which subunits co-assemble to form the receptors *in vivo*.

8.4.3. Neuropeptides and somatic muscle

As stated above, theories on somatic muscle coordination are largely based on current understanding of the physiological features and classical transmitters that regulate this system. However, these theories do not yet take account of the huge array of neuropeptides that have varied and potent actions on nematode somatic muscle. Neuropeptide physiology has been the subject of a number of recent reviews (Maule *et al.*, 2002, and references therein) and only key features will be dealt with here. Almost nothing is known about the effects of NLPs on nematode somatic muscle except that the peptide DYRPLQFamide, encoded on *nlp-12* (Nathoo *et al.*, 2001), causes a ventral coiling behaviour and a sustained contraction of dorsal body wall muscle segments of *A. suum* (Reinitz *et al.*, 2000). By far the most studied family of nematode neuropeptides is the FaRPs, which are of particular interest as they modulate sensory and motor function and, therefore, play a key role in nematode behaviour.

A variety of distinguishable mechanisms are employed by inhibitory neuropeptides to induce muscle relaxation in nematode body wall muscle (Bowman *et al.*, 1995; Maule *et al.*, 1995a, 1996). For example, SDPNFLRFamide (PF1), which is encoded on *C. elegans flp-1*, induces a slow relaxation through its interaction with a GPCR that directly or indirectly gates a hypodermal-based Ca^{2+} channel. The resultant Ca^{2+} influx activates an endogenous nitric oxide synthase that results in the production and release of NO, which can diffuse to the somatic muscle and induce relaxation by an unknown mechanism (Bowman *et al.*, 1995). More recent evidence suggests that released NO opens a K^+ channel in *A. suum* somatic muscle (Bowman *et al.*, 2002). Although a range of other FaRPs have qualitatively similar actions on somatic muscle, their mechanisms of action have not been investigated.

A second type of somatic muscle inhibition has been reported for the *P. redivivus* peptide KPNFIRFamide (PF4) (Maule *et al.*, 1995a,b). This peptide induces a rapid relaxation and hyperpolarization, which, at high concentrations (≥ 1 µM), reverses rapidly. The muscle tension profile generated by the addition of PF4 is markedly similar to that seen with additions of GABA, suggesting the involvement of the GABA channel in the peptide-induced effects; this was further supported by the finding that the effects of PF4 were Cl^--dependent, implicating a Cl^- channel in its action (Maule *et al.*, 1995a; Holden-Dye *et al.*, 1997). However, the inability of the GABA antagonist NCS 281-93 to abolish PF4-induced relaxation suggested that GABA and PF4 activated different Cl^- channels (Maule *et al.*, 1995a). Using the two-electrode current clamp technique, Purcell *et al.* (2002) revealed that PF4 directly gates a Cl^- channel on *A. suum* somatic muscle, the first report of a directly neuropeptide-gated Cl^- channel. The *flp-1*-encoded peptide KPNFLRFamide is the *C. elegans* homologue of PF4 and displays an identical functional profile. A notable feature common to these functionally distinct peptides is the C-terminal hexapeptide signature, PNFLRFamide. Structure–activity analyses of these peptides revealed that the Lys residue in positions 1 of PF4 and KPNFLRFamide is critical to activation of the Cl^- channel and prevents interaction with the PF1 receptor (Bowman *et al.*, 2002). Further complexity in neuropeptide–receptor interactions was revealed by the inactivity of the *flp-9* peptide KPSFVRFamide and the analogue KPSFIRFamide on dorsal or ventral muscle strips of *A. suum*; the actions of the chimera KPNFVRFamide, which only differs from KPSFVRFamide in position 3 (Asn^3 for Ser^3), were indistinguishable from those of PF4 (Marks *et al.*, 1999b). These data indicate that the replacement of the Asn^3 with Ser^3 is enough to abolish the biological activity of PF4. In this way, subtle amino acid changes to very similar neuropeptides can alter the receptor interaction, the downstream muscle effect and, presumably, the resultant behavioural response.

Two structurally similar FaRPs (KNEFIRFamide, AF1 and KHEYLRFamide, AF2), which are believed to act at different receptors, induce inhibition of locomotory waves when injected into the *A. suum* pseudocoel (Reinitz *et al.*, 2000). They also trigger distinctive biphasic responses in *A. suum* somatic muscle strips, comprising a transient relaxation followed by an increase in contraction amplitude and frequency. The excitatory phase has been linked to the inhibition of inhibitory neurones reported by Cowden *et al.* (1989) and is absent from denervated somatic muscle preparations, indicating a pre-synaptic site of action (Maule *et al.*, 1995a; Bowman *et al.*, 1996). One notable feature of AF1 and AF2 effects on *A. suum* somatic muscle is the dramatic and sustained increase in cAMP levels (Reinitz *et al.*, 2000; Thompson *et al.*, 2003). This increase in cAMP could be decoupled from the excitatory response (Thompson *et al.*, 2003) and may relate to the muscle relaxation phase of their actions.

Multiple FaRPs are known to induce myoexcitation in *A. suum* (Geary *et al.*, 1999; Maule *et al.*, 2001). One group of peptides that have excitatory effects on nematode somatic muscle are the PGVLRFamides, which comprise peptides that have variable N-terminal sequences and an invariant C-terminal signature. Multiple PGVLRFamides are encoded on the *afp-1* gene in *A. suum* and the *flp-18* gene in *C. elegans* (Edison *et al.*, 1997; Li *et al.*, 1999a). Following injection into the pseudocoel of *A. suum*, these peptides had profound behavioural effects, including an increase in erratic locomotory behaviour and a decrease in head searching activity (Davis and Stretton, 2001). The actions of AVPGVLRFamide (AF3) and/or GDVPGVLRFamide (AF4) on body wall muscle of *A. suum* and *A. galli* included sustained contraction that was independent of the cholinergic system and a depression of cAMP levels (Davis and Stretton, 1996; Trim *et al.*, 1997, 1998). As with the inhibitory peptides, a number of other peptides are known to have excitatory actions on somatic muscle, but little is known about their mechanisms of action (Maule *et al.*, 2001).

8.4.3.1. Neuropeptides and signalling heterogeneity

Unlike the situation with classical transmitters, the potential for multiple diverse signalling pathways associated with nematode neuropeptides is obvious. Current theory on the coordination of nematode locomotion struggles to accommodate the ever-increasing numbers of receptor and channel subunits and the consequent potential for signal complexity and subtlety within nematode neuromuscular systems. The somatic muscle alone displays a broad response range to structurally similar ligands, indicative of multiple receptors/receptor subtypes. Which of the published physiological data have relevance to *in vivo* behavioural activities is, for the most part, unknown. Indeed, the role played by the neuropeptide system in the coordination of locomotion is obscured by the lack of information on receptor–ligand interactions and expression profiles.

8.5. Specialized Muscular Systems

8.5.1. The pharynx

Specialized muscle systems in nematodes are non-striated and occur associated with the pharynx (and stylet protractor muscles in plant-parasitic nematodes), intestine (some species), vulva/ovijector, male reproductive system and rectum. Detailed accounts of these muscle systems have been published elsewhere (Albertson and Thomson, 1976; Wood, 1988; Wright and Perry, 1998). The component of the nematode nervous system that innervates these muscular structures has been designated the sympathetic nervous system (Baldwin and Perry, 2004).

The most highly developed muscular organ in most nematodes is the pharynx, which acts as a muscular pump and valve to concentrate and deliver food to the intestine. Its structure has been best defined in *C. elegans*, where it comprises 20 muscle cells and a network of 20 neurones and 18 other epithelial cells (Albertson and Thomson, 1976; White, 1988). The only other nematode for which there is detailed anatomical and functional information is *Ascaris*, which possesses a simple conical pharynx that acts as a two-stage pump (Mapes, 1965, 1966; Reger, 1966). In many plant-parasitic forms, the anterior portion is adapted to accommodate the stylet, which is used to suck fluid contents from plant cells and tissues (Wyss, 2002). The finely regulated muscle activity of the pharynx is dependent upon highly electrically coupled muscles, which have myogenic motor activity that can be precisely controlled by neuronal inputs.

In contrast to the trilateral symmetry of the pharynx, the pharyngeal nervous system displays bilateral symmetry (Chalfie and White, 1988). Of particular note is the fact that it possesses its own nerve ring and, in *C. elegans*, the only connections with the central nervous system are via two RIP interneurones. A single nerve fibre from each of the lateral pharyngeal nerves connects with the central nervous system in *A. suum*. The limited neuronal connectivity with the central nervous system for these nematodes suggests that the pharyngeal system is responsible for the coordinated motor activity, with the central nervous system simply relaying on or off signals based on sensory inputs.

Neuronal ablation studies in *C. elegans* revealed that feeding was controlled by two distinct muscle movements, isthmus peristalsis and pumping (Avery and Horvitz, 1989). The M4 motor neurone was necessary for the former while the marginal cell (MC) neurone was essential to the normal stimulation of pumping following a food trigger; pumping continued in MC-ablated worms. Twelve pharyngeal neurones were found to influence pharyngeal function (Avery and Thomas, 1997; Li *et al.*, 1997), with M3, M4 and MC being essential for normal pumping activity. Relaxation events are regulated by the inhibitory M3 motor neurones and some evidence suggests that they are activated by muscle contraction, a feature that would enable appropriate timing of relaxatory events. Somewhat surprisingly, laser ablation of the entire pharyngeal nervous system did not abolish pumping and the M4 neurone alone enabled the development of viable and fertile worms.

8.5.1.1. Pharyngeal physiology

In *Ascaris*, the pharyngeal nerves are predominantly serotoninergic and FaRPergic (Stretton and Johnson, 1985; Brownlee *et al.*, 1994; Johnson *et al.*, 1996), suggesting that both classical transmitters and neuropeptides play a role in the modulation of pharyngeal activity. The number of each neuronal subtype is unknown, although between five (Brownlee *et al.*, 1994) and 18 (Cowden *et al.*, 1993) FaRPergic and between three (Johnson *et al.*, 1996) and six (Brownlee *et al.*, 1994) serotoninergic neurones have been reported. FaRPergic nerves were identified immunocytochemically in innervation of the pharynx and stylet retractor muscles of *G. pallida* and *G. rostochiensis* (Kimber *et al.*, 2001). To date, only faint GABA immunoreactivity and no glutamate immunoreactivity has been identified in the pharynx (Guastella *et al.*, 1991).

A number of classical transmitters are known to modulate the activity of the pharynx in *A. suum* and *C. elegans*. Serotonin is needed to stimulate *A. suum* pharyngeal pumping and also enhances this activity in *C. elegans* (Brownlee *et al.*, 1995, 1997; Niacaris and Avery, 2003) and stylet protrusion in *G. rostochiensis* (Rolfe and Perry, 2001). Pharmacological studies have identified a 5-HT2-like receptor in the pharynx of *A. suum* (Trim *et al.*, 2001) and alternatively spliced 5-HT2-receptor isoforms have been characterized (Huang *et al.*, 2002). Both GABA and glutamate inhibit pharyngeal pumping in *A. suum* (Brownlee *et al.*, 1995, 1997). In *C. elegans*, glutamate was found to inhibit pharyngeal pumping via the M3 inhibitory neurone (Dent *et al.*, 1997) and to act on pm4 (metacorpus) muscle as well as the junction between pm5 and pm6 muscles (isthmus and terminal bulb) (Laughton *et al.*, 1997; Li *et al.*, 1997; Pemberton *et al.*, 2001). Glutamate-gated Cl⁻ channels are believed to be a primary site of action for the avermectin anthelmintics, which inhibit pharyngeal function in a range of nematode species (Geary *et al.*, 1993; Dent *et al.*, 1997; Laughton *et al.*, 1997; Sheriff *et al.*, 2002). Octopamine suppresses pharyngeal pumping in *C. elegans* and also reduces the activity of the M3 motor neurones (Rogers *et al.*, 2001; Niacaris and Avery, 2003). The neurotransmitter ACh has excitatory effects on the *C. elegans* pharynx (Avery and Horvitz, 1990), which are thought to be due to the M1, M2 and M5 neurones (Rand and Nonet, 1997). Also, an nAChR is believed to be required for MC signalling in the pharynx (Raizen *et al.*, 1995).

A few physiology studies have implicated FaRPs in the control of pharyngeal activity (Brownlee *et al.*, 1995; Brownlee and Walker, 1999; Rogers *et al.*, 2001). PF3 (KSAYMRFamide) inhibited serotonin-stimulated pharyngeal pumping activity in *A. suum* (Brownlee *et al.*, 1995) and increased pharyngeal action potential frequency in *C. elegans* (Rogers *et al.*, 2001); a number of other FaRPs have been shown to have excitatory or inhibitory effects on the *C. elegans* pharynx (Rogers *et al.*, 2001). The most potent of these peptides were found to exert their effects directly (APEASPFIRFamide; expressed in a neurone (I5) with synaptic input to the pharynx) or indirectly (AF1 and SAEPFGTMRFamide; expressed in extrapharyngeal neurones) on the pharynx. Delineation of the fine control of pharyngeal activity will depend on elucidation of all the signalling molecules that modulate its activity.

8.5.2. The ovijector/vulva

Most nematodes display a paired reproductive tract with paired uteri that unite to form a single duct (ovejector, ovijector or vagina) that facilitates the release of eggs or juvenile worms. This organ must coordinate the movement of luminal contents out of the worm while not compromising the hydrostatic skeleton. In *C. elegans*, 16 muscle cells control egg-laying, eight of which are associated with the uterus and eight with the vulva (White *et al.*, 1986; White, 1988). As with somatic muscles, these send arms to the ventral nerve cord and receive synaptic input from the hermaphrodite-specific neurone (HSN) and VCn motor neurones (White *et al.*, 1986).

In *Ascaris*, the paired vagina uteri accumulate eggs before the latter are passed into the muscular vagina vera (ovijector) and into the host intestinal lumen. The vagina vera comprises primarily circular muscle, which overlies a hypodermal layer, and an inner cuticle, which lines the lumen of the ovijector (Musso, 1930; Fellowes *et al.*, 1999). *Ascaris* eggs are released in bursts following peristaltic contractions of the ovijector (Terenina and Shishov, 1975), an activity that is myogenic and which occurs in both inactive and moving worms. This is in contrast to *C. elegans*, which releases eggs only during periods of inactivity (Horvitz *et al.*, 1982) or modified locomotory activity (Hardaker *et al.*, 2001). The vagina vera appears to have a well-developed nerve plexus that extends from the ventral nerve cord through the muscle layer (Smart *et al.*, 1992; Brownlee *et al.*, 1993).

8.5.2.1 Physiology of the ovijector/vulva

A number of classical transmitters (serotonin, octopamine and ACh) have been implicated in oviposition in *C. elegans* (Horvitz *et al.*, 1982; Trent *et al.*, 1983; Desai and Horvitz, 1989; Schinkmann and Li, 1992; Weinshenker *et al.*, 1995; Rand and Nonet, 1997; Waggoner *et al.*, 1998). Serotonin and ACh immunoreactivities have been localized to the HSNs and associated ventral nerve cord cells (VC4 and VC5). Serotonin has been shown to modulate the temporal pattern of egg-laying behaviour and the coordination of locomotory behaviour with egg-laying events (Hardaker *et al.*, 2001). The G-protein alpha subunit G_o modulates various serotonin-controlled behaviours, including egg-laying (Ségalat *et al.*, 1995). The involvement of ACh has been supported by the disruption of oviposition caused by mutations in nAChR subunit genes (Kim *et al.*, 2001). Furthermore, these authors proposed that nAChRs modulate serotonin response pathways in egg-laying neuromusculature in *C. elegans*.

Neuropeptides appear to play a significant role in the regulation of egg-laying in nematodes and have been localized to innervation of this region in a small number of species (Leach *et al.*, 1987; Atkinson *et al.*, 1988; Smart *et al.*, 1992; Brownlee *et al.*, 1993; Fellowes *et al.*, 1999). FaRP immunostaining in *A. suum* was localized to multiple parallel nerve tracts that run along the muscle layer of the ovijector and send fine fibres that are closely apposed to the circular muscle fibres (Fellowes *et al.*, 1999). FLP-1 peptides and serotonin have been proposed to function in concert to initiate the onset of egg-laying activity, probably due to the modulation of sensory cues transduced by these neuropeptides (Waggoner *et al.*, 2000). FaRPs have a variety of distinguishable effects on ovijector myoactivity in *A. suum* (Fellowes *et al.*, 1998, 2000; Moffett *et al.*, 2003; Fig. 8.2). These data point to the occurrence of multiple FaRP receptor subtypes in the ovijector of *A. suum* and, as with somatic muscle, reveal unexpected complexity in neuromuscular function.

8.5.3. Other specialized muscle systems

Although a number of other well-developed specialized muscle systems occur in nematodes (male reproductive system; defecation system), much less is known about the control of their activity. For details on the organization of these specialized neuromuscular systems the reader is directed to a number of detailed reviews (White *et al.*, 1986; Chalfie and White, 1988; Munn and Munn, 2002).

Defecation in *C. elegans* is controlled by three sets of muscles, the anal depressors, the sphincter and two intestinal muscles (White, 1988). The three muscles are electrically coupled via gap junctions and receive synaptic input from the DVB neurone, part of the GABAergic system. Well-fed *C. elegans* defecate every 45 s and mutational disruptions of the muscles in this region induce dysfunctional behaviour, with waste accumulation (Munn and Munn, 2002).

The posterior end of male nematodes is commonly specialized into a highly developed copulatory bursa, which houses retractable spicules that are inserted into the female vulva during copulation. The control of this structure requires additional specialized muscles and associated innervation as well as numerous sensory receptors to facilitate the mating process (Sulston *et al.*, 1980). A complex arrangement of 41 muscles and the cloacal ganglion, which projects nerve fibres into the pre-anal ganglion, are involved with male mating behaviour in *C. elegans* (White *et al.*, 1986; White, 1988). The complexity of the neuromuscular arrangement of this system is not surprising, considering that male mating is one of the most complex *C. elegans* behaviours (Chalfie and White, 1988; Liu and Sternberg, 1995). The role played by specific neurotransmitters in male mating behaviour is not known. Male *C. elegans* have six serotonin-immunoreactive neurones that innervate male-specific ventral muscles and serotonin signalling mutants display defective male tail-curling behaviour (Loer and Kenyon, 1993). Serotonin has been localized to five ventral cord neurones in the tail of male *A. suum* (Stretton and Johnson, 1985; Johnson *et*

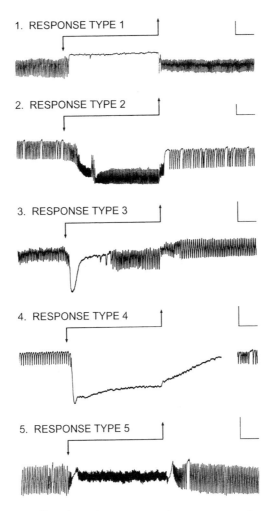

Fig. 8.2. Muscle tension recordings from the ovijector of *Ascaris suum* showing the five different response types seen following the addition of selected nematode FMRFamide-related peptides. Response type 1 comprises circular muscle relaxation and a shortening of the ovijector. Response type 2 comprises circular muscle contraction (lengthening of the ovijector) and a simultaneous increase in contraction frequency. Response type 3 consists of a transient contraction of the circular muscle. Response type 4 comprises transient contraction of the circular muscle and a subsequent slow relaxation. Response type 5 comprises muscle relaxation coupled with an increase in contraction frequency. Peptide was present during the period indicated by the arrows. Vertical scale represents 2 mg and horizontal scale represents 2 min. (For details see Moffett *et al.*, 2003.)

al., 1996). Furthermore, serotonin was found to depolarize the transverse ventral muscle cells and to influence male tail curling in *A. suum*, thereby mimicking the male copulatory posture (Reinitz and Stretton, 1996). These data clearly implicate serotonin in male mating behaviour.

8.6. Conclusions and Future Directions

Although the hard-wiring of the sensory apparatus and nerve–muscle motor system is essentially very simple in nematodes, it manages to coordinate a wide variety of complex and often quite subtle behavioural patterns, which undoubtedly contribute to the success of this phylum. From available biochemical and physiological evidence, it is clear that the intercellular communication systems that interlink sensory and motor function in the coordination of nematode behavioural traits are modulated by a complex array of signalling molecules, receptor proteins and ion channels. Even within the individual classes of receptor/channel subunits there are large numbers of small differences, which provide for a high level of heterogeneity in the protein complement associated with nematode behaviour. Unravelling the component parts within signalling pathways that influence individual behaviours and determining how small differences in the protein complement can modify them remains a significant challenge. Unfortunately, current understanding of the physiological and biochemical features of nematode behaviour has fallen far behind the rapidly expanding databases of molecular information. Probably the biggest challenge to the nematologist is finding ways to enable biochemical and physiological understanding to keep abreast of molecular progress. Advances are being made in the use of RNA interference (RNAi) techniques to examine the phenotypic effects of gene silencing at the organismal level, but measuring subtle changes in behaviour within the confines of transcriptome-wide knockout studies remains extremely difficult. Detailed physiological and biochemical assessment of individual behaviours will be needed to underpin and fully understand the molecular-based data.

The cornerstone of our understanding of nematode behaviour has been, and will continue to be, *C. elegans*. However, EST data and the expanding use of RNAi techniques on other nematodes are allowing researchers to dissect and contrast the role of individual gene products in distinct nematode species. This has particular significance for parasitologists. As well as numerous common behavioural traits, most parasitic nematodes display unique behaviours that are not evident in *C. elegans*. Indeed, it is not even known if common behaviours are triggered and modulated by the same mechanisms in different nematode species. To address these issues, a greater understanding of common behaviours/signalling mechanisms in different nematode species is needed. Once the basic components of individual signalling pathways have been determined, the mechanisms that integrate all the incoming signals and thereby result in specific behavioural responses will remain to be evaluated. A full appreciation of nematode behaviour will only be realized with the assimilation of a vast catalogue of complementary structural, biochemical, physiological and molecular information about these 'not so simple' animals.

References

Albertson, D.G. and Thomson, J.N. (1976) The pharynx of *Caenorhabditis elegans*. *Philosophical Transactions of the Royal Society B Biological Sciences* 275, 299–325.

Alexander, R.McN. (2002) Locomotion. In: Lee, D.L. (ed.) *The Biology of Nematodes*. Taylor and Francis, London, pp. 345–352.

Alfonso, A., Grundahl, K., Duerr, J.S., Han, H.P. and Rand, J.B. (1993) The *Caenorhabditis elegans unc-17* gene: a putative vesicular acetylcholine transporter. *Science* 261, 617–619.

Alfonso, A., Grundahl, K., McManus, J.R. and Rand, J.B. (1994) Cloning and characterization of the choline acetyltransferase structural gene (*cha-1*) from *C. elegans. Journal of Neuroscience* 14, 2290–2300.

Atchinson, W.D., Geary, T.G., Manning, B., Vandewaa, E.A. and Thompson, D.P. (1992) Comparative neuromuscular blocking actions of levamisole and pyrantel-type anthelmintics on rat and gastrointestinal nematode somatic muscle. *Toxicology and Applied Pharmacology* 112, 133–143.

Atkinson, H.J. (2002) Molecular approaches to novel crop resistance against nematodes. In: Lee, D.L. (ed.) *The Biology of Nematodes*. Taylor and Francis, London, pp. 569–598.

Atkinson, H.J., Isaac, R.E., Harris, P.D. and Sharpe, C.M. (1988) FMRFamide-like immunoreactivity within the nervous system of the nematodes *Panagrellus redivivus, Caenorhabditis elegans* and *Heterodera glycines. Journal of Zoology* 216, 663–671.

Aumann, J. (1993) Permeability of chemosensillum-associated exudates for lectins in a plant-parasitic and in a free-living nematode. *Fundamental and Applied Nematology* 16, 381–382.

Aumann, J., Ladehoff, H. and Rutencrantz, S. (1998) Gas chromatographic characterization of the female sex pheromone of *Heterodera schachtii* (Nematoda: Heteroderidae). *Fundamental and Applied Nematology* 21, 119–122.

Avery, L. and Horvitz, H.R. (1989) Pharyngeal pumping continues after laser killing of the pharyngeal nervous system of *C. elegans. Neuron* 3, 473–485.

Avery, L. and Horvitz, H.R. (1990) Effects of starvation and neuroactive drugs on feeding in *Caenorhabditis elegans. Journal of Experimental Zoology* 253, 263–270.

Avery, L. and Thomas, J.H. (1997) Feeding and digestion. In: Riddle, D.L., Blumenthal, T., Meyer, B.J. and Priess, J.R. (eds) *C. elegans II*. Cold Spring Harbor Laboratory Press, Cold Spring Harbor, New York, pp. 679–716.

Baldwin, J.G. and Perry, R.N. (2004) Nematode morphology, sensory structure and function. In: Cheng, Z., Cheng, S. and Dickson, D.W. (eds) *Nematology – Advances and Perspectives*, Vol. 1. CAB International, Wallingford, UK, pp. 175–256.

Ballivet, M., Alliod, C., Bertrand, S. and Bertrand, D. (1996) Nicotinic acetylcholine receptors in the nematode *Caenorhabditis elegans. Journal of Molecular Biology* 258, 261–269.

Bamber, B.A., Beg, A.A., Twyman, R.E. and Jorgensen, E.M. (1999) The *Caenorhabditis elegans unc-49* locus encodes multiple subunits heteromultimeric GABA receptor. *Journal of Neuroscience* 19, 5348–5359.

Bargmann, C.I. and Horvitz, H.R. (1991) Chemosensory neurons with overlapping functions direct chemotaxis to multiple chemicals in *C. elegans. Neuron* 7, 729–742.

Bargmann, C.I. and Mori, I. (1997) Chemotaxis and thermotaxis. In: Riddle, D.L., Blumenthal, T., Meyer, B.J. and Priess, J.R. (eds) *C. elegans II*. Cold Spring Harbor Laboratory Press, Cold Spring Harbor, New York, pp. 717–737.

Bargmann, C.I., Thomas, J.H. and Horvitz, H.R. (1990) Chemosensory cell function in the behaviour and development of *Caenorhabditis elegans. Cold Spring Harbor Symposium* 55, 529–538.

Bargmann, C.I., Hartwieg, E. and Horvitz, H.R. (1993) Odorant-selective genes and neurons mediate olfaction in *C. elegans*. *Cell* 74, 515–527.

Bascal, Z.A., Montgomery, A., Holden-Dye, L., Williams, R.G. and Walker, R.J. (1995) Histochemical mapping of NADPH diaphorase in nervous system of the parasitic nematode *Ascaris suum*. *Parasitology* 110, 625–637.

Bascal, Z.A., Holden-Dye, L., Willis, R.J., Smith, S.W. and Walker, R.J. (1996) Novel azole derivatives are antagonists at the inhibitory GABA receptor on the somatic muscle cells of the parasitic nematode *Ascaris suum*. *Parasitology* 112, 253–259.

Bird, A.F. and Bird, J. (1991) *The Structure of Nematodes*, 2nd edn. Academic Press, San Diego, 316 pp.

Blair, K.L., Barsuhn, C.L., Day, J.S., Ho, N.F.H., Geary, T.G. and Thompson, D.P. (1998) Biophysical model for organic acid excretion in *Ascaris suum*. *Molecular and Biochemical Parasitology* 93, 179–190.

Bowman, J.W., Winterrowd, C.A., Friedman, A.R., Thompson, D.P., Klein, R.D., Davis, J.P., Maule, A.G., Blair, K.L. and Geary, T.G. (1995) Nitric oxide mediates the inhibitory effects of SDPNFLRFamide, a nematode FMRFamide-related neuropeptide, in *Ascaris suum*. *Journal of Neurophysiology* 74, 1880–1888.

Bowman, J.W., Friedman, A.R., Thompson, D.P., Ichhpurani, A.K., Kellman, M.F., Marks, N.J., Maule, A.G. and Geary, T.G. (1996) Structure–activity relationships of KNEFIRFamide (AF1), a nematode FMRFamide-related peptide, on *Ascaris suum* muscle. *Peptides* 17, 381–387.

Bowman, J.W., Friedman, A.R., Thompson, D.P., Maule, A.G., Alexander-Bowman, S.J. and Geary, T.G. (2002) Structure–activity relationships of an inhibitory nematode FMRFamide-related peptide, SDPNFLRFamide (PF1), on *Ascaris suum* muscle. *International Journal for Parasitology* 32, 1765–1771.

Brading, A.F. and Caldwell, P.C. (1971) The resting membrane potential of the somatic muscle cells of *Ascaris lumbricoides*. *Journal of Physiology (London)* 217, 605–624.

Brockie, P.J., Madsen, D.M., Zheng, Y., Mellem, J.E. and Maricq, A.V. (2001) Differential expression of glutamate receptor subunits in the nervous system of *Caenorhabditis elegans* and their regulation by the homeodomain protein UNC-42. *Journal of Neuroscience* 21, 1510–1522.

Brownlee, D.J.A. and Walker, R.J. (1999) Actions of nematode FMRFamide-related peptides on the pharyngeal muscle of the parasitic nematode, *Ascaris suum*. *Annals of the New York Academy of Sciences* 897, 228–238.

Brownlee, D.J.A., Fairweather, I. and Johnston, C.F. (1993) Immunocytochemical demonstration of neuropeptides in the peripheral nervous system of the roundworm *Ascaris suum* (Nematoda, Ascaroidea). *Parasitology Research* 79, 302–308.

Brownlee, D.J.A., Fairweather, I. and Johnston, C.F. (1994) Immunocytochemical demonstration of peptidergic and serotoninergic components in the enteric nervous system of the roundworm *Ascaris suum* (Nematoda, Ascaroidea). *Parasitology* 108, 89–103.

Brownlee, D.J.A., Holden-Dye, L., Fairweather, I. and Walker, R.J. (1995) The action of serotonin and the nematode neuropeptide KSAYMRFamide on the pharyngeal muscle of the parasitic nematode, *Ascaris suum*. *Parasitology* 111, 379–384.

Brownlee, D.J.A., Holden-Dye, L. and Walker, R.J. (1997) Actions of the anthelmintic ivermectin on the pharyngeal muscle of the parasitic nematode, *Ascaris suum*. *Parasitology* 115, 553–561.

Burr, A.H. (1985) The photomovement of *Caenorhabditis elegans*, a nematode which lacks ocelli: proof that the response is to light not radiant heating. *Photochemistry and Photobiology* 41, 577–582.

Caldwell, P.C. and Ellory, J.C. (1968) Ion movement in the somatic muscle cells of *Ascaris lumbricoides. Journal of Physiology* 197, 75–76.

Chalfie, M. and Sulston, J.E. (1981) Developmental genetics of the mechanosensory neurones of *Caenorhabditis elegans. Developmental Biology* 82, 358–370.

Chalfie, M. and White, J.G. (1988) The nervous system. In: Wood, W.B. (ed.) *The Nematode* Caenorhabditis elegans. Cold Spring Harbor Press, Cold Spring Harbor, New York, pp. 337–391.

Chaudhuri, J. and Donahue, M.J. (1989) Serotonin receptors in the tissues of adult *Ascaris suum. Molecular and Biochemical Parasitology* 35, 191–198.

Clemens, C.D., Aumann, J., Spiegel, Y. and Wyss, U. (1994) Attractant-mediated behaviour of mobile stages of *Heterodera schachtii. Fundamental and Applied Nematology* 17, 569–574.

Colquhoun, L., Holden-Dye, L. and Walker, R.J. (1991) The pharmacology of cholinoceptors on the somatic muscle-cells of the parasitic nematode *Ascaris suum. Journal of Experimental Biology* 158, 509–530.

Combes, D., Fedon, Y., Grauso, M., Toutant, J.P. and Arpagaus, M. (2000) Four genes encode acetylcholinesterases in the nematodes *Caenorhabditis elegans* and *Caenorhabditis briggsae*: cDNA sequences, genomic structures, mutations and *in vivo* expression. *Journal of Molecular Biology* 300, 727–742.

Combes, D., Fedon, Y., Toutant, J.P. and Arpagaus, M. (2001) Acetylcholinesterase genes in the nematode *Caenorhabditis elegans. International Review of Cytology* 209, 207–239.

Cowden, C. and Stretton, A.O.W. (1993) AF2, an *Ascaris* neuropeptide: isolation, sequence and bioactivity. *Peptides* 14, 423–430.

Cowden, C. and Stretton, A.O.W. (1995) Eight novel FMRFamide-like neuropeptides isolated from the nematode *Ascaris suum. Peptides* 16, 491–500.

Cowden, C., Stretton, A.O.W. and Davis, R.E. (1989) AF1, a sequenced bioactive neuropeptide isolated from the nematode *Ascaris suum. Neuron* 2, 1465–1473.

Cowden, C., Sithigorngul, P., Brackley, P., Guastella, J. and Stretton, A.O.W. (1993) Localization and differential expression of FMRFamide-like immunoreactivity in the nematode *Ascaris suum. Journal of Comparative Neurology* 333, 455–468.

Davis, E.L., Aron, L.M., Pratt, L.H. and Hussey, R.S. (1992) Novel immunization procedures used to develop antibodies that bind to specific structures in *Meloidogyne* spp. *Phytopathology* 82, 1244–1250.

Davis, R.E. (1998a) Action of excitatory amino acids on hypodermis and the motornervous system of *Ascaris suum*: pharmacological evidence for a glutamate transporter. *Parasitology* 116, 487–500.

Davis, R.E. (1998b) Neurophysiology of glutamatergic signalling and anthelmintic action in *Ascaris suum*: pharmacological evidence for a kainate receptor. *Parasitology* 116, 471–486.

Davis, R.E. and Stretton, A.O.W. (1996) The motornervous system of *Ascaris*: electrophysiology and anatomy of the neurons and their control by neuromodulators. *Parasitology* 113 (Suppl.), S97–S117.

Davis, R.E. and Stretton, A.O.W. (2001) Structure–activity relationships of 18 endogenous neuropeptides on the motornervous system of the nematode *Ascaris suum. Peptides* 22, 7–23.

Del Castillo, J., De Mello, W.C. and Morales, T. (1963) The physiological role of acetylcholine in the neuromuscular system of *Ascaris lumbricoides. Archives Internationales de Physiologie et Biochimie* 71, 741–757.

Dent, J.A., Davis, M.W. and Avery, L. (1997) *avr-15* encodes a chloride channel subunit that mediates inhibitory glutamatergic neurotransmission and ivermectin sensitivity in *Caenorhabditis elegans. EMBO Journal* 16, 5867–5879.

Desai, C. and Horvitz, H.R. (1989) *Caenorhabditis elegans* mutants defective in the functioning of the motor neurons responsible for egg laying. *Genetics* 121, 703–721.

Desai, C., Garriga, G., McIntire, S.L. and Horvitz, H.R. (1988) A genetic pathway for the development of the *Caenorhabditis elegans* HSN motor neurons. *Nature* 336, 638–646.

Devine, K. and Jones, P.W. (2002) Investigations into the chemoattraction of the potato cyst nematodes *Globodera rostochiensis* and *G. pallida* towards fractionated potato root leachate. *Nematology* 5, 65–75.

Driscoll, M. and Kaplan, J. (1997) Mechanotransduction. In: Riddle, D.L., Blumenthal, T., Meyer, B.J. and Priess, J.R. (eds) *C. elegans II*. Cold Spring Harbor Laboratory Press, Cold Spring Harbor, New York, pp. 645–678.

Eckenhoff, R.G. and Yang, B.J. (1994) Absence of pressure antagonism of ethanol narcosis in *C. elegans. Neuroreport* 6, 77–80.

Edison, A.S., Messinger, L.A. and Stretton, A.O.W. (1997) *afp*-1: a gene encoding multiple transcripts of a new class of FMRFamide-like neuropeptides in the nematode *Ascaris suum. Peptides* 18, 929–935.

Emmons, S.W. and Sternberg, P.W. (1997) Male development and mating behaviour. In: Riddle, D.L., Blumenthal, T., Meyer, B.J. and Priess, J.R. (eds) *C. elegans II*. Cold Spring Harbor Laboratory Press, Cold Spring Harbor, New York, pp. 295–334.

Fellowes, R.A., Maule, A.G., Marks, N.J., Geary, T.G., Thompson, D.P., Shaw, C. and Halton, D.W. (1998) Modulation of the motility of the vagina vera of *Ascaris suum in vitro* by FMRFamide-related peptides. *Parasitology* 116, 277–287.

Fellowes, R.A., Dougan, P.M., Maule, A.G., Marks, N.J. and Halton, D.W. (1999) Neuromusculature of the ovijector of *Ascaris suum* (Ascaroidea, Nematoda): an ultrastructural and immunocytochemical study. *Journal of Comparative Neurology* 415, 518–528.

Fellowes, R.A., Maule, A.G., Marks, N.J., Geary, T.G., Thompson, D.P. and Halton, D.W. (2000) Nematode neuropeptide modulation of the vagina vera of *Ascaris suum*: *in vitro* effects of PF1, PF2, PF4, AF3 and AF4. *Parasitology* 120, 79–89.

Fioretti, L., Porter, A., Haydock, P.J. and Curtis, R.H.C. (2002) Monoclonal antibodies reactive with secreted-excreted products from the amphids and the cuticle surface of *Globodera pallida* affect nematode movement and delay invasion of potato roots. *International Journal for Parasitology* 32, 1709–1718.

Geary, T.G., Price, D.A., Bowman, J.W., Winterrowd, C.A., Mackenzie, C.D., Garrison, R.D., Williams, J.F. and Friedman, A.R. (1992) Two FMRFamide-like peptides from the free-living nematode *Panagrellus redivivus. Peptides* 13, 209–214.

Geary, T.G., Sims, S.M., Thomas, E.M., Vanover, L., Davis, J.P., Winterrowd, C.A., Klein, R.D., Ho, N.F. and Thompson, D.P. (1993) *Haemonchus contortus*: ivermectin-induced paralysis of the pharynx. *Experimental Parasitology* 77, 88–96.

Geary, T.G., Marks, N.J., Maule, A.G., Bowman, J.W., Alexander-Bowman, S.J., Larsen, M.J., Kubiak, T.M., Davis J.P. and Thompson, D.P. (1999) Pharmacology of FMRFamide-related peptides in helminths. *Annals of the New York Academy of Sciences* 897, 212–227.

Goh, S.L. and Davey, K.G. (1976) Selective uptake of noradrenaline, dopa and 5–hydroxytryptamine by the nervous system of *Phocanema decipiens* (Nematoda): a light autoradiographic and ultrastructural study. *Tissue and Cell* 8, 421–435.

Greet, D.N. and Perry, R.N. (1992) Sexual differentiation and behaviour of the Nematoda and Nematomorpha. In: Adiyodi, K.G. and Adiyodi, R.G. (eds) *Reproductive Biology of Invertebrates*, Vol. V. Oxford and IBH, New Delhi, pp. 148–173.

Guastella, J., Johnson, C.D. and Stretton, A.O. (1991) GABA-immunoreactive neurons in the nematode *Ascaris*. *Journal of Comparative Neurology* 307, 584–597.

Hamdan, F.F., Ungrin, M.D., Abramovitz, M. and Ribeiro, P. (1999) Characterization of a novel serotonin receptor from *Caenorhabditis elegans*: cloning and expression of two splice variants. *Journal of Neurochemistry* 72, 1372–1383.

Hardaker, L.A., Singer, E., Kerr, R., Zhou, G. and Schafer, W.R. (2001) Serotonin modulates locomotory behavior and coordinates egg-laying and movement in *Caenorhabditis elegans*. *Journal of Neurobiology* 49, 303–313.

Hart, A., Sims, S. and Kaplan, J. (1995) A synaptic code of sensory modalities revealed by analysis of the *C. elegans* GLR-1 glutamate receptor. *Nature* 378, 82–85.

Hobert, O. (2003) Behavioral plasticity in *C. elegans*: paradigms, circuits, genes. *Journal of Neurobiology* 54, 203–223.

Hoekstra, R., Visser, A., Wiley, L.J., Weiss, A.S., Sangster, N.C. and Roos, M.H. (1997) Characterization of an acetylcholine receptor gene of *Haemonchus contortus* in relation to levamisole resistance. *Molecular and Biochemical Parasitology* 84, 179–187.

Holden-Dye, L., Hewitt, G.M., Wann, K.T., Krogsgaard-Larsen, P. and Walker, R.J. (1988) Studies involving avermectin and the gamma-aminobutyric acid (GABA) receptor of *Ascaris suum* muscle. *Pesticide Science* 24, 231–245.

Holden-Dye, L., Krogsgaard-Larsen, P., Nielsen, L. and Walker, R.J. (1989) GABA receptors on the somatic muscle cells of the parasitic nematode, *Ascaris suum*: stereoselectivity indicates similarity to a GABA-type agonist recognition site. *British Journal of Pharmacology* 98, 841–850.

Holden-Dye, L., Brownlee, D.J.A. and Walker, R.J. (1997) The effects of the peptide KPNFIRFamide (PF4) on somatic muscle cells of the parasitic nematode *Ascaris suum*. *British Journal of Pharmacology* 120, 379–386.

Horvitz, H.R., Chalfie, M., Trent, J., Sulston, J. and Evans, P.D. (1982) Serotonin and octopamine in the nematode *Caenorhabditis elegans*. *Science* 216, 1012–1014.

Huang, X., Duran, E., Diaz, F., Xiao, H., Messer, W.S., Jr and Komuniecki, R.W. (1999) Alternative-splicing of serotonin receptor isoforms in the pharynx and muscle of the parasitic nematode, *Ascaris suum*. *Molecular and Biochemical Parasitology* 101, 95–106.

Huang, X., Xiao, H., Rex, E.B., Hobson, R.J., Messer, W.S., Komuniecki, P.R. and Komuniecki, R.W. (2002) Functional characterization of alternatively spliced 5-HT2 receptor isoforms from the pharynx and muscle of the parasitic nematode, *Ascaris suum*. *Journal of Neurochemistry* 83, 249–258.

Hussein, A.S., Kichenin, K. and Selkirk, M.E. (2002a) Suppression of secreted acetylcholinesterase expression in *Nippostrogylus brasiliensis* by RNA interference. *Molecular and Biochemical Parasitology* 122, 91–94.

Hussein, A.S., Harel, M. and Selkirk, M.E. (2002b) A distinct family of acetylcholinesterases is secreted by *Nippostrogylus brasiliensis*. *Molecular and Biochemical Parasitology* 123, 125–134.

Isaac, R.E., Siviter, R.J., Stancombe, P., Coates, D. and Shirras, A.D. (2000) Conserved roles for peptidases in the processing of invertebrate neuropeptides. *Biochemical Society Transactions* 28, 460–464.

Jaffe, H., Huettel, R.N., Demilo, A.B., Hayes, D.K. and Rebois, R.V. (1989) Isolation and identification of a compound from soybean cyst nematode, *Heterodera glycines*, with sex pheromone activity. *Journal of Chemical Ecology* 15, 2031–2043.

Jarman, M. (1959) Electrical activity in the muscle cells of *Ascaris lumbricoides*. *Nature* 184, 1244.

Jin, Y., Hoskins, R. and Horvitz, H.R. (1994) Control of type-D GABAergic neuron differentiation by *C. elegans* UNC-30 homeodomain protein. *Nature* 372, 780–783.

Johnson, C.D. and Stretton, A.O.W. (1987) GABA-immunoreactivity in inhibitory motor neurones of the nematode *Ascaris. Journal of Neuroscience* 7, 223–235.

Johnson, C.D., Reninitz, C.A., Sithigorngul, P. and Stretton, A.O.W. (1996) Neuronal localization of serotonin in the nematode *Ascaris suum. Journal of Comparative Neurology* 367, 352–360.

Jones, J.T. (2002) Nematode sense organs. In: Lee, D.L. (ed.) *The Biology of Nematodes.* Taylor and Francis, London, pp. 353–368.

Jones, J.T., Perry, R.N. and Johnston, M.R.L. (1991) Electrophysiological recordings of electrical activity and responses to stimulants from *Globodera rostochiensis* and *Syngamus trachea. Revue de Nématologie* 14, 467–473.

Jones, J.T., Perry, R.N., Johnston, M.R.L. and Burrows, P.R. (1992) Investigations of the secretions of nematode sense organs using molecular biological techniques. In: *European Society of Nematologists 21st International Symposium, Albufeira, Portugal*, p.49.

Jones, J.T., Perry, R.N. and Johnston, M.R.L. (1994) Changes in the ultrastructure of the amphids of the potato cyst nematode, *Globodera rostochiensis*, during development and infection. *Fundamental and Applied Nematology* 17, 369–382.

Kaplan, J.M. (1996) Sensory signaling in *Caenorhabditis elegans. Current Opinion in Neurobiology* 6, 494–499.

Kaplan, J.M. and Horvitz, H.R. (1993) A dual mechanosensory and chemosensory neuron in *Caenorhabditis elegans. Proceedings of the National Academy of Sciences, USA* 90, 2227–2231.

Kass, J., Jacob, J.C., Kim, P. and Kaplan, J.M. (2001) The EGL-3 proprotein convertase regulates mechanosensory responses of *Caenorhabditis elegans. Journal of Neuroscience* 21, 9265–9272.

Kim, J., Poole, D.S., Waggoner, L.E., Kempf, A., Ramirez, D.S., Treschow, P.D. and Schafer, W.R. (2001) Genes affecting the activity of nicotinic receptors involved in *Caenorhabditis elegans* egg-laying behavior. *Genetics* 157, 1599–1610.

Kimber, M.J., Fleming, C.C., Bjourson, A.J., Halton, D.W. and Maule, A.G. (2001) FMRFamide-related peptides in potato cyst nematodes. *Molecular and Biochemical Parasitology* 116, 199–208.

Komatsu, H., Mori, I., Rhee, J.S., Akaike, N. and Ohshima, Y. (1996) Mutations in a cyclic nucleotide-gated channel lead to abnormal thermosensation and chemosensation in *C. elegans. Neuron* 17, 707–718.

Laughton, D.L., Wheeler, S.V., Lunt, G.G. and Wolstenholme, A.J. (1995) The β-subunit of *Caenorhabditis elegans* avermectin receptor responds to glycine and is encoded by chromosome 1. *Journal of Neurochemistry* 64, 2354–2357.

Laughton, D.L., Lunt, G.G. and Wolstenholme, A.J. (1997) Reporter gene constructs suggest that the *Caenorhabditis elegans* avermectin receptor beta-subunit is expressed solely in the pharynx. *Journal of Experimental Biology* 200, 1509–1514.

Leach, L., Trudgill, D.L. and Gahan, P. (1987) Immunocytochemical localization of neurosecretory amines and peptides in the free-living nematode, *Goodeyus ulmi. Histochemical Journal* 19, 471–475.

Lee, D. L. and Atkinson, H.J. (1976) *The Physiology of Nematodes*, 2nd edn. Macmillan, London, 215 pp.

Lee, Y.S., Park, Y.S., Chang, D.J., Hwang, J.M., Min, C.K., Kaang, B.K. and Cho, N.J. (1999) Cloning and expression of a G protein-linked acetylcholine receptor from *Caenorhabditis elegans. Journal of Neurochemistry* 72, 58–65.

Lewis, J.A., and Hodgkin, J.A. (1977) Specific neuroanatomical changes in chemosensory mutants of the nematode *Caenorhabditis elegans. Journal of Comparative Neurology* 712, 489–510.

Li, C., Kyuhyung, K. and Nelson, S. (1999a) FMRFamide-related peptide gene family in *Caenorhabditis elegans. Brain Research* 848, 26–34.

Li, C., Nelson, L.S., Kim, K., Nathoo, A. and Hart, A.C. (1999b) Neuropeptide gene families in the nematode *Caenorhabditis elegans. Annals of the New York Academy of Sciences* 897, 239–253.

Li, H., Avery, L., Denk, W. and Hess, G.P. (1997) Identification of chemical synapses in the pharynx of *Caenorhabditis elegans. Proceedings of the National Academy of Sciences, USA* 94, 5912–5916.

Li, J., Zhu, X.D., Ashton, F.T., Gamble, H.R. and Shad, G.A. (2001) Sensory neuroanatomy of a passively ingested nematode parasite, *Haemonchus contortus*: amphidial neurons of the third stage larva. *Journal of Parasitology* 87, 65–72.

Liu, K.S. and Sternberg, P.W. (1995) Sensory regulation of male mating behavior in *Caenorhabditis elegans. Neuron* 14, 79–89.

Loer, C.M. and Kenyon, C.J. (1993) Serotonin-deficient mutants and male mating behavior in the nematode *Caenorhabditis elegans. Journal of Neuroscience* 13, 5407–5417.

McIntire, S.L., Garriga, G., White, J., Jacobson, D. and Horvitz, H.R. (1992) Genes necessary for directed axonal elongation or fasciculation in *C. elegans. Neuron* 8, 307–322.

McIntire, S.L., Jorgensen, E. and Horvitz, H.R. (1993a) Genes required for GABA function in *Caenorhabditis elegans. Nature* 364, 334–337.

McIntire, S.L., Jorgensen, E., Kaplan, J. and Horvitz, H.R. (1993b) The GABAergic nervous system of *Caenorhabditis elegans. Nature* 364, 337–341.

McIntire, S.L., Reimer, R.J., Schuske, K., Edwards, R.H. and Jorgensen, E. (1997) Identification and characterization of the vesicular GABA transporter. *Nature* 389, 870–876.

Mapes, C.J. (1965) Structure and function in the nematode pharynx I. The structure of the pharynges of *Ascaris lumbricoides, Oxyuris equi, Aplectana brevicaudata* and *Panagrellus silusiae. Parasitology* 55, 269–284.

Mapes, C.J. (1966) Structure and function in the nematode pharynx III: the pharyngeal pump of *Ascaris lumbricoides. Parasitology* 56, 137–149.

Maricq, A.V., Peckol, E., Driscoll, M. and Bargmann, C.I. (1995) Mechanosensory signalling in *C. elegans* mediated by the GLR-1 glutamate receptor. *Nature* 378, 78–81.

Marks, N.J., Shaw, C., Davis, J.P., Maule, A.G., Verhaert, P., Geary, T.G. and Thompson, D.P. (1996) Isolation of AF2 (KHEYLRFamide) from *Caenorhabditis elegans*: evidence for the presence of more than one FMRFamide-related peptide encoding gene. *Biochemical and Biophysical Research Communications* 217, 845–851.

Marks, N.J., Maule, A.G., Geary, T.G., Thompson, D.P., Davis, J.P., Halton, D.W., Verhaert, P. and Shaw, C. (1997) APEASPFIRFamide, a novel FMRFamide-related decapeptide from *Caenorhabditis elegans*: structure and myoactivity. *Biochemical and Biophysical Research Communications* 231, 591–595.

Marks, N.J., Maule, A.G., Li, C., Geary, T.G., Thompson, D.P., Halton, D.W. and Shaw, C. (1998) PF3/AF8 is present in the free-living nematode *Caenorhabditis elegans. Biochemical and Biophysical Research Communications* 248, 422–425.

Marks, N.J., Maule, A.G., Li, C., Nelson, L.S., Thompson, D.P., Alexander-Bowman, S., Geary, T.G., Halton, D.W., Verhaert, P. and Shaw, C. (1999a) Isolation, pharmacology and gene organization of KPSFVRFamide: a neuropeptide from *Caenorhabditis elegans. Biochemical and Biophysical Research Communications* 254, 222–230.

Marks, N.J., Sangster, N.C., Maule, A.G., Halton, D.W., Thompson, D.P., Geary, T.G. and Shaw, C. (1999b) Structural characterisation and pharmacology of KHEYLRFamide (AF2) and KSAYMRFamide (PF3/AF8) from *Haemonchus contortus*. *Molecular and Biochemical Parasitology* 100, 185–194.

Marks, N.J., Shaw, C., Halton, D.W., Li, C., Thompson, D.P., Geary, T.G. and Maule, A.G. (2001) Isolation and pharmacology of AADGAPLIRFamide and SV`PGVLRFamide from *Caenorhabditis elegans*. *Biochemical and Biophysical Research Communications* 286, 1170–1176.

Martin, R.J. (1982) Electrophysiological effects of piperazine and diethyl-carbamazine on *Ascaris suum* somatic muscle. *British Journal of Pharmacology* 77, 255–265.

Martin, R.J. (1985) γ-Aminobutyric acid- and piperazine-activated single-channel currents from *Ascaris suum* body muscle. *British Journal of Pharmacology* 84, 445–461.

Martin, R.J., Robertson, A.P., Bjorn, H. and Sangster, N.C. (1997) Heterogeneous levamisole receptors: a single-channels study of nicotinic acetylcholine receptors from *Oesophagostomum dentatum*. *European Journal of Pharmacology* 322, 249–257.

Martin, R.J., Purcell, J., Robertson, A.P. and Valkanov, M.A. (2002) Neuromuscular organisation and control in nematodes. In: Lee, D.L. (ed.) *The Biology of Nematodes*. Taylor and Francis, London, pp. 321–344.

Maule, A.G., Shaw, C., Bowman, J.W., Halton, D.W., Thompson, D.P., Geary, T.G. and Thim, L. (1994a) KSAYMRFamide: a novel FMRFamide-related heptapeptide from the free-living nematode, *Panagrellus redivivus*, which is myoactive in the parasitic nematode *Ascaris suum*. *Biochemical and Biophysical Research Communications* 200, 973–980.

Maule, A.G., Shaw, C., Bowman, J.W., Halton, D.W., Thompson, D.P., Geary, T.G. and Thim, L. (1994b) The FMRFamide-like neuropeptide AF2 (*Ascaris suum*) is present in the free-living nematode *Panagrellus redivivus* (Nematoda, Rhabdita). *Parasitology* 109, 351–356.

Maule, A.G., Geary, T.G., Bowman, J.W., Marks, N.J., Blair, K.L., Halton, D.W., Shaw, C. and Thompson, D.P. (1995a) Inhibitory effects of nematode FMRFamide-related peptides (FaRPs) on muscle strips from *Ascaris suum*. *Invertebrate Neuroscience* 1, 225–265.

Maule, A.G., Shaw, C., Bowman, J.W., Halton, D.W., Thompson, D.P., Thim, L., Kubiak, T.M., Martin, R.A. and Geary, T.G. (1995b) Isolation and preliminary biological characterization of KPNFIRFamide, a novel FMRFamide-related peptide from the free-living nematode, *Panagrellus redivivus*. *Peptides* 16, 87–93.

Maule, A.G., Geary, T.G., Bowman, J.W., Shaw, C., Halton, D.W. and Thompson, D.P. (1996) The pharmacology of nematode FMRFamide-related peptides. *Parasitology Today* 12, 351–357.

Maule, A.G., Marks, N.J. and Halton, D.W. (2001) Nematode neuropeptides. In: Kennedy, M.W. and Harnet, W. (eds) *Molecular Biology, Biochemistry and Immunology*. CAB International, Wallingford, UK, pp. 415–436.

Maule, A.G., Mousley, A., Marks, N.J., Day, T.A., Thompson, D.P., Geary, T.G. and Halton, D.W. (2002) Neuropeptide signalling systems – potential drug targets for parasite and pest control. *Current Topics in Medicinal Chemistry* 2, 733–758.

Mellem, J.E., Brockie, G.J., Zheng, Y., Madsen, D.M. and Maricq, A.V. (2002) Decoding of polymodal sensory stimuli by postsynaptic glutamate receptors in *C. elegans*. *Neuron* 36, 933–944.

Mendel, J.E., Korswagen, H.C., Liu, K.S., Hajdu-Cronin, Y.M., Simon, M.I., Plasterk, R.H.A. and Sternberg, P.W. (1995) Participation of the protein G_o in multiple aspects of behavior in *C. elegans*. *Science* 267, 1652–1655.

Moerman, D.G. and Fire, A. (1997) Muscle: structure, function and development. In: Riddle, D.L., Blumenthal, T., Meyer, B.J. and Priess, J.R. (eds) *C. elegans II*. Cold Spring Harbor Laboratory Press, Cold Spring Harbor, New York, pp. 417–470.

Moffett, C.L., Beckett, A.M., Mousley, A., Geary, T.G., Marks, N.J., Halton, D.W., Thompson, D.P. and Maule, A.G. (2003) The ovijector of *Ascaris suum*: multiple response types revealed by *C. elegans* FMRFamide-related peptides. *International Journal for Parasitology* 33, 859–876.

Mongan, N.P., Baylis, H.A., Adcock, C., Smith, G.R., Sansom, M.S. and Sattelle, D.B. (1998) An extensive and diverse gene family of nicotinic acetylcholine receptor alpha subunits in *Caenorhabditis elegans*. *Receptor Channels* 6, 213–228.

Mori, I. and Ohshima, Y. (1995) Neural regulation of thermotaxis in *Caenorhabditis elegans*. *Nature* 376, 344–348.

Morrison, G.E. and van der Kooy, D. (2001) A mutation in the AMPA-type glutamate receptor, *glr-1*, blocks olfactory associative and nonassociative learning in *Caenorhabditis elegans*. *Behavioural Neuroscience* 115, 640–649.

Munn, E.A. and Munn, P.D. (2002) Feeding and digestion. In: Lee, D.L. (ed.) *The Biology of Nematodes*. Taylor and Francis, London, pp. 211–232.

Musso, R. (1930) Die Genitalröhren von *Ascaris lumbricoides* und *megalocephala*. *Zeitschrift für Wissenschaftliche Zoologie* 137, 327–330.

Nathoo, A.N., Moeller, R.A., Westlund, B.A. and Hart, A.C. (2001) Identification of neuropeptide-like protein gene families in *Caenorhabditis elegans* and other species. *Proceedings of the National Academy of Sciences, USA* 98, 14000–14005.

Natoff, I.L. (1969) The pharmacology of the cholinoceptor in muscle preparations of *Ascaris lumbricoides* var. *suum*. *British Journal of Pharmacology* 37, 251–257.

Nelson, L.S., Rosoff, M.L. and Li, C. (1998) Disruption of a neuropeptide gene, *flp-1*, causes multiple behavioral defects in *Caenorhabditis elegans*. *Science* 281, 1686–1690.

Niacaris, T. and Avery, L. (2003) Serotonin regulates repolarization of the *C. elegans* pharyngeal muscle. *Journal of Experimental Biology* 206, 223–231.

Olde, B. and McCombie, W.R. (1997) Molecular cloning and functional expression of a serotonin receptor from *Caenorhabditis elegans*. *Journal of Molecular Neuroscience* 8, 53–62.

Opperman, C.H. and Chang, S. (1992) Nematode acetylcholinesterases: molecular forms and their potential role in nematode behaviour. *Parasitology Today* 8, 406–411.

Park, Y.S., Lee, Y.S., Cho, N.J. and Kaang, B.K. (2000) Alternative splicing of *gar-1*, a *Caenorhabditis elegans* G-protein-linked acetylcholine receptor gene. *Biochemical Biophysical Research Communications* 268, 354–358.

Pemberton, D.J., Franks, C.J., Walker, R.J. and Holden-Dye, L. (2001) Characterization of glutamate-gated chloride channels in the pharynx of wild-type and mutant *Caenorhabditis elegans* delineates the role of the subunit GluCl-alpha2 in the function of the native receptor. *Molecular Pharmacology* 59, 1037–1043.

Pennington, A.J. and Martin, R.J. (1990) A patch-clamp study of acetylcholine-activated ion channels in *Ascaris suum* muscle. *Journal of Experimental Biology* 154, 201–221.

Perry, R.N. (1996) Chemoreception in plant parasitic nematodes. In: Webster, R.K., Zentmyer, G. and Shaner, G. (eds) *Annual Review of Phytopathology*, Vol. 34. Annual Reviews, Palo Alto, California, pp. 181–199.

Perry, R.N. (1997) Plant signals in nematode hatching and attraction. In: Grundler, F.M.W., Ohl, S.A. and Fenoll, C. (eds) *Cellular and Molecular Aspects of Plant–Nematode Interactions*. Kluwer Academic Publishers, Dordrecht, The Netherlands, pp. 38–50.

Perry, R.N. (2001) Analysis of the sensory responses of parasitic nematodes using electro-physiology. *International Journal for Parasitology* 31, 908–917.

Perry, R.N. (2002) Hatching. In: Lee, D.L. (ed.) *The Biology of Nematodes*. Taylor and Francis, London, pp. 147–169.

Perry, R.N. and Aumann, J. (1998) Behaviour and sensory responses. In: Perry, R.N. and Wright, D.J. (eds) *The Physiology and Biochemistry of Free-living and Plant-parasitic Nematodes*. CAB International, Wallingford, UK, pp. 75–102.

Pritchard, D.I. (1993) Why do some parasitic nematodes secrete acetylcholinesterase (AChE)? *International Journal for Parasitology* 23, 549–550.

Purcell, J., Robertson, A.P., Thompson, D.P. and Martin, R.J. (2002) The time-course of the response to the FMRFamide-related peptide PF4 in *Ascaris suum* muscle cells indicates direct gating of a chloride ion-channel. *Parasitology* 124, 649–656.

Raizen, D.M., Lee, R.Y. and Avery, L. (1995) Interacting genes required for pharyngeal excitation by motor neuron MC in *Caenorhabditis elegans*. *Genetics* 141, 1365–1382.

Rand, J.B. and Nonet, M.L. (1997) Synaptic transmission. In: Riddle, D.L., Blumenthal, T., Meyer, B.J. and Priess, J.R. (eds) *C. elegans II*. Cold Spring Harbor Laboratory Press, Cold Spring Harbor, New York, pp. 611–643.

Ranganathan, R., Cannon, S.C. and Horvitz, H.R. (2000) MOD-1 is a serotonin-gated chloride channel that modulates locomotory behaviour in *C. elegans*. *Nature* 408, 470–475.

Reger, J.F. (1966) The fine structure of fibrillar components and plasma membrane contacts in esophageal myoepithelium of *Ascaris lumbricoides* (var. *suum*). *Journal of Ultrastructural Research* 14, 602–617.

Reinitz, C.A. and Stretton, A.O.W. (1996) Behavioural and cellular effects of serotonin on locomotion and male mating posture in *Ascaris suum* (Nematoda). *Journal of Comparative Physiology A: Sensory Neural and Behavioural Physiology* 178, 655–667.

Reinitz, C.A., Herfel, H.G., Messinger, L.A. and Stretton, A.O.W. (2000) Changes in locomotory behaviour and cAMP produced in *Ascaris suum* by neuropeptides from *Ascaris suum* of *Caenorhabditis elegans*. *Molecular and Biochemical Parasitology* 111, 185–197.

Rex, E. and Komuniecki, R.W. (2002) Characterization of a tyramine receptor from *Caenorhabditis elegans*. *Journal of Neurochemistry* 82, 1352–1359.

Richmond, J.E. and Jorgensen, E.M. (1999) One GABA and two acetylcholine receptors function at the *C. elegans* neuromuscular junction. *Nature Neuroscience* 2, 791–797.

Riddle, D.L. and Albert, P.S. (1997) Genetic and environmental regulation of dauer larva development. In: Riddle, D.L., Blumenthal, T., Meyer, B.J., and Priess, J.R. (eds) *C. elegans II*. Cold Spring Harbor Laboratory Press, Cold Spring Harbor, New York, pp. 739–768.

Riga, E., Perry, R.N., Barrett, J. and Johnston, M.R.L. (1995) Investigation of the chemosensory function of amphids of *Syngamus trachea* using electrophysiological techniques. *Parasitology* 111, 347–351.

Riga, E., Perry, R.N. and Barrett, J. (1996a) Electrophysiological analysis of the response of males of *Globodera rostochiensis* and *G. pallida* to their female sex pheromones and to potato root diffusate. *Nematologica* 42, 493–498.

Riga, E., Perry, R.N., Barrett, J. and Johnston, M.R.L. (1996b) Electrophysiological responses of the potato cyst nematodes, *Globodera rostochiensis* and *G. pallida*, to their sex pheromones. *Parasitology* 112, 239–246.

Riga, E., Holdsworth, D.R., Perry, R.N., Barrett, J. and Johnston, M.R.L. (1997a) Electrophysiological analysis of the responses of males of the potato cyst nematode, *Globodera rostochiensis*, to fractions of their homospecific sex pheromones. *Parasitology* 115, 311–316.

Riga, E., Perry, R.N., Barrett, J. and Johnston, M.R.L. (1997b) Electrophysiological responses of male potato cyst nematodes, *Globodera rostochiensis* and *G. pallida*, to some chemicals. *Journal of Chemical Ecology* 23, 417–428.

Robertson, A.P., Bjorn, H.E. and Martin, R.J. (1999) Resistance to levamisole resolved at the single-channel level. *FASEB Journal* 13, 749–760.

Robinson, A.F. (2002) Soil and plant interactions' impact on plant-parasitic nematode host finding and recognition. In: Lewis, E.E., Campbell, J.F. and Sukhdeo, M.V.K. (eds) *The Behavioural Ecology of Parasites*. CAB International, Wallingford, UK, pp. 89–110.

Rogers, C.M., Franks, C.J., Walker, R.J., Burke, J.F. and Holden-Dye, L. (2001) Regulation of the pharynx of *Caenorhabditis elegans* by 5-HT, octopamine and FMRFamide-like neuropeptides. *Journal of Neuroscience* 49, 235–244.

Rohrer, S.P. and Arena, J.P. (1995) Ivermectin interactions with invertebrate ion channels. *American Chemical Society Symposium Series* 591, 264–283.

Rohrer, W.H., Esch, H. and Saz, H.J. (1988) Neuromuscular electrophysiology of the filar-ial helminth *Dipetalonema viteae*. *Comparative Biochemistry and Physiology C* 91, 517–523.

Rolfe, R.N. (1999) Electrophysiological analysis of nematode responses. PhD thesis, University of Wales, Aberystwyth.

Rolfe, R.N. and Perry, R.N. (2001) Electropharyngeograms and stylet activity of second stage juveniles of *Globodera rostochiensis*. *Nematology* 3, 31–34.

Rolfe, R.N., Barrett, J. and Perry, R.N. (2000) Analysis of chemosensory responses of second stage juveniles of *Globodera rostochiensis* using electrophysiological techniques. *Nematology* 2, 523–533.

Rolfe, R.N., Barrett, J. and Perry, R.N. (2001a) Electrophysiological analysis of responses of adult females of *Brugia pahangi* to some chemicals. *Parasitology* 122, 347–357.

Rolfe, R.N., Barrett, J. and Perry, R.N. (2001b) Analysis of responses of *Leidynema appen-diculata* to acetylcholine using electrophysiological techniques. *Journal of Parasitology* 87, 917–919.

Rosenbluth, J. (1965a) Ultrastructure organisation of obliquely striated muscle fibres in *Ascaris lumbricoides*. *Journal of Cell Biology* 25, 495–515.

Rosenbluth, J. (1965b) Ultrastructure organisation of somatic muscle cells in *Ascaris lum-bricoides* II. Intermuscular junctions, neuromuscular junctions and glycogen stores. *Journal of Cell Biology* 26, 579–591.

Rosoff, M.L., Doble, K.E., Price, D.A. and Li, C. (1993) The *flp*-1 propeptide is processed into multiple, highly similar FMRFamide-like peptides in *Caenorhabditis elegans*. *Peptides* 14, 331–338.

Rozhova, E.K., Malyutina, T.A. and Shishov, B.A. (1980) Pharmacological characteristics if cholinoreception in somatic muscle of the nematode *Ascaris suum*. *General Pharmacology* 11, 141–146.

Sawin, E.R., Ranganthan, R. and Horvitz, H.R. (2000) *C. elegans* locomotory rate is mod-ulated by the environment through a dopaminergic pathway and by experience through a serotonergic pathway. *Neuron* 26, 619–631.

Schaeffer, J.M. and Bergstrom, A.R. (1988) Identification of gamma-aminobutyric acid and its binding sites in *Caenorhabditis elegans*. *Life Science* 43, 1701–1706.

Schinkmann, K. and Li, C. (1992) Localization of FMRFamide-like peptides in *Caenorhabditis elegans*. *Journal of Comparative Neurology* 316, 251–260.

Ségalat, L., Elkes, D.A. and Kaplan, J.M. (1995) Modulation of serotonin-controlled behav-iours by G$_o$ in *Caenorhabditis elegans*. *Science* 267, 1648–1651.

Segerberg, M.A. and Stretton, A.O. (1993) Actions of cholinergic drugs in the nematode

Ascaris suum: complex pharmacology of muscle and motor neurones. *Journal of General Physiology* 101, 271–296.

Sengupta, P., Chou, J.H. and Bargmann, C.I. (1996) *odr-10* encodes a seven transmembrane domain olfactory receptor required for the responses to the odorant diacetyl. *Cell* 84, 899–909.

Sheriff, J.C., Kotze, A.C., Sangster, N.C. and Martin, R.J. (2002) Effects of macrocyclic lactone anthelmintics on feeding and pharyngeal pumping in *Trichostrongylus colubriformis in vitro. Parasitology* 125, 477–484.

Shingai, R. (2000) Durations and frequencies of free locomotion in wild type and GABAergic mutants of *Caenorhabditis elegans. Neuroscience Research* 38, 71–83.

Smart, D., Johnston, C.F., Curry, W.J., Shaw, C, Halton, D.W., Fairweather, I. and Buchanan, K.D. (1992) Immunoreactivity to two specific regions of chromogranin A in the nervous system of *Ascaris suum*: an immunocytochemical study. *Parasitology Research* 78, 329–335.

Stanssens, P., Bergum, P.W., Gamoermans, Y., Jespeos, L., Laroche, Y., Huang, S. *et al.* (1996) Anticoagulant repertoire of the hookworm *Ancylostoma caninum. Proceedings of the National Academy of Sciences, USA* 93, 2149–2154.

Stewart, G.R., Perry, R.N., Alexander, J. and Wright, D.J. (1993a) A glycoprotein specific to the amphids of *Meloidogyne* species. *Parasitology* 106, 405–412.

Stewart, G.R., Perry, R.N. and Wright, D.J. (1993b) Studies on the amphid specific glycoprotein gp32 in different life-cycle stages of *Meloidogyne* species. *Parasitology* 107, 573–578.

Stewart, G.R., Perry, R.N. and Wright, D.J. (1994) Immunocytochemical studies on the occurrence of gamma-aminobutyric acid in the nervous system of the nematodes *Pangrellus redivivus, Meloidogyne incognita* and *Globodera rostochiensis. Fundamental and Applied Nematology* 17, 433–439.

Stewart, G.R., Perry, R.N. and Wright, D.J. (2001) Occurrence of dopamine in *Panagrellus redivivus* and *Meloidogyne incognita. Nematology* 3, 843–848.

Stretton, A.O.W. and Johnson, C.D. (1985) GABA and 5–HT immunoreactive neurones in *Ascaris. Society for Neuroscience Abstracts* 11, 626.

Sukhdeo, M.V.K., Sukhdeo, S.C. and Bansemir, A.D. (2002) Interactions between intestinal nematodes and their hosts. In: Lewis, E.E., Campbell, J.F. and Sukhdeo, M.V.K. (eds) *The Behavioural Ecology of Parasites*. CAB International, Wallingford, UK, pp. 223–242.

Sulston, J.E. and Horvitz, H.R. (1977) Post-embryonic cell lineages of the nematode, *Caenorhabditis elegans. Developmental Biology* 56, 110–156.

Sulston, J.E., Dew, M. and Brenner, S. (1975) Dopaminergic neurons in the nematode *Caenorhabditis elegans. Journal of Comparative Neurology* 163, 215–226.

Sulston, J.E., Albertson, D.G. and Thomson, J.N. (1980) The *Caenorhabditis elegans* male: postembryonic development of nongonadal structures. *Developmental Biology* 78, 542–576.

Suo, S., Sasagawa, N. and Ishiura, S. (2002) Identification of a dopamine receptor from *Caenorhabditis elegans. Neuroscience Letters* 319, 13–16.

Sze, J.Y., Victor, M., Loer, C., Shi, Y. and Ruvkun, G. (2000) Food and metabolic signalling defects in a *Caenorhabditis elegans* serotonin-synthesis mutant. *Nature* 403, 560–564.

Terenina, N.B. and Shishov, B.A. (1975) Mechanism of oviposition in the nematodes, *Ascaris suum* and *Ascaridia galli. Meditsinskaya Parazitalogiya i Parazitarnye Bolezni* 44, 322–325.

Thompson, D.P., Davis, J.P., Larsen, M.J., Thomas, M., Zinser, E.R., Bowman, J.W., Alexander-Bowman, S.J., Marks, N.J. and Geary, T.G. (2003) Effects of KHEYLRFamide and KNEFIRFamide on cAMP levels in *Ascaris suum* somatic muscle. *International Journal for Parasitology* 33, 199–208.

Treinen, M. and Chalfie, M. (1995) A mutated acetylcholine receptor subunit causes neuronal degeneration in *C. elegans*. *Neuron* 14, 871–877.

Trent, C., Tsuing, N. and Horvitz, H.R. (1983) Egg-laying defective mutants of the nematode *Caenorhabditis elegans*. *Genetics* 104, 619–647.

Trett, M.W. and Perry, R.N. (1985) Functional and evolutionary implications of the anterior sensory anatomy of species of root-lesion nematode (genus *Pratylenchus*). *Revue de Nématologie* 8, 341–355.

Trim, J.E., Holden-Dye, L., Wilson, J., Lockyer, M. and Walker, R.J. (2001) Characterization of 5-HT receptors in the parasitic nematode, *Ascaris suum*. *Parasitology* 122, 207–217.

Trim, N., Holden-Dye, L., Ruddell, R. and Walker, R.J. (1997) The effects of the peptides AF3 (AVPGVLRFamide) and AF4 (GDVPGVLRFamide) on the somatic muscle of the parasitic nematodes *Ascaris suum* and *Ascaridia galli*. *Parasitology* 115, 213–222.

Trim, N., Boorman, J.E., Holden-Dye, L. and Walker, R.J. (1998) The role of cAMP in the action of the peptide AF3 in the parasitic nematodes *Ascaris suum* and *Ascaridia galli*. *Molecular and Biochemical Parasitology* 93, 236–271.

Troemel, E.R., Kimmel, B.E. and Bargmann, C.I. (1995) Reprogramming chemotaxis responses: sensory neurones define olfactory preferences in *C. elegans*. *Cell* 91, 161–169.

Viney, M.E. (2002) Environmental control of nematode life cycles. In: Lewis, E.E., Campbell, J.F. and Sukhdeo, M.V.K. (eds) *The Behavioural Ecology of Parasites*. CAB International, Wallingford, UK, pp. 111–128.

Voinilo, V.A. (1976) The effect of organic acids on the issue from the cyst of potato nematode larva. *Zashchita Rastenii Minsk* 1, 147–150.

Waggoner, L.E., Zhou, G.T., Schafer, R.W. and Schafer, W.R. (1998) Control of alternative behavioral states by serotonin in *Caenorhabditis elegans*. *Neuron* 21, 203–214.

Waggoner, L.E., Hardaker, L.A., Golik, S. and Schafer, W.R. (2000) Effect of a neuropeptide gene on behavioral states in *Caenorhabditis elegans* egg-laying. *Genetics* 154, 1181–1192.

Walker, J., Hoekstra, R., Roos, M.H., Wiley, L.J., Weiss, A.S., Sangster, N.C. and Tait, A. (2001) Cloning and structural analysis of partial acetylcholine receptor subunit genes from the parasitic nematode *Teladorsagia circumcincta*. *Veterinary Parasitology* 97, 329–335.

Walker, R.J., Franks, C.J., Pemberton, D., Rogers, C. and Holden-Dye, L. (2000) Physiological and pharmacological studies on nematodes. *Acta Biologica Hungaria* 51, 379–394.

Walrond, J.P. and Stretton, A.O.W. (1985) Reciprocal inhibition in the motor nervous-system of the nematode *Ascaris*: direct control of ventral inhibitory neurones by dorsal excitatory motor neurones. *Journal of Neuroscience* 5, 9–15.

Wang, K.C. and Chen, T.A. (1985) Ultrastructure of the phasmids of *Scutellonema brachyurum*. *Journal of Nematology* 17, 175–186.

Wann, K.T. (1987) The electrophysiology of the somatic muscle cells of *Ascaris suum* and *Ascaridia galli*. *Parasitology* 94, 555–566.

Waterson, R.H. (1988) Muscle. In: Wood, W.B. (ed.) *The Nematode* Caenorhabditis elegans. Cold Spring Harbor Laboratory Press, Cold Spring Harbor, New York, pp. 281–335.

Weinshenker, D., Garriga, G. and Thomas, J.H. (1995) Genetic and pharmacological analysis of neurotransmitters controlling egg laying in *C. elegans*. *Journal of Neuroscience* 15, 6975–6985.

White, J.G. (1985) Neuronal connectivity in *Caenorhabditis elegans*. *Trends in Neuroscience*, 277–283.

White, J.G. (1988) The anatomy. In: Wood, W.B. (ed.) *The Nematode* Caenorhabditis elegans. Cold Spring Harbor Laboratory, Cold Spring Harbor, New York, pp. 81–122.

White, J.G., Southgate, E., Thomson, J.N. and Brenner, S. (1986) The structure of the nervous system of *Caenorhabditis elegans*. *Philosophical Transactions of the Royal Society of London B (Biological Sciences)* 314, 1–340.

Wikgren, M. and Fagerholm, H.P. (1993) Neuropeptides in sensory structures of nematodes. *Acta Biologica Hungaria* 44, 133–136.

Wiley, L.J., Weiss, A.S., Sangster, N.C. and Li, Q. (1996) Cloning and sequence analysis of the candidate nicotinic acetylcholine receptor alpha subunit gene *tar-1* from *Trichostrongylus colubriformis*. *Gene* 182, 97–100.

Williams, J.A., Shahkolahi, A.M., Abbassi, M. and Donahue, M.J. (1992) Identification of a novel 5–HTN (Nematoda) receptor from *Ascaris* muscle. *Comparative Biochemistry and Physiology* 101, 469–474.

Winter, M.D., McPherson, M.J. and Atkinson, H.J. (2002) Neuronal uptake of pesticides disrupts chemosensory cells of nematodes. *Parasitology* 125, 561–565.

Wood, W.B. (ed.) (1988) *The Nematode* Caenorhabditis elegans. Cold Spring Harbor Laboratory Press, Cold Spring Harbor, New York, 667 pp.

Wright, D.J. (1981) Nematicides: mode of action and new approaches to chemical control. In: Zuckerman, B.M. and Rhode, R.A. (eds) *Plant Parasitic Nematodes*, Vol. III. Academic Press, New York, pp. 421–449.

Wright, D.J. and Awan, F.A. (1978) Catecholaminergic structures in the nervous system of three nematode species, with observations on related enzymes. *Journal of Zoology* 185, 477–489.

Wright, D.J. and Perry, R.N. (1998) Musculature and neurobiology. In: Perry, R.N. and Wright, D.J. (eds) *The Physiology and Biochemistry of Free-living and Plant-parasitic Nematodes*. CAB International, Wallingford, UK, pp. 49–74.

Wright, D.J. and Perry, R.N. (2002) Physiology and biochemistry. In: Gaugler, R. (ed.) *Entomopathogenic Nematology*. CAB International, Wallingford, UK, pp. 145–168.

Wyss, U. (2002) Feeding behaviour of plant parasitic nematodes. In: Lee, D.L. (ed.) *The Biology of Nematodes*. Taylor and Francis, London, pp. 233–260.

Zhu, H., Duerr, J.S., Varoqui, H., McManus, J.R., Rand, J.B. and Erickson, J.D. (2001) Analysis of point mutants in the *Caenorhabditis elegans* vesicular acetylcholine transporter reveals domains involved in substrate translocation. *Journal of Biological Chemistry* 276, 41580–41587.

9 Molecular Basis for Behaviour

MAUREEN M. BARR AND JINGHUA HU

School of Pharmacy, University of Wisconsin-Madison, 777 Highland Avenue, Madison, WI 53705, USA

9.1. Introduction to *C. elegans*

Sydney Brenner selected *Caenorhabditis elegans* because of its rapid free-living life cycle, hermaphroditism, large brood size, genetic amenability, transparency and simple cellular complexity. The cell lineage as well as the entire connectivity and circuitry of the 302-cell hermaphrodite nervous system and partial reconstruction of the 381-cell male nervous system are known. The simple nervous system of *C. elegans* is capable of executing a wide range of complex behaviours. The general approach to studying *C. elegans* behaviours is to first describe the behaviour. Laser microsurgery is performed to determine what neurones are required for the behaviour. Genetic screens identify mutations in genes required for behaviour and culminate in the molecular cloning and characterization of behavioural genes. The organization of this chapter follows this classic forward genetics paradigm.

The molecular genetic tools available in *C. elegans* have enabled the study of numerous behaviours at the cellular, genetic and molecular levels. Genetic screens and cloning as well as functional genomics approaches have identified molecules required for nervous system development, differentiation and function. Techniques pioneered in *C. elegans*, including laser microbeam ablations, mutagenesis and genetic screening, RNA interference (RNAi), mRNA differential display and microarrays, transgenics, green fluorescence protein (GFP) reporters, expressed sequence tags (ESTs), genome annotation and data mining, may be extended to other nematode species. This chapter will discuss how classic and reverse genetic approaches may be employed to elucidate the molecular basis of behaviour.

Several *C. elegans* sourcebooks are available and discussion will not be reproduced here. Two outstanding books that function as '*C. elegans* bibles' have been published: *The Nematode* Caenorhabditis elegans (Wood and the Community of *C. elegans* Researchers, 1988), and C. elegans *II* (Riddle *et al.*, 1997). The former provides an excellent description of *C. elegans* developmental biology while the latter tackles *C. elegans* molecular biology. For technical aspects, C. elegans: *A Practical Approach* (Hope, 1999) is a useful reference. Perhaps the greatest community accomplishments since the publication of C. elegans *II* are the completion of the *C. elegans* genome (*C. elegans* Sequencing Consortium, 1998) and efficient data compilation at WormBase (www.wormbase.org) (Stein *et al.*, 2001; Harris *et al.*, 2003).

9.1.1. *C. elegans* behaviours

C. elegans exhibits a rich repertoire of sensory and motor behaviours and a large portion of C. elegans *II* is dedicated to behaviour. This chapter will not attempt to cover the molecular basis of every behaviour in great detail. For simplicity, we have categorized the behaviours by modality (sensory, motor or complex, using both sensory and motor programs) and provide a brief description of each behaviour.

9.1.1.1. Sensory behaviours

As a soil-dweller, *C. elegans* is exposed to soluble and volatile chemicals as well as temperature gradients and mechanical stimuli and modifies its behaviour based on these stimuli. All sensory behaviours result in a motor output. Chemotaxis, olfaction and osmotic avoidance have been studied intensively (Bargmann and Mori,

1997; Riddle and Albert, 1997). *C. elegans* is able to discriminate between a wide range of volatile and non-volatile (water-soluble) chemical attractants and repellents (Bargmann and Horvitz, 1991). The basic chemotaxis assay measures the ability of *C. elegans* to track a chemical gradient on an agar plate (Ward, 1973). Dauer juvenile formation requires two environmental signals, low food and high dauer pheromone, which trigger dauer development (Bargmann and Mori, 1997; Riddle and Albert, 1997). Nictation, a parasitic host-seeking behaviour whereby the juvenile stands on its tail and moves its head from side to side, is observed as a dauer-specific *C. elegans* behaviour. *C. elegans* prefers the temperature at which it was raised, will exhibit thermotaxis to that temperature and will alter movement so as to continuously stay at the cultivation temperature (Hedgecock and Russell, 1975; Bargmann and Mori, 1997).

    C. elegans sense and distinguish between several different forms of touch (tap, harsh, gentle, nose and head touch, substrate texture and proprioception) (Driscoll and Kaplan, 1997). Tapping the agar plate on which *C. elegans* are cultured evokes backing (Chiba and Rankin, 1990). A gentle touch with an eyelash anywhere along the length of the body induces *C. elegans* to move in the direction opposite to the force (Chalfie and Sulston, 1981; Chalfie *et al.*, 1985). Nose and head touch refers to the response elicited when the *C. elegans* nose or head bumps into an object and the animal reverses direction (Kaplan and Horvitz, 1993; Hart *et al.*, 1995). Substrate texture of the bacterial lawn on which *C. elegans* is cultured affects movement (Sawin *et al.*, 2000). *C. elegans* proprioception entails perception of movement and spatial orientation arising from stimuli within the body itself (Tavernarakis *et al.*, 1997) and may be important for coordinated locomotion and mating behaviour of males (Garcia *et al.*, 2001) (E.M. Peden and M.M. Barr, 2003, unpublished). For a more thorough discussion, excellent reviews on *C. elegans* mechanosensation have recently been published (Ernstrom and Chalfie, 2002; Goodman and Schwarz, 2003).

9.1.1.2. Motor behaviours

Wild-type *C. elegans* moves in a sinusoidal motion by alternating ventral and dorsal musculature turns (Driscoll and Kaplan, 1997). The neural circuit for locomotion has been described (White *et al.*, 1986) and involves the coordination of muscle movement by distinct classes of motor neurones. Egg-laying is driven by contraction of vulval and uterine muscles, which are innervated by two motor neurone classes, the hermaphrodite-specific neurones (HSNs) and ventral cord (VC) neurones (Trent *et al.*, 1983; Bargmann, 1993). Feeding behaviour is mediated by pharyngeal pumping (Avery and Thomas, 1997). The defecation motor programme (DMP) in the hermaphrodite occurs rhythmically every 45 s. The DMP entails a posterior body-wall contraction (pBoc) driving intestinal contents to an anterior body-wall contraction (aBoc), culminating in the expulsion (Exp) of gut contents (Avery and Thomas, 1997).

9.1.1.3. Complex behaviours

Complex and, at the time of this publication, least understood behaviours at the molecular level include aggregation or 'social' behaviour. Different natural isolates of *C. elegans* display either solitary (feeding as individuals) or social (feeding as aggregates) behaviour (de Bono, 2003). *C. elegans* exhibits behavioural plasticity in the forms of adaptation, habituation, learning and memory (Hobert, 2003). Ageing affects the rates of behaviour (Felkai *et al.*, 1999; Dillin *et al.*, 2002; Herndon *et al.*, 2002).

Male mating is the most complex behaviour executed by *C. elegans*. Male copulation is comprised of a stereotyped subseries of sensory and motor behaviours, namely, response, backing, turning, vulval location, spicule insertion and sperm transfer (Fig. 9.1A – D; Liu and Sternberg, 1995). Response behaviour entails the male tail touching the hermaphrodite, ceasing of forward movement and commencement of backing. When encountering the head or tail of the hermaphrodite, the male turns via a sharp ventral coil. He continues backing until he encounters, locates and stops at her vulva (also called Lov behaviour for location of vulva) (Barr and Sternberg, 1999). The male then inserts his copulatory spicules by periodic and prolonged muscle contractions (Garcia *et al.*, 2001). Once the spicules penetrate the vulva, sperm transfer into the hermaphrodite uterus ensues. In addition to copulatory behaviour, the *C. elegans* male exhibits two additional sex-specific behaviours. In response to a short-range diffusible signal from the adult hermaphrodite, the male increases the frequency of reversals (Simon and Sternberg, 2002). In a second male-specific locomotory behaviour, an isolated adult male will leave a food source. Emmons and Lipton (2003) propose that this 'leaving' behaviour is similar to sex drive. Throughout this chapter, we shall use *C. elegans* male mating behaviour as an example of how to study the molecular basis of behaviour.

9.1.2. Neuroanatomy

The nervous system is well described (Chalfie and White, 1988) and a few key features will be mentioned here. There are two distinct parts to the nervous system: 20 neurones form the pharyngeal nervous system while the non-pharyngeal neurones are located throughout the body. The nerve ring is located between the anterior and posterior bulbs of the pharynx, houses a majority of neurones and functions as the 'central nervous system' (CNS). The ventral nerve cord contains motor and interneurones that coordinate locomotion and act as a spinal cord. The 302 neurones in the hermaphrodite nervous system are categorized into 118 classes, of which about half are interneurones, a third are sensory neurones and a quarter are motor neurones. Some neurones appear to have multiple functions. For example, the interlabial (IL1) mechanoreceptors combine all three properties (Ward *et al.*, 1975). Using morphological criteria, White *et al.* (1986) determined that neurones are interconnected by approximately 5000 chemical synapses, 700 gap junctions and 2000 neuromuscular junctions.

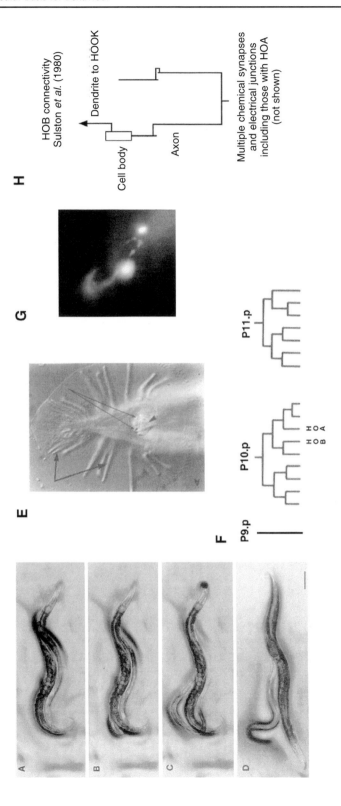

Fig. 9.1. Male mating behaviour. A–D, show the stereotyped mating subbehaviour of (A) backing, (B) turning, (C) location of vulva (Lov), and (D) spicule insertion and sperm transfer (reproduced from Liu and Sternberg, 1995). Hermaphrodite vulva is up and fertilized eggs are visible in the uterus. E. Differential interference contrast (DIC) image of ventral up male tail (reproduced from Liu and Sternberg, 1995). Thick arrows indicate sensory rays, thin arrows show post-cloacal sensilla and the arrowhead points to the hook structure. F. Hook lineage and equivalence group. G. GFP-labelled HOB neurone. H. HOB structure as determined by Sulston *et al.* (1980).

The male nervous system possesses 381 neurones and the complete reconstruction of the male nervous system via serial electron micrographs is under way (S. Emmons and D. Hall, New York, 2002, personal communication). Sexual dimorphism is reflected in behaviour: the two HSNs are required for egg-laying (Trent *et al.*, 1983; Desai and Horvitz, 1989) while many of the 87 male-specific neurones mediate male sensory behaviours (Liu and Sternberg, 1995). Most of the male-specific neurones are located in the tail (Figs 9.1E and 9.2b) and might

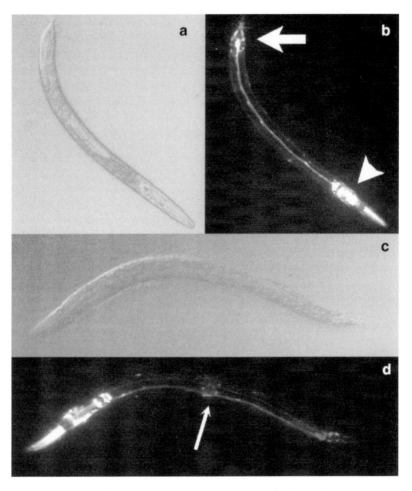

Fig. 9.2. The adult male and hermaphrodite nervous system labelled with pan-neural *Punc-119::GFP* (P is for promoter). The anterior nerve ring is indicated by the arrowhead, and the posterior tail ganglia by a thick arrow. The ventral nerve cord is visible as a long process running between the nerve ring and tail. (a) Differential interference contrast (DIC) image of adult male, ventral up. (b) Standard epifluorescence microscopy showing male nervous system. (c) DIC image of adult hermaphrodite. (d) Standard epifluorescence microscopy showing hermaphrodite nervous system. Note sexually dimorphic features, including the difference between the tail expression between male (b) and hermaphrodite (d) and hermaphrodite-specific neurones (HSNs) near the vulva (thin arrow) in (d).

be thought of as the male's second CNS. In Fig. 9.2, all of the neurones are visualized in living *C. elegans* adult males and hermaphrodites using the *unc-119* pan-neural promoter to drive expression of the GFP reporter.

9.1.3. Synaptic transmission

Classical neurotransmitters (acetylcholine ACh, GABA, the biogenic amines serotonin and dopamine), a few amino acids (glutamate and histamine) and modulatory neuropeptides (including FMRFamide-related peptides (FaRPs)) function in the *C. elegans* nervous system (Chalfie and White, 1988; Rand and Nonet, 1997; Li *et al.*, 1999; Thomas and Lockery, 1999; Nathoo *et al.*, 2001). Historically, pharmacological and histochemical staining were used to determine physiological function. ACh and GABA are the primary excitatory and inhibitory neurotransmitters, respectively, controlling motor functions. Exogenously applied ACh, ACh agonists (e.g. levamisole) or inhibitors of the ACh-degrading enzyme, acetylcholine esterase (AChE), penetrate the cuticle to cause muscle contraction, egg-laying, decreased pharyngeal pumping and male copulatory spicule protraction (Garcia *et al.*, 2001). GABA or GABA agonist (e.g. muscimol) application causes flaccid paralysis. Serotonin reserpine treatment induces egg-laying, pharyngeal pumping and male tail-curling while decreasing locomotion. Exogenous application of dopamine or dopamine antagonists increase or decrease high angled turn frequency, respectively (T.T. Hills, F. Adler, and A.V. Maricq, Utah, 2002, personal communication). Greater than 50 FaRP neuropeptides have been detected in nematodes and exogenous application of a subset increases the activity of the *C. elegans* pharynx in a cut preparation (Rogers *et al.*, 2001). The effects of *C. elegans* neuropeptides have also been explored using muscle preparations from other nematodes, including *Ascaris suum* (Brownlee *et al.*, 2000; Marks *et al.*, 2001).

9.1.4. Lineage and laser ablation

Cell lineage refers to the knowledge of the division patterns and fate decisions (Fig. 9.1F). Using differential interference contrast (DIC) or Nomarski microscopy to visualize cell divisions and migrations, Sulston (1988) determined the entire and largely invariant cell lineage from zygote to adult of both male and hermaphrodite. Sulston and Horvitz (1977) discovered that a set number of cells are programmed to die. For example, the HSNs die in the male embryo while the male-specific cephalic neurones, the CEMs, die in the hermaphrodite embryo. For characterizing the genetic and molecular mechanisms of programmed cell death, Horvitz shared the 2002 Nobel Prize in Medicine and Physiology with Brenner and Sulston (Bargmann and Hodgkin, 2002).

Knowledge of connectivity coupled with the ability to determine the function of single neurones in specific behaviours enables exploration of neural circuitry. With a transparent animal of known and invariant lineage, the function of individual neurones may be ascertained using laser microbeam studies (Sulston and White, 1980; Bargmann and Avery, 1995). In brief, a laser beam is focused on the nucleus of the cell or precursor cell of interest, using DIC, and pulsed until visible damage (seen as scarring or breakdown of the nucleus) is observed. Ablated animals are allowed to recover and subjected to a behavioural assay. Non-specific or collateral damage may result from laser surgery. To rule out this possibility, the nuclei of adjacent cells may be killed and examined for behavioural effects. Laser ablation may not completely eliminate the cell and therefore may not disrupt immediate interactions. Other limitations include the possibility that neurones may alter function when others are killed, that neurones may have functions in addition to the behaviour being assayed or that neurones may have redundant functions and killing one neurone may not be sufficient. Having said this, laser ablation has been used to define the cellular basis of several *C. elegans* behaviours, the first being mechanosensation (Chalfie *et al.*, 1985) and others including locomotion, egg-laying, feeding, chemotaxis to water-soluble chemicals, olfaction of airborne odorants and male mating.

Liu and Sternberg (1995) used laser ablation to assign function to many male-specific neurones. With the exception of the four CEM neurones in the head, the male sex-specific neurones are located in the tail (Sulston *et al.*, 1980; White *et al.*, 1986). The male tail consists of an elongated bursa, fan and proctodeum (Fig. 9.1E). The fan is composed of nine pairs of bilaterally arranged rays. The proctodeum houses two spicules and the gubernaculum. Additionally, there is a hook located anterior to the cloacal and a left–right pair of post-cloacal sensilla. Each ray and hook structure contains two ciliated sensory neurones, an A type neurone whose ciliated sensory ending is embedded in the cuticle and a predicted mechanosensor, and a B type neurone whose cilium is exposed to the environment and a predicted chemosensor. Each spicule has two ciliated sensory neurones (SPV and SPD) and one proprioceptive SPC motor neurone. Response to sensory contact and turning are mediated by the rays, CP motor neurones and sex-specific muscles (Loer and Kenyon, 1993; Liu and Sternberg, 1995). The PVY male-specific interneurone and some non-sex-specific VC neurones are required for backing (Liu, 1996; Liu and Sternberg, 1995). Lov behaviour requires three sensilla that are partially redundant in function. The hook is responsible for general location while the post-cloacal sensilla and spicules precisely locate the vulva (Liu and Sternberg, 1995). When either or both the HOA or HOB hook neurones are ablated, males fail to stop at the vulva, circle the hermaphrodite, and attempt to locate the vulva using an adaptive slow search behaviour that requires the post-cloacal sensilla and spicules. Spicule insertion requires the SPC motor neurones and spicule protractor muscles (Liu and Sternberg, 1995; Garcia *et al.*, 2001). The SPV sensory neurones inhibit (Liu and Sternberg, 1995) while the SPC or SPD spicule neurones may activate sperm transfer (Garcia *et al.*, 2001; E.M. Peden and M.M. Barr, 2003, unpublished results).

Nematodes share a generalized arrangement of four sense organs, or sensilla, in the head (Ward *et al.*, 1975; Ware *et al.*, 1975). Each sensillum has one or more ciliated neuronal endings and two non-neuronal support cells: a sheath cell and a socket cell. One of the left–right pair of *C. elegans* amphid sensilla is shown in Fig.

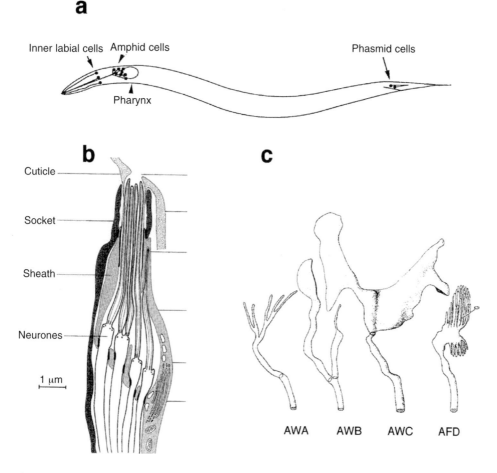

Fig. 9.3. (a) Illustration of cell body positions and dendrites of ciliated neurones in the *C. elegans* hermaphrodite (reproduced from Bargmann and Horvitz, 1991). (b) Transverse section of an amphid sensillum (reproduced from Ward *et al.*, 1975). Ciliated ends of neuronal dendrites are exposed to the environment through an opening in the cuticle. Sheath and socket cells provide structural support. (c) Complex cilium morphologies of amphid wing (AWA, AWB and AWC) and finger (AFD) neurones (reproduced from Ward *et al.*, 1975).

9.3. The amphid socket forms a pore with the cuticle and exposes the cilia of dendrites to the environment. Exposed amphid cilia (or any cilia that are exposed) are categorized as chemosensory. In contrast, if the pore does not open to the environment and the neuronal sensory endings are embedded in the cuticle, the sensillum is classified as mechanosensory. Studies in *C. elegans* reveal that anatomical descriptions may not always translate to functional descriptions. For example, the ASH amphid neurones open to the environment but have dual sensory functions (i.e. chemosensory and mechanosensory), which were discovered via laser ablations and behavioural analysis (Kaplan and Horvitz, 1993).

The laser-killing technique has been successfully applied to the study of thermotaxis and chemotaxis in animal-parasitic nematodes (Ashton *et al.*, 1999). EM reconstructions of the amphid of the dog hookworm *Ancylostoma caninum*, the human threadworm *Strongyloides stercoralis* and the sheep stomach worm *Haemonchus contortus* have been compared with that of *C. elegans*. The positions of corresponding amphid cell bodies have also been identified, mapped and labelled according to the homologous cell in *C. elegans* (Ashton *et al.*, 1998, 1999; Li *et al.*, 2000; Bhopale *et al.*, 2001). Using knowledge of the anatomy and function of individual *C. elegans* amphid neurones as a guide in laser ablation studies, Ashton *et al.* (1999) determined that the AFD finger cell amphid neurones mediate thermotaxis in *A. caninum* and *H. contortus*. The cellular basis of chemotaxis, thermotaxis and developmental switching in *S. stercoralis* was also studied, by killing selected neurones with a microbeam and assaying effects on behaviour. Hence, ablation may be a powerful tool suitable for studying nervous system function and the cellular basis of behaviour in other nematodes with characterized neuroanatomy.

9.1.5. Genetics: forward and reverse

Studies in *C. elegans* have revealed that mutation in a single gene may have profound effects on behaviour. In the first genetic screen, Brenner isolated 77 genes that when mutated cause uncoordinated (Unc) movement (Brenner, 1974). The molecular identity and functions of many of these *unc* genes have been determined using conventional genetic mapping and molecular cloning as described (Hodgkin, 1999). For example, mutations in *unc-6*, *unc-5* and *unc-40* cause defects in circumferential axon guidance and encode UNC-6/netrin secreted guidance factor and UNC-5 and UNC-40 netrin receptors, respectively (Wadsworth, 2002). The genes *unc-29* and *unc-38* encode subunits of an ionotropic acetylcholine receptor (Fleming *et al.*, 1997) while UNC-49 is an ionotropic GABA receptor (McIntire *et al.*, 1997).

Classic or 'forward' genetics, that is, identifying mutants defective in a behaviour and then determining the molecular identity of the mutated gene, has been a powerful approach to the study of behavioural genetics (for an elegant review of genetic screens see Jorgensen and Mango, 2002). A mutagen is used to induce mutations in the sperm and oocytes of wild-type hermaphrodites. Mutagens include chemicals, such as ethane methylsulphonate (EMS), to induce point mutations (GC-to-AT transitions) and small chromosomal deletions, and mobile transposable elements, such as Tc1, to generate chromosomal rearrangements, large deficiencies or small deletions (Johnsen and Baillie, 1997; Bessereau *et al.*, 2001). Mutagenized P0 hermaphrodites self-fertilize for one or two generations, and F2 or F3 homozygous mutant progeny are scored for the phenotype of interest. The primary advantage of a phenotype-driven, forward genetics mutant hunt is its unbiased nature. A disadvantage is the time-consuming process of positional mapping and gene identification, which excludes high-throughput, large-scale screening. Classic genetics has provided insight into the genetic basis of locomotion, egg-laying, dauer formation, ageing, feeding, defecation, thermotaxis, chemotaxis, olfaction, osmotic avoidance, touch, male mating, aggregation, circadian rhythms, behavioural plasticity and the effects of drugs on the nervous system.

Genetic epistasis studies may be used to study signalling pathways that specify behaviour. For example, a signalling network for dauer formation in *C. elegans* has been proposed based on genetic interactions (Riddle *et al.*, 1997). *C. elegans* forms dauer juveniles under the environmental conditions of low food and high dauer pheromone (an indicator of high population density). Over 30 genes are required for dauer formation. Mutations in *daf* (dauer formation) genes may result in the inability to form dauers (dauer-defective or Daf-d) or in the inappropriate formation of dauers in conditions of high food and low population density (dauer-constitutive or Daf-c). Genetic epistasis tests defined two signalling pathways for dauer formation. By examining the phenotypes of various daf-d; daf-c double mutants, it was determined that both pathways initiated at the *daf-22* pheromone signal and bifurcated at *daf-6*. Dauer formation and the role of select *daf* genes on adult longevity are discussed elsewhere (Lewis and Perez, Chapter 6, and Wharton, Chapter 13, this volume) and are not considered further here.

In addition to *C. elegans*, there are other nematodes, including *Heterorhabditis*, *Pristionchus pacificus* and *Oschieus* CEW1, that are amenable to genetic analysis: they have a free-living stage, are hermaphroditic and are easy to culture (Hashmi *et al.*, 1998; Sommer, 2000; Delattre and Felix, 2001). Genetic studies in these nematodes are providing insight into the evolution of developmental processes. As genetic and molecular tools are realized in *Heterorhabditis*, *P. pacificus* and *Oschieus* and/or new genetic nematode model systems are developed, comparative behavioural genetics and genomics may become a reality.

C. elegans was the first multicellular organism to have its genome completely sequenced (*C. elegans* Sequencing Consortium, 1998). The genome contains over 20,000 predicted genes, of which only 10% have been characterized by genetics or biochemistry and 10,000 by homology to characterized proteins or by literature observations, leaving 8000 novel genes of unknown function (Costanzo *et al.*, 2000, 2001). How do we determine the function of all these genes? As the *C. elegans* genome sequence project provided a platform for other genome sequencing projects, so will *C. elegans* functional genomics advance the understanding of other model systems.

Reverse genetics starts with a gene identified by its sequence and aims to elucidate function by eliminating (otherwise known as gene 'knockout') or altering the gene by mutation. A PCR-based approach may be used to identify chemically induced chromosomal deletions (Barstead, 1999). The *C. elegans* Gene Knockout Project aims to produce null alleles of all genes in the *C. elegans* genome (http://www.celeganskoconsortium.omrf.org). As a pilot study directed towards building a large-scale collection of mutants, over 1000 transposon insertions in or near 600 genes were generated and identified by anchored PCR, with some insertions resulting in a visible phenotype (Martin *et al.*, 2002). A third approach to knockdown gene function uses RNAi. RNAi has revolutionized *C. elegans* biology, allowing researchers to move from a gene-by-gene mode of characterization to high-throughput screening. RNAi techniques and applications are discussed in Section 9.3.

9.2. The Molecular Genetic Basis of C. *elegans* Behaviours

9.2.1. Cilium structure

Cilia and flagella are elongated structures defined by a characteristic microtubule arrangement and are found on a broad range of cell types mediating a plethora of biological processes. Sensory cilia are 9 + 0 microtubule-based structures that are located at the dendritic endings of sensory neurones and act as sensory receptor organs. In *C. elegans*, ciliated sensory neurones located in the head and tail sense an extensive variety of environmental signals and mediate a wide spectrum of behaviours (Fig. 9.3; Perkins *et al.*, 1986). In vertebrates, photoreceptor and olfactory neuroepithelial cells are well endowed with cilia. Motile cilia and flagella with a 9 + 2 microtubule ultrastructure are required for the swimming behaviour of ciliates, motility of sperm and beating of various mammalian cell types, including tracheal epithelial cells and embryonic organizer ventral node cells (Rosenbaum, 2002). Primary cilia dyskinesia, polycystic kidney disease and retinitis pigmentosa are human diseases affecting cilia (Pazour and Rosenbaum, 2002). Establishing and maintaining ciliary function is clearly essential for the well-being of an organism.

Intraflagellar transport (IFT), a ciliary/flagellar motility process, is necessary for the assembly and maintenance of all cilia and flagella, whether they are motile or sensory (Rosenbaum, 2002). IFT was first described in the biflagellated alga *Chlamydomonas* and visualized using video DIC microscopy as particle movement to and from the flagellar tip. IFT has also been observed using GFP tags in time-lapse fluorescence microscopy of sensory neurones of *C. elegans* as anterograde movement from the proximal end of dendrites to the distal tips of cilia and retrograde movement back to the cell body (Orozco *et al.*, 1999; Signor *et al.*, 1999a).

A kinesin-II motor complex drives anterograde IFT (from cell body to ciliated/flagellar endings) while cytoplasmic dynein directs retrograde movement (from cilia to cell body) at rates of 0.5–4 µm/s (Koushika and Nonet, 2000). A kinesin-II interacting multimolecular complex that is comprised of at least 16 different polypeptides subunits has been biochemically identified in *Chlamydomonas* (Cole *et al.*, 1998). IFT particles are comprised of two subunits, complex A and complex B. The genes coding for both IFT motors and particle polypeptides in *Chlamydomonas* have been shown to have homologues in *C. elegans* and in vertebrates (Cole, 2003). In *C. elegans*, mutations in these genes, identified as the Che, Osm and Daf mutants, cause defects in the development and function of cilia (Culotti and Russell, 1978; Perkins *et al.*, 1986). OSM-1, OSM-5 and OSM-6 are IFT172, IFT88 and IFT52 homologues, respectively, that localize to and move in ciliated sensory neurones via IFT (Cole, 2003).

In addition to a role in ciliogenesis, IFT may potentially function in the localization of signalling molecules to the ciliary membrane. Cilia act as sensory receptors in *C. elegans*. Cilium structure mutants of IFT genes are defective in multiple sensory processes: chemosensation, osmotic avoidance, dauer formation, nose touch mechanosensation and male mating behaviour, as well as the unlikely processes of axon morphology maintenance and ageing. What role, if any, do the IFT particles and polypeptides play in sensing and/or conveying environmental signals? Conditional loss of the kinesin-II subunit, KIF3A, results in opsin and arrestin transport defects in mammalian photoreceptors (Marszalek *et al.*, 2000). These studies indicate that kinesin-II-based transport is required for trafficking of sensory receptors to their proper subcellular compartment. In fact, IFT particle proteins are localized to the connecting cilia of vertebrate photoreceptors (Pazour *et al.*, 2000). We have proposed that IFT may play two distinct, evolutionarily conserved roles: a developmental role in ciliogenesis and a signalling role in trafficking of receptors and signalling proteins to the ciliary membrane (Qin *et al.*, 2001).

9.2.2. Cilium function: olfaction, mechanotransduction and male mating

C. elegans can sense and discriminate between a wide range of odorants. The amphid AWA, AWB and AWC sensory neurone pairs mediate volatile chemosensation (Fig. 9.3; Bargmann, 1993). AWA and AWC chemosensory neurones are required for chemoattraction, while AWB is required for chemorepulsion. Mutations in *odr* (odorant response) genes result in defective chemotaxis (Table 9.1).

The gene *odr-10* is required specifically for chemotaxis to diacetyl and encodes a G-protein-coupled receptor (GPCR) (Table 9.1; Sengupta *et al.*, 1996). This gene is expressed only in the AWA neurones, and ODR-10 protein localizes to AWA sensory cilia. Heterologous expression in mammalian cells demonstrated that ODR-10 is an odorant receptor (OR) that is activated by diacetyl (Zhang *et al.*, 1997). ORs were first discovered in rat (Buck and Axel, 1991), have been identified in invertebrates (*C. elegans* and the fruit fly *Drosophila melanogaster*) and vertebrates

Table 9.1. Genes required for olfaction.

Gene	Gene product	References
odr-10	GPCR	Sengupta *et al.*, 1996
osm-9	TRP channel	Colbert *et al.*, 1997
odr-3	Gα protein	Roayaie *et al.*, 1998
odr-1	Guanylate cyclase	L'Etoile and Bargmann, 2000
odr-4	Novel transmembrane protein	Dwyer *et al.*, 1998
odr-8	Not cloned	Dwyer *et al.*, 1998
odr-7	Nuclear hormone receptor transcription factor	Sengupta *et al.*, 1996
tax-2	Cyclic nucleotide gated channel (α subunit)	Coburn and Bargmann, 1998; Coburn *et al.*, 1996
tax-4	Cyclic nucleotide gated channel (β subunit)	Komatsu *et al.*, 1996
ocr-1	TRP cation channel	Tobin *et al.*, 2002
ocr-2	TRP cation channel	Tobin *et al.*, 2002

GPCR, G-protein-coupled receptor; TRP, transient receptor potential.

and represent the largest mammalian gene superfamily (~1500 in mouse, ~900 in human) (Young and Trask, 2002). Mammalian GPCR families have been categorized based on homology and/or expression pattern as ORs, taste receptors (Lindemann, 2001) and pheromone receptors (Barinaga, 1999). The *C. elegans* genome encodes 1000 predicted chemosensory GPCRs (Bargmann, 1998). Like ODR-10, many *C. elegans* GPCRs are expressed in sensory neurones and accumulate in ciliated endings of dendrites (Troemel *et al.*, 1995; Chou *et al.*, 1996). Ciliary localization in amphid neurones is observed for other chemosensory signalling molecules, including ODR-3 Gα protein (Roayaie *et al.*, 1998), TAX-2 and TAX-4 cyclic nucleotide gated channels (Coburn and Bargmann, 1996; Komatsu *et al.*, 1996; Coburn *et al.*, 1998), the guanylyl cyclases ODR-1 and DAF-11 (Birnby *et al.*, 2000; L'Etoile and Bargmann, 2000), the PEF-1 phosphatase (Ramulu and Nathans, 2001) and the transient receptor potential (TRP) cation channels TRP1, OSM-9 and OCR-2 (Colbert *et al.*, 1997; Tobin *et al.*, 2002; Table 9.1).

TRP channels and degenerin and epithelial Na$^+$ channels (DEG/ENaCs) are the only proteins demonstrated to act as mechanotransducers (Goodman and Schwarz, 2003). Large gene families encoding DEG/ENaCs and TRP channels have been identified in *C. elegans*, *Drosophila* and mammals. TRP channels were first identified in *D. melanogaster*, with the *trp* locus being required for photoreceptor light responses. All TRP proteins share a similar membrane topology: six transmembrane (tm) spanning domains with a pore region between tm5 and tm6. The TRP family may be divided into six subfamilies based on amino acid sequence analysis. DEG/ENaCs were first identified in *C. elegans* by gain of function mutations in channel proteins that result in cell swelling and neuronal degeneration. DEG/ENaC proteins all possess two tm domains separated by a large extracellular domain containing at least two cysteine-rich domains. A well-accepted model for mechanotransduction is that the heteromeric DEG/ENaC is tethered to extracellular matrix-associated proteins and intracellular cytoskeletal proteins. Mechanical force (whether it be push/pull, stretch or shear) gates the channel. Dominant degeneration mutations in MEC-4 and DEG-1 are located in a conserved region of the encoded protein that produces a hyperactive channel (i.e. activated independent of mechanical stimulation). The functions of only a few TRP and DEG/ENaC proteins are known in *C. elegans*. Reverse genetics approaches will probably reveal insight into the sensory functions of other mechanosensory channels.

Hodgkin (1983) was the first to systematically study the genetic basis for the development and function of male copulatory organs. Mating efficiency (measuring the number of cross-progeny of total progeny) was used as an indicator of anatomical or behavioural abnormalities. Hodgkin (1983) took two approaches: to examine mutant males of previously identified strains and to identify mutants with specific defects in males. Males appear in *C. elegans* hermaphrodite stocks at a low frequency (0.2%). Therefore, *him* (for high incidence of males) mutations are used to increase meiotic non-disjunction of X chromosomes, resulting in a larger male population (Hodgkin *et al.*, 1979). Males are observed at a frequency of 36% in self-fertilizing *him-5(e1490)* hermaphrodites and *him-5(e1490)* males mate efficiently. The *him-5* hermaphrodites were mutagenized and analysis of mutant male progeny identified male abnormal (Mab) strains as well as sensory-defective (*che-2* and *che-3*) strains.

Males with severe defects in all sensory neurone cilia (*osm-1*, *osm-5*, *osm-6* and *che-3*) exhibit pleiotropic male mating defects in response, Lov and sperm transfer (Barr and Sternberg, 1999; Qin *et al.*, 2001). The only ciliated cells in *C. elegans* are chemosensory and mechanosensory neurones (White *et al.*, 1986). The male has 46 predicted ciliated sensory neurones in his tail and four in his head (Sulston *et al.*, 1980). As mentioned above, *osm-1*, *osm-5* and *osm-6* are homologous to *Chlamydomonas* IFT genes (*che-3* encodes a cytoplasmic dynein motor (Signor *et al.*, 1999a)) and are required for ciliogenesis. Not surprisingly, *osm-5::gfp* and *osm-6::GFP* are expressed exclusively in ciliated neurones, with male-specific expression in four CEM head neurones and neurones of the hook, post-cloacal sensilla, rays and copulatory spicules (Collet *et al.*, 1998; Qin *et al.*, 2001). Gene expression pattern, protein subcellular localization and mutant phenotypes indicate that IFT genes are required for the structure and function of ciliated neurones in the adult male tail.

Few genes have been identified that are required solely for male sensory behaviours (Table 9.2). *lov-1* mutant males have response and Lov defects. The *lov-1* gene was cloned by genetic mapping and transformation rescue behavioural defects (refer to Section 9.3 for methodology). This gene is the *C. elegans* homologue of the human polycystic kidney disease gene PKD1 (Barr and Sternberg, 1999). Mutations in PKD1 or PKD2 account for 95% of autosomal dominant polycystic kidney disease (ADPKD), a human genetic disorder affecting 1 in 1000 individuals (Igarashi and Somlo, 2002). PKD-2 is the *C. elegans* homologue of the human polycystin 2 channel (encoded by the PKD2 gene) (Fig. 9.4). Polycystin 1 (encoded by PKD1) and polycystin 2 are proposed to form a receptor/channel complex. Polycystin 2 is a member of the TRP ion channel family, which has been implicated in mechanosensation in *D. melanogaster* and *C. elegans*. Like *lov-1*, *pkd-2* is required for response and Lov behaviours, indicating that *lov-1* and *pkd-2* act in the same genetic pathway (Barr *et al.*, 2001). Consistent with a role in sensation, *lov-1* and *pkd-2* are expressed in the male-specific B type sensory neurones of the hook and rays (with the exception of ray 6, which does not have an open ending) and also the CEMs (Fig. 9.5). Using a combination of full-length translational GFP fusions and antibodies, it was shown that LOV-1 and PKD-2 proteins are enriched in sensory cilia (Fig. 9.5).

Chemosensation and mechanosensation are probably involved in response and Lov behaviours (J. Wang and M.M. Barr, 2003, unpublished data). *C. elegans* sensory neurones can be polymodal: for example, by ultrastructural assignment, the ASH neurone appears to be chemosensory and yet functions in both mechanosensory (nose touch) and chemosensory (osmotic avoidance) modalities (Kaplan and Horvitz, 1993). HOB might similarly be a polymodal sensory neurone. Ablation of either HOA or HOB produces identical phenotypes (Liu and Sternberg, 1995) and HOA and HOB form multiple chemical synapses and electrical junctions (Fig. 9.1G, H; Sulston *et al.*, 1980), indicating extensive cross-talk between the two hook sensory neurones. LOV-1 has an extensive extracellular mucin-like domain, which

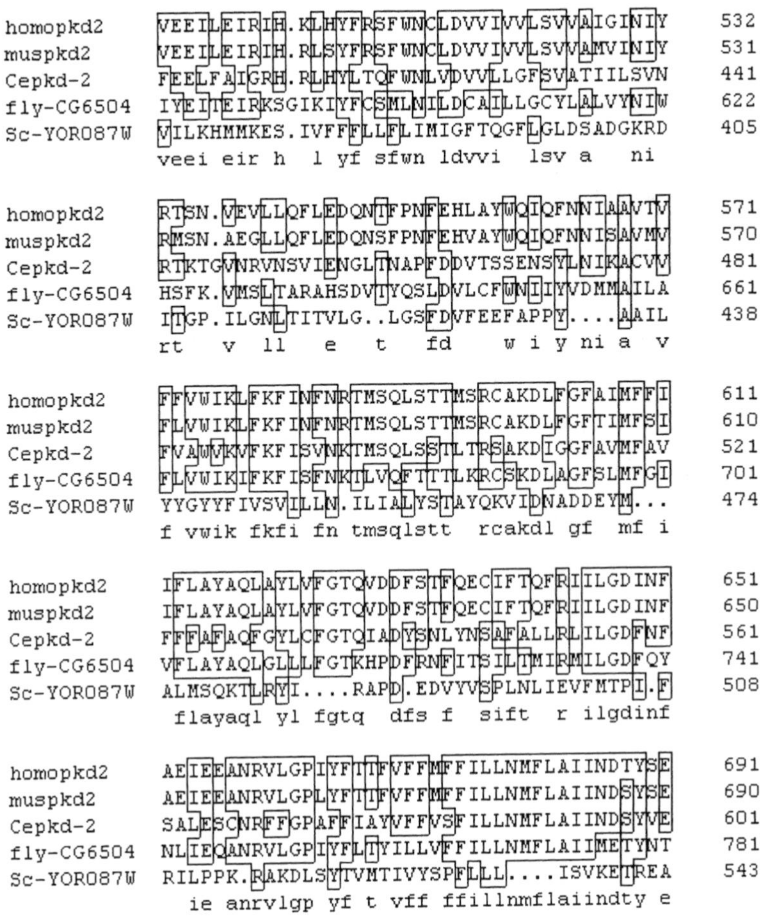

```
homopkd2   VEEILEIRIH.KLHYFRSFWNCLDVVIVVLSVVAIGINIY   532
muspkd2    VEEILEIRIH.RLSYFRSFWNCLDVVIVVLSVVAMVINIY   531
Cepkd-2    FEELFAIGRH.RLHYLTQFWNLVDVVLLGFSVATIILSVN   441
fly-CG6504 IYEITEIRKSGIKIYFCSMLNILDCAILLGCYLALVYNIW   622
Sc-YOR087W MILKHMMKES.IVFFFLLFLIMIGFTQGFLGLDSADGKRD   405
           veei eir h   l yf sfwn ldvvi   lsv a    ni

homopkd2   RTSN.VEVLLQFLEDQNTFPNFEHLAYWQIQFNNIAAVTV   571
muspkd2    RMSN.AEGLLQFLEDQNSFPNFEHVAYWQIQFNNISAVMV   570
Cepkd-2    RTKTGVNRVNSVIENGLTNAPFDDVTSSENSYLNIKMACVV   481
fly-CG6504 HSFK.VMSLTARAHSDVTYQSLDVLCFWNIIYVDMMAILA   661
Sc-YOR087W ITGP.ILGNLTITVLG..LGSFDVFEEFAPPY....AAIL   438
           rt    v  ll    e    t    fd    w iy ni a  v

homopkd2   FFVWIKLFKFINFNRTMSQLSTTMSRCAKDLFGFAIMFFI   611
muspkd2    FLVWIKLFKFINFNRTMSQLSTTMSRCAKDLFGFTIMFSI   610
Cepkd-2    FVAWVKVFKFISVNKTMSQLSSTLTRSAKDIGGFAVMFAV   521
fly-CG6504 FLVWIKIFKFISFNKTLVQFTTTLKRCSKDLAGFSLMFGI   701
Sc-YOR087W YYGYYFIVSVILLN.ILIALYSTAYQKVIDNADDEYM...   474
           f vwik fkfi fn tmsqlstt   rcakdl gf  mf i

homopkd2   IFLAYAQLAYLVFGTQVDDFSTFQECIFTQFRIILGDINF   651
muspkd2    IFLAYAQLAYLVFGTQVDDFSTFQECIFTQFRIILGDINF   650
Cepkd-2    FFFAFAQFGYLCFGTQIADYSNLYNSAFALLRLILGDFNF   561
fly-CG6504 VFLAYAQLGLLLFGTKHPDFRNFITSILTMIRMILGDFQY   741
Sc-YOR087W ALMSQKTLRYI....RAPD.EDVYVSPLNLIEVFMTPI.F   508
           flayaql yl fgtq   dfs f   sift   r ilgdinf

homopkd2   AEIEEANRVLGPIYFTTFVFFMFFILLNMFLAIINDTYSE   691
muspkd2    AEIEEANRVLGPLYFTTFVFFMFFILLNMFLAIINDSYSE   690
Cepkd-2    SALESCNRFFGPAFFIAYVFFVSFILLNMFLAIINDSYVE   601
fly-CG6504 NLIEQANRVLGPIYFLTYILLVFFILLNMFLAIIMETYNT   781
Sc-YOR087W RILPPK.RAKDLSYTVMTIVYSPFLLL....ISVKETREA   543
           ie anrvlgp yf t vff ffillnmflaiindty e
```

Fig. 9.4. Alignment of PKD-2 proteins from human PKD2, mouse PKD2, *C. elegans* PKD-2, *Drosophila* open reading frame (ORF) CG6504 and *S. cerevisiae* Y0R087W. Sequence names are shown next to the corresponding amino acid sequence.

could be involved in cell–cell or cell–matrix interaction. LOV-1 could physically link the hook and ray sensory endings to the cuticle and couple touch to intracellular PKD-2 channel-activated signalling, similar to hair cell mechanosensation or touch response in *C. elegans*. Stunningly, sensory function and ciliary localization of the *C. elegans* polycystins may be evolutionarily conserved. Polycystin 1 and 2 localize to renal cilia (Pazour *et al.*, 2002; Yoder *et al.*, 2002) and form a mechanosensitive channel in the primary cilium of cultured kidney cells (Nauli *et al.*, 2003). The powerful molecular genetic tools of *C. elegans* will enable simultaneous dissection of the molecular basis of male sensory behaviours, ciliary protein localization and PKD.

Turning behaviour is mediated by the neurotransmitters serotonin (Loer and Kenyon, 1993) and dopamine (DA) and unidentified neuropeptides (T. Liu and

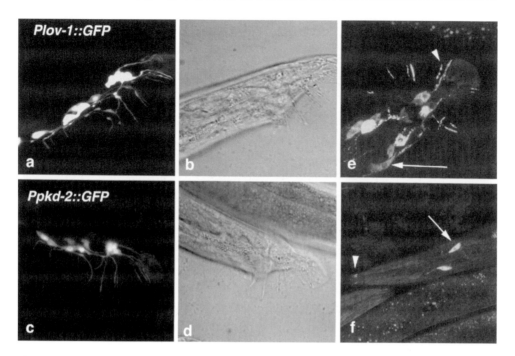

Fig. 9.5. GFP reporters may be used to study gene expression (a, c) and protein localization (e, f); *lov-1* and *pkd-2* are expressed in male-specific sensory neurones. (a, c) Transcriptional *Plov-1::GFP* and *Ppkd-2::GFP* reporters (P is for promoter) are expressed uniformly throughout all ray B neurones with the exception of ray 6. Expression is also observed in the CEMs and HOB, not shown. (b, d) DIC images of corresponding male tails in (a, c). (e, f) A translational, functional *Ppkd-2cDNA::GFP* (P is for promoter, cDNA is for complementary DNA) fusion is expressed in the same cells as *Plov-1::GFP* and *Ppkd-2::GFP* reporters. *Ppkd-2cDNA::GFP* is enriched in sensory cilia (arrowheads) and cell bodies (arrows) of the rays (e) and CEMs.

M. Barr, unpublished). R5A, R7A and R9A, the A type sensory neurones in rays 5, 7 and 9, express DA and the *cat-2* gene, which encodes a tyrosine hydroxylase for DA biosynthesis (Sulston and Horvitz, 1977; Sulston *et al.*, 1980; Lints and Emmons, 1999). R1, R3, R9 and male-specific CP motor neurones contain serotonin (Loer and Kenyon, 1993). Ablation studies indicate that it is the CP neurones and three most posterior rays (rays 7–9) that are essential for turning behaviour (Loer and Kenyon, 1993; Liu and Sternberg, 1995). How the male determines his position and executes well-timed and coordinated turning remains unknown. Intact sensory cilia are not required (M. Barr, unpublished results). Proprioception may be involved in turning behaviour and perhaps an unidentified DEG/ENaC mediates this mechanosensory behaviour.

Spicule insertion behaviour is a motor programme that entails periodic and prolonged spicule muscle contraction (Garcia *et al.*, 2001). When the PCB and PCC post-cloacal sensory neurones and SPC spicule motor neurones are activated, acetylcholine is released and triggers spicule protraction. The UNC-68 ryanodine receptor calcium channel is required for periodic contraction of spicule protractor muscles, while the EGL-19 L-type voltage-gated calcium channel α1 subunit is required for prolonged protractions (Table 9.2). The ACh agonists levamisole, nicotine and arecholine stimulate many muscle contractions, including spicule muscles. At least 42 ACh receptors (AChRs) are encoded by the *C. elegans* genome, many of which have unknown function (Bargmann, 1998). The UNC-38 nicotinic AChR is required for spicule protraction and it is likely that other AChRs mediate spicule insertion behaviour.

Sperm transfer follows spicule insertion. Ablation of the SPV sensory neurones results in premature sperm transfer (Liu and Sternberg, 1995) and phenocopies mutants defective in cilia formation (Qin *et al.*, 2001). A current model for sperm transfer behaviour is that the SPV neurones function to inhibit sperm transfer and that either the hermaphrodite uterus, full spicule distension during insertion or both provide a cue to the male to transfer sperm (Liu and Sternberg, 1995; Liu, 1996; Garcia *et al.*, 2001). Preliminary results suggest that *osm-9* and *ocr-2* are required for the sperm transfer step, are co-expressed in spicule neurones and produce proteins that localize to cilia of spicule neurones (Table 9.2; E.M. Peden and M. Barr, unpublished). The genes *osm-9* and *ocr-2* encode TRPV channels that have been demonstrated to function in *C. elegans* olfaction and nose touch (Tobin *et al.*, 2002). Whether or not *osm-9* and *ocr-2* mediate chemosensation or mechanosensation in the male remains to be determined.

Table 9.2. Genes required for male mating behaviour.

Behaviour	Gene	Gene product	Reference
Drive	*mod-5*	Serotonin reuptake transporter	Ranganathan *et al.*, 2001; Emmons and Lipton, 2003
	unc-77	Unknown	Emmons and Lipton, 2003
Response	*lov-1*	Membrane receptor	Barr and Sternberg, 1999
	pkd-2	TRP channel	Barr *et al.*, 2001
Turning	*egl-3*	Neuropeptide proprotein convertase	Kass *et al.*, 2001; T. Liu and M. Barr, unpublished
	cat-2	Tyrosine hydroxylase	Lints and Emmons, 1999; T. Liu and M. Barr, unpublished
	cat-1	Vesicular monoamine transporter	Loer and Kenyon, 1993; Duerr *et al.*, 1999
	cat-4	GTP cyclohydrolase I	Loer and Kenyon, 1993; C. Loer, unpublished
	bas-1	Aromatic amino acid decarboxylase	Loer and Kenyon, 1993; Hare and Loer, 2000
Lov			
HOB specification	*egl-44*	TEF factor	Yu *et al.*, 2003
HOB specification	*egl-46*	Zinc finger transcription factor	Yu *et al.*, 2003
HOB function	*lov-1*	Membrane receptor	Barr and Sternberg, 1999
HOB function	*pkd-2*	TRP channel	Barr *et al.*, 2001
Spicule insertion	*egl-19*	L-type voltage-gated calcium channel	Garcia *et al.*, 2001
	unc-29	Nicotinic acetylcholine receptor	Garcia *et al.*, 2001
	unc-38	Nicotinic acetylcholine receptor	Garcia *et al.*, 2001
	unc-68	Sarcoplasmic calcium channel	Garcia *et al.*, 2001
Sperm transfer	*osm-9*	TRP cation channel	Tobin *et al.*, 2002; E. Peden and M. Barr, unpublished
	ocr-2	TRP cation channel	

GTP, guanosine triphosphate.

9.3. Tools for Analysing *C. elegans* Behaviours

9.3.1. Transformation

Transformation of *C. elegans* involves injecting foreign DNA into the cytoplasm of the mitotically active hermaphrodite gonad (Mello and Fire, 1995). The injected DNA contains a mixture of the DNA of interest and a selectable marker to identify transformant progeny. Selectable markers include the dominant *rol-6(su1006)*, which results in a right-hand roller phenotype (Kramer *et al.*, 1990), *dpy-20(+)* to rescue a *dpy-20(e1282ts)* dumpy phenotype (Han and Sternberg, 1991), *pha-1(+)* to rescue *pha-1(e2123ts)* embryonic lethality at 25°C (Granato *et al.*, 1994) or the use of GFP reporter constructs as co-injection markers. Injected DNA forms long tandem arrays that are transmitted through the germ line as high-copy, extrachromosomal arrays. After selecting for the co-injection marker for several generations, the extrachromosomal arrays are stably transmitted in a non-Mendelian manner. Transformation in *C. elegans* has been typically employed to identify (or clone) a gene by rescuing a mutant phenotype with a wild-type gene copy, characterize gene expression patterns using lacZ or GFP reporter constructs (see below), determine structure–function relationships by overexpressing various forms of the gene, perform mosaic analysis to determine the site of gene action or interfere with gene function by RNAi. To integrate arrays into *C. elegans* chromosomes, γ-irradiation has been used (Mello and Fire, 1995).

Microparticle bombardment has been successful utilized as an alternative, less technically challenging approach to generating integrative transformants (Praitis *et al.*, 2001). *unc-119(ed3)* mutants are unable to form dauers and transformants are rescued by wild-type *unc-119(+)* DNA resulting in a non-Unc, non-Dauer phenotype. Using *unc-119* as a selection strategy, low-copy-number integrants are obtained. Transient transfection of parasitic worms, using injection and bombardment, has been described (Brooks and Isaac, 2002). The ability to generate transgenic animals by DNA transformation provides a basic tool essential to the dissection and understanding of gene function.

9.3.2. Cloning genes

Now that the *C. elegans* genome is complete, what exactly does it mean to 'clone a gene'? For a recessive, loss of function mutation, the gene of interest may be cloned by rescuing the mutant phenotype. The gene is first mapped by standard genetic methods and quickly refined by using a high-density single-nucleotide polymorphism (SNP) map at http://genome.wustl.edu/projects/celegans/index.php?snp=1 (Wicks *et al.*, 2001). It is estimated that SNPs occur at a rate of 1 in 1000 base pairs between individuals, whether the individuals be human or isogenic *C. elegans* strains. Interestingly, differences in aggregation behaviour in *C. elegans* are not due to complex polygenic traits but rather can be traced back to an SNP in the *npr-1* gene encoding the neuropeptide Y receptor. Analysis of 17 natural

C. elegans isolates revealed that social feeders encode a phenylalanine residue at position 215 of NPR-1, while solitary feeders encode a valine residue at this position (de Bono and Bargmann, 1998).

When the genetic mapping interval correlates to a manageable physical map and is covered by a reasonable number of cosmids (fewer than 50), pools of overlapping cosmids can be injected and transgenic lines scored for the rescue of mutant phenotype(s). Once rescue is obtained, then individual cosmids and finally subclones from the single rescuing cosmid are transformed to determine the single gene (or open reading frame (ORF)) that is responsible for the rescuing activity. To confirm correct gene identification, the mutant allele is sequenced and compared with the wild type to determine if the molecular DNA lesion or the rescuing fragment is altered. Alternatively, RNAi to the predicted ORF may be synthesized and used to phenocopy mutant defects.

9.3.3. Sequence analysis and data mining

With the completion of the *C. elegans* genome, sequence analysis is relatively straightforward. There are many online resources for ORF investigation, including genetics and genome information at http://elegans.swmed.edu/genome.shtml. At WormBase, a gene search of your favourite ORF will provide information on identification (including putative *C. briggsae* orthologues); genomic location; function (including RNAi phenotypes, expression pattern and protein domains); gene ontology classification according to biological process, cellular component, and molecular function; mutant alleles and strains available through the *Caenorhabditis* Genetics Center (CGC); similarities as determined by BLASTP matches to the longest protein product in *Homo sapiens, Mus musculus, Drosophila, Saccharomyces cerevisiae* and other species; reagents such as complementary DNAs (cDNAs) and transgenic animals; and a bibliography of published papers and unpublished abstracts from *C. elegans* meetings and the *Worm Breeder's Gazette* (Stein *et al.*, 2001; Harris *et al.*, 2003). WormBase provides both *C. elegans* and non-*C. elegans* researchers with the foundation to address problems in molecular genetics and functional and comparative genomics.

9.3.4. GFP

The jellyfish *Aequorea victoria* green fluorescent protein (GFP) was first used as a marker to study gene expression and protein localization in living *C. elegans* (Chalfie *et al.*, 1994). A promoter may be fused to GFP to generate a transcriptional fusion to examine gene expression patterns (Figs 9.2 and 9.5) in transgenic animals using standard epifluorescence, confocal or two-photon microscopy. A promoter driving expression of a full-length gene fused to GFP may be used to generate a translational fusion to examine subcellular protein localization. Andrew Fire has assembled an impressive series of expression vectors, and Fire Lab Vector Kits are available to academic laboratories at no charge (vector details and sequences can be downloaded from http://www.ciwemb.edu). Vectors include tissue-specific promoters, localization signals, and several mutant variants of GFP, including blue-shifted (BFP), yellow-shifted (YFP) and cyan-shifted fluorescence proteins (CFP). YFP and CFP demonstrate well-separated emission and excitation spectra, enabling double-labelling experiments (Miller *et al.*, 1999). Fluorescence resonance energy transfer (FRET) between YFP and CFP is observed in genetically encoded calcium sensors such as cameleon (Miyawaki *et al.*, 1997) and has been used to measure calcium transients in *C. elegans* (Kerr *et al.*, 2000), as discussed later in this chapter.

The GFP technique has provided insights into a wide range of biological phenomena, including embryogenesis (Bowerman, 2001), protein sorting and transport (Koushika and Nonet, 2000; Signor *et al.*, 2000) and cellular ultrastructure (using immuno-EM with fixed *C. elegans* and anti-GFP antibodies) (Paupard *et al.*, 2001). Based on published and unpublished GFP expression patterns, Shawn Lockery and Oliver Hobert have compiled a list of neuronal-specific promoters (http://chinook.uoregon.edu/promoters.html), which may be used to label your favourite neurones or drive expression of your favourite genes. GFP may be used in mutant screens looking for changes in expression or subcellular localization. This approach has been used to identify genes required for synapse formation, axon path-finding, endocytosis and olfactory adaptation (Jorgensen and Mango, 2002).

9.3.5. RNAi

RNAi is a rapid and simple reverse genetic technique to determine the cellular function of any predicted gene. RNAi refers to the introduction of exogenous double-stranded RNA (dsRNA) to knock down gene function by specific and rapid degradation of the corresponding mRNA(s). First demonstrated in *C. elegans*, RNAi has also proved effective in many other organisms, including *Drosophila*, trypanosomes, plants, zebrafish, mouse and mammalian cells. Recent articles about the mechanisms and applications of RNAi are available and discussion will not be repeated here (Grishok and Mello, 2002; Hannon, 2002; Tijsterman *et al.*, 2002). dsRNA can be introduced into *C. elegans* by direct injection into the hermaphrodite germline (Fire *et al.*, 1998), by soaking in dsRNA (Maeda *et al.*, 2001), by ingestion of bacteria engineered to produce dsRNA (Timmons and Fire, 1998; Timmons *et al.*, 2001) or by generating heritable inverted repeat (IR) genes (Tavernarakis *et al.*, 2000). As neurones are particularly resilient to the effects of RNAi, the primary advantage of the transgenic IR RNAi method is the ability to stably and heritably inactive genes that act in the nervous system. A major drawback of transgenic IR RNAi is that it is not a technique easily amenable to high-throughput screening.

RNAi was initially used in the *C. elegans* to investigate the function of single genes. RNAi has revolutionized *C. elegans* functional genomics research (Fraser *et al.*, 2000; Gonczy *et al.*, 2000; Dillin *et al.*, 2002; Lee *et al.*, 2002; Ashrafi *et al.*, 2003; Kamath *et al.*, 2003). Genome-wide screens looking for specific phenotypes caused by injected or ingested dsRNA have been executed, providing tremendous knowledge about the genes required for viability, development, ageing and fat metabolism. This systematic RNAi approach represents a major advance in *C. elegans* functional genomics.

While RNAi has become a powerful tool for analysing gene function, there are limitations. First, some nematode species appear to be resistant to RNAi (Brooks and Isaac, 2002). However, RNAi by soaking works in the animal parasite *Nippostrongylus brasiliensis* (Hussein *et al.*, 2002) and RNAi by ingestion is effective in the plant parasites *Heterodera glycines* and *Globodera pallida* (Urwin *et al.*, 2002). Secondly, a variety of factors can affect RNAi phenotypes, including dsRNA delivery method, developmental stage(s) at which dsRNA is applied, choice of dsRNA sequence, stability of the target protein, temperature and nature of the gene being tested (Maine, 2001). Finally, and most importantly for those researchers interested in nervous system development and function, postembryonic tissues, including neurones, are typically resistant to RNAi. Techniques to overcome this hurdle include the use of IR RNAi and the identification of a *C. elegans* mutant strain, *rrf-3(pk1426)*, that is hypersensitive to RNAi (Simmer *et al.*, 2002).

9.3.6. Functional genomics: microarray technologies

A great deal of information regarding nematode genomes is known from the completed *C. elegans* genome, and the ongoing genome sequencing efforts in several parasitic nematode species, including *Onchocerca volvulus*, *Brugia malayi* and *Schistosoma mansoni*, and EST projects in many other parasitic nematode species (www.nematodes.org and www.nematode.net). Ascertaining the function of all these genes remains a daunting challenge. The *C. elegans* Nematode Expression Pattern Database (NEXTDB at http://nematode.lab.nig.ac.jp), run by the laboratory of Yuji Kohara, is constructing an expression pattern map through EST analysis and systematic whole-mount *in situ* hybridization of ESTs. NEXTDB provides an important complement to systematic RNAi screens and DNA microarrays in *C. elegans* as well as a format for investigating functional genomics in other nematodes.

With genome information in hand, research priority is shifted from identification of genes to characterization of gene function (functional genomics). Functional genomics requires careful analysis of single genes combined with high-throughput technologies examining global gene expression and function. DNA microarrays ('DNA chips') are powerful tools that provide a genome-wide, high-throughput approach to the study of global gene expression. DNA microarrays have been applied to *C. elegans* research (Reinke, 2002; Reinke and White, 2002). Two sources of *C. elegans* microarrays exist. Stuart Kim's laboratory at Stanford University has developed full-genome cDNA chips with 1 kb pieces of exon-rich genomic DNA for every predicted gene (http://cmgm.stanford.edu/~kimlab/wmdirectorybig.html). Academic laboratories have free access to Kim laboratory microarrays and provide the RNA for collaborative microarray experiments that are performed at Stanford. Affymetrix manufactures commercially available oligonucleotide microarrays. The *C. elegans* full-genome DNA chips have been used to profile gene expression during development from eggs to adults (Hill *et al.*, 2000; Jiang *et al.*, 2001), to identify genes expressed in a tissue-specific manner, including those that function in touch neurones (Zhang *et al.*, 2002b) and muscle (Roy *et al.*, 2002), and to identify operons (Blumenthal *et al.*, 2002). Microarray data are accessible on the Internet and compiled at WormBase.

mRNA differential display is an attractive alternative for exploring functional genomics in nematode species without complete genome information (Liao and Freedman, 2002).

9.3.7. Proteomics

With the completion of the *C. elegans* genome project, research emphasis is shifting to functional genomics and proteomics. The term proteome refers to all the proteins expressed by a genome. This has given rise to the science of proteomics, the quantitative and qualitative study of proteomes in an organism to unravel biological processes. With the availability of *C. elegans* DNA microarrays and the ability to analyse expression of thousands of genes simultaneously, it may be asked why proteomics is so important. Proteins provide a structural and functional framework for cellular life and, with this, animal behaviour. Proteomics complements genomics in the study of physiological mechanisms of complex phenomena, such as nematode behaviours.

Animal behaviours may be considered as a complex set of chemical reactions. The study of *C. elegans* behaviours has proceeded in large part using genetic approaches. In vertebrates, biochemical approaches have been used to identify components that participate in a wide variety of cellular processes. Biochemical approaches have not been commonly applied to *C. elegans* behavioural research. An exception is that the abundance and repetitive organization of muscle have enabled biochemical analyses that extend molecular genetic approaches (Moerman and Fire, 1997). An extensive array of biochemically defined protein–protein interactions is essential to complement molecular genetics and provide the framework for understanding the nervous system.

By means of a variety of heterologous expression systems (e.g. *E. coli*, yeast, insect cells), cloned genes can be expressed *in vitro*. Expressed and purified proteins may be characterized using conventional biochemical methods, such as structural analysis, enzyme activity analysis or antibody generation. The determination of protein structure by physical methods such as X-ray crystallography or by computational prediction and modelling programmes reveals functional roles by means of comparison with proteins of known structure and function. Structural genomics of *C. elegans* at the University of Alabama, Birmingham, has the lofty goal of expressing, purifying, crystallizing and analysing every protein encoded by the *C. elegans* genome (http://sgce.cbse.uab.edu).

Perhaps most importantly, biochemical methods may be used to identify protein–protein interactions and this information is essential for understanding the basis of a specific behaviour. To examine protein–protein interactions, there are several possible approaches. As one method, researchers can use the yeast two-hybrid system with a protein or protein fragment as bait to screen a cDNA library to identify candidate interactions (Fields and Song, 1989). Proteome-wide interactions using the yeast two-hybrid system have been determined in the budding yeast *S. cerevisiae* (Kumar and Snyder, 2002; Uetz, 2002). In *C. elegans*, the yeast two-hybrid system was initially used on a protein-by-protein basis and has employed a systematic 'interactome' approach to vulva development (Walhout *et al.*, 2000a,b, 2002). In contrast to vertebrate systems, the physiological relevance of interactions is easily testable in living *C. elegans*.

A second approach to examining protein complexes involves affinity purification methods. With good antibodies against a particular protein, co-immunoprecipitation can be used to isolate associated proteins. A method related to co-immunoprecipitation but more useful for large-scale screening efforts is affinity chromatography utilizing immobilized protein fragments as bait. This method is technically simple, scalable and adaptable. However, this technique may not be amenable to detecting protein interactions that require recognition of overall protein conformations. Fusion protein affinity chromatography has been used successfully to identify a number of protein interactions in other model systems and may be applicable to nematode models (Hirabayashi and Kasai, 2002). The major limitation of biochemical methods is protein abundance. Poor heterologous expression or low endogenous expression levels are significant technical barriers. The emerging *C. elegans* primary cell culture system (see Section 9.3.10) may provide a source for ample native proteins (Christensen *et al.*, 2002).

9.3.8. Electrophysiological approaches

Understanding the properties of individual neurones and how neurones control behaviour is a common goal of all neurobiologists. Techniques to study the electrical properties of *C. elegans* neurones and muscles have only recently been developed (Francis *et al.*, 2003). The anatomy of *C. elegans* presents a number of difficulties for using traditional electrophysiological techniques. The adult is ~1 mm long with neuronal cell bodies 2–3 μm in diameter. Furthermore, the tough, pressurized cuticle can rupture when pierced, presenting a unique challenge to electrophysiologists.

These technical problems have been solved and the electrical properties of neurones and muscles can be recorded from *C. elegans* (Raizen and Avery, 1994; Davis *et al.*, 1995; Goodman *et al.*, 1998). A complete description of the method for patch clamping *C. elegans* neurones is available (Lockery and Goodman, 1998). In brief, worms are immobilized on an agarose pad, using cyanoacrylate glue, and dissected by nicking the cuticle near the target neurone. The neurone of interest may be labelled with GFP (for a list of neurone-specific promoters, see http://chinook.uoregon.edu/promotors.html). The patch pipette is applied to the neurone of choice and activity is recorded. In *C. elegans*, chemotaxis behaviour is controlled mainly by a single pair of chemosensory neurones, ASEL and ASER

(Bargmann and Horvitz, 1991). With the help of whole-cell patch-clamp electrophysiology, functional asymmetries and the molecular basis of distinguishing sodium from chloride were found to exist in this pair of neurones (Goodman *et al.*, 1998; Pierce-Shimomura *et al.*, 2001).

Most work on *C. elegans* muscle electrophysiology has focused on muscle cells of the pharynx because of their large size and easy accessibility. Two different methods, intracellular recording and the electropharyngeogram (EPG), have been used. In the first method, intracellular muscle recordings are taken from a severed worm head in which the posterior pharynx is exposed (Davis *et al.*, 1995). In the second method, the head of the worm is drawn into a suction electrode and the currents are measured at the whole-animal level (Avery *et al.*, 1995). Both methods give corroborative results. A preparation for making whole-cell patch-clamp recordings from the 50 μm body-wall muscles has also been developed (Richmond *et al.*, 1999). With this method it should be possible to study the mechanism of synaptic transmission in various *C. elegans* locomotion behaviours. For example, the properties of body wall muscle cells used for locomotion behaviour remained unclear. By using *in situ* patch-clamp techniques, it was found that the L-type voltage-dependent Ca^{2+} channel EGL-19 controls body-wall muscle function in *C. elegans* (Jospin *et al.*, 2002).

9.3.9. Calcium imaging

Ca^{2+} plays a critical role in many cellular processes, including muscle contraction, neuronal excitability, secretion, cell cycle, differentiation, apoptosis and gene transcription. It has been shown to be a key second messenger in a variety of regulatory and signalling processes. Most stimuli in nematode behaviour, such as neurotransmitters, hormones, odorants or electrical activity, can modulate the level of cellular calcium. Recent studies also suggest that calcium is the main carrier of the inward positive current, and neural calcium transients are ubiquitous in *C. elegans* (Bargmann and Kaplan, 1998; Goodman *et al.*, 1998). Calcium imaging is a powerful tool in analysing calcium dynamics. Furthermore, calcium imaging is an attractive alternative for monitoring the activity of excitable cells *in vivo*.

There are difficulties in using calcium imaging techniques in *C. elegans*. The anatomy of nematodes prevents the accessibility of neurones to traditional bath-applied dye, such as fura-2 and fluo-3. At the same time, reliable microinjection of calcium-sensitive dye into tiny *C. elegans* neurones or muscles is technically challenging but possible (Dal Santo *et al.*, 1999). To circumvent these limitations, the calcium indicator protein, cameleon, may be used for calcium imaging in intact *C. elegans* (Kerr *et al.*, 2000). Cameleon is a FRET-based, ratiometric calcium-sensitive protein that has many potential advantages for cell-specific imaging (Miyawaki *et al.*, 1997, 1999). Cell-specific promoters can be used to drive cameleon expression in target cells. Using cameleon in *C. elegans* is feasible due to the animal's transparency, the ease with which transgenic animals are generated and the numerous available neurone- and muscle-specific promoters.

By comparing the magnitude and duration of calcium influx in wild-type and mutant animals, Schaffer and co-workers (Kerr *et al.*, 2000) successfully determined the pharyngeal calcium transients and have provided significant insight into the function of calcium channel activity in feeding behaviour. This method can be potentially extended into the study of physiological characteristics of calcium dynamics in other specialized muscles that control particular nematode behaviours, such as body-wall muscles, that control locomotion, vulva muscles, which control egg-laying, male-specific muscles, which control turning, and spicule insertion steps of mating behaviour. Other potential uses of cameleon include the study of neuronal activity in response to application a stimulus, such as an odorant, a neurotransmitter or touch. The analysis of individual calcium channels is also possible. To be able to record neural calcium transients in intact *C. elegans* is a significant technological breakthrough for understanding the physiological and molecular basis of behaviour.

9.3.10. Primary cell culture system

Many experimental approaches have been developed to the study of nematode behaviours. However, for reasons discussed above, it is difficult to perform functional analysis of *C. elegans* neurones and muscle cells, and this, in turn, limits studies of cell- and tissue-specific profiling of gene expression patterns and protein analysis. A recently developed *C. elegans* embryonic cell culture system has provided an experimental solution to these technical hurdles (Christensen *et al.*, 2002). Using cell-specific promoters to label selected cells with GFP, it was shown that isolated embryonic cells could differentiate into the various neurones and muscle cell types that comprise newly hatched juveniles. Cultured cells could be used in electrophysiological assays, such as patch-clamp, and are sensitive to RNAi, dramatically knocking down gene expression. The primary *C. elegans* cell culture system provides a powerful foundation for a broad array of new experimental opportunities.

To gain molecular insight into cell excitability, signalling transduction and other important cellular processes, genes may be introduced via standard *C. elegans* transgenesis and resulting transgenic animals used to create primary cultured cells. The transfection method that is the norm for the introduction of exogenous genes in mammalian cells is still unavailable in *C. elegans* cell culture. Another drawback is that embryonic cells do not undergo postembryonic development, thereby limiting analysis to processes that act early on. However, those genes that are expressed later than in first-stage juveniles can be introduced into cultured cells easily by injecting DNA and generating transgenic animals to use in the establishment of cell cultures. Embryonic promoters may be used to drive expression of postembryonic genes.

Cell-specific GFP markers can also be used in enriching selected cell types via fluorescence-activated cell sorting (FACS) methods. This pool of GFP-expressing cells can be used as starting material for biochemical and molecular analysis. For example, 71 *mec-3*–dependent candidate genes were identified by culturing and sorting GFP-labelled touch cells of wild-type and *mec-3* mutant animals, isolating mRNA and performing microarray analysis (Zhang *et al.*, 2002b). Until the advent of cell culture, those genes that are differentially expressed in only a few cells (such as the six touch neurones) would not be identified using DNA microarrays because of sensitivity issues.

Routine biochemical and molecular approaches developed in mammalian cell culture systems can be easily transferred to *C. elegans* cell culture system. If gene X encodes a membrane ion channel, patch-clamp studies could be easily performed on cultured cells to study electrophysiological properties. Cultured cells can also be readily loaded with fluorescent probes that measure the levels of intracellular ions such as Ca^{2+}, H^+ and Cl^-, as well as the changes in membrane potential and current. RNAi assays can 'knock down' the expression of a specific gene in cultured cells. GFP variants, red fluorescent protein (RFP; also known as dsRED from *Discosoma*) (Campbell *et al.*, 2002; Zhang *et al.*, 2002a) and cell sorting methods facilitate co-culturing. Thus, in combination with the existing powerful *C. elegans* tools, the primary cell culture system provides the opportunity for analysing molecular mechanisms at a new level.

The neural basis of nematode behaviour lies in neuronal excitability and connectivity. Identifying and characterizing molecules involved in synaptogenesis and axon guidance could accelerate the understanding of basic molecular rules underlying neural activity. Cultured neurones and muscle cells from *Drosophila* to human species are capable of establishing functional synapses *in vitro* and, in many instances, retain synaptic target specificity. Likewise, cultured *C. elegans* neurones and muscle cells form synapses (Christensen *et al.*, 2002). This is an exciting finding: synapses are easily accessible for the functional analysis of their electrophysiological properties, which cannot be evaluated via existing tools.

9.4. Conclusion and Future Directions

C. elegans is an excellent model organism for examining behaviours at the cellular, genetic and molecular levels. Information gleamed from the completed *C. elegans* genome, ongoing nematode sequencing projects and other neurogenetic model systems will enable identification of conserved and species-specific molecules that regulate nervous system development, differentiation and function. In the post-genomic era, the major challenge is to understand gene interactions and environmental factors that generate complex behavioural patterns. The genetic amenability of *C. elegans* combined with the powerful armoury of tools suggests that this problem may be conquered at a molecular level.

Acknowledgements

We thank our laboratory colleagues for useful comments on the manuscript. We especially thank K. Knobel for images in Fig. 9.5 and T. Liu, E. Peden and J. Wang for sharing unpublished data. M.M.B. thanks D. Tilton for ongoing, thought-provoking discussion. Our research is supported by grants from the National Institutes of Health, the PKD Foundation and the American Heart Association.

References

Ashrafi, K., Chang, F.Y., Watts, J.L., Fraser, A.G., Kamath, R.S., Ahringer, J. and Ruvkun, G. (2003) Genome-wide RNAi analysis of *Caenorhabditis elegans* fat regulatory genes. *Nature* 421, 268–272.

Ashton, F.T., Bhopale, V.M., Holt, D., Smith, G. and Schad, G.A. (1998) Developmental switching in the parasitic nematode *Strongyloides stercoralis* is controlled by the ASF and ASI amphidial neurons. *Journal of Parasitology* 84, 691–695.

Ashton, F.T., Li, J. and Schad, G.A. (1999) Chemo- and thermosensory neurons: structure and function in animal parasitic nematodes. *Veterinary Parasitology* 84, 297–316.

Avery, L. and Thomas, J.T. (1997) Feeding and defecation. In: Riddle, D.L., Blumenthal, T., Meyer, B.J. and Priess, J. (eds) *C. elegans II.* Cold Spring Harbor Laboratory Press, Cold Spring Harbor, New York, pp. 679–715.

Avery, L., Raizen, D. and Lockery, S. (1995) Electrophysiological methods. *Methods in Cell Biology* 48, 251–269.

Bargmann, C. and Hodgkin, J. (2002) Accolade for C. elegans. *Cell* 111, 759–762.

Bargmann, C.I. (1993) Genetic and cellular analysis of behaviour in *C. elegans. Annual Review in Neuroscience* 16, 47–71.

Bargmann, C.I. (1998) Neurobiology of the *Caenorhabditis elegans* genome. *Science* 282, 2028–2033.

Bargmann, C.I. and Avery, L. (1995) Laser killing of cells in *Caenorhabditis elegans. Methods in Cell Biology* 48, 225–250.

Bargmann, C.I. and Horvitz, H.R. (1991) Chemosensory neurons with overlapping functions direct chemotaxis to multiple chemicals in *C. elegans. Neuron* 7, 729–742.

Bargmann, C.I. and Kaplan, J.M. (1998) Signal transduction in the *Caenorhabditis elegans* nervous system. *Annual Review in Neuroscience* 21, 279–308.

Bargmann, C.I. and Mori, I. (1997) Chemotaxis and thermotaxis. In: Riddle, D.L., Blumenthal, T., Meyer, B.J. and Priess, J. (eds) *C. elegans II.* Cold Spring Harbor, New York, Laboratory Press, pp. 717–737.

Barinaga, M. (1999) Mapping smells in the brain. *Science* 285, 508.

Barr, M.M. and Sternberg, P.W. (1999) A polycystic kidney-disease gene homologue required for male mating behaviour in *C. elegans. Nature* 401, 386–389.

Barr, M.M., DeModena, J., Braun, D., Nguyen, C.Q., Hall, D.H. and Sternberg, P.W. (2001) The *Caenorhabditis elegans* autosomal dominant polycystic kidney disease gene homologs *lov-1* and *pkd-2* act in the same pathway. *Current Biology* 11, 1341–1346.

Barstead, R. (1999) Reverse genetics. In: Hope, I.A. (ed.) *C. elegans: a Practical Approach.* Oxford University Press, New York, pp. 97–118.

Bessereau, J.L., Wright, A., Williams, D.C., Schuske, K., Davis, M.W. and Jorgensen, E.M. (2001) Mobilization of a *Drosophila* transposon in the *Caenorhabditis elegans* germ line. *Nature* 413, 70–74.

Bhopale, V.M., Kupprion, E.K., Ashton, F.T., Boston, R. and Schad, G.A. (2001) *Ancylostoma caninum*: the finger cell neurons mediate thermotactic behaviour by infective larvae of the dog hookworm. *Experimental Parasitology* 97, 70–76.

Birnby, D.A., Link, E.M., Vowels, J.J., Tian, H., Colacurcio, P.L. and Thomas, J.H. (2000) A transmembrane guanylyl cyclase (DAF-11) and Hsp90 (DAF-21) regulate a common set of chemosensory behaviours in *Caenorhabditis elegans*. *Genetics* 155, 85–104.

Blumenthal, T., Evans, D., Link, C.D., Guffanti, A., Lawson, D., Thierry-Mieg, J., Thierry-Mieg, D., Chiu, W.L., Duke, K., Kiraly, M. and Kim, S.K. (2002) A global analysis of *Caenorhabditis elegans* operons. *Nature* 417, 851–854.

Bowerman, B. (2001) Cytokinesis in the *C. elegans* embryo: regulating contractile forces and a late role for the central spindle. *Cell Structure and Function* 26, 603–607.

Brenner, S. (1974) The genetics of *Caenorhabditis elegans*. *Genetics* 77, 71–94.

Brooks, D.R. and Isaac, R.E. (2002) Functional genomics of parasitic worms: the dawn of a new era. *Parasitology International* 51, 319–325.

Brownlee, D., Holden-Dye, L. and Walker, R. (2000) The range and biological activity of FMRFamide-related peptides and classical neurotransmitters in nematodes. *Advances in Parasitology* 45, 109–180.

Buck, L. and Axel, R. (1991) A novel multigene family may encode odorant receptors: a molecular basis for odor recognition. *Cell* 65, 175–187.

Campbell, R.E., Tour, O., Palmer, A.E., Steinbach, P.A., Baird, G.S., Zacharias, D.A. and Tsien, R.Y. (2002) A monomeric red fluorescent protein. *Proceedings of the National Academy of Sciences, USA* 99, 7877–7882.

C. elegans Sequencing Consortium (1998) Genome sequence of the nematode *C. elegans*: a platform for investigating biology. *Science* 282, 2012–2018.

Chalfie, M. and Sulston, J. (1981) Developmental genetics of the mechanosensory neurons of *Caenorhabditis elegans*. *Developmental Biology* 82, 358–370.

Chalfie, M. and White, J. (1988) The nervous system. In: Wood, W.B. (ed.) *The Nematode Caenorhabditis elegans*. Cold Spring Harbor Laboratory Press, Cold Spring Harbor, New York, pp. 337–391.

Chalfie, M., Sulston, J.E., White, J.G., Southgate, E., Thomson, J.N. and Brenner, S. (1985) The neural circuit for touch sensitivity in *Caenorhabditis elegans*. *Journal of Neuroscience* 5, 956–964.

Chalfie, M., Tu, Y., Euskirchen, G., Ward, W.W. and Prasher, D.C. (1994) Green fluorescent protein as a marker for gene expression. *Science* 263, 802–805.

Chiba, C.M. and Rankin, C.H. (1990) A developmental analysis of spontaneous and reflexive reversals in the nematode *Caenorhabditis elegans*. *Journal of Neurobiology* 21, 543–554.

Chou, J.H., Troemel, E.R., Sengupta, P., Colbert, H.A., Tong, L., Tobin, D.M., Roayaie, K., Crump, J.G., Dwyer, N.D. and Bargmann, C.I. (1996) Olfactory recognition and discrimination in *Caenorhabditis elegans*. *Cold Spring Harbor Symposium on Quantitative Biology* 61, 157–164.

Christensen, M., Estevez, A., Yin, X., Fox, R., Morrison, R., McDonnell, M., Gleason, C., Miller, D.M., 3rd and Strange, K. (2002) A primary culture system for functional analysis of *C. elegans* neurons and muscle cells. *Neuron* 33, 503–514.

Coburn, C.M. and Bargmann, C.I. (1996) A putative cyclic nucleotide-gated channel is required for sensory development and function in *C. elegans*. *Neuron* 17, 695–706.

Coburn, C.M., Mori, I., Ohshima, Y. and Bargmann, C.I. (1998) A cyclic nucleotide-gated channel inhibits sensory axon outgrowth in larval and adult *Caenorhabditis elegans*: a distinct pathway for maintenance of sensory axon structure. *Development* 125, 249–258.

Colbert, H.A., Smith, T.L. and Bargmann, C.I. (1997) OSM-9, a novel protein with structural similarity to channels, is required for olfaction, mechanosensation, and olfactory adaptation in *Caenorhabditis elegans*. *Journal of Neuroscience* 17, 8259–8269.

Cole, D.G. (2003) The intraflagellar transport machinery of *Chlamydomonas reinhardtii*. *Traffic* 4, 435–442.

Cole, D.G., Diener, D.R., Himelblau, A.L., Beech, P.L., Fuster, J.C. and Rosenbaum, J.L. (1998) *Chlamydomonas* kinesin-II-dependent intraflagellar transport (IFT): IFT particles contain proteins required for ciliary assembly in *Caenorhabditis elegans* sensory neurons. *Journal of Cell Biology* 141, 993–1008.

Collet, J., Spike, C.A., Lundquist, E.A., Shaw, J.E. and Herman, R.K. (1998) Analysis of *osm-6*, a gene that affects sensory cilium structure and sensory neuron function in *Caenorhabditis elegans*. *Genetics* 148, 187–200.

Costanzo, M.C., Hogan, J.D., Cusick, M.E., Davis, B.P., Fancher, A.M., Hodges, P.E., Kondu, P., Lengieza, C., Lew-Smith, J.E., Lingner, C., Roberg-Perez, K.J., Tillberg, M., Brooks, J.E. and Garrels, J.I. (2000) The yeast proteome database (YPD) and *Caenorhabditis elegans* proteome database (WormPD): comprehensive resources for the organization and comparison of model organism protein information. *Nucleic Acids Research* 28, 73–76.

Costanzo, M.C., Crawford, M.E., Hirschman, J.E., Kranz, J.E., Olsen, P., Robertson, L.S., Skrzypek, M.S., Braun, B.R., Hopkins, K.L., Kondu, P., Lengieza, C., Lew-Smith, J.E., Tillberg, M. and Garrels, J.I. (2001) YPD, PombePD and WormPD: model organism volumes of the BioKnowledge library, an integrated resource for protein information. *Nucleic Acids Research* 29, 75–79.

Culotti, J.G. and Russell, R.L. (1978) Osmotic avoidance defective mutants of the nematode *Caenorhabditis elegans*. *Genetics* 90, 243–256.

Dal Santo, P., Logan, M.A., Chisholm, A.D. and Jorgensen, E.M. (1999) The inositol trisphosphate receptor regulates a 50–second behavioural rhythm in *C. elegans*. *Cell* 98, 757–767.

Davis, M.W., Somerville, D., Lee, R.Y., Lockery, S., Avery, L. and Fambrough, D.M. (1995) Mutations in the *Caenorhabditis elegans* Na,K-ATPase alpha-subunit gene, *eat-6*, disrupt excitable cell function. *Journal of Neuroscience* 15, 8408–8418.

de Bono, M. (2003) Molecular approaches to aggregation behavior and social attachment. *Journal of Neurobiology* 54, 78–92.

de Bono, M. and Bargmann, C.I. (1998) Natural variation in a neuropeptide Y receptor homolog modifies social behaviour and food response in *C. elegans*. *Cell* 94, 679–689.

Delattre, M. and Felix, M.A. (2001) Microevolutionary studies in nematodes: a beginning. *BioEssays* 23, 807–819.

Desai, C. and Horvitz, H.R. (1989) *Caenorhabditis elegans* mutants defective in the functioning of the motor neurons responsible for egg laying. *Genetics* 121, 703–721.

Dillin, A., Hsu, A.L., Arantes-Oliveira, N., Lehrer-Graiwer, J., Hsin, H., Fraser, A.G., Kamath, R.S., Ahringer, J. and Kenyon, C. (2002) Rates of behaviour and aging specified by mitochondrial function during development. *Science* 298, 2398–2401.

Driscoll, M. and Kaplan, J. (1997) Mechanotransduction. In: Riddle, D.L., Blumenthal, T., Meyer, B.J. and Priess, J. (eds) *C. elegans II*. Cold Spring Harbor Laboratory Press, Cold Spring Harbor, New York, pp. 645–677.

Duerr, J.S., Frisby, D.L., Gaskin, J., Duke, A., Asermely, K., Huddleston, D., Eiden, L.E. and Rand, J.B. (1999) The *cat-1* gene of *Caenorhabditis elegans* encodes a vesicular monoamine transporter required for specific monoamine-dependent behaviours. *Journal of Neuroscience* 19, 72–84.

Dwyer, N.D., Troemel, E.R., Sengupta, P. and Bargmann, C.I. (1998) Odorant receptor localization to olfactory cilia is mediated by ODR-4, a novel membrane-associated protein. *Cell* 93, 455–466.

Emmons, S.W. and Lipton, J. (2003) Genetic basis of male sexual behaviour. *Journal of Neurobiology* 54, 93–110.

Ernstrom, G.G. and Chalfie, M. (2002) Genetics of sensory mechanotransduction. *Annual Review of Genetics* 36, 411–453.

Felkai, S., Ewbank, J.J., Lemieux, J., Labbe, J.C., Brown, G.G. and Hekimi, S. (1999) CLK-1 controls respiration, behaviour and aging in the nematode *Caenorhabditis elegans*. *European Molecular Biology Organization Journal* 18, 1783–1792.

Fields, S. and Song, O. (1989) A novel genetic system to detect protein-protein interactions. *Nature* 340, 245–246.

Fire, A., Xu, S., Montgomery, M.K., Kostas, S.A., Driver, S.E. and Mello, C.C. (1998) Potent and specific genetic interference by double-stranded RNA in *Caenorhabditis elegans*. *Nature* 391, 806–811.

Fleming, J.T., Squire, M.D., Barnes, T.M., Tornoe, C., Matsuda, K., Ahnn, J., Fire, A., Sulston, J.E., Barnard, E.A., Sattelle, D.B. and Lewis, J.A. (1997) *Caenorhabditis elegans* levamisole resistance genes *lev-1*, *unc-29*, and *unc-38* encode functional nicotinic acetylcholine receptor subunits. *Journal of Neuroscience* 17, 5843–5857.

Francis, M.M., Mellem, J.E. and Maricq, A.V. (2003) Bridging the gap between genes and behaviour: recent advances in the electrophysiological analysis of neural function in *Caenorhabditis elegans*. *Trends in Neuroscience* 26, 90–99.

Fraser, A.G., Kamath, R.S., Zipperlen, P., Martinez-Campos, M., Sohrmann, M. and Ahringer, J. (2000) Functional genomic analysis of *C. elegans* chromosome I by systematic RNA interference. *Nature* 408, 325–330.

Garcia, L.R., Mehta, P. and Sternberg, P.W. (2001) Regulation of distinct muscle behaviours controls the *C. elegans* male's copulatory spicules during mating. *Cell* 107, 777–788.

Gonczy, P., Echeverri, G., Oegema, K., Coulson, A., Jones, S.J., Copley, R.R., Duperon, J., Oegema, J., Brehm, M., Cassin, E., Hannak, E., Kirkham, M., Pichler, S., Flohrs, K., Goessen, A., Leidel, S., Alleaume, A.M., Martin, S., Ozlu, N., Bork, P. and Hyman, A.A. (2000) Functional genomic analysis of cell division in *C. elegans* using RNAi of genes on chromosome III. *Nature* 408, 331–336.

Goodman, M.B. and Schwarz, E.M. (2003) Transducing touch in *Caenorhabditis elegans*. *Annual Review in Physiology* 65, 429–452.

Goodman, M.B., Hall, D.H., Avery, L. and Lockery, S.R. (1998) Active currents regulate sensitivity and dynamic range in *C. elegans* neurons. *Neuron* 20, 763–772.

Granato, M., Schnabel, H. and Schnabel, R. (1994) *pha-1*, a selectable marker for gene transfer in *C. elegans*. *Nucleic Acids Research* 22, 1762–1763.

Grishok, A. and Mello, C.C. (2002) RNAi (Nematodes: *Caenorhabditis elegans*). *Advances in Genetics* 46, 339–360.

Han, M. and Sternberg, P.W. (1991) Analysis of dominant-negative mutations of the *Caenorhabditis elegans let-60* ras gene. *Genes and Development* 5, 2188–2198.

Hannon, G.J. (2002) RNA interference. *Nature* 418, 244–251.

Hare, E. and Loer, C.M. (2000) Characterization of a serotonin-synthetic aromatic acid decarboxylase gene in the nematode *C. elegans*. *Society of Neuroscience Abstracts* 26, 564.

Harris, T.W., Lee, R., Schwarz, E., Bradnam, K., Lawson, D., Chen, W., Blasier, D., Kenny, E., Cunningham, F., Kishore, R., Chan, J., Muller, H.M., Petcherski, A., Thorisson, G., Day, A., Bieri, T., Rogers, A., Chen, C.K., Spieth, J., Sternberg, P., Durbin, R. and Stein, L.D. (2003) WormBase: a cross-species database for comparative genomics. *Nucleic Acids Research* 31, 133–137.

Hart, A.C., Sims, S. and Kaplan, J.M. (1995) Synaptic code for sensory modalities revealed by *C. elegans* GLR-1 glutamate receptor. *Nature* 378, 82–85.

Hashmi, S., Hashmi, G., Glazer, I. and Gaugler, R. (1998) Thermal response of *Heterorhabditis bacteriophora* transformed with the *Caenorhabditis elegans* hsp70-encoding gene. *Journal of Experimental Zoology* 281, 164–170.

Hedgecock, E.M. and Russell, R.L. (1975) Normal and mutant thermotaxis in the nematode *Caenorhabditis elegans. Proceedings of the National Academy of Sciences, USA* 72, 4061–4065.

Herndon, L.A., Schmeissner, P.J., Dudaronek, J.M., Brown, P.A., Listner, K.M., Sakano, Y., Paupard, M.C., Hall, D.H. and Driscoll, M. (2002) Stochastic and genetic factors influence tissue-specific decline in ageing *C. elegans. Nature* 419, 808–814.

Hill, A.A., Hunter, C.P., Tsung, B.T., Tucker-Kellogg, G. and Brown, E.L. (2000) Genomic analysis of gene expression in *C. elegans. Science* 290, 809–812.

Hirabayashi, J. and Kasai, K. (2002) Separation technologies for glycomics. *Journal of Chromatography B Analytical Technological Biomedical Life Science* 771, 67–87.

Hobert, O. (2003) Behavioral plasticity in *C. elegans*: paradigms, circuits, genes. *Journal of Neurobiology* 54, 203–223.

Hodgkin, J. (1983) Male phenotypes and mating efficiency in *Caenorhabditis elegans. Genetics* 103, 43–64.

Hodgkin, J. (1999) Conventional genetics. In: Hope, I. (ed.) C. elegans: *a Practical Approach*. Oxford University Press, New York, pp. 245–269.

Hodgkin, J., Horvitz, H.R. and Brenner, S. (1979) Nondisjunction mutants of the nematode *Caenorhabditis elegans. Genetics* 91, 67–94.

Hope, I.A. (ed.) (1999) C. elegans: *a Practical Approach*. Oxford University Press, New York, 300 pp.

Hussein, A.S., Kichenin, K. and Selkirk, M.E. (2002) Suppression of secreted acetylcholinesterase expression in *Nippostrongylus brasiliensis* by RNA interference. *Molecular and Biochemical Parasitology* 122, 91–94.

Igarashi, P. and Somlo, S. (2002) Genetics and pathogenesis of polycystic kidney disease. *Journal of the American Society of Nephrology* 13, 2384–2398.

Jiang, M., Ryu, J., Kiraly, M., Duke, K., Reinke, V. and Kim, S.K. (2001) Genome-wide analysis of developmental and sex-regulated gene expression profiles in *Caenorhabditis elegans. Proceedings of the National Academy of Sciences USA* 98, 218–223.

Johnsen, R.C. and Baillie, D.L. (1997) Mutation. In: Riddle, D.L., Blumenthal, T., Meyer, B.J. and Priess, J. (eds) C. elegans *II*. Cold Spring Harbor Laboratory Press, Cold Spring Harbor, New York, pp. 79–95.

Jorgensen, E.M. and Mango, S.E. (2002) The art and design of genetic screens: *Caenorhabditis elegans. Nature Reviews Genetics* 3, 356–369.

Jospin, M., Jacquemond, V., Mariol, M.C., Segalat, L. and Allard, B. (2002) The L-type voltage-dependent Ca^{2+} channel EGL-19 controls body wall muscle function in *Caenorhabditis elegans. Journal of Cell Biology* 159, 337–348.

Kamath, R.S., Fraser, A.G., Dong, Y., Poulin, G., Durbin, R., Gotta, M., Kanapin, A., Le Bot, N., Moreno, S. and Sohrmann, M. (2003) Systematic functional analysis of the *Caenorhabditis elegans* genome using RNAi. *Nature* 421, 231–237.

Kaplan, J.M. and Horvitz, H.R. (1993) A dual mechanosensory and chemosensory neuron in *Caenorhabditis elegans. Proceedings of the National Academy of Sciences, USA* 90, 2227–2231.

Kass, J., Jacob, T.C., Kim, P. and Kaplan, J.M. (2001) The egl-3 proprotein convertase regulates mechanosensory responses of *Caenorhabditis elegans. Journal of Neuroscience* 21, 9265–9272.

Kerr, R., Lev-Ram, V., Baird, G., Vincent, P., Tsien, R.Y. and Schafer, W.R. (2000) Optical imaging of calcium transients in neurons and pharyngeal muscle of *C. elegans*. *Neuron* 26, 583–594.

Komatsu, H., Mori, I., Rhee, J.S., Akaike, N. and Ohshima, Y. (1996) Mutations in a cyclic nucleotide-gated channel lead to abnormal thermosensation and chemosensation in *C. elegans*. *Neuron* 17, 707–718.

Koushika, S.P. and Nonet, M.L. (2000) Sorting and transport in *C. elegans*: a model system with a sequenced genome. *Current Opinions in Cell Biology* 12, 517–523.

Kramer, J.M., French, R.P., Park, E.C. and Johnson, J.J. (1990) The *Caenorhabditis elegans rol-6* gene, which interacts with the *sqt-1* collagen gene to determine organismal morphology, encodes a collagen. *Molecular Cell Biology* 10, 2081–2089.

Kumar, A. and Snyder, M. (2002) Protein complexes take the bait. *Nature* 415, 123–124.

Lee, S.S., Lee, R.Y., Fraser, A.G., Kamath, R.S., Ahringer, J. and Ruvkun, G. (2002) A systematic RNAi screen identifies a critical role for mitochondria in *C. elegans* longevity. *Nature Genetics* 25, 25.

L'Etoile, N.D. and Bargmann, C.I. (2000) Olfaction and odor discrimination are mediated by the *C. elegans* guanylyl cyclase ODR-1. *Neuron* 25, 575–586.

Li, C., Kim, K. and Nelson, L.S. (1999) FMRFamide-related neuropeptide gene family in *Caenorhabditis elegans*. *Brain Research* 848, 26–34.

Li, J., Zhu, X., Boston, R., Ashton, F.T., Gamble, H.R. and Schad, G.A. (2000) Thermotaxis and thermosensory neurons in infective larvae of *Haemonchus contortus*, a passively ingested nematode parasite. *Journal of Comparative Neurology* 424, 58–73.

Liao, V.H. and Freedman, J.H. (2002) Differential display analysis of gene expression in invertebrates. *Cellular and Molecular Life Sciences* 59, 1256–1263.

Lindemann, B. (2001) Receptors and transduction in taste. *Nature* 413, 219–225.

Lints, R. and Emmons, S.W. (1999) Patterning of dopaminergic neurotransmitter identity among *Caenorhabditis elegans* ray sensory neurons by a TGFbeta family signaling pathway and a Hox gene. *Development* 126, 5819–5831.

Liu, K.S. (1996) Male mating behaviour of *Caenorhabditis elegans*. PhD thesis, California Institute of Technology, Pasadena.

Liu, K.S. and Sternberg, P.W. (1995) Sensory regulation of male mating behaviour in *Caenorhabditis elegans*. *Neuron* 14, 79–89.

Lockery, S.R. and Goodman, M.B. (1998) Tight-seal whole-cell patch clamping of *Caenorhabditis elegans* neurons. *Methods in Enzymology* 293, 201–217.

Loer, C.M. and Kenyon, C.J. (1993) Serotonin-deficient mutants and male mating behaviour in the nematode *Caenorhabditis elegans*. *Journal of Neuroscience* 13, 5407–5417.

McIntire, S.L., Reimer, R.J., Schuske, K., Edwards, R.H. and Jorgensen, E.M. (1997) Identification and characterization of the vesicular GABA transporter. *Nature* 389, 870–876.

Maeda, I., Kohara, Y., Yamamoto, M. and Sugimoto, A. (2001) Large-scale analysis of gene function in *Caenorhabditis elegans* by high-throughput RNAi. *Current Biology* 11, 171–176.

Maine, E.M. (2001) RNAi As a tool for understanding germline development in *Caenorhabditis elegans*: uses and cautions. *Developmental Biology* 239, 177–189.

Marks, N.J., Shaw, C., Halton, D.W., Thompson, D.P., Geary, T.G., Li, C. and Maule, A.G. (2001) Isolation and preliminary biological assessment of AADGAPLIRFamide and SVPGVLRFamide from *Caenorhabditis elegans*. *Biochemical and Biophysical Research Communications* 286, 1170–1176.

Marszalek, J.R., Liu, X., Roberts, E.A., Chui, D., Marth, J.D., Williams, D.S. and Goldstein, L.S. (2000) Genetic evidence for selective transport of opsin and arrestin by kinesin-II in mammalian photoreceptors. *Cell* 102, 175–187.

Martin, E., Laloux, H., Couette, G., Alvarez, T., Bessou, C., Hauser, O., Sookhareea, S., Labouesse, M. and Segalat, L. (2002) Identification of 1088 new transposon insertions of *Caenorhabditis elegans*: a pilot study toward large-scale screens. *Genetics* 162, 521–524.

Mello, C. and Fire, A. (1995) DNA transformation. *Methods in Cell Biology* 48, 451–482.

Miller, D.M., 3rd, Desai, N.S., Hardin, D.C., Piston, D.W., Patterson, G.H., Fleenor, J., Xu, S. and Fire, A. (1999) Two-color GFP expression system for *C. elegans*. *Biotechniques* 26, 914–918, 920–911.

Miyawaki, A., Llopis, J., Heim, R., McCaffery, J.M., Adams, J.A., Ikura, M. and Tsien, R.Y. (1997) Fluorescent indicators for Ca^{2+} based on green fluorescent proteins and calmodulin. *Nature* 388, 882–887.

Miyawaki, A., Griesbeck, O., Heim, R. and Tsien, R.Y. (1999) Dynamic and quantitative Ca^{2+} measurements using improved cameleons. *Proceedings of the National Academy of Sciences, USA* 96, 2135–2140.

Moerman, D.G. and Fire, A. (1997) Muscle: structure, function, and development. In: Riddle, D.L., Blumenthal, T., Meyer, B.J. and Priess, J. (eds) *C. elegans II*. Cold Spring Harbor Laboratory Press, Cold Spring Harbor, New York, pp. 417–500.

Nathoo, A.N., Moeller, R.A., Westlund, B.A. and Hart, A.C. (2001) Identification of neuropeptide-like protein gene families in *Caenorhabditis elegans* and other species. *Proceedings of the National Academy of Sciences USA* 98, 14000–14005.

Nauli, S.M., Alenghat, F.J., Luo, Y., Williams, E., Vassilev, P., Li, X., Elia, A.E., Lu, W., Brown, E.M., Quinn, S.J., Ingber, D.E. and Zhou, J. (2003) Polycystins 1 and 2 mediate mechanosensation in the primary cilium of kidney cells. *Nature Genetics* 33, 129–137.

Orozco, J.T., Wedaman, K.P., Signor, D., Brown, H., Rose, L. and Scholey, J.M. (1999) Movement of motor and cargo along cilia. *Nature* 398, 674.

Paupard, M.C., Miller, A., Grant, B., Hirsh, D. and Hall, D.H. (2001) Immuno-EM localization of GFP-tagged yolk proteins in *C. elegans* using microwave fixation. *Journal of Histochemistry and Cytochemistry* 49, 949–956.

Pazour, G. and Rosenbaum, J. (2002) Intraflagellar transport and cilia-dependent diseases. *Trends in Cell Biology* 12, 551–555.

Pazour, G.J., Dickert, B.L., Vucica, Y., Seeley, E.S., Rosenbaum, J.L., Witman, G.B. and Cole, D.G. (2000) *Chlamydomonas* IFT88 and its mouse homologue, polycystic kidney disease gene *tg737*, are required for assembly of cilia and flagella. *Journal of Cell Biology* 151, 709–718.

Pazour, G.J., San Agustin, J.T., Follit, J.A., Rosenbaum, J.L. and Witman, G.B. (2002) Polycystin-2 localizes to kidney cilia and the ciliary level is elevated in *orpk* mice with polycystic kidney disease. *Current Biology* 12, R378–R380.

Perkins, L.A., Hedgecock, E.M., Thomson, J.N. and Culotti, J.G. (1986) Mutant sensory cilia in the nematode *Caenorhabditis elegans*. *Developmental Biology* 117, 456–487.

Pierce-Shimomura, J.T., Faumont, S., Gaston, M.R., Pearson, B.J. and Lockery, S.R. (2001) The homeobox gene *lim-6* is required for distinct chemosensory representations in *C. elegans*. *Nature* 410, 694–698.

Praitis, V., Casey, E., Collar, D. and Austin, J. (2001) Creation of low-copy integrated transgenic lines in *Caenorhabditis elegans*. *Genetics* 157, 1217–1226.

Qin, H., Rosenbaum, J.L. and Barr, M.M. (2001) An autosomal recessive polycystic kidney disease gene homolog is involved in intraflagellar transport in *C. elegans* ciliated sensory neurons. *Current Biology* 11, 457–461.

Raizen, D.M. and Avery, L. (1994) Electrical activity and behaviour in the pharynx of *Caenorhabditis elegans*. *Neuron* 12, 483–495.

Ramulu, P. and Nathans, J. (2001) Cellular and subcellular localization, N-terminal acylation, and calcium binding of *Caenorhabditis elegans* protein phosphatase with EF- hands. *Journal of Biological Chemistry* 276, 25127–25135.

Rand, J.B. and Nonet, M.L. (1997) Synaptic transmission. In: Riddle, D.L., Blumenthal, T., Meyer, B.J. and Priess, J. (eds) C. elegans *II*. Cold Spring Harbor Laboratory Press, Cold Spring Harbor, New York, pp. 611–643.

Ranganathan, R., Sawin, E.R., Trent, C. and Horvitz, H.R. (2001) Mutations in the *Caenorhabditis elegans* serotonin reuptake transporter MOD-5 reveal serotonin-dependent and -independent activities of fluoxetine. *Journal of Neuroscience* 21, 5871–5884.

Reinke, V. (2002) Functional exploration of the *C. elegans* genome using DNA microarrays. *Nature Genetics* 32 (Suppl. 2), 541–546.

Reinke, V. and White, K.P. (2002) Developmental genomic approaches in model organisms. *Annual Review of Genomics and Human Genetics* 3, 153–178.

Richmond, J.E., Davis, W.S. and Jorgensen, E.M. (1999) UNC-13 is required for synaptic vesicle fusion in *C. elegans*. *Nature Neuroscience* 2, 959–964.

Riddle, D.L. and Albert, P.S. (1997) Genetic and environmental regulation of dauer larva development. In: Riddle, D.L., Blumenthal, T., Meyer, B.J. and Priess, J. (eds) C. elegans *II*. Cold Spring Harbor Laboratory Press, Cold Spring Harbor, New York, pp. 739–768.

Riddle, D.L., Blumenthal, T., Meyer, B.J. and Priess, J. (eds) (1997) C. elegans *II*. Cold Spring Harbor Laboratory Press, Cold Spring Harbor, New York, 1222 pp.

Roayaie, K., Crump, J.G., Sagasti, A. and Bargmann, C.I. (1998) The G alpha protein ODR-3 mediates olfactory and nociceptive function and controls cilium morphogenesis in *C. elegans* olfactory neurons. *Neuron* 20, 55–67.

Rogers, C.M., Franks, C.J., Walker, R.J., Burke, J.F. and Holden-Dye, L. (2001) Regulation of the pharynx of *Caenorhabditis elegans* by 5–HT, octopamine, and FMRFamide-like neuropeptides. *Journal of Neurobiology* 49, 235–244.

Rosenbaum, J. (2002) Intraflagellar transport. *Current Biology* 12, R125.

Roy, P.J., Stuart, J.M., Lund, J. and Kim, S.K. (2002) Chromosomal clustering of muscle-expressed genes in *Caenorhabditis elegans*. *Nature* 418, 975–979.

Sawin, E.R., Ranganathan, R. and Horvitz, H.R. (2000) *C. elegans* locomotory rate is modulated by the environment through a dopaminergic pathway and by experience through a serotonergic pathway. *Neuron* 26, 619–631.

Sengupta, P., Chou, J.H. and Bargmann, C.I. (1996) *odr-10* encodes a seven transmembrane domain olfactory receptor required for responses to the odorant diacetyl. *Cell* 84, 899–909.

Signor, D., Wedaman, K.P., Orozco, J.T., Dwyer, N.D., Bargmann, C.I., Rose, L.S. and Scholey, J.M. (1999) Role of a class DHC1b dynein in retrograde transport of IFT motors and IFT raft particles along cilia, but not dendrites, in chemosensory neurons of living *Caenorhabditis elegans*. *Journal of Cell Biology* 147, 519–530.

Signor, D., Rose, L.S. and Scholey, J.M. (2000) Analysis of the roles of kinesin and dynein motors in microtubule-based transport in the *Caenorhabditis elegans* nervous system. *Methods* 22, 317–325.

Simmer, F., Tijsterman, M., Parrish, S., Koushika, S.P., Nonet, M.L., Fire, A., Ahringer, J. and Plasterk, R.H. (2002) Loss of the putative RNA-directed RNA polymerase RRF-3 makes *C. elegans* hypersensitive to RNAi. *Current Biology* 12, 1317–1319.

Simon, J.M. and Sternberg, P.W. (2002) Evidence of a mate-finding cue in the hermaphrodite nematode *Caenorhabditis elegans*. *Proceedings of the National Academy of Sciences, USA* 99, 1598–1603.

Sommer, R.J. (2000) Comparative genetics: a third model nematode species. *Current Biology* 10, 879–881.

Stein, L., Sternberg, P., Durbin, R., Thierry-Mieg, J. and Spieth, J. (2001) WormBase: network access to the genome and biology of *Caenorhabditis elegans*. *Nucleic Acids Research* 29, 82–86.

Sulston, J. (1988). Cell lineage. In: Wood, W.B. (ed.) *The Nematode* Caenorhabditis elegans. Cold Spring Harbor Laboratory Press, Cold Spring Harbor, New York, pp. 123–155.

Sulston, J.E. and Horvitz, H.R. (1977) Post-embryonic cell lineages of the nematode, *Caenorhabditis elegans*. *Developmental Biology* 56, 110–156.

Sulston, J.E. and White, J.G. (1980) Regulation and cell autonomy during postembryonic development of *Caenorhabditis elegans*. *Developmental Biology* 78, 577–597.

Sulston, J.E., Albertson, D.G. and Thomson, J.N. (1980) The *Caenorhabditis elegans* male: postembryonic development of nongonadal structures. *Developmental Biology* 78, 542–576.

Tavernarakis, N., Shreffler, W., Wang, S. and Driscoll, M. (1997) *unc-8*, a DEG/ENaC family member, encodes a subunit of a candidate mechanically gated channel that modulates *C. elegans* locomotion. *Neuron* 18, 107–119.

Tavernarakis, N., Wang, S.L., Dorovkov, M., Ryazanov, A. and Driscoll, M. (2000) Heritable and inducible genetic interference by double-stranded RNA encoded by transgenes. *Nature Genetics* 24, 180–183.

Thomas, J.H. and Lockery, S. (1999) Neurobiology. In: Hope, I.A. (ed.) C. elegans: *a Practical Approach*. Oxford University Press, New York, pp. 143–179.

Tijsterman, M., Ketting, R.F. and Plasterk, R.H. (2002) The genetics of RNA silencing. *Annual Review in Genetics* 36, 489–519.

Timmons, L. and Fire, A. (1998) Specific interference by ingested dsRNA. *Nature* 395, 854.

Timmons, L., Court, D.L. and Fire, A. (2001) Ingestion of bacterially expressed dsRNAs can produce specific and potent genetic interference in *Caenorhabditis elegans*. *Gene* 263, 103–112.

Tobin, D., Madsen, D., Kahn-Kirby, A., Peckol, E., Moulder, G., Barstead, R., Maricq, A. and Bargmann, C. (2002) Combinatorial expression of TRPV channel proteins defines their sensory functions and subcellular localization in *C. elegans* neurons. *Neuron* 35, 307–318.

Trent, C., Tsuing, N. and Horvitz, H.R. (1983) Egg-laying defective mutants of the nematode *Caenorhabditis elegans*. *Genetics* 104, 619–647.

Troemel, E.R., Chou, J.H., Dwyer, N.D., Colbert, H.A. and Bargmann, C.I. (1995) Divergent seven transmembrane receptors are candidate chemosensory receptors in *C. elegans*. *Cell* 83, 207–218.

Uetz, P. (2002) Two-hybrid arrays. *Current Opinion in Chemical Biology* 6, 57–62.

Urwin, P.E., Lilley, C.J. and Atkinson, H.J. (2002) Ingestion of double-stranded RNA by preparasitic juvenile cyst nematodes leads to RNA interference. *Molecular Plant Microbe Interactions* 15, 747–752.

Walhout, A.J., Boulton, S.J. and Vidal, M. (2000a) Yeast two-hybrid systems and protein interaction mapping projects for yeast and worm. *Yeast* 17, 88–94.

Walhout, A.J., Sordella, R., Lu, X., Hartley, J.L., Temple, G.F., Brasch, M.A., Thierry-Mieg, N. and Vidal, M. (2000b) Protein interaction mapping in *C. elegans* using proteins involved in vulval development. *Science* 287, 116–122.

Walhout, A.J., Reboul, J., Shtanko, O., Bertin, N., Vaglio, P., Ge, H., Lee, H., Doucette-Stamm, L., Gunsalus, K.C., Schetter, A.J., Morton, D.G., Kemphues, K.J., Reinke, V., Kin, S.K., Piano, F. and Vidal, M. (2002) Integrating interactome, phenome, and transcriptome mapping data for the *C. elegans* germline. *Current Biology* 12, 1952–1958.

Ward, S. (1973) Chemotaxis by the nematode *Caenorhabditis elegans*: identification of attractants and analysis of the response by use of mutants. *Proceedings of the National Academy of Sciences, USA* 70, 817–821.

Ward, S., Thomson, N., White, J.G. and Brenner, S. (1975) Electron microscopical reconstruction of the anterior sensory anatomy of the nematode *Caenorhabditis elegans*. *Journal of Comparative Neurology* 160, 313–337.

Ware, R.W., Clark, D.V., Crossland, K. and Russell, R.L. (1975) The nerve ring of the nematode *Caenorhabditis elegans*: sensory input and motor output. *Journal of Comparative Neurology* 162, 71–110.

White, J.G., Southgate, E., Thomson, J.N. and Brenner, S. (1986) The structure of the nervous system of the nematode *Caenorhabditis elegans*: the mind of a worm. *Philosophical. Transactions of the Royal Society of London* 314, 1–340.

Wicks, S.R., Yeh, R.T., Gish, W.R., Waterston, R.H. and Plasterk, R.H. (2001) Rapid gene mapping in *Caenorhabditis elegans* using a high density polymorphism map. *Nature Genetics* 28, 160–164.

Wood, W.B. and the Community of *C. elegans* researchers (1988) *The Nematode* Caenorhabditis elegans. Cold Spring Harbor Laboratory Press, Cold Spring Harbor, New York, 667 pp.

Yoder, B.K., Hou, X. and Guay-Woodford, L.M. (2002) The polycystic kidney disease proteins, polycystin-1, polycystin-2, polaris, and cystin, are co-localized in renal cilia. *Journal of the American Society of Nephrology* 13, 2508–2516.

Young, J.M. and Trask, B.J. (2002) The sense of smell: genomics of vertebrate odorant receptors. *Human Molecular Genetics* 11, 1153–1160.

Yu, H., Pretot, R.F., Burglin, T.R. and Sternberg, P.W. (2003) Distinct roles of transcription factors EGL-46 and DAF-19 in specifying the functionality of a polycystin-expressing sensory neuron necessary for *C. elegans* male vulva location behavior. *Development* 130, 5217–5227.

Zhang, J., Campbell, R.E., Ting, A.Y. and Tsien, R.Y. (2002a) Creating new fluorescent probes for cell biology. *Nature Reviews Molecular and Cellular Biology* 3, 906–918.

Zhang, Y., Chou, J.H., Bradley, J., Bargmann, C.I. and Zinn, K. (1997) The *Caenorhabditis elegans* seven-transmembrane protein ODR-10 functions as an odorant receptor in mammalian cells. *Proceedings of the National Academy of Sciences, USA* 94, 12162–12167.

Zhang, Y., Ma, C., Delohery, T., Nasipak, B., Foat, B.C., Bounoutas, A., Bussemaker, H.J., Kim, S.K. and Chalfie, M. (2002b) Identification of genes expressed in *C. elegans* touch receptor neurons. *Nature* 418, 331–335.

10 Biotic Interactions

PATRICIA TIMPER[1] AND KEITH G. DAVIES[2]

[1]USDA-ARS, Crop Protection and Management Research Unit, PO Box 748, Tifton, GA 31793, USA; [2]Nematode Interactions Unit, Rothamsted Research, Harpenden, Hertfordshire AL5 2JQ, UK

10.1. Introduction

Nematodes encounter other organisms as they forage for food, feeding sites and hosts or as they seek more hospitable environments and mates. Some of these encounters are beneficial to the nematode, but most encounters are neutral or antagonistic. How nematodes interact with beneficial and antagonistic organisms, including members of their own species, is the subject of this chapter. Because neutral encounters are unlikely to shape nematode behaviour, they will not be dealt with here.

There are two types of interactions in which a nematode benefits but causes no harm to other organisms: mutualism and commensalism (Table 10.1). Mutualism results in enhanced growth, fecundity or survival of both organisms. There are few examples of mutualistic associations involving nematodes. One classic example is the relationship between entomopathogenic nematodes in the genera *Steinernema* and *Heterorhabditis* and their symbiotic bacteria. Bacteria in the genera

Table 10.1. Interactions of nematodes with other organisms which have a positive (+) or negative (–) effect on nematode survival, growth or fecundity.

Type of interaction	Nematode	Other organism	Description of interaction (example involving nematodes)
Mutualism	+	+	Both organisms benefit (entomopathogenic nematodes and their bacterial partners)
Commensalism	+	0	Nematode benefits but other organism is unharmed (phoretic relationships between nematodes and arthropods)
Antagonism			
Predation	–	+	Nematode preyed on by another invertebrate (soil nematodes and predatory mites)
Parasitism	–	+	Nematode parasitized by a microorganism (soil nematodes and parasitic fungi)
Amensalism	–	0	Nematode is inhibited while other organism is unaffected (bacteria that repel plant-parasitic nematodes)
Competition	–	–	Interaction harms both organisms, shared resource (reduced fecundity when nematodes co-occur in vertebrate intestine)

Xenorhabdus and *Photorhabdus*, respectively, require these nematodes to gain entry into host insects and the bacteria, in turn, provide essential nutrients for nematode development and reproduction. There are also instances when population densities of two plant-parasitic species are greater when they co-occur on the same host plant than when they occur separately (Eisenback, 1993). The mechanism of this mutual stimulation is unknown but may involve suppressed host resistance. The nature of the interaction between the pine wood nematode *Bursaphelenchus xylophilus* and longhorn beetles in the genus *Monochamus* may also be considered mutualism. This nematode is dependent on the beetle for transport to new tree hosts, whereas the beetle benefits from the increased abundance of weakened trees (Mamiya, 1984). The use of a transport host is not unique to *B. xylophilus*. Many nematodes use larger organisms for transport to a new food source or a more suitable habitat (phoresis). However, most phoretic relationships are a form of commensalism, in which the growth, fecundity or survival of one organism is enhanced while the other is unaffected. With the exception of *B. xylophilus* and *Monochamus* spp., specific nematode behaviours related to mutualistic associations have not been described. We shall therefore focus our attention on the many diverse and fascinating behaviours that nematodes exhibit in their phoretic associations.

There are four types of interactions in which the nematode is harmed, through decreased survival, growth or fecundity, by another organism: predation, parasitism, amensalism and competition (Table 10.1). Nematodes themselves can be predators and parasites; however, behaviours related to food acquisition, including predator–prey and parasite–host interactions, are discussed in Bilgrami and Gaugler (Chapter 4, this volume). The concepts of predation and parasitism require little description. Predators tend to be larger than their prey and consume or partially consume multiple individuals in their life cycle. Parasites tend to be smaller than their hosts and live on or in a single individual either part or all of their life cycle. Fungi in the genera *Arthrobotrys*, *Dactylella*, *Duddingtonia* and *Monacrosporium*, to name a few, form traps that ensnare nematodes. Such trapping fungi are treated as predators in this chapter because an individual fungal colony can trap multiple nematodes. The production of a substance that reduces the survival, growth or fecundity of another organism is termed antibiosis. Antibiosis is a form of amensalism in which one organism is harmed while the other organism (the antibiotic producer) is unaffected. Antibiosis is sometimes considered a form of interference competition because in some cases the antibiotic is produced to protect a resource from competitors (e.g. allelopathy in plants). However, in many other situations, particularly those involving nematodes, it is not clear that the antibiotic is protecting a shared resource (space or nutrition). Competition between two species occurs when they utilize a shared, limited resource such as space or nutrition. The effect of competition is detrimental to both organisms, though the effects are sometimes asymmetrical in that one organism is harmed more than the other. Of course, competition is not limited to interspecific interactions. Individuals within a species also compete for limited resources.

From the above discussion, it is clear that antagonistic interactions involving nematodes are diverse. However, not all of these interactions have well-described behavioural components. For example, there has been considerable research done on biological control of plant-parasitic nematode using microbial antagonists, primarily predators and parasites (Stirling, 1991; Kerry, 2000). Nevertheless, little is known of the behavioural interactions between nematodes and microbial antagonists and, as a consequence, these sections are briefer than the sections dealing with competition.

10.2. Phoresy

Because of their small size and inability to move long distances, many nematodes have formed phoretic relationships with larger, more motile organisms that share their habitat or food source. The association can be either external or internal (e.g. reproductive tract or trachea), or a combination of both. Phoretic hosts provide nematodes with reliable transport to fresh resources and protection from the biotic and abiotic environments. Three orders, Rhabditida, Diplogasterida and Aphelenchida, contain most of the nematodes known to form phoretic relationships. However, there are a few genera in Tylenchida that are phoretic on bark beetles (Massey, 1974) and one species in Strongylida that is phoretic on the fungal genus *Pilobolus* (Robinson, 1962). A key preadaptation for phoresy in Nematoda is the formation of dauer juveniles (Sudhaus, 1976, cited in Kiontke, 1996; Maggenti, 1981). Dauer juveniles are specialized for dispersal and survival and, while not all dauer juveniles become phoretic, all phoretic nematodes are dauer juveniles. Another adaptation that facilitates phoresy is a behaviour termed nictation or *Winken* (waving), in which the dauer juvenile lifts its body off the substrate and stands on its tail tip in a straight posture or performs waving motions. This behaviour increases the chance of being picked up by a passing host (Campbell and Gaugler, 1993). In most cases, these hosts are invertebrates. Phoretic relationships can be either facultative or obligate; they are often obligate when the nematode is adapted to aerial parts of plants or to highly dispersed, ephemeral resources, such as dung pats, carrion and arthropod brood chambers.

10.2.1. Facultative transport

Phoresy may be an important means of dispersal for many nematode species, but is not essential to the species survival. These nematodes tend to be adapted to a wide range of habitats and are able to locate new resources without aid from another organism.

Entomopathogenic nematodes in the genera *Steinernema* and *Heterorhabditis* (Rhabditida) are sometimes transported by larger invertebrates that are not hosts for the nematode. Epsky *et al.* (1988) observed nictating dauer juveniles of *Steinernema carpocapsae* attaching to the backs of predatory mites. Several dozen nematodes would accumulate in parallel formation on the larger mites. Individuals that mounted the mites avoided consumption and were in a position for greater dispersal than those that did not mount. *Heterorhabditis bacteriophora*, which do not nictate, were unable to attach themselves to mites and were readily consumed. The presence of the earthworm *Lumbricus terrestris* in soil increased upward dispersal of *S. carpocapsae* and *Steinernema feltiae* compared with soil without the earthworm, even when remnant burrows were present (Shapiro *et al.*, 1995). Viable dauer juveniles of *Steinernema* spp. were found on the surface, in the casts and within the body of *L. terrestris*, suggesting that these nematodes were transported upward by earthworms.

Nematodes in the genus *Caenorhabditis* (Rhabditidae) are adapted to decomposing organic matter containing abundant nutrients and microorganisms (Sudhaus and Kiontke, 1996). All species except one (*Caenorhabditis drosophilae*) display nictating behaviour, a sign that they are phoretically associated with another animal (Sudhaus and Kiontke, 1996). Although some species are dependent on phoresy because they are specialized for specific resources such as dung or rotting cacti, others may be less dependent because they utilize a variety of resources or renewable resources, such as plant litter. One example of facultative transport in this group is *Caenorhabditis remanei*, an associate of terrestrial isopods and snails (Baird *et al.*, 1994; Baird, 1999). In woodland habitats of Ohio, this nematode was found on several species of isopods collected from under rotting logs, rocks and leaf litter (Baird, 1999). Dauer juveniles of *C. remanei* were located on the underside of the dorsal plates and ventral appendages of the isopods. On agar plates, the dauers responded to the presence of isopod hosts or their frass by nictating. Within 5 min of exposing isopods to *C. remanei*, dauers were observed moving up the legs, antennae and dorsal plates. When the isopods were subsequently transferred to a moist environment, the nematodes dismounted their host, whereas they remained on the host in a dry environment.

Perhaps the most unusual and unexpected phoretic relationship between a nematode and another organism is that of the bovine lungworm, *Dictyocaulus viviparus* (Trichostrongylidae), and the fungus *Pilobolus* spp. Both the nematode and the fungus are deposited in ruminant dung and must be ingested to complete their life cycle. However, grazing animals do not normally feed near their dung pats. Consequently both *D. viviparus* and *Pilobus* spp. must disperse away from the pats to increase the probability of ingestion. *Pilobus* spp. accomplish this by forcible discharge of their sporangia (a structure containing spores) up to 3 m from the dung pat (Doncaster, 1981). The nematode has limited dispersal capabilities and must rely on passive dispersal by heavy rain or on the hooves of cattle (Robinson, 1962; Eysker and De Coo, 1988). Robinson (1962) first described another mechanism by which the nematode is passively dispersed from dung pats, by being catapulted along with the sporangium of *Pilobolus* spp. On the dung surface, the normally sluggish third-stage infective juveniles of *D. viviparus* become active, occasionally adopting an erect posture similar to the nictation behaviour of Rhabditida (Robinson, 1962). When a sporangiophore of *Pilobolus* spp. is contacted, they climb up and accumulate around the base of the sporangium. Just before discharge, the nematodes burrow into the gelatinous matrix surrounding the spores and are shot away from the dung pat with the sporangium (Doncaster, 1981). After landing, the nematodes may gradually emerge from the sporangia deposited on foliage or emerge more rapidly following ingestion by a ruminant (Somers, 1985).

10.2.2. Obligate transport

Nematodes adapted to patchy, ephemeral resources may depend on a variety of arthropods for transport. For example, *Diplogaster cerea* (Diplogasteridae) is found in decaying cacti in Arizona (Kiontke, 1996). In this dry climate, the nematode is dependent on several beetle and fly species for transport to new locations of rotting vegetation. However, more commonly, nematodes that rely exclusively on phoresy for their survival have developed synchronized life cycles with specific transport hosts, as displayed by another cactophilic nematode, *C. drosophilae*. In tissues of decaying saguaro cactus, second-stage juveniles of this nematode accumulate around pupae of *Drosophila nigrospiracula*, where they develop into non-nictating dauers (Kiontke, 1997). When a fly emerges from its pupa, the dauers crawl on to the head and enter the inflatable sac used to open the puparium. The nematodes are then transported to a new rotting cactus. Similar synchronized life cycles occur in dung pats between *Rhabditis dubia* and moth flies (Psychodidae) and between *Diplogaster coprophila* and scavenger flies (Sepsidae); in brood balls between *Diplogaster* spp. and dung-burying beetles (Scarabaeidae); and in underground brood cells between *Bursaphelenchus seani* (Aphelenchoididae) and helictid wasps and between *Aduncospiculum halicti* (Diplogasteroididae) and anthophorid wasps (Giblin and Kaya, 1983, 1984; Sudhaus and Kuhne, 1989; Giblin-Davis *et al.*, 1990; Kiontke, 1996; Kuhne, 1996). In many cases, the association of the nematode with the insect is obligatory; without host contact, the dauer juvenile will not develop even when placed on a suitable substrate. When the nematode is delivered by the insect to a new decaying cactus, dung pat or brood chamber, it disembarks and completes its life cycle by feeding on fungi and bacteria.

Two plant parasites, the pine wood nematode *B. xylophilus* and the red ring nematode *Bursaphelenchus cocophilus*, are also highly synchronized with their phoretic hosts, beetles in the genera *Monochamus* (Cerambycidae) and *Rhynchophorus* (Curculionidae) respectively (Giblin-Davis, 1993). Because of the economic importance of pine wilt disease, the interaction between *B. xylophilus* and *Monochamus* is the best-studied phoretic association between a nematode and another organism. The nematodes are carried to susceptible pine trees as fourth-stage dauer juveniles in the trachea of the beetle (Mamiya, 1984). The dauers exit the host and invade the wood tissue through feeding or oviposition wounds created by the beetle. Inside the resin canals of the tree, the nematode moults and enters its propagative phase, during which it completes several generations, feeding on parenchyma and epithelial cells as well as on fungi. After the death of the tree and depletion of food resources, *B. xylophilus* switches to the dispersal phase by forming third-stage dispersal juveniles (J3). This stage is morphologically and physiologically distinct from the propagative J3 (Kondo and Ishibashi, 1978). The dispersal J3 accumulate around the pupal chambers of *Monochamus* and moult to the dauer juvenile near the time of beetle eclosion (Mamiya, 1984; Necibi and Linit, 1998). After eclosion, the dauers mount the adults and enter the tracheal system.

A number of cues are involved in guiding the nematode through the dispersal phase of its life cycle. The first cue, which initiates formation of the dispersal juveniles, is some unknown indicator of low food supply (Ishibashi and Kondo, 1977) and is independent of the presence or absence of a beetle host (Warren and Linit, 1993). Also unknown is the cue resulting in aggregation of the dispersal J3 around the pupal chamber of *Monochamus*. This apparently host-specific cue may be a product of the last-instar beetle or the blue stain fungi, which are more abundant in the chambers of *Monochamus* than in the chambers of other beetles (Maehara and Futai, 2002). Both propagative juveniles and dispersal juveniles are attracted to the unsaturated fatty acids that are deposited by the beetle larvae in the walls of the chamber (Miyazaki *et al.*, 1977a,b; cited in Maehara and Futai, 2002). Juveniles of *B. xylophilus* were also attracted to lipid extracts of *Monochamus* larvae but not to lipid extracts of cabbage looper larvae (Bolla *et al.*, 1989). The moult from dispersal J3 to dauer juvenile, occurs only in the presence of a beetle host (Warren and Linit, 1993; Necibi and Linit, 1998). Although some dauers form soon after pupation, the greatest number form soon after eclosion of the adult beetle (Necibi and Linit, 1998; Maehara and Futai, 2001). Pulverized callow adults also stimulate the moult from dispersal J3 to dauer juvenile, indicating the involvement of one or more substances related to insect eclosion rather than an increase in CO_2 production (Necibi and Linit, 1998). Toluene, a component of the adult beetle cuticle, is attractive to dauers of *B. xylophilus* and may play a role in attachment of the nematode to the insect. Once on the adult beetle, the dauer juveniles may be directed to the spiracles by following a CO_2 gradient (Miyazaki *et al.*, 1978, cited in Mamiya, 1984). Volatile monoterpenes from pine trees, particularly β-myrcene, also attract dauer juveniles and are thought to stimulate exit from the beetle to the feeding wound (Ishikawa *et al.*, 1986). However, the dauers do not immediately exit the beetle upon visiting a susceptible pine. Intrinsic factors such as the level of neutral storage lipid in the dauer appear to modify the response of the nematode to extrinsic factors such as plant and insect volatiles (Stamps and Linit, 1998a,b). Dauer juveniles low in neutral lipids were attracted to β-myrcene, while those with high lipid content were attracted to toluene (Stamps and Linit, 2001). Stamps and Linit (2001) adapted the rolling fulcrum model of Miller and Strickler (1984) to explain the nematode's decision to stay within or leave the beetle during visits to susceptible trees (Fig. 10.1). In this model, the fulcrum is pushed to the right as the lipid content of the nematode is depleted, leading to a greater sensitivity of the lever to pine volatiles, such as β-myrcene, and a greater tendency to exit the beetle in response to these cues.

10.2.3. Necromeny

Necromeny is a special type of phoresy in which dauer juveniles colonize a living host and wait until the host dies. After death, the nematodes develop on bacteria

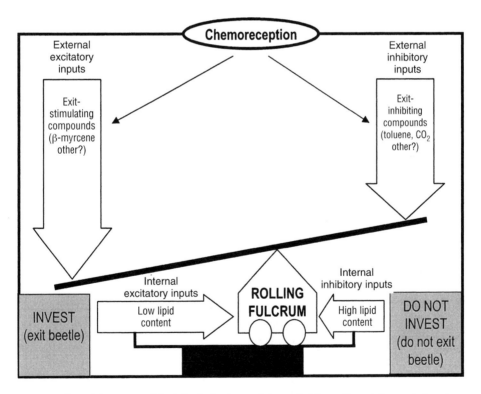

Fig. 10.1. Rolling fulcrum model of the behaviour of dauer (J4) juveniles of *Bursaphelenchus xylophilus* and their response to internal and external stimuli related to their exit behaviour from beetle hosts (redrawn from Stamps and Linit, 2001).

from the decomposing corpse. In other words, necromeny is phoresy over time rather than space. Several rhabditid species (Schulte, 1989; Baird *et al.*, 1994) and at least two diplogasterid species (Manegold and Kiontke, 2001) are known to practise necromeny. The nematode behaviours involved in establishing and maintaining a necromenic relationship are similar to those described for phoretic relationships. Except in the case of *C. remanei*, which is also phoretic on isopods, it is unclear whether or not necromeny is an essential part of the life cycle (Baird, 1999).

10.3. Antagonism

10.3.1. Predation and parasitism

Predation is a relationship between two different populations of organisms at different trophic levels in which the one at the higher trophic level, the predator, 'lives off' the one at the lower trophic level, the prey, to the detriment of the latter. In predation, individual prey are killed. Parasitism is also a relationship between two populations at different trophic levels; however in parasitism, the parasite, although detrimental to its host, does not usually kill it. At least this is true in highly co-evolved forms of host–parasite interaction. The range of antagonists is highly diverse, including fungi, bacteria and other invertebrates, such as amoebae, insects, mites, tardigrades, turbellaria and, indeed, even other nematodes. Even a brief description of each group and their interaction with nematodes would be a huge undertaking and readers are referred to the comprehensive review of Stirling (1991) for greater detail.

The most prolific research area on antagonists has been in the use of microbial control agents for the management of animal- and plant-parasitic nematode pests (Stirling, 1991; Sikora, 1992; Dickson *et al.*, 1994; Waller and Faedo, 1996; Kerry 2000; Kerry and Hominick, 2000; Larsen, 2000). Studies using antagonists for the biological control of plant-parasitic nematodes are dominated by reports of experiments designed to assess whether a particular organism can be used as a control agent. Nevertheless, there are few nematode biological control agents on the market compared with those for plant pathogens (Whipps and Davies, 2000), suggesting that their use is at present too inconsistent for development as commercial products. This emphasizes the need to increase our understanding of the interactions between nematodes and the various other biotic components of the system.

10.3.1.1. Effects of predation

Nematode behaviour can be investigated at the individual and population level; indeed, for nematode populations to coexist with their antagonists, be they predators or parasites, different behavioural strategies have developed. The importance of this can be seen by recent studies demonstrating that *Caenorhabditis elegans* can switch from a solitary to a gregarious type of behaviour by the change in a single amino acid (de Bono and Bargmann, 1998; de Bono *et al.*, 2002). However the modification in behaviour brought about by such a change and its effect on nematode interaction with antagonists have yet to be determined. Both sedentary and active stages of nematodes have been reported to be subject to predation by various groups of invertebrates, but the role of predators in the regulation of nematode populations remains unclear. Nematode biomass in a modelled grassland ecosystem, where the soil faunal community was manipulated, remained unaffected; however, their numbers were significantly affected by the different soil faunal treatments (Bradford *et al.*, 2002). This suggests that, although overall nematode biomass was not changing, their community structure had changed considerably.

Whole groups of nematodes, including *Aporcelaimus, Nygolaimus, Sectonema, Labronema, Dorylaimus* and *Actinolaimus* spp., are predacious. Esser (1963) reported that the stylet thrusting behaviour of *Dorylaimus* was altered in the presence of eggs of plant-parasitic nematodes. He suggested that this was in response to chemical cues released from the eggs. Rates of predation by predatory nematodes were found to increase with increasing number of prey, and excised or injured prey attracted predators (Yeates, 1969; Bilgrami *et al.*, 1985; Shafqat *et al.*, 1987; Khan *et al.*, 1995). Escape from predation was thought to be related to various physical, chemical and behavioural characteristics of the prey that conferred differing degrees of resistance to predators (Khan *et al.*, 1995). Different groups of nematodes appear to adopt different behavioural strategies to immobilize their prey: aphelenchids paralyse their prey by injecting saliva (Hechler, 1963), diplogasterids hold their prey by oesophageal suction (Bilgrami and Jairajpuri, 1989) and nygolaim predators paralyse their prey by the disruption of internal body organs (Bilgrami *et al.*, 1985). In some nematodes this is assisted by extracorporeal digestion using oesophageal secretions (Grootaert *et al.*, 1977; Bilgrami and Jairajpuri, 1989; Khan *et al.*, 1991). However, initial wounding appears critical to prey susceptibility. Puncturing the cuticle leads to a loss of hydrostatic pressure, affecting mobility and making these individuals more vulnerable to secondary infections by saprophytic microorganisms such as bacteria and fungi. However, some groups of fungi, the nematode-trapping fungi, can be predatory in their own right.

Nematode-trapping fungi are important predators of nematodes and, although their inundative application to control plant parasites has met with limited success, some species have recently been found to be excellent tools for the reduction of animal-parasitic nematodes (Wolstrup *et al.*, 1996; Larsen, 1999, 2000). These nematode-trapping fungi form a diverse collection from various fungal taxonomic groups and are capable of the predation of vermiform life stages of nematodes using various types of trapping devices (Barron, 1977). Several studies investigating the interaction between nematodes and trapping fungi have shown that mycelia, conidia and the traps of some fungi can attract various mycophagous and bacteriophagous nematodes on agar plates (Balan *et al.*, 1976; Field and Webster, 1977; Jansson and Nordbring-Hertz, 1979; 1980; Jansson, 1982a,b). Carbon dioxide has the ability to attract different nematode species (Johnson and Viglierchio, 1961; Croll and Matthews, 1977; Gaugler *et al.*, 1980; Robinson, 1995). Researchers have suggested that CO_2 produced by trapping fungi also attracts nematodes (Balan and Gerber, 1972). Moreover, mycelium that had been stimulated to produce traps was more attractive than mycelium that had not been stimulated to produce traps (Field and Webster, 1977). The presence of nematodes can act as a stimulus for the induction of trap formation (Bartnicki-Garcia *et al.*, 1964; Nansen, 1988) and trapping activity appears related to the nutrient status of the fungal substrate (Cooke, 1962).

Not all fungi are equally successful at attracting nematodes. In a test of 14 nematophagous fungi, ten were attractive to *Panagrellus redivivus*, three were neutral and one was repellent (Jansson and Nordbring-Hertz, 1979). The attractiveness of nematodes to fungi is also dependent on the ecological state of the fungi. Fungi, as discussed above, can be attractive to nematodes but biological control studies in which *Monacrosporium* spp. and *Hirsutella rhossiliensis* had been incorporated into alginate pellets were found to be repellent, leading to reduced biological control. As might be expected, the injection of CO_2 into the assay increased nematode attraction and fungal-induced mortality (Robinson and Jaffee, 1996). Similarly, not all nematode species respond in the same manner, and sialic acid has been implicated in chemotaxis and subsequent infection (Jansson and Nordbring-Hertz, 1980, 1984). The stage of the nematode has also been found to be important; e.g. in studies of animal- and entomopathogenic nematodes, whether or not the nematode is sheathed or exsheathed has a significant effect on the interaction between the trapping fungus and the nematode (Timper and Kaya, 1989). Changes in the cuticle may also be expected to affect mobility in the soil; however, surface mutants of *C. elegans* that were differentially trapped by different fungi were not shown to differ significantly in their ability to migrate on agar plates (Mendoza de Gives *et al.*, 1999a).

10.3.1.2. Effects of parasitism

There are several types of microbial parasites that infect nematodes. These can be facultative parasites, such as the fungi *Paecilomyces lilacinus* and *Pochonia chlamydosporia* (= *Verticillium chlamydosporium*), or obligate parasites, such as *Nematophthora gynophila*, *H. rhossiliensis* and the Gram-positive bacterium *Pasteuria penetrans*. Further details and descriptions of most of these organisms, their life cycles, and how they interact with nematodes can be found in the following reviews: Stirling (1991), Kerry and Jaffee (1997), Kerry (2000) and Kerry and Hominick (2000). Recently other organisms have caught the attention of researchers. Rickettsia-like organisms have been observed in some J2 plant-parasitic nematodes (Shepherd *et al.*, 1973) and shortly afterwards intracellular bacteria were observed in filarial nematodes (McLaren *et al.*, 1975; Vincent *et al.*, 1975; Kozek and Marroquin, 1977). The application of molecular techniques has revolutionized work with Rickettsia-like organisms and they have been shown to be close relatives of *Wolbachia*, a group of bacteria known to infect a wide range of arthropods and filarial nematodes (O'Neill *et al.*, 1992; Sironi *et al.*, 1995; Werren, 1997; Werren and O'Neill, 1997; Bandi *et al.*, 1998). These bacteria, thought originally to be commensals, can now be characterized as parasites because they appear to affect the reproduction and fecundity of their nematode hosts (Werren, 1997; Casiraghi *et al.*, 2002; Rao *et al.*, 2002; Taylor, 2002). Within insect groups, *Wolbachia* can have a large effect on the behaviour of insect hosts (Stevens *et al.*, 2001), including sex-role reversal, changes in mating systems and mate choice behaviour in butter-flies. We suggest that similar behavioural changes may well occur in nematodes and understanding this interaction may be helpful in controlling filarial parasites.

Another group of endoparasites that has recently received attention is the *Pasteuria* group of Gram-positive bacteria. Four species are recognized: *Pasteuria ramosa*, a parasite of *Daphnia* spp. (water-fleas) and *Pasteuria penetrans*, *Pasteuria nishizawae* and *Pasteuria thornei*, parasites of root-knot, cyst and lesion nematodes, respectively. The morphologies, life cycles and ecologies of these parasites have been extensively reviewed (Sayre and Starr, 1988; Chen and Dickson, 1998). *Pasteuria* spp. may affect nematode behaviour in a number of ways, including acquisition of spores as nematodes migrate through soil. There are several reports indicating that *Meloidogyne* spp. juveniles with seven to 50 adhering spores appear less able to locate and invade a root system (Stirling, 1984; Brown and Smart, 1985; Davies *et al.*, 1988, 1991). It is not unreasonable to assume that the adhesion of other microorganisms may also affect movement. Studies of host specificity between *Pasteuria penetrans* and its nematode host (Davies *et al.*, 1988; Stirling, 1991; Sharma and Davies, 1996; Espanol *et al.*, 1997; Mendoza de Gives, 1999b) indicate that individual populations of the bacterium do not adhere to and recognize all nematode populations. Indeed, it has recently been shown that cuticle heterogeneity, as exhibited by *Pasteuria* endospore attachment, is not linked to the phylogeny of root-knot nematodes (Davies *et al.*, 2001). Cuticle heterogeneity can play an important role in influencing nematode behaviour in terms of acquisition of endospores. Because surface antigenicity of nematodes is under environmental control (Grenache *et al.*, 1996), such factors may play an important role in the interactions between the surface of nematodes and any microorganisms. In turn, these interactions may affect the general mobility and behaviour of the nematodes as they move through their environment.

Little is known about the genetics behind the interactions of nematode hosts and their parasites. The recently completed genome of *C. elegans* has stimulated research on its interaction with various pathogens. There are some 20 species of bacteria known to be pathogens of *C. elegans*, of which six are Gram-positive while the remainder are Gram-negative (Ewbank, 2002). Thus, *C. elegans* appears to be poorly protected when compared with other invertebrates such as insects. For example *Drosophila* and, indeed, most other invertebrates have well-developed immune systems, including coelomocytes, which are important in cellular defence mechanisms (Hoffmann and Reichhart, 1997). The best-characterized innate system of immunity concerns the Toll pathway, which is activated by insect expo-sure to fungi and results in the expression of antifungal peptides that protect the insect from infection. This pathway has been conserved throughout evolution and is also important in vertebrate immunity (Imler and Hoffmann, 2000). Although there is homology between genes in the Toll pathway of insects and *C. elegans* (Fallon *et al.*, 2001) it would appear that there has been functional divergence. For example, a mutation in *tol-1, nr2033*, which in *Drosophila* appears to affect resistance against bacterial infection, does not have the same effect in *C. elegans*. *Caenorhabditis elegans* is initially highly attracted to *Serratia marcescens* strain Db11, but over time Db11 has a strong tendency to repel wild-type worms. With *tol-1, nr2033* mutants this response is significantly reduced and they are never repelled by the pathogen (Pujol *et al.*, 2001). Expression of *tol-1* is restricted to the nervous system and it has been proposed that these genes contribute to the recognition of specific bacterial components that result in changes in behaviour (Pujol *et al.*, 2001).

10.3.2. Amensalism

Amensalism is the interaction between two organisms in which one organism is suppressed by another without detriment to the one doing the suppressing. In amensalism, the unaffected population usually releases antibiotics or toxins that restrict the growth of the other population. Whether this is without cost to the organism producing the antibiotic or toxin can be difficult to measure, for in the economy of nature toxin production uses energy which otherwise could be used for other purposes, including growth and reproduction. Therefore it is arguable that amensalism in its pure form may not exist.

10.3.2.1. The role of antibiosis and possible tritrophic interaction

The use of rhizobacteria and endophytic bacteria as a potential method for controlling plant-parasitic nematodes has led to an increased understanding of antibiosis. *In vitro* experiments have shown that some rhizosphere bacteria and fungi produce compounds that inhibit egg hatch and J2 mobility, but whether these compounds are produced *in vivo* and in sufficient quantity to be effective in the rhizosphere is unclear (Kerry, 2000). Microbial agents that are stimulated to grow in the rhizosphere and that are not necessarily parasites may adhere to the cuticle of migrating nematodes, thus reducing nematode mobility. For example, the endophytic bacteria *Enterobacter asburiae* was found adhering to the cuticle of *Meloidogyne incognita* (Hallmann *et al.*, 1998). Rhizobacteria, such as *Pseudomonas fluorescens* and *Pseudomonas putida*, may directly influence the migration of infective juvenile nematodes towards roots (Aalten *et al.*, 1998); however, indirect effects are also possible. For example, reduced root penetration by *Globodera pallida* in the presence of *Bacillus aphaericus* or *Agrobacterium radiobacter* was shown to be related to the development of systemic resistance induced by the presence of the bacteria in the rhizosphere (Hallmann *et al.*, 1998; Hasky-Günther *et al.*, 1998).

Entomopathogenic nematodes have been shown to suppress plant-parasitic nematodes (Bird and Bird, 1986). Recently the inundative application of entomopathogenic nematodes, with their associated symbiotic bacteria, was found to reduce the number of genera and abundance of plant-parasitic but not free-living nematodes (Somasekhar *et al.*, 2002). Although the mechanism for this effect is unclear, it has been suggested that allelochemicals produced by the nematodes and/or their symbiotic bacteria may be selectively active against the plant-parasitic nematode community (Grewal *et al.*, 1999; Jagdale *et al.*, 2002). Applications of just the symbiotic bacteria, *Xenorhabdus* and *Photorhabdus* spp., associated with these entomopathogenic nematodes were also found to reduce plant-parasitic nematodes (Samaliev *et al.*, 2000). The compounds responsible for these effects are thought to be bacterial toxins (Bowen *et al.*, 1998) or other secondary metabolites (Hu *et al.*, 1999).

Bacteria are not the only organisms that produce compounds antagonistic to nematodes; plants and fungi also produce compounds that play a role in antibiosis. Probably the best-known case for nematode antibiosis by plants is that of marigolds, *Tagetes* spp. The mode of action of these plants on nematodes involves polythienyls, and *in vitro* tests show that different groups of nematodes reveal differential susceptibility to these compounds. The free-living nematode *C. elegans* was ten times more sensitive than the plant-parasitic nematode *Pratylenchus penetrans* (Kyo *et al.*, 1990) and α-terthienyl from marigold roots affected the host-finding ability of the entomopathogenic nematode *Steinernema glaseri* (Kanagy and Kaya, 1996). Other compounds produced by plants have recently been reviewed with respect to their efficacy for plant-parasitic nematode control (Chitwood, 2002), but whether or not they affect nematode behaviour in soil is unknown.

Clearly plants can protect themselves directly by producing toxins. Plants also protect their leaves from herbivores indirectly by producing chemical signals that attract the herbivore's natural enemies, indicating that a tritrophic interaction is possible between a plant, an insect herbivore and an insect-parasitic wasp (Dicke *et al.*, 1990; Turlings *et al.*, 1995; Takabayashi and Dicke, 1996). Roots are also prone to herbivore attack and, therefore, they too might be expected to release compounds that attract natural enemies. Interestingly, a similar tritrophic interaction has been established between the roots of a coniferous plant (*Thuja occidentalis*), a root weevil larvae (*Otiorhynchus sulcatus*) and an entomopathogenic nematode (*Heterorhabditis megidis*) (van Tol *et al.*, 2001). In the future, new subtle signals may be discovered that affect nematode behaviour.

10.3.3. Competition

This is not intended to be a comprehensive review of competition, as the primary focus of this chapter is nematode behaviour, not ecology. However, it is necessary to provide some background on the importance of competition in nematode communities before discussing behaviours that may result in reduced competition and coexistence with competitors. There is controversy within parasitology about the role of competition in structuring communities of parasites, including nematodes, within vertebrate hosts (Poulin, 2001). Holmes and Price (1986) suggested that there are two types of parasitic communities within a host (infracommunities): isolationist communities, which have discrete niches and are unaffected by interspecific competition, and interactive communities, which have overlapping niches and are affected by interspecific competition. These are extreme ends of a spectrum, with many infracommunities displaying elements of isolation and interaction (Stock and Holmes, 1988; Moore and Simberloff, 1990). The role of competition in structuring communities of free-living, plant-parasitic and insect-parasitic nematodes has not been challenged. Scientists studying these nematodes have documented the existence of competition, but have generally been cautious in speculating about its importance in shaping nematode communities or in evolution.

   Competition among species is determined by either experimentation or field observation. In experiments, population parameters are measured when species are alone against when they co-occur. In field observation studies, communities are sampled and patterns of species richness and co-occurrences are determined. In both types of studies, reduced population numbers, fecundity and size are used as evidence for competitive interactions. With parasites of vertebrates, spatial displacement (when one species causes a shift in the position of another species) is also considered evidence of competition (Poulin, 2001). Competitive interactions among plant-parasitic nematodes (Eisenback, 1993), gastrointestinal nematodes of vertebrates (Poulin, 2001) and entomopathogenic nematodes (Kaya and Koppenhofer, 1996) have been reviewed. The consensus is that competition is often asymmetrical, with one species being harmed more than the other. Asymmetrical interactions may arise because the shared host is more susceptible to one species than the other or one species may be an intrinsically superior competitor. What makes one nematode a superior competitor to another depends on the interacting species. Timing of resource colonization also affects the outcome of competition with the initial colonizer having the competitive advantage.

10.3.3.1. Exploitation and interference competition

Exploitation competition, defined as depletion of the resource to a level that is insufficient for optimal growth and reproduction, is the most common type of intra- and interspecific competition experienced by nematodes. This form of competition can lead to changes in sex ratio and reduced body size, fecundity and survival. All nematodes experience some degree of intraspecific competition when resources are limited, but it has been associated more with parasitic than with free-living species (Hominick and Tingley, 1984; Fleming, 1988; Selvan *et al.*, 1993; Koppenhofer and Kaya, 1995; Dezfuli *et al.*, 2002). Examples of interspecific exploitation competition involving nematodes are too numerous to cover and only a few will be highlighted. The first and simplest example occurs in cultures of the bacterial-feeding nematodes *Panagrolaimus detritophagus* and *P. superbus* (Rhabditida). When grown on agar plates, the density of both species is greater when they are cultured alone than when cultured together (Sohlenius, 1988). The second example occurs in wax worm larvae concurrently infected with two species of *Steinernema*. At low inoculation rates, progeny production of *S. carpocapsae* was reduced whereas that of *S. glaseri* was not (Koppenhofer *et al.*, 1995). At high inoculation rates, progeny production of both species was reduced, but the reduction was greater for *S. carpocapsae* than for *S. glaseri*. The intrinsic superiority of *S. glaseri* appears to be due to faster utilization of the food resources and compatibility with its competitor's symbiotic bacterium. The third and clearest example of exploitation competition occurs in barley infected with two nematode species. Here, the presence of the root-lesion nematode *Pratylenchus neglectus* reduced the rate of development and fecundity of the root-knot nematode *Meloidogyne chitwoodi* (Umesh *et al.*, 1994). Analysis of carbon assimilation suggested that feeding and migration of the lesion nematode reduced the availability of nutrients to *M. chitwoodi*.

Interactions in which one species is prevented from access to a resource by physical or chemical means is termed interference competition. This type of competition can result in reduced establishment and survival, and in niche displacement. One of the most interesting cases of intraspecific interference competition is that of pinworms (Oxyuroidea: Thelastomatidae), which live in the hind-gut of certain arthropods. Hosts typically contain a single male–female pair (Zervos, 1988a,b; Adamson *et al.*, 1992). It is rare to find more than two males within the host, though sometimes multiple females are present. When more than one female occurs, female size and fecundity are reduced, suggesting exploitation competition. Even when hosts are fed multiple male eggs, only one male develops and the others are expelled in the faeces after hatching (Zervos, 1988b). A similar expulsion occurs with females, but not as intensely as with males. The hypothesis is that males and females produce a sex-specific substance that eliminates intraspecific competitors (Zervos, 1988a,b).

More commonly, interference competition occurs between two species. Formation of root galls from the root-knot nematode *M. incognita* was suppressed when large numbers of the mycophagous nematode *Aphelenchus avenae* or entomopathogenic nematodes *Steinernema* spp. were applied to soil (Bird and Bird, 1986; Ishibashi and Choi, 1991; Ishibashi and Matsunaga, 1995). Entomopathogenic nematodes and *A. avenae* aggregate at root tips, perhaps orienting to CO_2 produced there, and possibly interfere with root penetration of the plant parasites. With entomopathogenic nematodes, toxic metabolites from their symbiotic bacterium may also play a role in suppression of *Meloidogyne* spp. root penetration (Grewal *et al.*, 1999; Samaliev *et al.*, 2000).

Colonization of a host by one parasitic species often leads to conditions unsuitable for the establishment of a second parasitic species. With gastrointestinal parasites of vertebrates, this can occur as a consequence of the host reaction to the initial colonizer. Acanthocephalans can cause severe destruction of the host tissue, thus preventing parasitic nematodes from establishing in that region of the intestine (Vidal-Martinez and Kennedy, 2000). The initial colonizing species may also induce a specific immune reaction in the host that adversely affects the establishment of a second nematode species (i.e. cross-immunity) (Moqbel and Wakelin, 1979). With entomopathogenic nematodes, antibiotics produced by the symbiotic bacterium prevent the establishment of a second symbiotic bacterium in the host (Akhurst, 1982). Because *Heterorhabditis* spp. and *Steinernema* spp. cannot utilize each others bacterial symbiont, colonization of a host by the symbiont of one genus effectively bars access to host resources by the other genus (Alatorre-Rosas and Kaya, 1991).

10.3.3.2. Mechanisms of coexistence

Both exploitation and interference competition can eventually lead to local extinction of the weaker competitor. However, nematodes frequently coexist with competitors in the same habitat or host (Kaya and Koppenhofer, 1996; Ettema, 1998; Poulin, 2001). Clearly, there are mechanisms that reduce competitive interactions, thus allowing coexistence. In the following sections, we shall describe possible mechanisms of reducing both intra- and interspecific competition, many of which involve behavioural responses from the nematode.

10.3.3.2.1. Adjustments in resource utilization

A fundamental niche is the niche that a species could potentially occupy, whereas the realized niche is a subset of the fundamental niche that a species is restricted to in the presence of a competitor (Begon and Mortimer, 1986). Although a niche can occupy a number of dimensions (space, time, temperature, food type, etc.), we shall limit our discussion to a single niche dimension or resource, either nutrition or space. Resource partitioning occurs when one or both species restrict their fundamental niche to minimize overlap and reduce competition. It is often difficult to separate resource partitioning from interference competition. For example, is the nematode avoiding resources where it will probably perform poorly or is it being excluded from the resource by a competitor (interference competition)? A prerequisite for resource partitioning is the ability to detect the presence of competitors and select a resource that will maximize fitness.

Resource partitioning has been documented several times with intestinal nematode parasites of vertebrates (Stock and Holmes, 1988; Moore and Simberloff, 1990; Patrick, 1991; Haukisalmi and Henttonen, 1993). In each of these cases, evidence for resource partitioning is a shift in position of attachment when another species is present (Fig. 10.2). However, because these nematodes cannot be directly observed during site selection, it is not always clear whether shifts in attachment sites were a consequence of interference competition or prudent site selection by the nematodes. The biology of the nematode will determine its ability to actively select sites. Some gastrointestinal nematodes lack the mobility to move from site to site, while others are able to actively migrate to a preferred site. Nematodes lacking mobility would be unlikely to participate in resource partitioning; therefore, shifts in site of attachment of these nematodes may be evidence of interference competition.

In other parasitic nematodes, there is strong evidence of detection and avoidance of previously infected hosts. Given a choice, more entomopathogenic nematodes will penetrate uninfected hosts rather than hosts infected with a conspecific or heterospecific nematode (Glazer, 1997; Wang and Ishibashi, 1998; 1999). Apparently, these nematodes are able to detect volatile cues emanating from

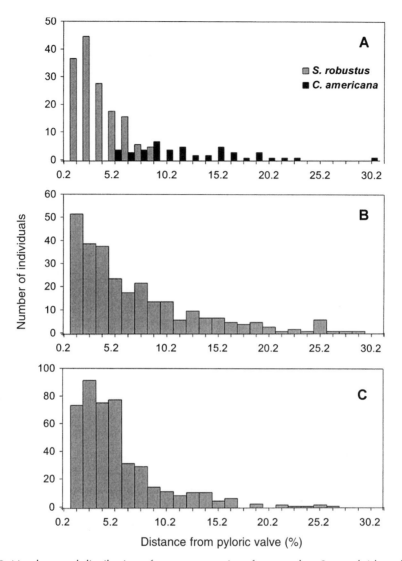

Fig. 10.2. Numbers and distribution of two core species of nematodes, *Strongyloides robustus* and *Capillaria americana*, found in the small intestine of the southern flying squirrel, *Glaucomys volans*. A. *S. robustus* and *C. americana* in naturally infected squirrels. B. *S. robustus* after experimental infection using 50 juveniles. C. *S. robustus* after experimental infection with 200 juveniles. (From Patrick, 1991.)

infected hosts (Grewal *et al.*, 1997). However, the duration post-infection is critical. Infective stages of *S. carpocapsae* were attracted to hosts recently infected by conspecifics but were repelled from hosts with later infections. The ability to detect heterospecific infections depends on the species involved. For example, infective juveniles of *S. carpocapsae* were repelled from hosts infected with most heterospecific nematodes (four of five species), whereas juveniles of *S. glaseri* were attracted to hosts infected with most heterospecific nematodes and only repelled by hosts infected with *Steinernema riobrave*. This differential response to previously infected hosts may be related to the inherent competitive ability of the nematode species, with a weak competitor showing greater sensitivity to a heterospecific infection than a strong competitor. In co-infections of wax worm larvae, *S. carpocapsae* was a poor competitor for host resources compared with *S. glaseri* (Koppenhofer *et al.*, 1995). Interestingly, *S. carpocapsae* appears to react to the presence of *S. glaseri* infective juveniles by increasing nictation behaviour and rates of host penetration (Wang and Ishibashi, 1999). It is unclear whether this is a specific response to potential competition or a general response to agitation from another organism. Entomopathogenic nematodes may also detect the presence of other insect pathogens. In soil containing a healthy host and a host infected with the fungus *Beauveria bassiana*, more *S. carpocapsae* and *H. bacteriophora* accumulated around the healthy than around the infected host (Barbercheck and Kaya, 1991).

Some plant-parasitic nematodes also have the ability to detect the presence of competitors within roots. Fewer lesion nematodes (*Pratylenchus penetrans*) entered roots when either root-knot (*M. incognita*) or stunt nematodes (*Tylenchorhynchus claytoni*) were present in another part of the root system (Estores and Chen, 1972; Miller and McIntyre, 1975). Although this could be a case of interference competition via induced host resistance, the lesion nematode's capability of detecting the presence of other nematodes prior to invasion also suggests that a decision is being made as to the quality of the resource.

10.3.3.2.2. Niche differentiation

Resource partitioning and avoidance of previously infected hosts are flexible responses to competition. If competitors are removed, the resource will be fully utilized. Niche differentiation, on the other hand, is an inflexible, genetically determined response that may or may not be a result of competition. In niche differentiation, a species is specialized for using a particular resource, or part of a resource, such has host type or region of the host intestine. If competitors are removed, resource utilization will be unaffected. Niche differentiation may arise from co-evolution of competing species or from independent evolution resulting in adaptation to different environmental conditions (Connell, 1980). Regardless of the mechanism, niche differentiation results in coexistence of species utilizing similar resources. Although there are many examples of niche differentiation in the nematology literature, we shall focus on foraging strategies because they are directly related to nematode behaviour.

Different foraging strategies can lead to niche differentiations in space and type of food/host utilized. The best-described foraging strategies are those exhibited by entomopathogenic nematodes. These nematodes display a continuum of foraging behaviours between a sit-and-wait (ambusher) strategy and a widely foraging (cruiser) strategy (Lewis *et al.*, 1992; Campbell and Gaugler, 1993; Campbell and Kaya, 2002). Nematodes utilizing a sit-and-wait strategy, such as *Steinernema carpocapsae* and *S. scapterisci*, stay near the soil surface without dispersing from their original location (Moyle and Kaya, 1981), and spend a significant amount of time standing on their tails (nictating) on top of surface particles (Campbell and Gaugler, 1993). Nictation increases contact with mobile insects (Campbell and Gaugler, 1993). Nematodes utilizing a widely foraging strategy, such as *H. bacteriophora* and *S. glaseri*, readily disperse below the soil surface (Georgis and Poinar, 1983a,b; Schroeder and Beavers, 1987) and do not nictate. Although entomopathogenic nematodes have broad and overlapping host ranges, the divergence of foraging strategies leads to the utilization of different hosts. Sit-and-wait foragers are adapted to infection of mobile hosts on the soil surface, while widely foraging species are adapted to infection of sedentary subterranean hosts (Campbell and Gaugler, 1993). In host-finding competition studies, *S. carpocapsae* was superior to the widely foraging species *H. bacteriophora* and *S. glaseri* when hosts were on the soil surface, but was inferior to these species when hosts were below the surface (Alatorre-Rosas and Kaya, 1990; Koppenhofer *et al.*, 1996). When both surface and subterranean hosts were available, *S. carpocapsae* and *S. glaseri* were able to coexist in laboratory microcosms; however, when only subterranean hosts were present, *S. carpocapsae* went extinct in some microcosms (Koppenhofer and Kaya, 1996).

10.3.3.2.3. Aggregation and dispersal

Aggregation and dispersal are responses to environmental stress. Here we are concerned with how aggregation and dispersal may relieve the stress imposed by competition. For a broader discussion of nematode behaviours related to environmental stress, see Wharton, Chapter 13, this volume. Aggregation of potential competitors on a patchy or divided resource may promote coexistence (Atkinson and Shorrocks, 1981). In a recent study of insect communities of mushrooms, spatial aggregation and not resource partitioning explained the coexistence of species with overlapping resource requirements (Wertheim *et al.*, 2000). Nematodes are similar to these mycophagous insects in that their food resources are often patchily distributed (e.g. microbial colonies, plant roots and insect hosts). Moreover, aggregation behaviours are common among nematodes (Croll, 1970). On bacterial cultures, *P. detritophagus* tended to aggregate near the point of inoculation, whereas *P. superbus* dispersed over the agar surface (Sohlenius, 1988). Aggregation of the competitively superior *P. detritophagus* may have created refuges for the inferior competitor *P. superbus*, thus contributing to the coexistence of these two species during 5 years of culture, an outcome predicted by the simulation model of Atkinson and Shorrocks *(1981)*.

Some natural populations of *C. elegans* exhibit either solitary or social feeding behaviour (de Bono and Bargmann, 1998). Solitary strains will aggregate when food becomes limiting, but social strains will aggregate even when food is abundant. In the absence of food, neither strain type forms aggregates. Aggregation is a response to noxious chemicals (de Bono *et al.*, 2002). Social strains lack a functioning *npr-1* gene (de Bono and Bargmann, 1998), which appears to suppress the response to aversive stimuli in solitary strains. The question is why is the stress response kept 'on' in some populations of *C. elegans* even in apparently low-stress environments (i.e. unlimited food supply)? Perhaps social populations persist, in part, because their aggregation on food resources reduces competitive interactions with other species.

While aggregation may play a role in reducing interspecific competition, dispersal is primarily a means of reducing intraspecific competition, though it may reduce interspecific competition as well. When resources become limited, mobile nematode stages leave in search of new resources. In many microbivorous and parasitic species, specialized stages are produced that are adapted for survival, dispersal, attachment to phoretic hosts and infection in the case of parasitic species.

The best studied of these specialized stages is the dauer juvenile of *C. elegans* and other rhabditid nematodes (Riddle, 1988). In *C. elegans*, dauer juveniles develop in response to a pheromone, low nutrient supply and elevated temperatures (Golden and Riddle, 1984). The concentration of pheromone serves as an indicator of population density. The ratio of pheromone to nutrient level influences dauer juvenile formation and recovery such that dauers develop when population densities are high and food is limited, whereas dauers moult to the fourth-stage juvenile (i.e. recover) when population densities are low and food is abundant (Golden and Riddle, 1982). Similarly, the formation of infective juveniles (dauers) of entomopathogenic nematodes within the host is induced by high population densities and low nutrient levels (Popiel *et al.*, 1989). Thus, with *C. elegans* and entomopathogenic nematodes, formation of the dauer juvenile is directly related to intraspecific competition.

Dispersal behaviour of some entomopathogenic nematodes may depend on the time of emergence from the insect cadaver. Early-emerging infective juveniles of *H. megidis* did not disperse far, but they showed a strong response to host cues (O'Leary *et al.*, 1998). Late-emerging individuals tended to disperse further but did not readily respond to host cues. This greater tendency for dispersal in late-emerging individuals would facilitate locating hosts that have not already been parasitized by the early-emerging individuals, thereby reducing intraspecific competition.

10.4. Conclusions

Nematode behaviours related to interactions with beneficial and deleterious organisms are in general poorly understood. This is because most encounters are hidden within soil or hosts, where nematode behaviour cannot be directly observed. Consequently, *in vitro* arenas (e.g. Petri dishes) tend to be used to study behavioural interactions of nematodes and other organisms. Such arenas are useful for indicating a potential response; however, they are limiting in that they cannot mimic the complexity of cues that the nematode is exposed to in its natural habitat. This level of complexity is aptly illustrated by the interaction between *B. xylophilus* and its beetle host *Monochamus*, where a number of different cues, both exogenous and endogenous, guide the dispersal behaviour of the nematode. Many of the studies contributing to our knowledge of this interaction were conducted in simplified arenas where compounds were tested for their behavioural effects on different stages of *B. xylophilus*. This knowledge, along with observations from natural habitats, was integrated into an elegant behavioural model to describe the interplay of exogenous and endogenous cues in the dispersal of *B. xylophilus* by *Monochamus*.

Another tool that could advance our understanding of the role behaviour plays in nematode interactions with other organisms is the use of behavioural phenotypes. For example, solitary and social feeding strains of *C. elegans* could be used to test the hypothesis that aggregation promotes coexistence of competing species, or that widely foraging solitary strains are more prone to encounters with antagonists and experience higher rates of mortality than are less mobile social strains.

Although there have been many exciting advances in our understanding of the behavioural ecology of nematodes in the years since Croll (1970) published his book on nematode behaviour, there remains much to be learned. In particular, there is little known regarding the behavioural components of competition among competing plant parasites or interactions of all nematode groups with their antagonists.

References

Aalten, P.M., Vitour, D., Blanvillain, D., Gowen, S.R. and Sutra, L. (1998) Effect of rhizosphere fluorescent *Pseudomonas* strains on plant-parasitic nematodes *Radopholus similis* and *Meloidogyne* sp. *Letters on Applied Microbiology* 27, 357–361.

Adamson, M.L., Buck, A. and Noble, S. (1992) Transmission pattern and intraspecific competition as determinants of population structure in pinworms (Oxyurida: Nematoda). *Journal of Parasitology* 78, 420–426.

Akhurst, R.J. (1982) Antibiotic activity of *Xenorhabdus* spp., bacteria symbiotically associated with insect pathogenic nematodes of the families Heterorhabditidae and Steinernematidae. *Journal of General Microbiology* 128, 3061–3066.

Alatorre-Rosas, R. and Kaya, H.K. (1990) Interspecific competition between entomopathogenic nematodes in the genera *Heterorhabditis* and *Steinernema* for an insect host in sand. *Journal of Invertebrate Pathology* 55, 179–188.

Alatorre-Rosas, R. and Kaya, H.K. (1991) Interaction between two entomopathogenic nematode species in the same host. *Journal of Invertebrate Pathology* 57, 1–6.

Atkinson, W.D. and Shorrocks, B. (1981) Competition on a divided and ephemeral resource: a simulation model. *Journal of Animal Ecology* 50, 461–471.

Baird, S.E. (1999) Natural and experimental associations of *Caenorhabditis remanei* with *Trachelipus rathkii* and other terrestrial isopods. *Nematology* 1, 471–475.

Baird, S.E., Fitch, D.H.A. and Emmons, S.W. (1994) *Caenorhabditis vulgaris* sp. n. (Nematoda: Rhabditidae): a necromenic associate of pill bugs and snails. *Nematologica* 40, 1–11.

Balan, J. and Gerber, N.N. (1972) Attraction and killing of the nematode *Panagrellus redivivus* by the predaceous fungus *Arthrobotrys dactyloides*. *Nematologica* 18, 163–173.

Balan, J., Krizkov, L., Nemec, P. and Kolozsvary, A. (1976) A qualitative method for detection of nematode-attracting substances and proof of production of three different attractants by the fungus *Monacrosporium rutgeriensis*. *Nematologica* 22, 306–311.

Bandi, C., Anderson, T.J.C., Genchi, C. and Blaxter, M.L. (1998) Phylogeny of *Wolbachia* in filarial nematodes. *Proceedings of the Royal Society of London, Series B: Biological Sciences* 265, 2407–2413.

Barbercheck, M.E. and Kaya, H.K. (1991) Effect of host condition and soil texture on host finding by the entomogenous nematodes *Heterorhabditis bacteriophora* (Rhabditida, Heterorhabditidae) and *Steinernema carpocapsae* (Rhabditida, Steinernematidae). *Environmental Entomology* 20, 582–589.

Barron, G.L. (1977) *The Nematode Destroying Fungi.* Canadian Biological Publication Ltd, Guelph, 140 pp.

Bartnicki-Garcia, S., Eren, J. and Pramer, D. (1964) Carbon dioxide dependent morphogenesis in *Arthrobotrys conoides. Nature* 204, 804.

Begon, M. and Mortimer, M. (1986) *Population Ecology: a Unified Study of Animals and Plants,* 2nd edn. Blackwell Scientific Publications, Oxford, 220 pp.

Bilgrami, A.L. and Jairajpuri, M.S. (1989) Resistance of prey to predation and strike rate of predators, *Mononchoides longicaudatus* and *M. fortidens* (Nematoda: Diplogasterida). *Revue de Nematologie* 12, 45–49.

Bilgrami, A.L., Ahmad, I. and Jairajpuri, M.S. (1985) Predatory behaviour of *Aquatides thornei* (Nygolaimina : Nematoda). *Nematologica* 30, 457–462.

Bird, A.F. and Bird, J. (1986) Observations on the use of insect parasitic nematodes as a means of biological control of root-knot nematodes. *International Journal for Parasitology* 16, 511–516.

Bolla, J.A., Bramble, J. and Bolla, R.I. (1989) Attraction of *Bursaphelenchus xylophilus,* pathotype MPSy-1, to *Monochamus carolinensis* larvae. *Japanese Journal of Nematology* 19, 32–37.

Bowen, D., Rocheleau, T.A., Blackburn, M., Andreev, O., Golubeva, E., Bhartia, R. and French-Constant, R.H. (1998) Insecticidal toxins from the bacterium *Photorhabdus luminescens. Science* 280, 2129–2132.

Bradford, M.A., Jones, T.H., Bardgett, R.D., Black, H.I.J., Boag, B., Bonkowski, M., Cook, R., Eggers, T., Grange, A.C., Grayston, S.J., Kandeler, E., McCaig, A.E., Newington, J.E., Prosser, J.I., Setätä, H., Staddon, P.L., Tordoff, G.M., Tscherko, D. and Lawton, J.H. (2002) Impacts of soil faunal community composition on model grassland ecosystems. *Science* 298, 615–618.

Brown, S.M. and Smart, G.C., Jr (1985) Root penetration by *Meloidogyne incognita* juveniles infected with *Bacillus penetrans. Journal of Nematology* 17, 123–126.

Campbell, J.F. and Gaugler, R. (1993) Nictation behaviour and its ecological implications in the host search strategies of entomopathogenic nematodes (Heterorhabditidae and Steinernematidae). *Behaviour* 126, 155–169.

Campbell, J.F. and Kaya, H.K. (2002) Variation in entomopathogenic nematode (Steinernematidae and Heterorhabditidae) infective-stage jumping behaviour. *Nematology* 4, 471–482.

Casiraghi, M., McCall, J.W., Simoncini, L., Kramer, L.H., Sacchi, L., Genchi, C., Werren, J.H. and Bandi, C. (2002) Tetracycline treatment and sex-ratio distortion: a role for *Wolbachia* in the moulting of filarial nematodes? *International Journal of Parasitology* 32, 1457–1468.

Chen, Z.X. and Dickson, D.W. (1998) Review of *Pasteuria penetrans:* biology, ecology and biological control potential. *Journal of Nematology* 30, 313–340.

Chitwood, D.J. (2002) Phytochemical based strategies for nematode control. *Annual Review of Phytopathology* 40, 221–249.

Connell, J.H. (1980) Diversity and the coevolution of competitors, or the ghost of competition past. *Oikos* 35, 131–138.

Cooke, R.C. (1962) Behaviour of nematode-trapping fungi during the decomposition of organic matter in soil. *Transactions of the British Mycological Society* 45, 314–320.

Croll, N.A. (1970) *The Behaviour of Nematodes*. Edward Arnold, London, 117 pp.

Croll, N.A. and Matthews, B.E. (1977) *Biology of Nematodes*. John Wiley & Sons, New York, 201 pp.

Davies, K.G., Kerry, B.R. and Flynn, C.A. (1988) Observations on the pathogenicity of *Pasteuria penetrans*, a parasite of root-knot nematodes. *Annals of Applied Biology*, 112, 491–501.

Davies, K.G., Laird, V. and Kerry, B.R. (1991) The motility, development and infection of *Meloidogyne incognita* encumbered with spores of the obligate hyperparasite *Pasteuria penetrans*. *Revue de Nématologie* 14, 611–618.

Davies, K.G., Fargette, M., Balla, G., Daudi, A., Duponnois, R., Gowen, S.R., Mateille, T., Phillips, M.S., Sawadogo, A., Trivino, C., Vouyoukalou, E. and Trudgill, D.L. (2001) Cuticle heterogeneity as exhibited by *Pasteuria* spore attachment is not linked to the phylogeny of parthenogenetic root-knot nematodes (*Meloidogyne* spp.). *Parasitology* 122, 111–120.

de Bono, M. and Bargmann, C.I. (1998) Natural variation in a neuropeptide Y receptor homolog modifies social behavior and food response in *C. elegans*. *Cell* 94, 679–689.

de Bono, M., Tobin, D.M., Davis, M.W., Avery, L. and Bargmann, C.I. (2002) Social feeding in *Caenorhabditis elegans* is induced by neurons that detect aversive stimuli. *Nature* 419, 899–903.

Dezfuli, B.S., Volponi, S., Beltrami, I. and Poulin, R. (2002) Intra- and interspecific density-dependent effects on growth in helminth parasites of the cormorant, *Phalacrocorax carbo sinensis*. *Parasitology* 124, 537–544.

Dicke, M., Sabelis, M.W., Takabayashi, J., Bruin, J. and Posthumus, M.A. (1990) Plant strategies of manipulating predator-prey interactions through allelochemicals: prospects for application in pest control. *Journal of Chemical Ecology* 16, 3091–3118

Dickson, D.W., Oostendorp, M., Giblin-Davis, R.M. and Mitchell, D.J. (1994) Control of plant-parasitic nematodes by biological antagonists. In: Rosen, D., Bennett, F.D. and Capinera, J.L. (eds) *Pest Management in the Subtropics: Biological Control – a Florida Perspective*. Intercept, Andover, UK, pp. 575–601.

Doncaster, C.C. (1981) Observations on relationships between infective juveniles of bovine lungworm, *Dictyocaulus viviparus* (Nematoda, Strongylida) and the fungi, *Pilobolus kleinii* and *Pilobolus crystallinus* (Zygomycotina, Zygomycetes). *Parasitology* 82, 421–428.

Eisenback, J.D. (1993) Interactions between nematodes in cohabitance. In: Khan, A.W. (ed.) *Nematode Interactions*. Chapman and Hall, London, pp. 134–174.

Epsky, N.D., Walter, D.E. and Capinera, J.L. (1988) Potential role of nematophagous microarthropods as biotic mortality factors of entomogenous nematodes (Rhabditida: Steinernematidae and Heterorhabditidae). *Journal of Economic Entomology* 81, 821–825.

Espanol, M., Verdejo-Lucas, S., Davies, K.G. and Kerry, B.R. (1997) Compatibility between *Pasteuria penetrans* and *Meloidogyne* populations from Spain. *Biocontrol Science and Technology* 7, 219–230.

Esser, R.P. (1963) Nematode interactions in plates on non-sterile water agar. *Proceedings of the Soil Crop Science Society of Florida* 23, 121–128.

Estores, R.A. and Chen, T.A. (1972) Interactions of *Pratylenchus penetrans* and *Meloidogyne incognita* as coinhabitants in tomato. *Journal of Nematology* 4, 170–174.

Ettema, C.H. (1998) Soil nematode diversity: species coexistence and ecosystem function. *Journal of Nematology* 30, 159–169.

Ewbank, J.J. (2002) Tackling both sides of the host-pathogen equation with *Caenorhabditis elegans*. *Microbes and Infection* 4, 247–256.

Eysker, M. and De Coo, F.A.M. (1988) *Pilobolus* species and rapid translation of *Dictyocaulus viviparus* from cattle faeces. *Research in Veterinary Science* 44, 178–182.

Fallon, P.G., Allen, R.L. and Rich, T. (2001) Toll signaling: bugs, flies, worms and man. *Trends in Immunology* 22, 63–66.

Field, J.I. and Webster, J. (1977) Traps of predacious fungi attract nematodes. *Transactions of the British Mycological Society* 68, 467–469.

Fleming, M.W. (1988) Size of inoculum dose regulates in part worm burdens, fecundity, and lengths in ovine *Haemonchus contortus* infections. *Journal of Parasitology* 74, 975–978.

Gaugler, R., LeBeck, L., Nakagaki, B. and Boush, G.M. (1980) Orientation of the entomogenous nematode *Neoaplectana carpocapsae* to carbon dioxide. *Environmental Entomology* 9, 649–652.

Georgis, R. and Poinar, G.O. (1983a) Effects of soil texture on the distribution and infectivity of *Neoaplectana glaseri* (Nematoda: Steinernematidae). *Journal of Nematology* 15, 329–332.

Georgis, R. and Poinar, G.O. (1983b) Vertical migration of *Heterorhabditis bacteriophora* and *H. heliothidis* (Nematoda: Heterorhabditidae) in sandy loam soil. *Journal of Nematology* 15, 652–654.

Giblin, R.M. and Kaya, H.K. (1983) Field observations on the association of *Anthophora bomboides stanfordiana* (Hymenoptera: Anthophoridae) with the nematode *Bursaphelenchus seani* (Aphelenchida: Aphelenchoididae). *Annals of the Entomological Society of America* 76, 228–231.

Giblin, R.M. and Kaya, H.K. (1984) Associations of halictid bees with the nematodes, *Aduncospiculum halicti* (Diplogasterida: Diplogasteroididae) and *Bursaphelenchus kevini* (Aphelenchida: Aphelenchoididae). *Journal of the Kansas Entomological Society* 57, 92–99.

Giblin-Davis, R.M. (1993) Interactions of nematodes with insects. In: Khan, A.W. (ed.) *Nematode Interactions*. Chapman and Hall, London, pp. 303–344.

Giblin-Davis, R.M., Norden, B.B., Batra, S.W.T. and Eickwort, G.E. (1990) Commensal nematodes in the glands, genitalia, and brood cells of bees (Apoidea). *Journal of Nematology* 22, 150–161.

Glazer, I. (1997) Effects of infected insects on secondary invasion of steinernematid entomopathogenic nematodes. *Parasitology* 114, 597–604.

Golden, J.W. and Riddle, D.L. (1982) A pheromone influences larval development in the nematode *Caenorhabditis elegans*. *Science* 218, 578–580.

Golden, J.W. and Riddle, D.L. (1984) The *Caenorhabditis elegans* dauer larva: developmental effects of pheromone, food, and temperature. *Developmental Biology* 102, 368–378.

Grenache, D.G., Caldicott, L., Albert, P.S., Riddle, D.L. and Politz, S.M. (1996) Environmental induction and genetic control of surface antigen switching in the nematode *Caenorhabditis elegans*. *Proceedings of the National Academy of Sciences, USA* 93, 12388–12393.

Grewal, P.S., Lewis, E.E. and Gaugler, R. (1997) Response of infective stage parasites (Nematoda: Steinernematidae) to volatile cues from infected hosts. *Journal of Chemical Ecology* 23, 503–515.

Grewal, P.S., Lewis, E.E. and Venkatachari, S. (1999) Allelopathy: a possible mechanism of suppression of plant-parasitic nematodes by entomopathogenic nematodes. *Nematology* 1, 735–743.

Grootaert, P., Jaques, A. and Small, R.W. (1977) Prey selection in *Butlerius* sp. (Rhabditida: Diplogasteridae). *Mededelingen van de Faculteit Landbouwwetenschappen Universiteit Gent* 42, 1559–1562.

Hallmann, J., Quadt-Hallmann, A., Rodriguez-Kabana, R. and Kloepper, J.W. (1998) Interactions between *Meloidogyne incognita* and endophytic bacteria in cotton and cucumber. *Soil Biology and Biochemistry* 30, 925–937.

Haskey-Günther, K., Hoffmann-Hergarten, S. and Sikora, R. (1998) Resistance against potato cyst nematode *Globodera pallida* systemically induced by rhizobacteria *Agrobacterium radiobacter* (G12) and *Bacillus sphaericusi* (B43). *Fundamental and Applied Nematology* 21, 511–517.

Haukisalmi, V. and Henttonen, H. (1993) Coexistence in helminths of the bank vole *Clethrionomys glareolus*. II. Intestinal distribution and interspecific interactions. *Journal of Animal Ecology* 62, 230–238.

Hechler, H.C. (1963) Description, developmental biology and feeding habits of *Seinura tenuicaudata* (de Man). J.B. Goodey, 1960 (Nematoda : Aphelenchida), a nematode predator. *Proceedings of the Helminthological Society of Washington* 30, 182–195.

Hoffmann, J.A. and Reichhart, J.M. (1997) *Drosophila* immunity. *Trends in Cell Biology* 7, 309–316.

Holmes, J.C. and Price, P.W. (1986) Communities of parasites. In: Anderson, D.J. and Kikkawa, J. (eds) *Community Ecology: Pattern and Process*. Blackwell Scientific, Oxford, pp. 187–213.

Hominick, W.M. and Tingley, G.A. (1984) Mermithid nematodes and the control of insect vectors of human disease. *Biocontrol News and Information* 5, 7–20.

Hu, K.J., Li, J.X. and Webster, J.M. (1999) Nematicidal metabolites produced by *Photorhabdus luminescens* (Enterobacteriaceae), bacterial symbiont of entomopathogenic nematodes. *Nematology* 1, 457–469.

Imler, J.L. and Hoffmann, J.A. (2000) Toll and Toll-like proteins: an ancient family of receptors signaling infection. *Review of Immunogenetics* 2, 294–635

Ishibashi, N. and Choi, D.R. (1991) Biological control of soil pests by mixed application of entomopathogenic and fungivorous nematodes. *Journal of Nematology* 23, 175–181.

Ishibashi, N. and Kondo, E. (1977) Occurrence and survival of the dispersal forms of pine wood nematode, *Bursaphelenchus lignicolus*. *Applied Entomology and Zoology* 12, 293–302.

Ishibashi, N. and Matsunaga, T. (1995) Interference by entomopathogenic and fungivorous nematodes of root invasion by plant-parasitic nematodes. *Journal of Nematology* 27, 503–504.

Ishikawa, M., Shuto, Y. and Watanabe, H. (1986) B-myrcene, a potent attractant component of pine wood for the pine wood nematode, *Bursaphelenchus xylophilus*. *Agricultural and Biological Chemistry* 50, 1863–1866.

Jagdale, G.B., Somasekhar, N., Grewal, P.S. and Klein, M.G. (2002) Suppression of plant-parasitic nematodes by application of live and dead infective juveniles of an entomopathogenic nematode *Steinernema carpocapsae*, on boxwood (*Buxus* spp.). *Biological Control*, 24 42–49.

Jansson, H.-B. (1982a) Predacity by nematophagous fungi and its relation to the attraction of nematodes. *Microbial Ecology* 8, 233–240.

Jansson, H.-B. (1982b) Attraction of nematodes to endoparasitic nematophagous fungi. *Transactions of the British Mycological Society* 79, 25–29.

Jansson, H.-B. and Nordbring-Hertz, B. (1979) Attraction of nematodes to living mycelium of nematophagous fungi. *Journal of General Microbiology* 112, 89–93.

Jansson, H.-B. and Nordbring-Hertz, B. (1980) Interactions between nematophagous fungi and plant-parasitic nematodes: attraction, induction of trap formation and capture. *Nematologica* 26, 383–389.

Jansson, H.-B. and Nordbring-Hertz, B. (1984) Involvement of sialic acid in nematode chemotaxis and infection by an endoparasitic nematophagous fungus. *Journal of General Microbiology* 130, 39–43.

Johnson, R.N. and Viglierchio, D.R. (1961) The accumulation of plant parasitic nematode larvae around carbon dioxide and oxygen. *Proceedings of the Helminthological Society of Washington* 28, 171–174.

Kanagy, J.M.N. and Kaya, H.K. (1996) The possible role of marigold roots and α-terthienyl in mediating host-finding by steinernematid nematodes. *Nematologica* 42, 220–231.

Kaya, H.K. and Koppenhofer, A.M. (1996) Effects of microbial and other antagonistic organism and competition on entomopathogenic nematodes. *Biocontrol Science and Technology* 6, 357–371.

Kerry, B.R. (2000) Rhizosphere interactions and the exploitation of microbial agents for the biological control of plant-parasitic nematodes. *Annual Review of Phytopathology* 38, 423–441.

Kerry, B.R. and Hominick, W.B. (2000) Biological control. In: Lee, D.L. (ed.) *Biology of Nematodes*. Harwood Academic, Reading, UK, pp. 483–509.

Kerry, B.R. and Jaffee, B.A. (1997) Fungi as biological control agents for plant parasitic nematodes. In: Wicklow, D.T. and Söderström, B. (eds) *The Mycota IV: Environmental and Microbial Relationships*. Springer Verlag, Berlin, pp. 203–218.

Khan, Z., Bilgrami, A.L. and Jairajpuri, M.S. (1991) Observations on the predation ability of *Aporcelaimellus nivalis* (Altherr, 1952) Heyns, 1966 (Nematoda : Dorylaimida) *Nematologica,* 37, 333–342.

Khan, Z., Bilgrami, A.L. and Jairajpuri, M.S. (1995) A comparative study on the predation by *Allodorylaimus americanus* and *Discolaimus silvicolus* (Nematoda : Dorylaimida) on different species of plant parasitic nematodes. *Fundamental and Applied Nematology* 18, 99–108.

Kiontke, K. (1996) The phoretic association of *Diplogaster coprophila* Sudhaus & Rehfeld, 1990 (Diplogasteridae) from cow dung with its carriers, in particular flies of the family Sepsidae. *Nematologica* 42, 354–366.

Kiontke, K. (1997) Description of *Rhabditis (Caenorhabditis) drosophilae* n. sp. and *R. (C.) sonorae* n. sp. (Nematoda: Rhabditida) from saguaro cactus rot in Arizona. *Fundamental and Applied Nematology* 20, 305–315.

Kondo, E. and Ishibashi, N. (1978) Ultrastructural differences between the propagative and dispersal forms of the pine wood nematode, *Bursaphelenchus lignicolus*. *Applied Entomology and Zoology* 13, 1–11.

Koppenhofer, A.M. and Kaya, H.K. (1995) Density-dependent effects on *Steinernema glaseri* (Nematoda: Steinernematidae) within an insect host. *Journal of Parasitology* 81, 797–799.

Koppenhofer, A.M. and Kaya, H.K. (1996) Coexistence of entomopathogenic nematode species (Steinernematidae and Heterorhabditidae) with different foraging behavior. *Fundamental and Applied Nematology* 19, 175–183.

Koppenhofer, A.M., Kaya, H.K., Shanmugam, S. and Wood, G.L. (1995) Interspecific competition between steinernematid nematodes within an insect host. *Journal of Invertebrate Pathology* 66, 99–103.

Koppenhofer, A.M., Baur, M.E. and Kaya, H.K. (1996) Competition between two stein-ernematid nematode species for an insect host at different soil depths. *Journal of Parasitology* 82, 34–40.

Kozek, W.J. and Marroquin, H.F. (1977) Intracytoplasmic bacteria in *Onchocerca volvulus*. *American Journal of Tropical Medicine and Hygiene* 26, 663–678.

Kuhne, R. (1996) Relations between free-living nematodes and dung-burying *Geotrupes* spp. (Coleoptera: Geotrupini). *Fundamental and Applied Nematology* 19, 263–271.

Kyo, M., Miyauchi, Y., Fujimoto, T. and Mayama, S. (1990) Production of nematicidal compounds by hairy root cultures of *Tagetes patula* L. *Plant Cell Reporter* 9, 393–397.

Larsen, M. (1999) Biological control of helminths. *International Journal of Parasitology* 29, 139–146.

Larsen, M. (2000) Prospects for controlling animal parasitic nematodes by predacious micro fungi. *Parasitology* 120, S121–S131.

Lewis, E.E., Gaugler, R. and Harrison, R. (1992) Entomopathogenic nematode host finding: response to host contact cues by cruise and ambush foragers. *Parasitology* 105, 309–315.

McLaren, D.J., Worms, M.J., Laurence, B.R. and Simpson, M.G. (1975) Micro-organisms in filarial larvae (Nematoda). *Transactions of Royal Society of Tropical Medicine and Hygiene* 69, 509–514.

Maehara, N. and Futai, K. (2001) Presence of the cerambycid beetles *Psacothea hilaris* and *Monochamus alternatus* affecting the life cycle strategy of *Bursaphelenchus xylophilus*. *Nematology* 3, 455–461.

Maehara, N. and Futai, K. (2002) Factors affecting the number of *Bursaphelenchus xylophilus* (Nematoda: Aphelenchoididae) carried by several species of beetles. *Nematology* 4, 653–658.

Maggenti, A. (1981) *General Nematology*. Springer-Verlag, New York, 372 pp.

Mamiya, Y. (1984) The pine wood nematode. In: Nickle, W.R. (ed.) *Plant and Insect Nematodes*. Marcel Dekker, New York, pp. 589–626.

Manegold, A. and Kiontke, K. (2001) The association of two *Diplogasteroides* species (Secernentea: Diplogastrina) and cockchafers (*Melolontha* spp., Scarabaeidae). *Nematology* 3, 603–606.

Massey, C.L. (1974) *Biology and Taxonomy of Nematode Parasites and Associates of Bark Beetles in the United States*. US Government Printing Office, Washington, DC, 233 pp.

Mendoza de Gives, P., Davies, K.G., Clark, S.J. and Behnke, J.M. (1999a) Predatory behaviour of trapping fungi against *srf* mutants of *Caenorhabditis elegans* and different plant and animal parasitic nematodes. *Parasitology* 119, 95–104.

Mendoza de Gives, P., Davies, K.G., Morgan, M. and Behnke, J.M. (1999b) Attachment tests of *Pasteuria penetrans* to the cuticle of plant and animal parasitic nematodes, free living nematodes and *srf* mutants of *Caenorhabditis elegans*. *Journal of Helminthology* 73, 67–71.

Miller, J.R. and Strickler, K.L. (1984) Finding and accepting host plants. In: Bell, W.J and Carde, R.T. (eds) *Chemical Ecology of Insects*. Sinauer Associates, Sunderland, Massachusetts, pp. 127–157.

Miller, P.M. and McIntyre, J.L. (1975) *Tylenchorhynchus claytoni* feeding on tobacco roots inhibits entry of *Pratylenchus penetrans*. *Journal of Nematology* 17, 327.

Miyazaki, M., Oda, K. and Yamaguchi, A. (1977a) Behavior of *Bursaphelenchus ligicolus* to unsaturated fatty acids. *Journal of the Japan Wood Research Society* 23, 255–261.

Miyazaki, M., Oda, K. and Yamaguchi, A. (1977b) Deposit of fatty acids in the wall of pupal chamber made by *Monochamus alternatus*. *Journal of the Japan Wood Research Society* 23, 307–311.

Miyazaki, M., Yamaguchi, A. and Oda, K. (1978) Behaviour of *Bursaphelenchus lignicolus* in response to carbon dioxide released by respiration of *Monochamus alternatus* pupa. *Journal of the Japan Forestry Society* 60, 249–254.

Moore, J. and Simberloff, D. (1990) Gastrointestinal helminth communities of bobwhite quail. *Ecology* 71, 344–359.

Moqbel, R. and Wakelin, D. (1979) *Trichinella spiralis* and *Strongyloides ratti*: immune interaction in adult rats. *Experimental Parasitology* 47, 65–72.

Moyle, P.L. and Kaya, H.K. (1981) Dispersal and infectivity of the entomogenous nematode, *Neoaplectana carpocapsae* Weiser (Rhabditida: Steinernematidae), in sand. *Journal of Nematology* 13, 295–300.

Nansen, P., Grøvold, J., Henriksen, S.A. and Wolstrup, J. (1988) Interactions between the predacious fungus *Arthrobotys oligospora* and third stage larvae of a series of animal parasitic nematodes. *Veterinary Parasitology* 26, 327 -337.

Necibi, S. and Linit, M.J. (1998) Effect of *Monochamus carolinensis* on *Bursaphelenchus xylophilus* dispersal stage formation. *Journal of Nematology* 30, 246–254.

O'Leary, S.A., Stack, C.M., Chubb, M.A. and Burnell, A.M. (1998) The effect of day of emergence from the insect cadaver on the behavior and environmental tolerances of infective juveniles of the entomopathogenic nematode *Heterorhabditis megidis* (strain UK211). *Journal of Parasitology* 84, 665–672.

O'Neill, S.L., Giordano, R., Colbert, A.M.E., Karr, T.L. and Robertson, H.M. (1992) 16s rRNA phylogenetic analysis of bacterial endosymbionts associated with cytoplasmic incompatibility in insects. *Proceedings of the National Academy of Sciences, USA* 89, 2699–2702.

Patrick, M.J. (1991) Distribution of enteric helminths in *Glaucomys volans* L. (Sciuridae): a test of competition. *Ecology* 72, 755–758.

Popiel, I., Grove, D.L. and Friedman, M.J. (1989) Infective juvenile formation in the insect parasitic nematode *Steinernema feltiae*. *Parasitology* 99, 77–81.

Poulin, R. (2001) Interactions between species and the structure of helminth communities. *Parasitology* 122, S3–S11.

Pujol, N., Link, E.M., Liu, L.X., Kurz, L.C., Alloing, G., Tan, M.W., Ray, K.P., Solari, R., Johnson, C.D. and Ewbank, J.J. (2001) A reverse genetic analysis of components of the Toll signalling pathway in *Caenorhabditis elegans*. *Current Biology* 11, 809–821.

Rao, R.U., Moussa, H. and Weil, G.J. (2002) *Brugia malayi*: effects of antibacterial agents on larval viability and development *in vitro*. *Experimental Parasitology* 101, 77–81.

Riddle, D.L. (1988) The dauer larva. In: Wood, E.B. (ed.) *The Nematode* Caenorhabditis elegans. Cold Spring Harbor Laboratory, Cold Spring Harbor, New York, pp. 393–412.

Robinson, A.F. (1995) Optimal release rates for attracting *Meloidogyne incognita*, *Rotylenchus reniformis* and other nematodes to carbon dioxide in sand. *Journal of Nematology* 27, 42–50.

Robinson, A.F. and Jaffee, B.A. (1996) Repulsion of *Meloidogyne incognita* by alginate pellets containing hyphae of *Monacrosporium cionopagum*, *M. ellipsosporum*, or *Hirsutella rhossiliensis*. *Journal of Nematology* 28, 133–147.

Robinson, J. (1962) *Pilobolus* spp. and translation of infective larvae of *Dictyocaulus viviparus* from faeces to pastures. *Nature* 193, 353–354.

Samaliev, H.Y., Andreoglou, F.I., Elawad, S.A., Hague, N.G.M. and Gowen, S.R. (2000) The nematicidal effects of the bacteria *Pseudomonas oryzihabitans* and *Xenorhabdus nematophilus* on the root-knot nematode *Meloidogyne javanica*. *Nematology* 2, 507–514.

Sayre, R.M. and Starr, M.P. (1988) Bacterial diseases and antagonists of nematodes. In: Poinar, G.O. and Jansson, H.-B. (eds) *Diseases of Nematodes*. Vol II. CRC Press, Boca Raton, Florida, pp. 69–101.

Schroeder, W.J. and Beavers, J.B. (1987) Movement of the entomogenous nematodes of the families Heterorhabditidae and Steinernematidae in soil. *Journal of Nematology* 19, 257–259.

Schulte, F. (1989) The association between *Rhabditis necromena* Sudhaus & Schulte, 1989

(Nematoda: Rhabditidae) and native and introduced millipedes in South Australia. *Nematologica* 35, 82–89.

Selvan, S., Campbell, J.F. and Gaugler, R. (1993) Density-dependent effects on entomopathogenic nematodes (Heterorhabditidae and Steinernematidae) within an insect host. *Journal of Invertebrate Pathology* 62, 278–284.

Shafqat, S., Bilgrami, A.L. and Jairajpuri, M.S. (1987) Evaluation of the predation ability of *Dorylaimus stagnalis* Dujardin, 1845 (Nematoda : Dorylaimida). *Revue de Nématologie* 10, 455–461.

Shapiro, D.I., Tylka, G.L., Berry, E.C. and Lewis, L.C. (1995) Effects of earthworms on the dispersal of *Steinernema* spp. *Journal of Nematology* 27, 21–28.

Sharma, S.B. and Davies, K.G. (1996) A comparison of two sympatric species of *Pasteuria* isolated from a tropical vertisol soil. *World Journal of Microbiology and Biotechnology* 12, 361–366.

Shepherd, A.M., Clark, S.A. and Kempton, A. (1973) An intracellular microorganism associated with the tissues of *Heterodera* spp. *Nematologica* 19, 31–34.

Sikora, R.A. (1992) Management of the antagonistic potential in agricultural ecosystems for the biological control of plant parasitic nematodes. *Annual Review of Phytopathology* 30, 245–270.

Sironi, M., Bandi, C., Sacchi, L., Di Sacco, B., Damiani, G. and Genchi, C. (1995) A close relative of the arthropod endosymbiont *Wolbachia* in a filarial worm. *Molecular Biochemical Parasitology* 74, 223–227.

Sohlenius, B. (1988) Interactions between two species of *Panagrolaimus* in agar cultures. *Nematologica* 34, 208–217.

Somasekhar, N., Grewel, P.S., De Nardo, E.A.B. and Stinner, B.R. (2002) Non-target effects of entomopathogenic nematodes on the soil nematode community. *Journal of Applied Ecology* 39, 735–744

Somers, C.J. (1985) Viability and pattern of emergence of *Dictyocaulus viviparus* larvae in sporangia of the fungus *Pilobolus kleinii*. *Research in Veterinary Science* 39, 124–126.

Stamps, W.T. and Linit, M.J. (1998a) Chemotactic response of propagative and dispersal forms of the pinewood nematode *Bursaphelenchus xylophilus* to beetle and pine derived compounds. *Fundamental and Applied Nematology* 21, 243–250.

Stamps, W.T. and Linit, M.J. (1998b) Neutral storage lipid and exit behavior of *Bursaphelenchus xylophilus* fourth-stage dispersal juveniles from their beetle vectors. *Journal of Nematology* 30, 255–261.

Stamps, W.T. and Linit, M.J. (2001) Interaction of intrinsic and extrinsic chemical cues in the behaviour of *Bursaphelenchus xylophilus* (Aphelenchida: Aphelenchoididae) in relation to its beetle vector. *Nematology* 3, 295–301.

Stevens, L., Giordano, R. and Fialho, R.E. (2001) Male-killing, nematode infections, bacteriophage infection, and virulence of cytoplasmic bacteria in the genus *Wolbachia*, *Annual Review of Ecology and Systematics* 32, 519–545.

Stirling, G. (1984) Biological control of *Meloidogyne javanica* with *Bacillus penetrans*. *Phytopathology* 74, 55–60.

Stirling, G. (1991) *Biological Control of Plant Parasitic Nematodes*. CAB International, Wallingford, UK, 282 pp.

Stock, T.M. and Holmes, J.C. (1988) Functional relationships and microhabitat distributions of enteric helminths of grebes (Podicipedidae): the evidence for interactive communities. *Journal of Parasitology* 74, 214–227.

Sudhaus, W. (1976) Vergleichende untersuchungen zur Phylogenie, Systematik, Ökologie, Biologie und Ethologie der Rhabditidae (Nematoda). *Zoologica* 43, 1–229.

Sudhaus, W. and Kiontke, K. (1996) Phylogeny of *Rhabditis* subgenus *Caenorhabditis* (Rhabditidae, Nematoda). *Journal of Zoological Systematics and Evolutionary Research* 34, 217–233.

Sudhaus, W. and Kuhne, R. (1989) Nematodes associated with Psychodidae: description of *Rhabditis berolina* sp. n. and redescription of *R. dubia* Bovien, 1937 (Nematoda: Rhabditidae), with biological and ecological notes, and a phylogenetic discussion. *Nematologica* 35, 305–320.

Takabayashi, J. and Dicke, M. (1996) Plant–carnivore mutualism through herbivore-induced carnivore attractants. *Trends in Plant Science* 1, 109–113.

Taylor, M.J. (2002) *Wolbachia* endosymbiotic bacteria of filarial nematodes: a new insight into disease pathogenesis and control. *Archives of Medical Research* 33, 422–424.

Timper, P. and Kaya, H.K. (1989) Role of the second-stage cuticle of entomogenous nematodes in preventing infection by nematophagous fungi. *Journal of Invertebrate Pathology* 54, 314–321.

Turlings, T.C.J., Loughrin, J.H., McCall, P.J., Röse, U.S.R., Lewis, W.J. and Tumlinson, J.H. (1995) How caterpillar-damaged plants protect themselves by attracting parasitic wasps. *Proceedings of the National Academy of Sciences, USA* 92, 4169–4174.

Umesh, K.C., Ferris, H. and Bayer, D.E. (1994) Competition between the plant-parasitic nematodes *Pratylenchus neglectus* and *Meloidogyne chitwoodi*. *Journal of Nematology* 26, 286–295.

van Tol, R.W.H.M., van der Sommen, A.T.C., Boff, M.I.C., van Bezooijen, J., Sabelis, M.W. and Smits, P.H. (2001) Plants protect their roots by alerting the enemies of grubs. *Ecology Letters* 4, 292–294.

Vidal-Martinez, V.M. and Kennedy, C.R. (2000) Potential interactions between the intestinal helminths of the cichlid fish *Cichlasoma synspilum* from southeastern Mexico. *Journal of Parasitology* 86, 691–695.

Vincent, A.L., Portaro, J.K. and Ash, L.R. (1975) A comparison of the body wall ultrastructure of *Brugia pahangi* with that of *Brugia malayi*. *Journal of Parasitology* 63, 567–570.

Waller, P.J. and Faedo, M. (1996) The prospects for biological control of free-living stages of nematode parasites of livestock. *International Journal of Parasitology* 26, 915–925.

Wang, X.D. and Ishibashi, N. (1998) Effects of precedent infection of entomopathogenic nematodes (Steinernematidae) on the subsequent invasion of infective juveniles. *Japanese Journal of Nematology* 28, 8–16.

Wang, X.D. and Ishibashi, N. (1999) Infection of the entomopathogenic nematode, *Steinernema carpocapsae*, as affected by the presence of *Steinernema glaseri*. *Journal of Nematology* 31, 207–211.

Warren, J.E. and Linit, M.J. (1993) Effect of *Monochamus carolinensis* on the life history of the pinewood nematode, *Bursaphelenchus xylophilus*. *Journal of Nematology* 25, 703–709.

Werren, J.H. (1997) Biology of *Wolbachia*. *Annual Review of Entomology* 42, 587–609.

Werren, J.H. and O'Neill, S.L. (1997) The evolution of heritable symbionts. In: O'Neill, S.L., Hoffman, A.A. and Werren, J.H. (eds) *Influential Passengers: Inherited Microorganisms and Arthropod Reproduction*. Oxford University Press, Oxford, UK, pp. 1–41.

Wertheim, B., Sevenster, J.G., Eijs, I.E. and Van Alphen, J.J.M. (2000) Species diversity in a mycophagous insect community: the case of spatial aggregation vs resource partitioning. *Journal of Animal Ecology* 69, 335–351.

Whipps, J.M. and Davies, K.G. (2000) Success in biological control of plant pathogens and nematodes by microorganisms. In: Gurr, G. and Wratten, S. (eds) *Biological Control:*

Measures of Success. Kluwer Academic Publishers, Dordrecht, The Netherlands, pp. 231–269.

Wolstrup, J., Nansen P., Gronvold, J., Henriksen, S.A. and Larsen, M. (1996) Toward practical biological control of parasitic nematodes in domestic animals. *Journal of Nematology* 28, 129–132.

Yeates, G.W. (1969) Predation by *Mononchoides potohikus* (Nematoda: Diplogasterida) in laboratory culture. *Nematologica* 15, 1–9.

Zervos, S. (1988a) Population dynamics of a thelastomatid nematode of cockroaches. *Parasitology* 96, 353–368.

Zervos, S. (1988b) Evidence for population self-regulation, reproductive competition and arrhenotoky in a thelastomatid nematode of cockroaches. *Parasitology* 96, 369–379.

11 Abiotic Factors

MARY E. BARBERCHECK[1] AND LARRY DUNCAN[2]

[1]Department of Entomology, Pennsylvania State University, University Park, PA 16802, USA; [2]Department of Entomology and Nematology, University of Florida, IFAS, Citrus Research and Education Center, 700 Experiment Station Road, Lake Alfred, FL 33850, USA

11.1. Introduction

This chapter concentrates on the behavioural responses of nematodes to abiotic factors in their environment. Environmental factors considered here are broadly categorized as energy, chemical, physical and mechanical. We shall focus on exogenous (e.g. directed and non-directed movement) as opposed to endogenous (e.g. pharyngeal pumping, uterine or intestinal contractions) behaviours (Croll and Matthews, 1977). Behavioural responses to exogenous stimuli are not fixed, but can be affected by many factors, including individual characteristics, e.g. age, physiological stage, condition and sex, as well as population characteristics, such as density

(Stanton, 1990; Barbercheck and Kaya, 1991; Grewal *et al.*, 1993; Lewis *et al.*, 1997; O'Leary *et al.*, 1998; Dempsey and Griffin, 2000; Peckol *et al.*, 2001; de Bono *et al.*, 2002). Reviews on nematode response to abiotic factors can be found in Croll and Matthews (1977) and Croll and Sukhdeo (1981) and on endogenous behaviours in Burr and Robinson (Chapter 2, this volume), Riga (Chapter 3, this volume) and Wharton (Chapter 13, this volume).

Several types of assays are used to study nematode behaviour. Because most nematodes are colourless and microscopic in size, they are difficult to observe on opaque surfaces. Therefore, studies of behaviour often require destruction of habitat or substrate to determine movement or aggregation. To avoid this limitation, other approaches for observing specific nematode behaviours have included assays on or in substrates, such as filter-paper, solutions, transparent gels thin layers of solid substrates (e.g. sand) on transparent gels and assay arenas containing glass or gel beads. Whereas the use of these methodologies greatly reduces the complexity of research, they may not accurately reproduce the dynamics of abiotic factors as they occur in natural substrates (Wallace, 1971; Barbercheck and Kaya, 1991). For example, structural heterogeneity and chemical gradients interact so that *Caenorhabditis elegans* has more linear movement compared with a homogeneous environment (Anderson *et al.*, 1997a,b). When structure was present, foraging became an avoidance strategy, thus allowing the nematode to escape structural traps, such as 'dead-end' pores, and then continue to react to attractant gradients.

Although laboratory assays have limitations, they provide information to form hypotheses about behaviours of individual nematodes. Comparative advantages and limitations of assay techniques have been reviewed (Dusenbery, 1980; Lee, 2002; Lewis, 2002). The extensive use of *C. elegans* as a model in biology has contributed greatly to the understanding of the genetic bases of behaviour in nematodes (see Barr and Hu, Chapter 9, this volume). Mutational analysis, reverse genetic analyses and laser ablation of structures have been used to study coordinated movement, chemotaxis, thermotaxis, osmotic avoidance and mechanosensation in *C. elegans* (Sulston and White, 1980; Chalfie and White, 1988; Bargmann and Avery, 1995; Liu *et al.*, 1999).

11.2. Energy

11.2.1. Temperature

More than any other ambient factor, temperature modulates all physiological processes and behaviours of nematodes. Nematode metabolism is reasonably well modelled as a temperature-dependent function of body size (Atkinson, 1980; Moens and Vincx, 2000). The rate of development of a species is a linear function of temperature, within physiological limits. The rates of individual development and population growth are directly related to numbers of accumulated heat units above a base temperature at which development is not detected (Ferris *et al.*, 1996; Gao and Becker, 2002). Predicted climate change in response to elevated CO_2 in the atmosphere may shorten the generation time of nematodes. This effect may increase the population density and dominance of species better adapted to warm habitats at the expense of species near the edge of their geographical range (McSorley and Porazinska, 2001). Ruess *et al.* (1999) and Sohlenius and Bostrom (1999) found that elevating temperature in subarctic sites increased the abundance of bacterial- and fungal-feeding nematodes but reduced species richness. However, the effects of species immigration and long-term changes to plant communities and soil physical properties on soil food webs and nematode communities at those sites are unknown. Although effects of interacting factors involved in climate change are numerous and largely ill-defined, the damaging effects of many parasitic nematodes may increase in importance with increased temperature. For example, in Germany cutaneous *larva migrans* (caused by endemic hookworms) is rare but increases dramatically during years with unusually warm summers (Klose *et al.*, 1996). Studies of the likely range expansion of some phytoparasitic species suggest that the new areas would be large enough to warrant efforts at exclusion by regulatory agencies (Boag and Neilson, 1996; McSorley and Porazinska, 2001).

The direct effect of temperature on the development and population growth of nematodes also gives it major adaptive significance; the abilities to acclimate, orient towards optimal temperatures and anticipate the advent of extreme low and high temperatures are profoundly selected behaviours. Acclimation to non-lethal temperature change is exhibited by nematodes and other poikilotherms as a temporal adaptation of physiological processes to function optimally at the new temperature (Bullock, 1955; Jagdale and Grewal, 2003). Oxygen consumption of *Panagrellus redivivus* and *Turbatrix aceti* decreased by as much as 70% when culture temperature was reduced from 20 to 10°C. Subsequent respiration increased gradually over several days until it equalled or exceeded that at 20°C (Cooper and Ferguson, 1973). After similar acclimation regimes, *Ditylenchus dipsaci* exposed to a thermal gradient aggregated at the acclimation temperature (Croll, 1970). Assuming a favourable environment during acclimation, this taxic behaviour should be adaptive since it is likely to achieve a continuation of normal activity without the need to reacclimate. Other components of the environment interact with acclimation temperature. *Caenorhabditis elegans* displayed positive or negative thermotaxis with respect to acclimation temperature, depending on whether the worms were fed or starved, respectively (Hedgecock and Russell, 1975; Mori and Oshima, 1995; Li *et al.*, 2000). Whether this behaviour represents associative learning or suppression of thermotaxis by starvation is unresolved (Bargmann and Mori, 1997).

A model of nematode movement in diurnally dynamic vertical thermogradients that occur in soil predicted that nematode migration patterns change during the day, with unequal distances being traversed in either direction (Dusenbery, 1989). The predicted patterns resulted in an accumulation of nematodes at a constant soil depth, regardless of initial depth. Such behaviour could lead nematodes to an optimal soil depth for encountering plant roots, but thermal gradients emanating from the rhizosphere are shallow compared with those created by sunlight, so that orientation towards roots probably involves other stimuli (Pline *et al.*, 1988).

Nematode responses to gradients in controlled experiments invariably raise questions about the importance of these behaviours in natural systems with complex gradients of physical and chemical stimuli. Klingler (1972) found strong interactions between the effects of ambient temperature and gradients of heat and CO_2 on *Pratylenchus penetrans* and *D. dipsaci* movement. Nematodes responded more strongly to heat at low ambient temperatures and CO_2 at high ambient temperatures. To test nematode responses to dynamic gradients in three-dimensional space, Robinson (1994) created a diurnal temperature gradient flux in sand cylinders that approximated historical data obtained in cotton fields. He distinguished between thermal and gravitational effects on nematode movement by creating a horizontal temperature gradient. Following a 3-day exposure to 'natural' gradients, *Meloidogyne incognita* moved downward and towards the thermal gradient. These responses and a relatively slow response time for this nematode are consistent with the model of Dusenbery (1989). However, *Rotylenchulus reniformis* moved downward but away from the heat source, in contrast to its behaviour in static temperature gradients (Robinson and Heald, 1989). In contrast, two species of entomopathogenic nematodes and a foliar nematode, *Ditylenchus phyllobius*, moved randomly in this system. The responses of these nematodes, with the exception of *M. incognita*, would not be predicted from their behaviour in heat gradients on surfaces, illustrating the complexity of competing stimuli that influence nematode orientation (see Riga, Chapter 3, this volume).

Acclimation to low or high temperature is required by many nematodes for survival, either through avoidance or tolerance of even more extreme conditions (see Wharton, Chapter 13, this volume). Lacking the capacity for thermoregulation, nematodes must anticipate future environmental conditions that require an altered (tolerant) physiological state that cannot be achieved under the new conditions. *Steinernema carpocapsae* can survive in a desiccated state at 5°C if dehydration first occurs at 23°C. However, nematodes dehydrated at 5°C perish because they are unable to synthesize the glycerol necessary to balance the internal osmotic pressure during dehydration (Qiu and Bedding, 2002). Plant-parasitic (Forge and MacGuidwin, 1992a), entomopathogenic (Brown and Gaugler, 1996), vertebrate-parasitic (Ash and Atkinson, 1986) and free-living nematodes (Wharton and Brown, 1991; Wharton and Ferns, 1995; Wharton, 1998) overwintering in frozen soil often require preconditioning at low, but above-freezing temperatures for physiological changes, which may include desiccation (Evans and Perry, 1976) and/or production of cryoprotectants, such as trehalose (Ash and Atkinson, 1983; Behm, 1997; Jagdale and Grewal, 2003). Preconditioning above ambient temperature increases tolerance to lethal temperatures in free-living and insect- and vertebrate-parasitic species (Snutch and Baillie, 1983; Grewal *et al.*, 1994; Raghavan *et al.*, 1999) by stimulating the overexpression of heat-shock proteins, which prevent and repair denaturation of proteins (Linquist and Craig, 1988). The heat-shock protein hsp70 is one of several such proteins overexpressed during the preconditioning period in many nematodes. *Heterorhabditis bacteriophora* transformed with the hsp70A gene from *C. elegans* was 18 times more heat-tolerant than wild-type nematodes (Gaugler and Hashmi, 1996).

Temperature acts as a cue to synchronize the development and activity of nematodes to coincide with optimal environmental conditions and avoid or survive harsh conditions (see Wharton, Chapter 13, this volume). If infective third-stage juveniles of the ascarid *Obeliscoides cuniculi* and some trichostrongylids, such as *Haemonchus contortus*, are exposed to cold temperature prior to being ingested by a host, the normal time for development from fourth stage to adult within the host increases, delaying oviposition and avoiding excretion of eggs from the host during winter (Waller and Thomas, 1975; Watkins and Fernando, 1984; Eysker, 1997). Latent infection is exhibited by some entomopathogenic nematodes that infect insects at low temperature, wherein host death does not occur until the advent of warmer temperatures (Brown *et al.*, 2002).

Obligate diapause involving temperature is a common adaptation in nematodes (Evans, 1987). The plant parasite *Heterodera avenae* hatches in the spring but requires a period of low temperature before hatch occurs in response to elevated temperature (Banyer and Fisher, 1971). Climatic differences and seasonal variation in the time of egg production appear to modulate the temperature regimes necessary to break diapause in *H. avenae* (Rivoal, 1983; Evans, 1987). Rhabditid dauers are third-stage juveniles in developmental arrest that are morphologically and physiologically adapted to survive adverse conditions. Induction of the dauer stage in *C. elegans* involves stimulation by a pheromone that increases with population density (Golden and Riddle, 1984). The response to the pheromone is modulated by food availability and temperature (Riddle and Albert, 1997). A high ratio of food to pheromone reduces dauer formation and maintenance in the dauer state, while higher temperature favours development and maintenance of dauer juveniles.

11.2.2. Electromagnetic

Electromagnetic waves are actually radiation – a form of energy. All electromagnetic radiation is comprised of photons, massless particles each travelling in a wave-like pattern and moving at the speed of light. Each photon contains a certain amount of energy. The electromagnetic spectrum includes, in order of decreasing wavelength, radio, microwave, infrared, visible light, ultraviolet and ionizing emissions. The only difference between the various types of electromagnetic radiation is the amount of energy found in the photons. Radio waves have photons with low energies and have not been studied with regard to nematode behaviour, while infrared radiation (heat) was considered in Section 11.2.1.

11.2.2.1. Microwaves

Microwaves are electromagnetic waves having a frequency range from 300 MHz to 300 GHz and have a little more energy than radio waves. They are the principal carriers of television, telephone and data transmissions between stations on earth and between earth and satellites. Early investigations of the effect of microwaves on nematodes focused on control (O'Bannon and Good, 1971). Exposure of *M.*

incognita in soil to 2450 MHz for 30 s was lethal due to substrate heating. *Panagrellus redivivus* exposed to modulated low-energy microwave radiation at a non-thermal carrier frequency (2.45 GHz) showed behavioural changes (Samoiloff *et al.*, 1973). The 'activated' behaviour was characterized by rapid changes in orientation and increased amplitude of head and tail movements. Prolonged exposure to weak microwave fields (750–1000 MHz, 0.5 W) at 25°C induced a heat-shock response in transgenic *C. elegans* strains carrying *hsp-16* reporter genes (de Pomerai *et al.*, 2002). Both the growth rate and the proportion of worms later maturing into egg-bearing adults increased following microwave exposure, whereas both decreased after mild heat treatment at 28°C for the same period. The authors considered that the biological consequences of microwave exposure are opposite to those attributable to mild heating and suggested a non-thermal mechanism for the effects of microwaves (de Pomerai *et al.*, 2002). Lower power levels tended to induce larger responses, which is opposite to the trend anticipated for any simple heating effect. The evidence suggests that microwave radiation causes nematode stress, presumably reflecting increased levels of protein damage within cells. The response levels observed were comparable to those observed with moderate concentrations of metal ions such as Zn^{2+} and Cu^{2+} (Daniells *et al.*, 1998).

11.2.2.2. Visible light

Electromagnetic radiation that the human eye can see is perceived as colours ranging from violet (shorter wavelengths, 400 nm) to red (longer wavelengths, 700 nm). Our understanding of nematode behaviour responses to visible light is still rudimentary. Photoreceptive organs are present in many nematodes that reside in habitats exposed to light. As many as half the species of marine nematodes and many freshwater nematodes possess putative photosensory organs consisting of paired ocelli or pigmented spots in the region of the pharynx (McLaren, 1976). Croll (1970) reviewed an extensive literature on nematode photoreception with the caveat that much of the research failed to control for heat gradients caused by the light source. Demonstrating a response to light by nematodes requires elaborate experimental control because *C. elegans* and *M. incognita* display klinotaxis to gradients as shallow as 0.1 and 0.001°C/cm, respectively (Pline *et al.*, 1988; Bargmann and Mori, 1997).

    Mermis nigrescens females ascend moist vegetation to deposit eggs. This movement is accompanied by lifting of the head to produce a scanning movement that causes the photosensory organ to be alternately shaded and illuminated by the pigment. Croll (1966a) confirmed that oviposition in *M. nigrescens* is either arrested or greatly reduced in the dark. The nematode exhibited spectral sensitivity by aggregating in areas of green light when given a choice between red and green wavelengths (Croll, 1966b) and exhibits phototaxis (Burr *et al.*, 1990).

    Some nematodes without pigment spots respond to visible light, including *C. elegans* (Burr, 1985). Two mermithid species exhibit transverse orientation – perpendicular to the direction of light (Robinson *et al.*, 1990). Oxygen consumption of *H. contortus* increases in light (Wilson, 1996). Although inactive in the dark,

Anguina agrostis, *Rhabdias bufonis* and *Trichonema* responded to cold light with increased movement, which in some cases continued for several hours (Croll, 1970). In contrast, *Pellodera* sp. and *T. aceti* were very active in darkness and showed no response to light. Little is known about the mechanism for these responses; however, Li and Baehr (1998) identified an enzyme (CEδ) in *C. elegans* similar in activity to one in humans (PDEδ) that regulates membrane binding by proteins during photoreception.

11.2.2.3. Ultraviolet radiation

Electromagnetic energy at the ultraviolet (UV) wavelengths (750 nm to 1 mm) is invisible to humans although it is detectable by some other organisms, notably bees. UV is harmful and even lethal to many nematodes, so it is not surprising that negative phototaxis is displayed by many soil-borne species (Croll, 1970). *Globodera rostochiensis* and *Heterodera schachtii* exposed to UV light developed transparent blisters beneath the cuticle, and infection of potato was reduced in a dose-dependent manner (Green and Plumb, 1967). Development and reproduction by juvenile *Nippostrongylus muris* and *Trichinella spiralis* exposed to UV light were impaired (Stowens, 1942; Keeling, 1960). UV radiation also induces chromosomal rearrangements, and short wavelengths (254 nm, a wavelength absorbed by the atmosphere) have been used to induce *C. elegans* mutants with heritable traits (Stewart *et al.*, 1991).

Development affects UV sensitivity. *Rhabditis tokai* egg hatch decreased and juvenile morphological abnormalities increased the earlier eggs were exposed to UV radiation. *Caenorhabditis elegans* accumulates fluorescent pigment (lipofuscin) as it ages, and sensitivity to UV radiation decreases with the age of the nematode (Klass, 1977).

Short but not long (366 nm) wavelength UV light reduced the infectivity and eventually killed *S. carpocapsae* after exposures of just a few minutes (Gaugler and Boush, 1978). At 302 nm, a UV wavelength approximating sunlight, reproductive capacity and virulence of *H. bacteriophora* were degraded at exposure times that were less than half those needed to cause similar effects in *S. carpocapsae* (Gaugler *et al.*, 1992). Several stilbene fluorescent brighteners have shown UV-protective properties for formulating nematodes and other biological control organisms (Nickle and Shapiro, 1994).

11.2.2.4. Ionizing

Ionizing radiation is radiation with sufficient energy to remove electrons from atoms or molecules (including protein and DNA) when it passes through or collides with material. The loss of an electron is called ionization. The principal forms of ionizing radiation that have an impact on biological systems are X-rays and gamma rays (terms that indicate origin rather than different kinds of radiation) and alpha and beta particles.

X-rays have extremely short wavelengths, lying between gamma rays and ultraviolet radiation on the electromagnetic spectrum. The minimum dose to reduce reproduction among several free-living and plant-parasitic nematodes was 20,000 R, and some required >160,000 R (Myers, 1960). No hatch of plant-parasitic *G. rostochiensis* occurred after cysts were irradiated with a dose of 640,000 R, and exposure to 720,000 R prevented hatch of *Meloidogyne* eggs (Fassuliotis, 1958). *Rhabditis tokai* show a marked resistance to X-rays, but malformations occurred when immature juveniles were irradiated with X-rays. Data suggest that X-rays induce single strand breaks in DNA, which can be rapidly and efficiently rejoined by a repair mechanism (Ishii and Suzuki, 1980). Radiation resistance decreased slightly throughout the first, proliferative phase of *C. elegans* embryogenesis (Ishii and Suzuki, 1990). Nematodes were 40-fold more resistant to the lethal effects of X-rays when irradiated in the second half of embryogenesis as compared with the first half. This is probably due to the absence of cell divisions during this stage. Radiation resistance increased with advancing juvenile stages. A radiation-hypersensitive mutant, *rad-1*, irradiated in the first half of embryogenesis, is about 30-fold more sensitive than the wild type, but in the second half its X-ray sensitivity is the same as that of the wild type (Ishii and Suzuki, 1990).

Gamma radiation caused four harmful effects to the entomopathogenic nematode *S. carpocapsae* (Gaugler and Boush, 1979). Exposure to 10,000 rad completely inhibited reproduction, 100,000 rad inhibited maturation and 300,000 rad reduced pathogenicity by 50%, whereas much higher exposures were required for death. The observed effects were attributed to enzyme inactivation and nucleoprotein damage.

Some nematodes appear to be resistant to gamma radiation (Samoiloff *et al.*, 1973; Samoiloff, 1980; Rinaldo *et al.*, 2002). The behaviour of *Panagrellus silusiae* was normal following irradiation up to 200,000 rad, with uncoordinated behaviour following doses between 400,000 and 500,000 rad, and a cessation of activity beyond 500,000 rad. Hartman *et al.* (1996) determined survival after gamma irradiation (generated from either a ^{137}Cs or a ^{60}Co source) for two strains of *C. elegans*. Nematodes were between 1.3 and 39 times more sensitive to caesium than to cobalt. The magnitude of this differential sensitivity was dependent upon strain, developmental stage and sex. Because cobalt- and caesium-generated gamma radiation have nearly identical energy depositions, the differential sensitivity probably reflects different mechanisms of processing the slightly different spectra of DNA damage induced by these two sources. Sex-specific differences in radiation sensitivity were probably due to male possession of a single X-chromosome rather than two, as for hermaphrodites. In gravid *C. elegans* irradiated with gamma rays, the frequency of aberrations in the early embryonic cells increased proportionally with the dosage of gamma rays (Sadaie and Sadaie, 1989). Aberrations decreased greatly following incubation of the irradiated gravid worms for 2 days. These results suggest a mechanism for gonadal repair of chromosomal aberrations, including chromosomal non-separation, and that gamma-induced chromosome aberrations do not stop hatching behaviour of *C. elegans*.

The biological effects of high-energy ionized-particle radiation are of particular interest for long-duration space flights where exposure levels represent a potential health hazard (Johnson and Nelson, 1991; Nelson *et al.*, 1992). *C. elegans* is being used as a model to address this question. The STS-42 mission carried the first International Microgravity Laboratory payload on the space shuttle *Discovery* on 22 January 1992. In this experiment, the generation of mutations in *C. elegans* was ten times greater during space flight than on earth; however, no irregularities of reproduction and development were evident. No obvious differences were seen in the development, behaviour and chromosome mechanics of *C. elegans* as a function of microgravity (http://lifesci.arc.nasa.gov/lis2/Chapter4_Programs /IML/IML_1.html). Hartman *et al.* (1996) employed the *fem-3* gene of *C. elegans* to determine the mutation frequency as well as the nature of mutations induced by low earth orbit space radiation ambient to space shuttle flight STS-76. Mutation frequency was 3.3-fold higher than the spontaneous rate. Their data provide evidence that a mutagenic component of ambient space radiation is the high linear energy transfer charged particles, such as iron ions (Hartman *et al.*, 1996).

11.2.3. Electrical field/charge

Electricity is a phenomenon associated with stationary or moving electric charges. The various manifestations of electricity are the result of the accumulation or motion of electrons. Electrophysiological measurements demonstrate that an electrochemical potential exists across the body wall of parasitic nematodes. The cuticle/hypodermis complex is differentially permeable to both inorganic and organic ions and it is likely that active transport of ions or outward diffusion of metabolic end-products contributes extensively to the maintenance of transmural electrochemical gradients (Pax *et al.*, 1995). The free-living nematode, *P. silusiae*, exposed to short pulses of electric current displayed changes in swimming behaviour (Samoiloff *et al.*, 1973), including rapid alterations in orientation, with anterior and posterior ends sweeping over a maximum area with each cycle of movement.

The electrotactic threshold, the current and potential required to evoke a response, has been determined for some nematodes. The potential gradient rather than magnitude of the current is important in electrotactic orientation. Nematodes showing electrotaxis usually move towards or accumulate at the cathode in an electric field (Croll and Matthews, 1977). In an electric field established in a wet sand bed at 10 V/cm, migrating *H. schachtii* move preferentially to the cathode, whereas *Meloidogyne javanica* move towards the anode. The pH of the migrating bed modified the proportions of nematodes moving towards the electrodes, and treatment of nematode populations with cationic or anionic detergents, protein solutions and enzymes also modified the behaviour of these nematodes in an electric field (Viglierchio and Yu, 1983). The mechanism of the response modifications was attributed to electrophoresis. The electric field induces a net charge on the sensitive nematode, with the concurrent generation of a directional field force that becomes superimposed upon the random movement of the nematode (Viglierchio and Yu, 1983). Sukul and Croll (1978) observed that *C. elegans* responded directionally to a direct current (DC), but that anodal or cathodal migrations depended on the combination of potential difference, current and KCl concentration in the assay media. Nematodes did not move all the way to the electrodes; about 2 mm from the electrodes individuals reversed to a critical distance. The authors suggested that ions, especially cations, mediate electrotaxis (Sukul *et al.*, 1975, 1977; Sukul and Croll, 1978). Sukul *et al.* (1975) suggested that the amphids of *Anguina tritici* contain largely negatively charged organic molecules, which attract cations in solution. They hypothesized that nematodes sense electrical potential change from the disturbance of the ionic atmosphere of the anionic amphidial molecules under an applied electrical field.

As with electric charges, the effect of this magnetic force acting at a distance is expressed in terms of a field of force. The electromagnetic force is more than 1000 times as strong as the earth's gravitational force. Despite Croll and Matthews's (1977) report of an absence of research on magnetic fields and nematode behaviour, little relevant work on the subject has been conducted since then. Orientation of *A. tritici* was lost when a uniform magnetic field of 750 Oe was applied in a direction perpendicular to the electrical field in ferro- and paramagnetic salt solutions having high magnetic susceptibilities (Sukul *et al.*, 1975). *Meloidgyre incognita, A. tritici, Rhabditis aberrans* and *Diploscapter coronata* were exposed to high-intensity magnetic fields (1000 or 2000 gauss) for 6, 24 or 48 h. Exposure for 24 or 48 h resulted in significant nematode mortality (Sukul *et al.*, 1975, 1977). Intermittent exposure slightly inhibited the reproduction and postembryonic development of *C. elegans*, and caused a marked but transient derangement in locomotory behaviour. Alternating high magnetic fields can elicit chronic and acute biological effects but the effects may be well tolerated or compensated (Bessho *et al.*, 1995). Expression of the heat-shock protein gene *hsp-16–lacZ* was enhanced in transgenic *C. elegans* exposed to magnetic fields up to 0.5 T at 60 Hz, indicating a stress response to exposure to magnetic fields (Miyakawa *et al.*, 2001). The *hsp-16* promoter was more efficiently expressed at the embryonic than at the postembryonic stage, irrespective of exposure, suggesting that *hsp-16* induction occurs at the transcriptional step. Gutzeit (2001) also reported an enhanced stress response after exposure to extremely low-frequency electromagnetic fields in transgenic *C. elegans* and *Drosophila melanogaster*. Gutzeit (2001) also suggested that this response was controlled by the *hsp-16* or *hsp-70* promoter.

11.2.4. Gravity

Gravity, or gravitation, is the force in operation between the earth and other bodies or the force acting to draw objects towards the earth. Gravity is unique in that its direction and intensity are constant, and therefore may be useful to nematodes as an orientation reference (Croll and Matthews, 1977). The ability to detect and respond in respect of gravity is through the action of mechanoreceptors in many animals. Little is known about the mechanisms underlying gravitropic or gravitactic signal transduction. One important question in signal transduction is how mechanical signals, such as pressure or mechanical force delivered to a cell, are interpreted to direct biological responses. Many nematodes will move up a vertical substrate and this has been related to gravity but also to random movement. Some authors consider that gravity is of little importance to soil-dwelling nematodes because the forces due to the water films in which they live are many times stronger than gravitational forces acting on them (Dusenbery, 1980).

Gravity may mediate vertical movement of nematodes in or on a substrate. Vertical migration is especially common in animal-parasitic forms that migrate from dung pats or soil on to vegetation, e.g. the insect parasite *M. nigrescens* (Gans and Burr, 1994). The plant parasite *D. dipsaci*, which attacks plants at or above the soil surface, moves up and down the soil with rainfall. Other plant-parasites showing vertical above-ground migration include *Aphelenchoides ritzemabosi*, *A. agrostis* and *A. tritici*. The vinegar eelworm, *T. aceti*, moves upwards in a vinegar culture medium.

Gravitropism allows plants to direct their growth in response to gravity, promoting upward growth for shoots and downward growth for roots. There is evidence from *Arabidopsis thaliana* that the ARG1 locus participates in a gravity-signalling process involving the plant cytoskeleton. The expression of ARG1 in *Arabidopsis* is ubiquitous and an orthologue in *C. elegans* has been identified (Sedbrook *et al.*, 1999).

Genetic, molecular and electrophysiological studies in organisms ranging from nematodes to mammals have highlighted members of the recently discovered degenerin/epithelial sodium channel (DEG/ENaC) family of ion channels as strong candidates for the metazoan mechanotransducer. Tavernarakis and Driscoll (2001) suggested that degenerins, a large family of proteins, function as channel subunits and mediators of mechanosensitive behaviours of *C. elegans*.

11.3. Chemicals

11.3.1. Oxygen

Nematodes generate energy via aerobic respiration or anaerobic fermentation. Most nematodes can use both metabolic processes; however, the ability to utilize fermentation pathways for development may depend largely on whether oxygen is normally absent during some life stages. Animal parasites that inhabit microaerobic or anaerobic environments in their hosts often have a facultative capacity for more efficient aerobic energy generation during stages when oxygen is available. However, the response to anoxia by bacterivorous and fungivorous free-living nematodes is to ferment existing glycogen reserves to exhaustion (and death) or to obtain energy necessary to transform to a cryptobiotic, non-metabolizing state until oxygen becomes available (Cooper and Van Gundy, 1970). The response to anoxia is stage-specific in many nematodes, e.g. dauer juveniles tend to be better able than adults at surviving anaerobic conditions (Anderson, 1978). Eggs of the bird parasite *Heterakis gallinarum* can remain unembryonated for up to 60 days in the anaerobic bird gut and deep in faecal pats. Embryonation proceeds normally with the availability of oxygen following faecal decomposition (Saunders *et al.*, 2000).

The requirement for anaerobic respiration may be related to the relatively large size of many animal-parasitic nematodes. Lacking an efficient circulatory system, the diffusion of ambient oxygen to internal organs of large nematodes is insufficient for aerobic metabolism (Behm, 2002). Conversely, the small size of many soil-borne nematodes permits them to maintain aerobic respiration at relatively low oxygen concentrations (see below). Nematodes that normally develop anaerobically accumulate glycogen but relatively little neutral lipid (1–8% dry weight) because it cannot be used for fermentation. In contrast, nematodes that can develop only in aerobic conditions accumulate energy-rich neutral lipid (30–40% dry weight) at the expense of glycogen (3–8%) (Cooper and Van Gundy, 1970).

Oxygen in soil air ranges from approximately that in the atmosphere (20%) to as low as 5% or less, depending on soil texture, moisture and biological activity. Nematodes in roots and soil that become waterlogged or that experience significant decomposition of organic matter experience reduced oxygen tensions. *Aphelenchus avenae* and *Caenorhabditis* sp. reproduce normally at oxygen concentrations as low as 10% (Cooper *et al.*, 1970). Reproduction was reduced at 5% and inhibited at 4% oxygen. Intermittent anoxia inhibited oviposition and reproduction by *Hemicycliophora arenaria* (Cooper *et al.*, 1970). This is consistent with the observation that population density of *H. arenaria* in the field was positively correlated with length of time between irrigation (Van Gundy *et al.*, 1968). Infective stages of entomopathogenic nematodes also rely on stored carbohydrate reserves to survive at low oxygen potential. Survival of *Steinernema glaseri* in soil over 2 weeks was directly proportional to oxygen concentrations between 1 and 20%, whereas *S. carpocapsae*, a species that is adapted to relative immobility in the absence of host cues, was less affected by oxygen concentration (Kung *et al.*, 1990a,b).

11.3.2. Carbon dioxide

CO_2 in soil air tends to be several hundred times the concentration in the atmosphere (0.03%), and the oxygen concentration decreases accordingly (Brady, 1974). At levels above 5% CO_2 has an anaesthetic or toxic effect on nematodes (Pline and Dusenbery, 1987; Sciacca *et al.*, 2002). Because CO_2 is a metabolic by-product secreted by most organisms, it has received a great deal of attention as a potential kairomone mediating nematode food-searching behaviour (see Riga, Chapter 3, this volume, for a discussion on CO_2 and nematode chemotaxis).

Some mammal parasites respond to CO_2 with important kinetic behaviours. CO_2 or undissociated carbonic acid stimulates exsheathment (loss of the second-stage cuticle retained by infective third-stage juveniles) of *H. contortus*, *Trichostrongylus colubriformis* and *Nematospiroides dubius* upon penetration of the host gut (Rogers and Sommerville, 1963; Croll, 1970). Eggs of several *Ascaris* species, *Heterakis gallinae* and two species of *Trichuris* can be induced to hatch at elevated levels of CO_2 in combination with other conditions found in mammalian intestines (Fairbairn, 1961; Hass and Todd, 1962). Human breath and CO_2 concentrations in the range found in mammalian breath (3–4%) stimulated non-directional movement of *Ancylostoma caninum* and *Strongyloides stercoralis* and caused a cessation of movement and coiling by *H. contortus*; human breath scrubbed of CO_2 did not affect movement (Sciacca *et al.*, 2002). Selection for these behaviours should favour host encounters by these species, because *A. caninum* and *S. stercoralis* enter the host via skin penetration whereas *H. contortus* enters via ingestion.

Above-ambient CO_2 induces nictation in *A. caninum*, *Ancylostoma duodenale* and *Necator americanus* (Granzer and Haas, 1991). Slightly higher levels (5%), typical of those found within the host, can induce quiescence and feeding behaviours typical of transformation to the parasitic phase (Sciacca *et al.*, 2002). Microfilariae of *Wuchereria bancrofti* and other filarial nematodes respond to diurnal cues in host physiology that coincide with insect vector feeding behaviour to enhance the probability of transmission (Croll, 1970; Muller and Wakelin, 1998; Lee, 2002). Increased CO_2 tension in the pulmonary veins at night causes microfilariae of some species to cease swimming against the blood flow in the lungs, releasing the nematodes into the peripheral blood, where they can be acquired by vectors. Elevating the ambient oxygen level of patients during this period causes the microfilariae to rapidly disappear from the peripheral blood (Hawking, 1962).

Atmospheric CO_2 has increased by 18% since 1959 and the annual amplitude, which is governed by seasonal plant growth, increased by a similar amount (Keeling *et al.*, 1996). These trends suggest increased plant biomass since 1959, possibly due to climate warming and a lengthening of the growing season. Cotton (Runion *et al.*, 1994) and pine seedlings (Markkola *et al.*, 1996) are among several plants shown to support generally greater numbers of bacterivorous and fungivorous nematodes when grown at elevated CO_2 levels. However, the decomposition characteristics of plant material vary depending on the levels of atmospheric CO_2, and short-term experiments are unlikely to distinguish between responses to this material to and prior organic resources. To study long-term responses to elevated CO_2, Yeates *et al.* (1999) examined nematode communities in sites near natural CO_2 vents. They found that the abundance and diversity of nematodes decreased with elevated CO_2 and that bacterivorous nematodes increased as a proportion of the community. However, variation in soil texture may have caused the relationship; when soils were analysed separately, microbial carbon was the best predictor of bacterivorous nematode data and the relationships were positive in some soils and negative in others. These results suggest that nematode responses to elevated CO_2 will be strongly affected by local edaphic conditions.

11.3.3. Other chemicals

Chemical gradients have been created in a variety of substrates by delivering or placing specific chemicals, chemical blends, hosts, host tissues or products on, in or near an assay arena to study the behaviour of nematodes (Dusenbery, 1983; Terrill and Dusenbery, 1996; Riga *et al.*, 1997). The earliest sensory behaviour studied in *C. elegans* was chemotaxis (Dusenbery, 1980; Barr and Hu, Chapter 9, this volume). A common set of chemosensory mechanisms and principles is emerging as a result of multidisciplinary research using a variety of model organisms, including nematodes and other invertebrates (Krieger and Breer, 1999). Recognition and discrimination of chemicals, as well as chemo-electrical transduction and processing of olfactory signals, appear to be mediated by similar mechanisms in phylogenetically diverse animals (Riga, Chapter 3, this volume; Perry and Maule, Chapter 8, this volume). The common approach of organisms to surviving in and interpreting chemical stimuli appears to be a phylogenetically conserved strategy.

Numerous studies of nematode responses to volatile and non-volatile chemical stimuli have been conducted. *C. elegans* moves up gradients of numerous attractive compounds and down gradients of repellent ones. The sensory neurones for many of these responses have been identified (Davis *et al.*, 1986; Perkins *et al.*, 1986; Chalfie and White, 1988; Riga, Chapter 3, this volume; Perry and Maule, Chapter 8, this volume). Ferree and Lockery (1999) derived a linear neural network model of the chemotaxis control circuit in *C. elegans* and demonstrated that this model is capable of producing nematode-like chemotaxis. Based on simple computational rules, they found that optimized linear networks typically control chemotaxis by computing the first time derivative of the chemical concentration and modulating the body turning rate in response to this derivative. The authors argue that their linear neural network model is consistent with behavioural studies and is a plausible mechanism for at least some components of chemotaxis in real nematodes.

Avoidance of an abrupt exposure to repellent stimuli is distinct from gradient chemotaxis in *C. elegans*. Avoidance is activated by potentially lethal stimuli, such as high osmolarity, heavy metal ions, detergents and strong acids (Culotti and Russell, 1978; Thomas and Lockery, 1999). When *C. elegans* comes into contact with this type of stimulus, it abruptly reverses direction, makes a hairpin turn and moves away from the toxic stimulus (Bargmann *et al.*, 1993). Hilliard *et al.* (2002) suggested that the antagonistic activity of head and tail sensory neurones is integrated to generate appropriate escape behaviours: detection of a repellent by head neurones mediates reversals, which are suppressed by antagonistic inputs from tail neurones. Their results support the hypothesis that *C. elegans* senses repellents by defining a head-to-tail spatial map of the chemical environment.

Chemosensation is critical for nematode survival (see Wharton, Chapter 13, this volume), host finding (see Bilgrami and Gaugler, Chapter 4, this volume), reproduction (Huettel, Chapter 5, this volume) and orientation (see Riga, Chapter 3, this volume). Nematodes, even terrestrial species, are aquatic organisms and must be able to detect potentially lethal conditions in the soil solution (Prot, 1979). Behaviours related to osmotic stress have been recorded for several nematodes. During osmotic stress, juvenile *P. redivivus* coil (Pollock and Samoiloff, 1976). In greenhouse studies, infectivity of *M. incognita* on tomato was impaired by increasing soil solution concentration (Edongali *et al.*, 1982). Free-living and animal-, plant- and insect-parasitic nematodes prefer different NaCl concentrations on a gradient, and these differences may reflect survival requirements in their respective natural environments (Dusenbery, 1980, 1983; Pye and Burman, 1981; Riddle and Bird, 1985; Thurston *et al.*, 1994; Tobato-Kudo *et al.*, 2000). Rutherford and Croll (1979) observed two behavioural thresholds for *C. elegans* in response to changes in NaCl concentrations. At 10^{-5} M, an increase in reversal rate and decrease in forward movement was noted. At 10^{-4}–10^{-3} M, wave-form activity was significantly retarded. This second threshold concentration corresponds to that reported for an accumulation response. NaCl acts as a repellent at concentrations above 300 mM and may elicit an escape response. Infective juveniles of the rodent parasite, *Strongyloides ratti*, preferred NaCl concentrations below 80 mM on agarose gel – a unidirectional avoidance movement was observed at higher concentrations and random dispersal was observed when placed initially in a favourable area (Tobata-Kudo *et al.*, 2000). At high concentrations, NaCl, KCl and $CaCl_2$ inhibited the ability of the entomopathogen, *S. glaseri* to move through a soil column and locate and infect a susceptible host. Calcium chloride and KCl had no effect on *H. bacteriophora* survival, infection efficiency or movement through a soil column, but moderate concentrations of these salts enhanced *H. bacteriophora* virulence. NaCl at high salinities (> 16 dS/m), however, adversely affected each of these parameters (Thurston *et al.*, 1994).

Nematode response to host-related chemicals depends on a wide range of biological and environmental stimuli (Zuckerman and Jansson, 1984; Gaugler *et al.*, 1990; Ishibashi and Kondo, 1990; Lewis *et al.*, 1992, 1993, 1995a,b; Lipton and Emmons, 1999; Campbell and Kaya, 2000). For example, various invertebrate and vertebrate parasites demonstrate chemotaxis on agar to host plasma/serum gradients (Khlibsuwan *et al.*, 1992; Tobata-Kudo *et al.*, 2000; Rolfe *et al.*, 2001). Skin-penetrating zooparasitic nematodes, e.g. the dog hookworm, *A. caninum*, and the threadworm *S. stercoralis*, use thermal and chemical signals for host-finding, whereas the passively ingested sheep stomach worm, *H. contortus*, uses environmental signals to position itself for ingestion (Ashton *et al.*, 1999).

Bacterial-feeding nematodes are colonizers of high populations of bacteria, as found on decaying organic materials (Dusenbery, 1980). Attractive bacteria produce an alkaline environment attractive to *C. elegans*. Cyclic adenosine monophosphate (cAMP), which attracts *C. elegans*, is also produced by bacteria. Evidence for the role of complex antagonistic mechanisms for feeding behaviours is illustrated by aggregation in *C. elegans*. De Bono *et al.* (2002) and Coates and de Bono (2002) recently reported that aggregation behaviour associated with fresh food or high population density in *C. elegans* is induced by neurones that detect adverse or stressful stimuli. They suggested a model for regulation of social feeding by opposing sensory inputs, with aversive inputs to nociceptive neurones promoting social feeding and antagonistic inputs from an unidentified neurone inhibiting aggregation. It was surprising that in this case food (*Escherichia coli*) was the source of aversive stimuli, but others have also found that many bacteria that serve as food for *C. elegans* can be lethal under some conditions (Grewal, 1991; Marroquin *et al.*, 2000). De Bono *et al.* (2002) suggested that aggregation provides a defence for nematodes by secretion of enzymes that inactivate bacterial toxins in the food source or stimulate dauer formation and dispersal.

Nitrogen compounds are produced during the decomposition of organic materials and may be an important cue for nematodes. Infective juveniles of the entomopathogen *H. bacteriophora* on agar plates were attracted to 16 and 160 µg of nitrogen and repelled by 1600 and 8000 µg (Shapiro *et al.*, 2000). The authors hypothesized that nitrogen released from infected insects is attractive early in infection and repellent from longer-infected cadavers, which may prevent the secondary invasion of unsuitable hosts by conspecific nematodes. A similar mechanism (odour-mediated host recognition) was proposed for avoidance of inter- and intraspecific competition by insect-parasitic steinernematid nematodes (Grewal *et al.*, 1997).

Experience and habituation also affect nematode response to chemical gradients (Rutherford and Croll, 1979; Croll and Sukhdeo, 1981; Grewal *et al.*, 1993; Lewis *et al.*, 1995b; Thomas and Lockery, 1999; Peckol *et al.*, 2001; Cohen *et al.*, 2002). Infective juveniles of *S. glaseri*, a ranging forager or cruiser, are strongly attracted to host volatiles and switch their movement from ranging search to localized search after contact with host cuticle, whereas *S. carpocapsae*, a sit-and-wait forager or ambusher, does not (Lewis *et al.*, 1992; Lewis, 2002). Campbell *et al.* (2003) proposed that the behaviour sequences comprising the cruising and ambushing host-finding strategies each evolved only once in the genus *Steinernema*, with the ancestral species being an intermediate forager (Fig. 11.1).

Changes in the environment cause both short- and long-term changes in the behaviour of *C. elegans*. Peckol *et al.* (2001) demonstrated that specific sensory experiences cause changes in chemosensory receptor gene expression that may alter sensory perception. Continuous presentation of an olfactory stimulus causes a decrement of the chemotaxis response in *C. elegans* (Bernhard and van der Kooy, 2000). They found a difference between the learning process of habituation (a readily reversible decrease in behavioural response) and other types of olfactory plasticity, such as adaptation, which cannot be dishabituated. Pre-exposing *C. elegans* to a high concentration of diacetyl vapours caused a chemotaxis decrement that was not reversible despite the presentation of potentially dishabituating stimuli. Pre-exposure to low diacetyl concentrations produced habituation of the chemotaxis response (a dishabituating stimulus could reverse the response decrement back to baseline levels). The distinct behavioural effects produced by diacetyl vapour pre-exposure highlighted a concentration-dependent dissociation between two decremental olfactory processes: adaptation at high diacetyl concentrations versus habituation at low vapour concentrations (Bernhard and van der Kooy, 2000).

11.3.4. pH

The effects of pH on nematode behaviour have been examined in several species. *C. elegans* is attracted to high pH and avoids pH lower than 4 formed by either organic or inorganic acids (Dusenbery, 1980; Sambongi *et al.*, 2000). Entomopathogenic nematode survival and activity is greatest at non-acid pH (Fischer and Führer, 1990; Kung *et al.*, 1990a). Acidic pH is known to cause pain

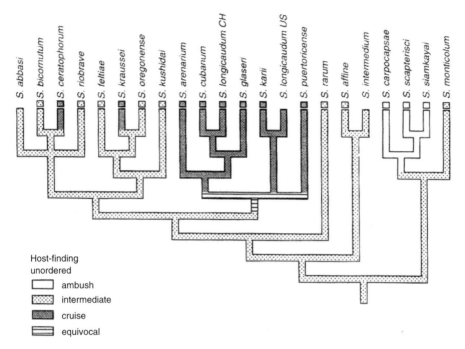

Fig. 11.1. Hypothetical evolutionary history of host-finding behaviours in a genus of entomopathogenic nematodes. Foraging strategies are mapped on to a phylogeny of *Steinernema* (from Stock *et al.*, 2001). Foraging strategy categories were based on the ratio between the number of infective juveniles finding a mobile host (unrestrained *Galleria mellonella* larva on the surface of a sand-filled 90 mm Petri dish) minus the number finding a non-mobile host (*G. mellonella* larva buried beneath the surface of a sand-filled 90 mm Petri dish) and the total number of nematodes finding both types of hosts. Cruise foragers had ratios close to –1.0, intermediate foragers had ratios close to 0.0 and ambush foragers had ratios close to 1.0 (reprinted with permission from Campbell *et al.*, 2003).

sensation, through nociceptive neurones, as well as taste transduction in mammals. Sambongi *et al.* (2000) provided experimental evidence that *C. elegans* recognizes protons as a nociceptive stimulus, through multiple neurones, which elicits avoidance behaviour. A system similar to that of mammalian signal transduction is responsible for acid avoidance by nematodes (Sambongi *et al.*, 2000).

In laboratory experiments, acid pH reduced the efficacy of the entomopathogens *S. carpocapsae*, *Steinernema feltiae* and *H. bacteriophora* against diapausing larvae of *Cephalcia abietis*, and it was speculated that raising soil pH may induce nematode epizootics in herbivorous insects by increasing the activity of entomopathogenic nematodes (Jaworska, 1993). Mortality of the cotton leaf worm, *Spodoptera littoralis*, from *H. bacteriophora* and *S. carpocapsae* was higher and more rapid at pH 6.9 and 8.0 than at pH 5.6 (Ghally, 1995). Kung *et al.* (1990a) demonstrated reduced survival of steinernematid nematodes at pH 10, but observed no differences in a pH range of 4–8. Infections of mosquitoes by *Romanomermis culicivorax* have been observed over a pH range of 3.6–8.6 (Brown and Platzer, 1978). Petersen (1984) concluded that, under normal conditions, pH does not appear to be a limiting factor for *R. culicivorax* or most other aquatic nematodes.

11.4. Physical Factors

11.4.1. Mechanoreception

Nematodes are exquisitely sensitive to a variety of mechanical stimuli, including touch, vibration and pressure. The behavioural responses to these sensations are important for many activities, such as foraging, avoidance of enemies, orientation (see Riga, Chapter 3, this volume), mating, feeding and penetration of host tissue.

Most nematodes respond to gentle touch by withdrawal or reversal of movement. *Rhabditis* sp. moves forward when touched at the posterior end and backward when touched at the anterior end (Croll and Smith, 1970). Diffuse touch (vibration or a tap on a culture dish) causes *C. elegans* to either accelerate forward movement or initiate backward movement, and the frequency of either response is age-dependent (Rankin, 1991). Most mechanotransduction information has been derived from the study of wild-type and mutant *C. elegans* and of worms with laser-ablated neurones (Driscoll and Kaplan, 1997). At least six types of neurones, located in different parts of the body, are required for normal responses to gentle touch in *C. elegans* (Duggan *et al.*, 1996). The responses by the different neurone types are non-redundant and their combined function increases the sensitivity of nematodes to mechanostimuli (Kaplan and Horvitz, 1993; Driscoll and Kaplan, 1997). The tap response is mediated primarily by an integrated response of the anterior and posterior touch cells, which have dendrites with an intracellular mantle embedded in the epidermis along either the anterior or posterior length of the body (Kaplan and Horvitz, 1993; Wicks and Rankin, 1995). Anterior cells respond to tap with reversal movements and posterior cells with accelerations. Lack of either type of cell results in exaggeration of the opposite behaviour. A number of inter-acting mechanosensory (*mec*) genes and their products have been identified and rudimentary models of their role in transduction of stimuli into behaviour have been proposed (Gu *et al.*, 1996; Welsh *et al.*, 2002). These models describe protein-mediated linkages between the nematode cuticle, extracellular mantel and move-ment-sensitive ion channels. Similar protein linkages between microtubules in the receptor cells and the ion channels are envisioned (Driscoll and Kaplan, 1997).

Behavioural responses to touch are modulated by the structure of the environ-ment. Anderson *et al.* (1997a,b) simulated host-seeking behaviour of *C. elegans* in a 'complex environment' consisting of sand grains sprinkled on an agar surface. Their model was based on experiments comparing the behaviour of nematodes in environments of increasing complexity. Typical foraging behaviour in the absence of sand (wide head movement and eventual directed movement in response to chemical gradients) was restricted by the presence of sand particles. In the hetero-geneous environment the movements become more linear within the sand pores, but these were interrupted frequently by the touch withdrawal movement as the nematode anterior encountered a blocked pore, followed by a random change in direction. Thus, the typical response to touch of withdrawal appears to be adaptive in the sense of avoiding blind ends and in modifying the typical klinotactic foraging movement to one that is dictated by physical structure.

11.4.2. Soil matrix

Most studies on matrix effects on nematode behaviour have focused on soil texture rather than on soil structure, even though the latter may be of greater significance in the environment. Structural pore space is largely determined by size and arrangement of aggregates and affects the movement of water, air, chemicals and organisms in soil. Prevalence and virulence of many plant-parasitic nematodes of economic importance vary predictably in soils of different texture (Barker *et al.*, 1998) because texture is related to soil porosity, water potential and other chemical and physical properties affected by particle size. For example, reproduction of *M. incognita* on cotton was much greater in coarse than in fine-textured soil, whereas *R. reniformis* reproduction was highest in soils with moderate levels of clay and silt (Koenning *et al.*, 1996). Cotton yields during 3 years were inversely proportional to nematode population densities at planting for both species in all soils; however, the effects of soil texture on nematode reproduction resulted in higher density ranges and crop damage in those soils favouring nematode reproduction. Often soil texture affects the fundamental relationship between yield and nematode population density at planting and can interact with other variables that affect nematodes and yield. Soybean yield suppression as a function of *Heterodera glycines* population density and nematode population growth was greater in sand and sandy loam textures than in finer-textured soils (Koenning and Barker, 1995). Moreover, compared with non-irrigated plots, yield suppression by nematodes was greater in irrigated plots in the coarse soils, but was not affected by irrigation in fine-textured soils.

The importance of soil structure as opposed to texture is illustrated by Workneh *et al.* (1999), who examined 1462 soybean fields and found an inverse relationship between *H. glycines* and clay content in no-till fields, but no relationship in tilled fields. In fine-textured soil, *H. glycines* numbers were greater in tilled than in no-till fields, but population density exhibited no relationship with tillage practices in textures coarser than silty clay loam. Greater structure (bulk density, porosity, aeration) alteration by tillage of fine-textured compared with coarse-textured soil demonstrates the importance of texture to the relative effects of cultural practices that affect nematode communities. Similar relationships have been found when considering the important role of free-living nematodes in mineralization of soil nutrient elements. Grazing pressure on bacteria by bacterivorous nematodes and subsequent mineralization of organic N was higher in sandy than in finer-textured soils (Hassink *et al.*, 1993a). Bacterial biomass was correlated with pore sizes between 0.2 and 1.2 μm, which excluded nematodes in the finer-textured soils, whereas nematode numbers, but not bacteria, were correlated with pore sizes of 30–90 μm in the sandy soils. The amount of N mineralized per bacterium was related to grazing pressure (Hassink *et al.*, 1993b). Hunt *et al.* (2001) noted the importance of fine soil particles in increasing species richness by restricting nematode movement.

Virulence and survival of entomopathogenic nematodes often differ somewhat in their relationships with soil texture. Presumably this is because persistence is favoured by soil properties that induce quiescence (Kung *et al.*, 1990a; Glazer, 2002), whereas active movement is an important component of virulence (Portillo-Aguilar *et al.*, 1999). Virulence of *S. carpocapsae* increased with sand content of soil, whereas survival was highest in sandy loam compared with sand, clay loam or clay soils (Kung *et al.*, 1990b). Similar effects of texture on virulence and pathogenicity of *S. glaseri*, *H. bacteriophora* and *S. carpocapsae* were reported by Portillo-Aguilar *et al.* (1999), who also varied the bulk density of each soil studied to demonstrate the key role of soil structure in nematode movement and survival. The proportion of pores in each treatment equal to or greater in size than the nematode body diameter explained more than 60% of the variation in virulence of the three species. Spatial patterns detected in surveys of endemic entomopathogenic nematodes are reviewed by Hominick (2002), and several but not all surveys revealed a direct relationship between nematode prevalence and sand content of the soil (e.g. Hara *et al.*, 1991; Zhang *et al.*, 1992).

Soil texture may be less important for many vertebrate-parasitic nematodes that contact hosts at the soil surface. For example, human infections by *Toxocara* spp. were proportional to the prevalence of eggs in soil samples, which was unrelated to soil texture (Mizgajska, 2001).

Little attention has been given to the role of soil texture in the regulation of nematodes by natural enemies. Predatory mites were found to have highest activity against *M. javanica* in sandy soil (Walia and Mathur, 1997). Worldwide surveys of *Pasteuria penetrans* spores attached to cuticles of *Meloidogyne* showed increased disease prevalence with increasing clay content (Mateille *et al.*, 2002). The authors suggested that the spores in sandy soils are more likely to be washed into deeper strata and to adsorb to surfaces of the finer, more highly charged particles. They further suggested that smaller pore sizes favour contact with the cuticles of migrating nematodes.

11.4.3. Water potential

Movement, development and survival of nematodes in soil are regulated by the interaction between soil porosity and water potential. Physical principles governing the effect of this interaction on nematode biology have been comprehensively reviewed (Wallace, 1971) and modelled (Hunt *et al.*, 2001). Depending on pore size and hydrology, a nematode may reside in a pore that is filled with water or in a film of water adhering to a soil particle or root surface. Large, water-filled soil pores require nematodes to swim, an inefficient form of locomotion for most species. Reduced aeration in saturated soil is often detrimental to nematodes

(Sotomayor *et al.*, 1999; Soriano *et al.*, 2000), although it can induce quiescence to extend longevity in some species (Van Gundy *et al.*, 1967). In most soils, the developmental rate (Rebois, 1973; Towson and Apt, 1983) and movement (Wallace, 1971) of many nematodes is highest near the inflection point of the soil moisture release curve where soil water content changes from saturated to slightly drier states.

As water drains or evaporates from the soil pore, aeration increases, nematodes become increasingly constrained within a film of water and locomotion is more efficient due to the resistance to nematode thrusting provided by the surface tension of the water film. The optimum water potential for infection of *Galleria mellonella* by *S. carpocapsae* was lower than that for the larger *S. glaseri*, perhaps due to correspondence between the thickness of the respective water films and nematode body diameters (Koppenhöfer *et al.*, 1995). With increasing dryness, the thickness of water films decreases to the point that increasing surface tension forces and friction from the substrate impede motility; nematodes may lose internal water to the environment and either die or initiate behaviours, such as coiling and aggregation, required for drought survival (see Wharton, Chapter 13, this volume). Coiling reduces the surface area of individual nematodes, thereby reducing the rate of water loss and providing time to complete metabolic processes required for desiccation survival. Some nematodes that produce large numbers of dauer juveniles exhibit an aggregation behaviour in drying soil that results in 'nematode wool'. For example, fourth-stage juvenile *D. dipsaci* in onion aggregate in dense mats as the plant tissue dries. Nematodes in the centre of the mat survive much longer than those at the surface which desiccate quickly. Many entomopathogenic nematodes exhibit the same behaviour and mats consisting of millions of third-stage infective juveniles can be stored at high relative humidity for up to 6 months without mortality (Lindegren *et al.*, 1993).

Many nematodes have physiological or behavioural adaptations that allow resumption of activity after quiescence induced by moisture-limiting conditions (Glazer, 2002). Survival of the entomopathogenic nematode *Steinernema riobrave* may be enhanced by quiescence induced by moisture deficits (Duncan *et al.*, 1996a). In experiments to simulate a rainfall or irrigation event, virulence of entomopathogenic nematodes in low-moisture conditions could be restored by rehydrating the soil. Insect mortality was generally low in low-moisture, nematode-infested soils before rehydration, but increased to high levels post-hydration (Grant and Villani, 2003). Low water potential can also increase winter survival of *Meloidogyne hapla* by reducing the pore space filled with ice and inducing physiological changes that improve survival of freezing (Forge and MacGuidwin, 1992b).

The degree to which closely related nematodes are adapted to low water potential is often reflected in their life histories. In contrast to the human pinworm, *Enterobius vermicularis*, transmission of the mouse pinworm, *Syphacia obvelata*, requires intimate contact between its hosts. Eggs of the former species are desiccation-tolerant, whereas those of the latter perish quickly in water or when desiccated (Grice and Prociv, 1993). Entomopathogenic nematode survival in dry soil is improved when infective juveniles are contained within an insect cadaver

(Koppenhöfer *et al.*, 1997). The mechanism is more effective for *S. riobrave* and *S. carpocapsae*, species adapted to surface soil horizons or geographical regions that experience frequent desiccation, than for *S. glaseri*, which typically forages deeper in the soil. Infection by *H. bacteriophora* produces a gummy consistency of the host cadaver that helps retain moisture.

Migration along moisture gradients is a behaviour critical for survival for many nematodes. Numerous species of plant-parasitic nematodes have been shown to orient to water gradients as shallow as 1% change in water content over 10 cm (Wallace, 1971). Gouge *et al.* (2000) showed apparent migration and accumulation of *S. riobrave* at the same hydration level along moisture gradients in soil columns that were either rehydrated daily or permitted to continuously dehydrate.

Spatial and temporal heterogeneity of water potential in soil affects nematode populations in a variety of ways and can often confound comparison of nematode behaviour in nature with that exhibited under controlled conditions. *Tylenchulus semipenetrans* and *S. riobrave* exhibit poor drought tolerance (Tsai and Van Gundy, 1988; Duncan *et al.*, 1996a). However, survival of *S. riobrave* and population growth of *T. semipenetrans* were far greater in extremely dry surface soil horizons (< −150 kPa) than in surface horizons irrigated normally (at −15 kPa) provided that plant roots extended into deeper soil horizons with high water potential (Duncan and El-Morshedy, 1996; Duncan and McCoy, 2001). Hydraulic lift of the deeper water via the root xylem apparently resulted in a thin water film at the rhizoplane in the dry soil horizon, which provided a niche with ideal water potential for nematode development and survival (Fig. 11.2). Savin *et al.* (2001) proposed that population growth of microbivorous and predacious nematodes was favoured by low water potential (−50 kPa) because nematodes become trapped within water 'enclosures', placing them in closer proximity to their sources of food. Young citrus trees that are interplanted with mature trees often grow at continuously high, rather than fluctuating, soil water potential because irrigation must be scheduled to accommodate the greater amount of water transpired by the mature trees. Population densities of *Belonolaimus longicaudatus* in the rhizospheres of the young and old trees are directly related to soil water potential at the dry point of the irrigation cycle, thereby causing significant damage to roots of young, but not mature, trees (Duncan *et al.*, 1996b).

11.5. Conclusions and Future Directions

Clearly, recent research on nematode behaviour has focused heavily on *C. elegans*. This research focus has contributed greatly to a deeper understanding of the molecular and biochemical mechanisms of behaviour, which may apply generally to nematodes and other organisms. However, because nematodes are extremely diverse, are ubiquitous and vary tremendously in their lifestyles, we shall only gain a full understanding of nematode behaviour and biology by adding a comparative approach with other nematode taxa studied under more realistic environmental conditions. Knowledge about differences in sensitivities, tolerances and behavioural

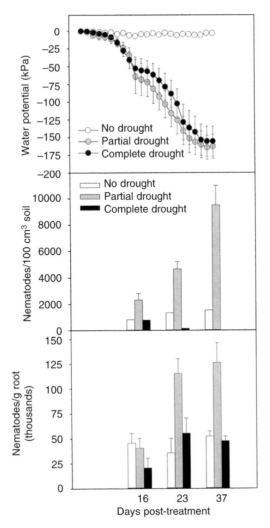

Fig. 11.2. The effect of hydraulic lift on population growth of *Tylenchulus semipenetrans*. Root systems of citrus seedlings were grown in double vertical pots, in which the top pot was infested with nematodes. Three treatments consisted of irrigating both pots (no drought), only the bottom pot (partial drought) or neither pot (complete drought). Hydraulic lift could occur only under the partial drought treatment. Despite similar soil water potential in the upper pot (top panel) under complete and partial drought, nematode population growth (lower two panels) was favoured by dry soil combined with hydraulic lift. For clarity, the increments between measurement dates on the bottom two panels are not drawn to scale (redrawn from Duncan and El-Morshedy, 1996).

responses of nematodes to various environmental challenges is important for fundamental science, but may also be useful for practical reasons, e.g. to enhance the activities of beneficial nematodes and manage the activities of pest nematodes and to help predict the effects that major perturbations, e.g. global climate change, will have on ecosystem function as well as plant, animal and human health. Comparative analyses can be used to elucidate patterns, understand complex relationships and better describe the variation in nematode populations and behaviours in complex environments.

References

Anderson, A.R.A., Young, I.M., Sleeman, B.D., Griffiths, B.S. and Robertson, W.M. (1997a) Nematode movement along a chemical gradient in a structurally heterogeneous environment. 1. Experiment. *Fundamental and Applied Nematology* 20, 157–163.

Anderson, A.R.A., Sleeman, B.D, Young, I.M., Griffiths, B.S. and Robertson, W.M. (1997b) Nematode movement along a chemical gradient in a structurally heterogeneous environment. 2. Theory. *Fundamental and Applied Nematology* 20, 165–172.

Anderson, G.L. (1978) Responses of dauer larvae of *Caenorhabditis elegans* (Nematoda: Rhabditidae) to thermal stress and oxygen deprivation. *Canadian Journal of Zoology* 56, 1786–1791.

Ash, C.P.J. and Atkinson, H.J. (1983) Evidence for a temperature-dependent conversion of lipid reserves to carbohydrate in quiescent eggs of the nematode, *Nematodirus battus*. *Comparative Biochemistry and Physiology* 76B, 603–610.

Ash, C.P.J. and Atkinson, H.J. (1986) *Nematodirus battus*: development of cold hardiness in dormant eggs. *Experimental Parasitology* 62, 24–28.

Ashton, F.T., Li, J. and Schad, G.A. (1999) Chemo- and thermosensory neurons: Structure and function in animal parasitic nematodes. *Veterinary Parasitology* 84, 297–316.

Atkinson, H.J. (1980) Respiration in nematodes. In: Zuckerman, B.M. (ed.) *Nematodes as Biological Models*, Vol. 2. Academic Press, London, pp. 101–142.

Banyer, R.J. and Fisher, J.M. (1971) Effect of temperature on hatching of eggs of *Heterodera avenae*. *Nematologica* 17, 519–534.

Barbercheck, M.E. and Kaya, H.K. (1991) Effect of host condition and soil texture on host finding by the entomogenous nematodes *Heterorhabditis bacteriophora* (Rhabditida Heterorhabditidae) and *Steinernema carpocapsae* (Rhabditida Steinernematidae). *Environmental Entomology* 20, 582–589.

Bargmann, C.I. and Avery, L. (1995) Laser killing of cells in *Caenorhabditis elegans*. *Methods in Cell Biology* 48, 225–250.

Bargmann, C.I. and Mori, I. (1997) Chemotaxis and thermotaxis. In: Riddle, D.L., Blumenthal, T., Meyer, B.J. and Priess, J.R. (eds) *C. elegans II*. Cold Spring Harbor Laboratory Press, Cold Spring Harbor, New York, pp. 717–737.

Bargmann, C.I., Hartwieg, E. and Horvitz, B.R. (1993) Odorant-selective genes and neurons mediate olfaction in *C. elegans*. *Cell* 74, 515–527.

Barker, K.R., Pederson, G.A. and Windham, G.L. (eds) (1998) *Plant and Nematode Interactions*. ASA Headquarters, Madison, Wisconsin, 771 pp.

Behm, C.A. (1997) The role of trehalose in the physiology of nematodes. *International Journal for Parasitology* 27, 215–229.

Behm, C.A. (2002) Metabolism. In: Lee, D.L. (ed.) *The Biology of Nematodes*. Taylor and Francis, London and New York, pp. 261–290.

Bernhard, N. and van der Kooy, D. (2000) Behavioral and genetic dissection of two forms of olfactory plasticity in *Caenorhabditis elegans*: adaptation and habituation. *Learning and Memory* 7, 199–212.

Bessho, K., Yamada, S., Kunitani, T., Nakamura, T., Hashiguchi, T., Tanimoto, Y., Harada, S., Yamamoto, H. and Hosono, R. (1995) Biological responses in *Caenorhabditis elegans* to high magnetic fields. *Experientia* 51, 284–288.

Boag, B. and Neilson, R. (1996) Effects of potential climatic changes on plant-parasitic nematodes. *Aspects of Applied Biology* 45, 331–334.

Brady, N.C. (1974) *The Nature and Properties of Soils*, 8th edn. Macmillan Publishing, New York, 639 pp.

Brown, B.J. and Platzer, E.G. (1978) Salts and the infectivity of *Romanomermis culicivorax*. *Journal of Nematology* 9, 166–172.

Brown, I.M. and Gaugler, R. (1996) Cold tolerance of steinernematid and heterorhabditid nematodes. *Journal of Thermal Biology* 21, 115–121.

Brown, I.M., Lovett, B.J., Grewal, P.S. and Gaugler, R. (2002) Latent infection: a low temperature survival strategy in steinernematid nematodes. *Journal of Thermal Biology* 27, 531–539.

Bullock, T.H. (1955) Compensation for temperature in the metabolism and activity of poikilotherms. *Biological Reviews* 30, 311–342.

Burr, A.H.J. (1985) The photomovement of *Caenorhabditis elegans*, a nematode which lacks ocelli: proof that the response is to light not radiant heating. *Photochemistry and Photobiology* 41, 577–582.

Burr, A.H.J., Babbinski, C.P.F. and Ward, A.J. (1990) Components of phototaxis of the nematode *Mermis nigrescens*. *Journal of Comparative Physiology* 167A, 245–255.

Campbell, J.F. and Kaya, H.K. (2000) Influence of insect associated cues on the jumping behavior of entomopathogenic nematodes (*Steinernema* spp.). *Behaviour* 137, 591–609.

Campbell, J.F., Lewis, E.E., Stock, S.P., Nadler, S. and Kaya, H.K. (2003) Evolution of host search strategies in entomopathogenic nematodes. *Journal of Nematology* 35, 142–145.

Chalfie, M. and White, J. (1988) The nervous system. In: Wood, W.B. (ed.) *The Nematode* Caenorhabditis elegans. Cold Spring Harbor Laboratory, Cold Spring Harbor, New York, pp. 337–391.

Coates, J.C. and de Bono, M. (2002) Antagonistic pathways in neurons exposed to body fluid regulate social feeding in *Caenorhabditis elegans*. *Nature* 419, 925–929.

Cohen, N.E., Brown, I.M. and Gaugler, R. (2002) Physiological ageing and behavioural plasticity of *Heterorhabditis bacteriophora* infective juveniles. *Nematology* 4, 81–87.

Cooper, A.F., Jr and Van Gundy, S.D. (1970) Metabolism of glycogen and neutral lipids by *Aphelenchus avenae* and *Caenorhabditis* sp. in aerobic, microaerobic, and anaerobic environments. *Journal of Nematology* 2, 305–315.

Cooper, A.F., Jr, Van Gundy, S.D. and Stolzy, L.H. (1970) Nematode reproduction in environments of fluctuating aeration. *Journal of Nematology* 2, 182–188.

Cooper, S.C. and Ferguson, J.H. (1973) The effects of cold acclimation upon the oxygen consumption of two species of free-living nematodes. *Journal of Nematology* 5, 241–245.

Croll, N.A. (1966a) The phototactic response and spectral sensitivity of *Chromadorina viridis* (Nematoda, Chromadorida) with a note on the nature of the paired pigment spots. *Nematologica* 12, 610–614.

Croll, N.A. (1966b) A contribution to the light sensitivity of the *Mermis subnigrescens* chromatrope. *Helminthology* 40, 33–38.

Croll, N.A. (1970) *The Behaviour of Nematodes: Their Activity, Senses and Responses*. St Martin's Press, New York, 117 pp.

Croll, N.A. and Matthews, B.E. (1977) *Biology of Nematodes*. John Wiley & Sons, New York, 201 pp.

Croll, N.A. and Smith, J.M. (1970) The sensitivity and responses of *Rhabditis* sp. to peripheral mechanical stimulation. *Proceedings of the Helminthological Society of Washington* 37, 1–5.

Croll, N.A. and Sukhdeo, M.V.K. (1981) Hierarchies in nematode behavior. In: Zuckerman, B.M. and Rohde, R.A. (ed.) *Plant Parasitic Nematodes*, Vol. III. Academic Press, New York, pp. 227–251.

Culotti, J.G. and Russell, R.L. (1978) Osmotic avoidance defective mutants of the nematode *Caenorhabditis elegans*. *Genetics* 90, 243–256.

Daniells, C., Duce, I., Thomas, D., Sewell, P., Tattersall, J. and de Pomerai, D. (1998) Transgenic nematodes as biomonitors of microwave-induced stress. *Mutation Research* 399, 55–64.

Davis, B.O., Jr, Goode, M. and Dusenbery, D.B. (1986) Laser microbeam studies of role of the amphid receptors in chemosensory behavior of the nematode *Caenorhabditis elegans*. *Journal of Chemical Ecology* 12, 1339–1348.

de Bono, M., Tobin, D.M., Davis, M.W., Avery, L. and Bargmann, C.I. (2002) Social feeding in *Caenorhabditis elegans* is induced by neurons that detect aversive stimuli. *Nature* 419, 899–903.

Dempsey, C.M. and Griffin, C.T. (2000) Phased activity in *Heterorhabditis megidis* infective juveniles. *Parasitology* 124, 605–613.

de Pomerai, D.I., Dawe, A., Djerbib, L., Allan, J., Brunt, G. and Daniells, C. (2002) Growth and maturation of the nematode *Caenorhabditis elegans* following exposure to weak microwave fields. *Enzyme and Microbial Technology* 30, 73–79.

Driscoll, M. and Kaplan, J. (1997) Mechanotransduction. In: Riddle, D.L., Blumenthal, T., Meyer, B.J. and Priess, J.R. (eds) *C. elegans II*. Cold Spring Harbor Laboratory Press, Cold Spring Harbor, New York, pp. 645–677.

Duggan, A., Ma, C. and Chalfie, M. (1996) Regulation of touch receptor differentiation by the *Caenorhabditis elegans mec*-3 and *unc*-86 genes. *Development* 125, 4107–4119.

Duncan, L.W. and El-Morshedy, M.M. (1996) Population changes of *Tylenchulus semipenetrans* under localized versus uniform drought in the citrus root zone. *Journal of Nematology* 28, 360–368.

Duncan, L.W. and McCoy, C.W. (2001) Hydraulic life increases herbivory by *Diaprepes abbreviatus* larvae and persistence of *Steinernema riobrave* in dry soil. *Journal of Nematology* 33, 142–146.

Duncan, L.W., Dunn, D.C. and McCoy, C.W. (1996a) Spatial patterns of entomopathogenic nematodes in microcosms: implications for laboratory experiments. *Journal of Nematology* 28, 252–258.

Duncan, L.W., Noling, J.W., Inserra, R.N. and Dunn, D. (1996b) Spatial patterns of *Belonolaimus* spp. among and within citrus orchards on Florida's central ridge. *Journal of Nematology* 28, 352–359.

Dusenbery, D.B. (1980) Behavior of free-living nematodes. In: Zuckerman, B.M. (ed.) *Nematodes as Biological Models*, Vol. 1. *Behavioral and Developmental Models*. Academic Press, New York, pp. 127–158.

Dusenbery, D.B. (1983) Chemotactic behavior of nematodes. *Journal of Nematology* 15, 168–173.

Dusenbery, D.B. (1989) A simple animal can use a complex stimulus pattern to find a location: nematode thermotaxis in soil. *Biological Cybernetics* 60, 431–437.

Edongali, E.A., Duncan, L. and Ferris, H. (1982) Influence of salt concentration on infectivity and development of *Meloidogyne incognita* on tomato. *Revue de Nématologie* 5, 111–118.

Evans, A.A.F. (1987) Diapause in nematodes as a survival strategy. In: Veech, J.A. and

Dickson, D.W. (eds) *Vistas on Nematology: a Commemoration of the Twenty-fifth Anniversary of the Society of Nematologists.* Society of Nematologists, Hyattsville, Maryland, pp. 180–187.

Evans, A.A.F. and Perry, R.N. (1976) Survival strategies in nematodes. In: Croll, N.A. (ed.) *The Organization of Nematodes.* Academic Press, London and New York, pp. 383–424.

Eysker, M. (1997) Some aspects of inhibited development of trichostrongylids in ruminants. *Veterinary Parasitology* 72, 265–283.

Fairbairn, D. (1961) The *in vitro* hatching of *Ascaris lumbricoides* eggs. *Canadian Journal of Zoology* 39, 153–162.

Fassuliotis, G. (1958) Observations on the biological effects of ionizing radiation on the life cycle of *Heterodera rostochiensis* Wollenweber, 1923. PhD thesis, New York University.

Ferree, T.C. and Lockery, S.R. (1999) Computational rules for chemotaxis in the nematode *C. elegans. Journal of Computational Neuroscience* 6, 263–277.

Ferris, H., Eyre, M., Venette, R.C. and Lau, S.S. (1996) Population energetics of bacterial-feeding nematodes: stage-specific development and fecundity rates. *Soil Biology and Biochemistry* 28, 271–280.

Fischer, P. and Führer, E. (1990) Effect of soil acidity on the entomophilic nematode *Steinernema kraussei* Steiner. *Biology and Fertility of Soils* 9, 174–177.

Forge, T.A. and MacGuidwin, A.E. (1992a) Impact of thermal history on tolerance of *Meloidogyne hapla* second-stage juveniles to external freezing. *Journal of Nematology* 24, 262–268.

Forge, T.A. and MacGuidwin, A.E. (1992b) Effects of water potential and temperature on survival of the nematode *Meloidogyne hapla* in frozen soil. *Canadian Journal of Zoology* 70, 1553–1560.

Gans, C. and Burr, A.H. (1994) Unique locomotory mechanisms of *Mermis nigrescens*, a large nematode that crawls over soil and climbs through vegetation. *Journal of Morphology* 222, 133–148.

Gao, X. and Becker, J.O. (2002) Population development of both sexes of *Heterodera schachtii* is diminished in a beet cyst nematode-suppressive soil. *Biological Control* 25, 187–194.

Gaugler, R. and Boush, G.M. (1978) Effects of ultraviolet radiation and sunlight on the entomogenous nematode, *Neoaplectana carpocapsae. Journal of Invertebrate Pathology* 32, 291–296.

Gaugler, R. and Boush, G.M. (1979) Effects of gamma radiation on the entomogenous nematode, *Neoaplectana carpocapsae. Journal of Invertebrate Pathology* 33, 121–123.

Gaugler, R. and Hashmi, S. (1996) Genetic engineering of an insect parasite. In: Setlow, J. (ed.) *Genetic Engineering Principles and Methods.* Plenum, New York, pp. 135–157.

Gaugler, R., Campbell, J.F. and McGuire, T. (1990) Fitness of a genetically improved entomopathogenic nematode. *Journal of Invertebrate Pathology* 56, 106–116.

Gaugler, R., Bednarek, A. and Campbell, J.F. (1992) Ultraviolet inactivation of heterorhabditid and steinernematid nematodes. *Journal of Invertebrate Pathology* 59, 155–160.

Ghally, S.E. (1995) Some factors affecting the activity and pathogenicity of *Heterorhabditis heliothidis* and *Steinernema carpocapsae* nematodes. *Journal of the Egyptian Society of Parasitology* 25, 125–135.

Glazer, I. (2002) Survival biology. In: Gaugler, R. (ed.) *Entomopathogenic Nematology.* CAB International, Wallingford, UK, pp. 169–187

Golden, J.W. and Riddle, D.L. (1984) The *Caenorhabditis elegans* dauer larva: developmental effects of pheromone, food, and temperature. *Developmental Biology* 102, 368–378.

Gouge, D.H., Smith, K.A., Lee, L.L. and Henneberry, T.J. (2000) Effect of soil depth and moisture on the vertical distribution of *Steinernema riobrave* (Nematoda: Steinernematidae). *Journal of Nematology* 32, 223–228.

Grant, J.A. and Villani, M.G. (2003) Soil moisture effects on entomopathogenic nematodes. *Environmental Entomology* 32, 80–87.

Granzer, M. and Haas, W. (1991) Host-finding and host recognition of infective *Ancylostoma caninum* larvae. *International Journal for Parasitology* 21, 429–440.

Green, C.D. and Plumb, S. (1967) The effect of ultra-violet radiation on the invasion survival and fertility of larvae of *Heterodera rostochiensis*. *Nematologica* 13, 186–190.

Grewal, P.S. (1991) Influence of bacteria and temperature on the reproduction of *Caenorhabditis elegans* (Nematoda: Rhabditidae) infesting mushrooms (*Agaricus bisporus*). *Nematologica* 37, 72–82.

Grewal, P.S., Selvan, S., Lewis, E.E. and Gaugler, R. (1993) Male insect-parasitic nematodes: a colonizing sex. *Experimentia* 49, 605–608.

Grewal, P.S., Selvan, S. and Gaugler, R. (1994) Thermal adaptation of entomopathogenic nematodes: niche breadth for infection, establishment, and reproduction. *Journal of Thermal Biology* 19, 245–253.

Grewal, P.S., Lewis, E.E. and Gaugler, R. (1997) Response of infective stage parasites (Nematoda: Steinernematidae) to volatile cues from infected hosts. *Journal of Chemical Ecology* 23, 503–515.

Grice, R.L. and Prociv, P. (1993) *In vitro* embryonation of *Syphacia obvelata* eggs. *International Journal for Parasitology* 23, 257–260.

Gu, G., Caldwell, G.A. and Chalfie, M. (1996) Genetic interactions affecting touch sensitivity in *Caenorhabditis elegans*. *Proceedings of the National Academy of Sciences, USA* 93, 6577–6582.

Gutzeit, H.O. (2001) Biological effects of ELF-EMF enhanced stress response: new insights and new questions. *Electro- and Magnetobiology* 20, 15–26.

Hara, A.H., Gaugler, R., Kaya, H.K. and LeBeck, L.M. (1991) Natural populations of entomopathogenic nematodes (Rhabditida: Heterorhabditidae, Steinernematidae) from the Hawaiian Islands. *Environmental Entomology* 20, 211–216.

Hartman, P., Goldstein, P., Algarra, M., Hubbard, D. and Mabery, J. (1996) The nematode *Caenorhabditis elegans* is up to 39 times more sensitive to gamma radiation generated from ^{137}Cs than from ^{60}Co. *Mutation Research* 363, 201–208.

Hass, D.K. and Todd, A.C. (1962) Extension of a technique for hatching ascarid eggs *in vitro*. *American Journal of Veterinary Research* 23, 169–170.

Hassink, J., Bouwman, L.A., Zwart, K.B. and Brussaard, L. (1993a) Relationships between habitable pore space, soil biota and mineralization rates in grassland soils. *Soil Biology and Biochemistry* 25, 47–55.

Hassink, J., Bouwman, L.A., Zwart, K.B., Bloem, J. and Brussaard, L. (1993b) Relationships between soil texture, physical protection of organic matter, soil biota, and C and N mineralization in grassland soil. *Geoderma* 57, 105–128.

Hawking, F. (1962) Microfilarial infestation as an instance of periodic phenomena seen in a host-parasite relationship. *Annals of the New York Academy of Sciences* 98, 940–953.

Hedgecock, E.M. and Russell, R.L. (1975) Normal and mutant thermotaxis in the nematode *Caenorhabditis elegans*. *Proceedings of the National Academy of Sciences, USA* 72, 4061–4065.

Hilliard, M.A., Bargmann, C.I. and Bazzicalupo, P. (2002) *C. elegans* responds to chemical repellents by integrating sensory inputs from the head and the tail. *Current Biology* 12, 730–734.

Hominick, W.M. (2002) Biogeography. In: Gaugler, R. (ed.) *Entomopathogenic Nematology.* CAB International, Wallingford, UK, pp. 115–143.

Hunt, H.W., Wall, D.H., Decrappeo, N.M. and Brenner, J.S. (2001) A model for nematode locomotion in soil. *Nematology* 3, 705–716.

Ishibashi, N. and Kondo, E. (1990) Behavior of infective juveniles. In: Gaugler, R. and Kaya, H.K. (eds) *Entomopathogenic Nematodes in Biological Control.* CRC Press, Boca Raton, Florida, pp. 139–150.

Ishii, N. and Suzuki, K. (1980) Killing effects of UV light and X-rays on free living nematode *Rhabditis tokai. Journal of Radiation Research* 21, 137–147.

Ishii, N. and Suzuki, K. (1990) X-ray inactivation of *Caenorhabditis elegans* embryos or larvae. *International Journal of Radiation Biology* 58, 827–834.

Jagdale, G.B. and Grewal, P.S. (2003) Acclimation of entomopathogenic nematodes to novel temperatures: trehalose accumulation and the acquisition of thermotolerance. *International Journal for Parasitology* 33, 145–152.

Jaworska, M. (1993) Investigations on the possibility of using entomophilic nematodes in reduction of *Cephalcia abietis* (L.) (Hym., Pamhillidae) population. *Polskie Pismo Entomologiczne* 62, 201–213.

Johnson, T.E. and Nelson, G.A. (1991) *Caenorhabditis elegans*: a model system for space biology studies. *Experimental Gerontology* 26, 299–309.

Kaplan, J.M. and Horvitz, H.R. (1993) A dual mechanosensory and chemosensory neuron in *Caenorhabditis elegans. Proceedings of the National Academy of Sciences, USA* 90, 2227–2231.

Keeling, C.D., Chin, J.F.S. and Whorf, T.P. (1996) Increased activity of northern vegetation inferred from atmospheric CO_2 measurements. *Nature* 382, 146–149.

Keeling, J.E.D. (1960) The effects of ultra-violet radiation on *Nippostrongylas muris.* 1. Irradiation of infective larvae, lethal and sublethal effects. *Annals of Tropical Medicine and Parasitology* 54, 182–191.

Khlibsuwan, W., Ishibashi, N. and Kondo, E. (1992) Response of *Steinernema carpocapsae* infective juveniles to the plasma of three insect species. *Journal of Nematology* 24, 156–159.

Klass, M.R. (1977) Ageing in the nematode *Caenorhabditis elegans*: major biological and environmental factors influencing life span. *Mechanisms of Ageing and Development* 6, 413–429.

Klingler, J. (1972) The effect of single and combined heat and CO_2 stimuli at different ambient temperatures on the behavior of two plant-parasitic nematodes. *Journal of Nematology* 4, 95–100.

Klose, C., Mravak, S., Geb, M., Bienzle, U. and Meyer, C.G. (1996) Autochthonous cutaneous larva migrans in Germany. *Tropical Medicine and International Health* 1, 503–504.

Koenning, S.R. and Barker, K.R. (1995) Soybean photosynthesis and yield as influenced by *Heterodera glycines*, soil type and irrigation. *Journal of Nematology* 27, 51–62.

Koenning, S.R., Walters, S.A. and Barker, K.R. (1996) Impact of soil texture on the reproductive and damage potentials of *Rotylenchulus reniformis* and *Meloidogyne incognita* on cotton. *Journal of Nematology* 28, 527–536.

Koppenhöfer, A.M., Kaya, H.K. and Taormino, S.P. (1995) Infectivity of entomopathogenic nematodes (Rhabditida: Steinernematidae) at different soil depths and moistures. *Journal of Invertebrate Pathology* 65, 193–199.

Koppenhöfer, A.M., Baur, M.E., Stock, S.P., Choo, H.Y., Chinnasri, B. and Kaya, H.K. (1997) Survival of entomopathogenic nematodes within host cadavers in dry soil. *Applied Soil Ecology* 6, 231–240.

Krieger, J. and Breer, H. (1999) Olfactory reception in invertebrates. *Science* 286, 720–723.

Kung, S.-P., Gaugler, R. and Kaya, H.K. (1990a) Influence of soil pH and oxygen on persistence of *Steinernema* spp. *Journal of Nematology* 22, 440–445.

Kung, S.-P., Gaugler, R. and Kaya, H.K. (1990b) Soil type and entomopathogenic nematode persistence. *Journal of Invertebrate Pathology* 55, 401–406.

Lee, D.L. (2002) Behaviour. In: Lee, D.L. (ed.) *The Biology of Nematodes*. Taylor and Francis, London, pp. 369–387.

Lewis, E.E. (2002) Behavioural ecology. In: Gaugler, R. (ed.) *Entomopathogenic Nematology*. CAB International, Wallingford, UK, pp. 205–223.

Lewis, E.E., Gaugler, R. and Harrison, R. (1992) Entomopathogenic nematode host finding: response to host contact cues by cruise and ambush foragers. *Parasitology* 105, 309–316.

Lewis, E.E., Gaugler, R. and Harrison, R. (1993) Response of cruiser and ambusher entomopathogenic nematodes (Steinernematidae) to host volatile cues. *Canadian Journal of Zoology* 71, 765–769.

Lewis, E.E., Grewal, P.S. and Gaugler, R. (1995a) Hierarchical order of host cues in parasite foraging strategies. *Parasitology* 110, 207–213.

Lewis, E.E., Selvan, S., Campbell, J.F. and Gaugler, R. (1995b) Changes in foraging behaviour during the infective stage of entomopathogenic nematodes. *Parasitology* 110, 583–590.

Lewis, E.E., Campbell, J.F. and Gaugler, R. (1997) The effects of aging on the foraging behavior of *Steinernema carpocapsae* (Rhabditida: Steinernematidae). *Nematologica* 43, 355–362.

Li, N. and Baehr, W. (1998) Expression and characterization of human PDEδ and its *Caenorhabditis elegans* ortholog CEδ. *FEBS Letters* 440, 454–457.

Li, N., Ashton, F.T., Gamble, H.R. and Schad, G.A. (2000) Sensory neuroanatomy of a passively ingested nematode parasite, *Haemonchus contortus*: amphidial neurons of the first stage larva. *Journal of Comparative Neurology* 417, 299–314.

Lindegren, J.E., Valero, K.A. and Mackey, B.E. (1993) Simple *in vivo* production and storage methods for *Steinernema carpocapsae* infective juveniles. *Journal of Nematology* 25, 193–197.

Linquist, S. and Craig, E.A. (1988) The heat-shock proteins. *Annual Review of Genetics* 22, 631–677.

Lipton, J. and Emmons, S.W. (1999) An appetitive behavior is sexually dimorphic in *C. elegans*. *Worm Breeder's Gazette* 16, 56.

Liu, L.X., Spoerke, J.M., Mulligan, E.L., Chen, J., Reardon, B. and Westlund, B. (1999) High throughput isolation of *Caenorhabditis elegans* deletion mutants. *Genome Research* 9, 859–867.

McLaren, D.J. (1976) Nematode sense organs. In: Dawes, B. (ed.) *Advances in Parasitology*. Academic Press, London, pp. 195–265.

Markkola, A.M., Ohtonen, A., Ahonen-Jonnarth, U. and Ohtonen, R. (1996) Scots pine responses to CO_2 enrichment. 1. Ectomycorrhizal fungi and soil fauna. *Environmental Pollution* 94, 309–316.

Marroquin, L.D., Elyassnia, D., Griffitts, J.S., Feitelson, J.S. and Aroian, R.V. (2000) *Bacillus thuringiensis* (Bt) toxin susceptibility and isolation of resistance mutants in the nematode *Caenorhabditis elegans*. *Genetics* 155, 1693–1699.

Mateille, T., Trudgill, D.L., Trivino, C., Bala, G., Abdoussalam, S. and Vouyoukalou, E. (2002) Multisite survey of soil interactions with infestation of root-knot nematodes (*Meloidogyne* spp.) by *Pasteuria penetrans*. *Soil Biology and Biochemistry* 34, 1417–1424.

McSorley, R. and Porazinska, D.L. (2001) Elements of sustainable agriculture. *Nematropica* 31, 1–2.

Miyakawa, T., Yamada, S., Harada, S.I., Ishimori, T., Yamamoto, H. and Hosono, R. (2001) Exposure of *Caenorhabditis elegans* to extremely low frequency high magnetic fields induces stress responses. *Bioelectromagnetics* 22, 333–339.

Mizgajska, H. (2001) Eggs of *Toxocara* spp. in the environment and their public health implications. *Journal of Helminthology* 75, 147–151.

Moens, T. and Vincx, M. (2000) Temperature, salinity and food thresholds in two brackish-water bacterivorous nematode species: assessing niches from food absorption and respiration experiments. *Journal of Experimental Marine Biology and Ecology* 243, 137–154.

Mori, I. and Ohshima, Y. (1995) Neural regulation of thermotaxis in *Caenorhabditis elegans*. *Nature* 376, 344–348.

Muller, R. and Wakelin, D. (1998) Lymphatic filariasis. In: Collier, L., Balows, A. and Sussma, M. (eds) *Topley and Wilson's Microbiology and Microbial Infections*, 9th edn, Vol 5. Arnold, London, pp. 610–619.

Myers, R.F. (1960) The sensitivity of some plant-parasitic and free-living nematodes to gamma and X-irradiation. *Nematologica* 5, 56–63.

Nelson, G.A., Schubert, W.W. and Marshall, T.M. (1992) Radiobiological studies with the nematode *Caenorhabditis elegans*: genetic and developmental effects of high LET radiation. *Nuclear Tracks and Radiation Measurements* 20, 227–232.

Nickle, W.R. and Shapiro, M. (1994) Effects of eight brighteners as solar radiation protectants for *Steinernema carpocapsae*, All strain. *Journal of Nematology* 26, 782–784.

O'Bannon, J.H. and Good, J.M. (1971) Applications of microwave energy to control nematodes in soil. *Journal of Nematology* 3, 93–94.

O'Leary, S.A., Stack, C.M., Chubb, M.A. and Burnell, A.M. (1998) The effect of day of emergence from the insect cadaver on the behavior and environmental tolerances of infective juveniles of the entomopathogenic nematode *Heterorhabditis megidis* (strain UK211). *Journal of Parasitology* 84, 665–672.

Pax, R.A., Geary, T.G., Bennett, J.L. and Thompson, D.P. (1995) *Ascaris suum*: characterization of transmural and hypodermal potentials. *Experimental Parasitology* 80, 85–97.

Peckol, E.L. and Troemel, E.R. and Bargmann, C.I. (2001) Sensory experience and sensory activity regulate chemosensory receptor gene expression in *Caenorhabditis elegans*. *Proceedings of the National Academy of Sciences, USA* 98, 11032–11038.

Perkins, L.A., Hedgecock, E.M., Thompson, J.N. and Culotti, J.G. (1986) Mutant sensory cilia in the nematode *Caenorhabditis elegans*. *Developmental Biology* 117, 456–486.

Petersen, J.J. (1984) Nematode parasites of mosquitoes. In: Nickle, W.R. (ed.) *Plant and Insect Nematodes*. Marcel Dekker, New York, pp. 797–820

Pline, M. and Dusenbery, D.B. (1987) Responses of plant-parasitic nematode *Meloidogyne incognita* to carbon dioxide determined by video camera-computer tracking. *Journal of Chemical Ecology* 13, 873–888.

Pline, M., Diez, J.A. and Dusenbery, D.B. (1988) Extremely sensitive thermotaxis of the nematode *Meloidogyne incognita*. *Journal of Nematology* 20, 605–608.

Pollock, C. and Samoiloff, M.R. (1976) The development of nematode behaviour: stage-specific behaviour in *Panagrellus redivivus*. *Canadian Journal of Zoology* 54, 674–679.

Portillo-Aguilar, C., Villani, M.G., Tauber, M.J., Tauber, C.A. and Nyrop, J.P. (1999) Entomopathogenic nematode (Rhabditida: Heterorhabditidae and Steinernematidae) response to soil texture and bulk density. *Environmental Entomology* 28, 1021–1035.

Prot, J.-C. (1979) Influence of concentration gradients of salts on the behaviour of four plant parasitic nematodes. *Revue de Nématologie* 2, 11–16.

Pye, A.E. and Burman, M. (1981) *Neoaplectana carpocapsae*: nematode accumulation on chemical and bacterial gradients. *Experimental Parasitology* 51, 13–20.

Qiu, L. and Bedding, R.A. (2002) Characteristics of protectant synthesis of infective juveniles of *Steinernema carpocapsae* and importance of glycerol as a protectant for survival of the nematodes during osmotic dehydration. *Comparative Biochemistry and Physiology Part B: Biochemistry and Molecular Biology* 131, 757–765.

Raghavan, N., Ghosh, I., Eisinger, W.S., Pastrana, D. and Scott, A.L. (1999) Developmentally regulated expression of a unique small heat shock protein in *Brugia malayi*. *Molecular and Biochemical Parasitology* 104, 233–246.

Rankin, C.H. (1991) Interactions between two antagonistic reflexes in the nematode *C. elegans*. *Journal of Compartive Physiology* 169, 59–67.

Rebois, R.V. (1973) Effect of soil water on infectivity and development of *Rotylenchulus reniformis* on soybean, *Glycine max*. *Journal of Nematology* 5, 246–248.

Riddle, D.L. and Albert, P.S. (1997) Genetic and environmental regulation of dauer larva development. In: Riddle, D.L., Blumenthal, T., Meyer, B.J. and Priess, J.R. (eds) *C. elegans II*. Cold Spring Harbor Laboratory Press, Cold Spring Harbor, New York, pp. 739–768.

Riddle, D.L. and Bird, A.F. (1985) Responses of the plant parasitic nematodes *Rotylenchulus reniformis*, *Anguina agrostis* and *Meloidogyne javanica* to chemical attractants. *Parasitology* 91, 185–195.

Riga, E., Perry, R.N., Barrett, J. and Johnston, M.R.L. (1997) Electrophysiological responses of male potato cyst nematodes, *Globodera rostochiensis* and *G. pallida*, to some chemicals. *Journal of Chemical Ecology* 23, 417–428.

Rinaldo, C., Bazzicalupo, P., Ederle, S., Hilliard, M. and La Volpe, A. (2002) Roles for *Caenorhabditis elegans* rad-51 in meiosis and in resistance to ionizing radiation during development. *Genetics* 160, 471–479.

Rivoal, R. (1983) Biologie d'*Heterodera avenae* Wollen-weber en France. 3. Evolution des diapauses des races Fr_1 and Fr_4 au cours de plusiers années consécutives; influence de la température. *Revue de Nématologie* 6, 157–164.

Robinson, A.F. (1994) Movement of five nematode species through sand subjected to natural temperature gradient fluctuations. *Journal of Nematology* 26, 46–58.

Robinson, A.F. (1995) Optimal release rates for attracting *Meloidogyne incognita*, *Rotylenchulus reniformis*, and other nematodes to carbon dioxide in sand. *Journal of Nematology* 27, 42–50.

Robinson, A.F. and Heald, C.M (1989) Accelerated movement of nematodes from soil in Baermann funnels with temperature gradients. *Journal of Nematology* 21, 370–378.

Robinson, A.F., Baker, G.L. and Heald, C.M. (1990) Transverse phototaxis by infective juveniles of *Agamermis* sp. and *Hexamermis* sp. *Journal of Parasitology* 76, 147–152.

Rogers, W.P. and Sommerville, R.I. (1963) The infective stage of nematode parasites, and its significance in parasitism. *Advances in Parasitology* 1, 109–178.

Rolfe, R.N., Barrett, J. and Perry, R.N. (2001) Electrophysiological analysis of responses of adult females of *Brugia pahangi* to some chemicals. *Parasitology* 122, 347–357.

Ruess, L., Michelsen, A., Schmidt, I.K., Jonasson, S. and Griffiths, B.S. (1999) Simulated climate change affecting microorganisms, nematode density and biodiversity in subarctic soils. *Plant and Soil* 212, 63–73.

Runion, G.B., Curl, E.A., Rogers, H.H., Backman, P.A., Rodriguez-Kabana, R. and Helms, B.E. (1994) Effects of free-air CO_2 enrichment on microbial populations in the rhizosphere and phyllosphere of cotton. *Agricultural and Forest Meteorology* 70, 117–130.

Rutherford, T.A. and Croll, N.A. (1979) Wave forms of *Caenorhabditis elegans* in a hemical attractant and repellent and in thermal gradients. *Journal of Nematology* 11, 232–240.

Sadaie, T. and Sadaie, Y. (1989) Rad-2–dependent repair of radiation-induced chromosomal aberrations in *Caenorhabditis elegans*. *Mutation Research* 218, 25–32.

Sambongi, Y., Takeda, K., Wakabayashi, T., Ueda, I., Wada, Y. and Futai, M. (2000) *Caenorhabditis elegans* senses protons through amphid chemosensory neurons: proton signals elicit avoidance behavior. *Neuroreport* 11, 2229–2232.

Samoiloff, M.R. (1980) Action of physical and chemical agents on free-living nematodes. In: Zuckerman, B.M. (ed.) *Nematodes as Biological Models*, Vol. 2. *Aging and Other Model Systems*. Academic Press, New York, pp. 81–98.

Samoiloff, M.R., McNicholl, P., Cheng, R. and Balakanich, S. (1973) Regulation of nematode behavior by physical means. *Experimental Parasitology* 33, 253–262.

Saunders, L.M., Tompkins, D.M. and Hudson, P.J. (2000) The role of oxygen availability in the embryonation of *Heterakis gallinarum* eggs. *International Journal for Parasitology* 30, 1481–1485.

Savin, M.C., Görres, J.H., Neher, D.A. and Amador, J.A. (2001) Uncoupling of carbon and nitrogen mineralization: role of microbivorous nematodes. *Soil Biology and Biochemistry* 33, 1463–1472.

Sciacca, J., Forbes, W.M., Ashton, F.T., Lombardini, E., Gamble, H.R. and Schad, G.A. (2002) Response to carbon dioxide by the infective larvae of three species of parasitic nematodes. *Parasitology International* 51, 53–62.

Sedbrook, J.C., Chen, R. and Masson P.H. (1999) ARG1 (altered response to gravity) encodes a DnaJ-like protein that potentially interacts with the cytoskeleton. *Proceedings of the National Academy of Sciences, USA* 96, 1140–1145.

Shapiro, D.I., Lewis, E.E., Paramasivam, S. and McCoy, C. (2000) Nitrogen partitioning in *Heterorhabditis bacteriophora*-infected hosts and the effects on nitrogen attraction/repulsion. *Journal of Invertebrate Pathology* 76, 43–48.

Snutch, T.P. and Baillie, D.L. (1983) Alterations in the pattern of gene expression following heat shock in the nematode *Caenorhabditis elegans*. *Canadian Journal of Biochemistry and Cell Biology* 61, 480–487.

Sohlenius, B. and Bostrom, S. (1999) Effects of global warming on nematode diversity in a Swedish tundra soil – a soil transplantation experiment. *Nematology* 1, 695–709.

Soriano, I.R.S., Prot, J.C. and Matias, D.M. (2000) Expression of tolerance for *Meloidogyne graminicola* in rice cultivars as affected by soil type and flooding. *Journal of Nematology* 32, 309–317.

Sotomayor, D., Allen, L.H., Jr, Chen, Z., Dickson, D.W. and Hewlett, T. (1999) Anaerobic soil management practices and solarization for nematode. *Nematropica* 29, 153–170.

Stanton, J.M. (1990) Behavior of *Pterotylenchus cecidogenus* (Nematoda) in soil and on *Desmodium ovalifolium* as related to infection and host resistance. *Nematologica* 36, 448–456.

Stewart, H.I., Rosenbluth, R.E. and Baillie, D.L. (1991) Most ultraviolet irradiation induced mutations in the nematode *Caenorhabditis elegans* are chromosomal rearrangements. *Mutation Research/Fundamental and Molecular Mechanisms of Mutagenesis* 249, 37–54.

Stock, S.P., Campbell, J.F. and Nadler, S.A. (2001) Phylogeny of *Steinernema* Travasso, 1927 (Cephalobina: Steinernematidae). *Journal of Parasitology* 87, 877–889.

Stowens, D. (1942) The effect of ultraviolet radiation on *Trichinella spiralis*. *American Journal of Hygiene* 36, 264.

Sukul, N.C. and Croll, N.A. (1978) Influence of potential difference and current on the electrotaxis of *Caenorhabditis elegans*. *Journal of Nematology* 10, 314–317.

Sukul, N.C., Das, P.K. and Ghosh, S.K. (1975) Cation-mediated orientation of nematodes under electrical fields. *Nematologica* 21, 145–150.

Sukul, N.C., Das, P.K. and Ghosh, S.K. (1977) The influence of pre-incubation of *Labronema digitatum* at different ionic concentrations to subsequent electrotaxis. *Nematologica* 23, 24–28.

Sulston, J.E. and White, J.G. (1980) Regulation and cell anatomy during postembryonic development of *Caenorhabditis elegans*. *Developmental Biology* 78, 577–597.

Tavernarakis, N. and Driscoll, M. (2001) Mechanotransduction in *Caenorhabditis elegans*: the role of DEG/ENaC ion channels. *Cell Biochemistry and Biophysics* 35, 1–18.

Terrill, W.F. and Dusenbery, D.B. (1996) Threshold chemosensitivity and hypothetical chemoreceptor function of the nematode *Caenorhabditis elegans*. *Journal of Chemical Ecology* 22, 1463–1475.

Thomas, J.H. and Lockery, S. (1999) Neurobiology. In: Hope, I.A. (ed.) *C. elegans: a Practical Approach*. Oxford University Press, New York, pp. 143–179.

Thurston, G.S., Ni, Y. and Kaya, H.K. (1994) Influence of salinity on survival and infectivity of entomopathogenic nematodes. *Journal of Nematology* 26, 345–351.

Tobata-Kudo, H., Higo, H., Koga, M. and Tada, I. (2000) Chemokinetic behavior of the infective third-stage larvae of *Strongyloides ratti* on a sodium chloride gradient. *Parasitology International* 49, 183–188.

Towson, A.J. and Apt, W.J. (1983) Effect of soil water potential on survival of *Meloidogyne javanica* in fallow soil. *Journal of Nematology* 15, 110–114.

Tsai, B.Y. and Van Gundy, S.D. (1988) Comparison of anhydrobiotic ability of the citrus nematode with other plant-parasitic nematodes. In: Goren, R. and Mendel, K. (eds) *Proceedings of the Sixth International Citrus Congress*, Vol. 2, Balabin, Publishers, Philadelphia, pp, 983–992.

Van Gundy, S.D., Bird, A.F. and Wallace, H.R. (1967) Aging and starvation in larvae of *Meloidogyne javanica* and *Tylenchulus semipenetrans*. *Phytopathology* 57, 559–571.

Van Gundy, S.D., McElroy, F.D., Cooper, A.F. and Stolzy, L.H. (1968) Influence of soil temperature, irrigation and aeration on *Hemicycliophora arenaria*. *Soil Science* 106, 270–274.

Viglierchio, D.R. and Yu, P.K. (1983) On nematode behavior in an electric field. *Revue de Nématologie* 6, 171–178.

Walia, K.K. and Mathur, S. (1997) Influence of soil texture on the predacious activity of two nematophagous mites against root-knot nematode, *Meloidogyne javanica*. *Annals of Plant Protection Sciences* 5, 168–170.

Wallace, H.R. (1971) Abiotic influences in the soil environment. In: Zuckerman, B.M., Mai, W.F. and Rohde, R.A. (eds) *Plant Parasitic Nematodes*, Vol. 1. Academic Press, New York, pp. 257–280.

Waller, P.J. and Thomas, R.J. (1975) Field studies on inhibition of *Haemonchus contortus* in sheep. *Parasitology* 71, 285–291.

Watkins, A.R.J. and Fernando, M.A. (1984) Arrested development of the rabbit stomach worm *Obelescoides cuniculi*: manipulation of the ability to arrest through processes of selection. *International Journal of Parasitology* 14, 559–570.

Welsh, M.J., Price M.P. and Xie, J. (2002) Biochemical basis of touch perception: mechanosensory function of degenerin/epithelial Na$^+$ channels. *Journal of Biological Chemistry* 277, 2369 -2372.

Wharton, D.A. (1998) Comparison of the biology and freezing tolerance of *Panagrolaimus davidi*, an Antarctic nematode, from field samples and cultures. *Nematologica* 44, 643–653.

Wharton, D.A. and Brown, I.M. (1991) Cold tolerance mechanisms of the Antarctic nematode *Panagrolaimus davidi*. *Journal of Experimental Biology* 155, 629–641.

Wharton, D.A. and Ferns, D.J. (1995) Survival of intracellular freezing by the Antarctic nematode *Panagrolaimus davidi*. *Journal of Experimental Biology* 198, 1381–1387.

Wicks, S.R. and Rankin, C.H. (1995) Integration of mechanosensory stimuli in *Caenorhabditis elegans*. *Journal of Neuroscience* 15, 2434–2444.

Wilson, P.A.G. (1996) The light sense in nematodes. *Science* 151, 337–338.

Workneh, F., Yang, X.B. and Tylka, G.L. (1999) Soybean brown stem rot, *Phytophthora sojae*, and *Heterodera glycines* affected by soil texture and tillage relations. *Phytopathology* 89, 844–850.

Yeates, G.W., Newton, P.D. and Ross, D.J. (1999) Response of soil nematode fauna to naturally elevated CO_2 levels influenced by soil pattern. *Nematology* 1, 285–293.

Zhang, G.Y., Yang, H.W., Zhang, S.G. and Jiang, H. (1992) Survey on the natural occurrence of entomophilic nematodes (Steinernematidae and Heterorhabditidae) in the Beijing area. *Chinese Journal of Biological Control* 8, 157–159.

Zuckerman, B.M. and Jansson, H.B. (1984) Nematode chemotaxis and possible mechanisms of host/prey recognition. *Annual Review of Phytopathology* 22, 95–113.

12 Population Dynamics

BRIAN BOAG[1] AND GREGOR W. YEATES[2]

[1]Scottish Crop Research Institute, Invergowrie, Dundee DD2 5DA, Scotland, UK; [2]Landcare Research, Private Bag 11052, Palmerston North, New Zealand

12.1. Introduction

Population dynamics has variously been defined as 'the numerical changes in a population that occur within a stated period of time' (Gregorich *et al.*, 2001) or 'the changes in the structure of a population over time' (Lawrence, 1995). These definitions encompass numerical changes in nematode numbers within a year (seasonal variation) (Kendall and Buckland, 1971) and between years (population cycle) (Goodman and Payne, 1979) and both these components are included in this chapter. As it is essential to understanding the population dynamics of nematodes, an outline of their spatial distribution is also presented.

12.2. Spatial Distribution

The distribution of nematodes in soils is three-dimensional and it is imperative that there is information on both horizontal and vertical distributions if the population dynamics of these organisms is to be fully understood. The reason for the non-uniform distribution of nematodes in soils, sediments and plant and animal tissues and organs is poorly understood but reasons vary considerably, depending on the nematode species as well as external factors. Unfortunately, data on nematode aggregation are time-consuming and expensive to collect and, therefore, such data sets are rare and usually available only for species of economic importance.

12.2.1. Soil and plant-parasitic nematodes

12.2.1.1. Vertical distribution

The vertical distribution of nematodes has been studied in a number of habitats, including the tundra (Kuzmin, 1976), sand dunes (Yeates, 1967) and coniferous forests (Wasilewska, 1974). In a forest nursery, Boag (1981) found that the distribution of plant-parasitic nematodes belonging to the genus *Trichodorus* was not correlated with the vertical depth distribution of tree roots and that there were significant differences between species, e.g. *T. velatus* was more abundant between 0 and 29 cm while *T. primitivus* was more numerous between 30 and 49 cm. Similarly, Flegg (1968a) found for *Longidorus* and *Xiphinema* in a pear orchard and Forge *et al.* (1998) for *Pratylenchus penetrans* in a raspberry plantation that nematode numbers were not correlated with root distribution. Yeates (1980) studied the vertical distribution of nematode genera in pasture at 11 sites and found most genera at 0–2.5 cm and least at 20–30 cm. Boag *et al.* (1987) studied the vertical distribution of nematodes in 37 arable fields in eastern Scotland and found that significant differences existed among the depths at which different virus-vector nematode species were found and suggested that to detect them the optimum sampling depth was the 10–19 cm increment. These findings were supported by Feil *et al.* (1997), who studied the depth distribution of *Xiphinema index* among vineyards.

There appear to be a number of reasons why nematodes differ in their vertical distribution. It has been suggested that nematode populations are related to root distribution (Yeates, 1980; Rawsthorne and Brodie, 1986), but even for migratory and endoparasitic plant-parasitic nematodes this relationship does not always hold (Boag *et al.*, 1977; Forge *et al.*, 1998). Soil type has been implicated as a major reason for the distribution of some species, as have soil moisture and temperature (Wallace, 1971; Brodie, 1976; Griffiths *et al.*, 1996; Feil *et al.*, 1997; Young *et al.*, 1998). However, there are many other factors affecting the vertical distribution of nematodes in soil, e.g. competition between species (Boag and Alphey, 1988), the presence or absence of earthworms (Shapiro *et al.*, 1995), the tillage regime (Whitehead, 1977) and nematode migration (Rossner, 1972). In most cases probably no single reason is responsible for the vertical distribution of nematodes (MacGuidwin and Stanger, 1991). Sohlenius and Sandor (1987) suggested that the differences they observed in the distribution of nematodes in arable soils were that at greater depths the nematodes suffered from a shortage of food while nematodes near the surface suffered greater predation pressure. Different genera seem to have genetically programmed preferred depths at which they tend to reside, e.g. *Trichodorus* and *Paratrichodorus* are usually found at greater depth than *Rotylenchus* or *Paratylenchus* (Rossner, 1970; Boag and Alphey, 1988).

12.2.1.2. Horizontal distribution

The horizontal distribution of nematodes has received considerable attention since it is of ecological significance and could be a species characteristic (Taylor, 1979), as well as being of economic importance in determining sampling procedures for detecting and estimating the size of plant-parasitic populations (Noe and Campbell; 1985; Been and Schomaker, 1996). For some crops (e.g. coffee) nematode distribution is determined by the distance of the roots from the plant (Zhang and Schmitt, 1995), while other factors affect the distribution of nematodes when the roots of the crop are more uniformly distributed. While originally nematode distribution in the soil was considered to be random (Cotton, 1979), it is now considered that the distribution of soil nematodes is usually aggregated and not randomly or uniformly distributed. One of the first and most comprehensive studies on the horizontal distribution of nematodes, which demonstrated their aggregated distribution, was that of Goodell and Ferris (1980, 1981), which resulted from collecting 1936 samples on a 44 × 44 matrix (Fig. 12.1). Subsequent studies have used similar square grids (Ruess, 1995; Marshall *et al.*, 1998), modified grids (Boag *et al.*, 1987), nested sampling (Webster and Boag, 1992) and nested linear samples (Ettema and Yeates, 2003).

There have been a number of approaches to analysing the aggregated distribution of nematodes in the field, e.g. Morisita's index of dispersion (Wheeler *et al.*, 1987) and the negative binomial distribution. The negative binomial distribution has been used to describe the non-random distribution of nematodes in the soil (Goodell and Ferris, 1981; McSorley, 1982; Ruess, 1995). However, it has been found to have some ecological limitations, e.g. the *k* value (a measure of aggregation) is dependent upon the population mean (Taylor *et al.*, 1979). To overcome this problem some nematologists preferred to use Taylor's power law to measure aggregation (Taylor, 1961). This index of aggregation is independent of the mean and can be used to produce a transformation that 'normalizes' data (Taylor, 1970) and it has been used extensively to measure nematode aggregation (Caubel *et al.*, 1972; Boag and Topham, 1984; Abd-Elgawad and Hasabo, 1995) and to devise sampling programmes (McSorley and Dickson, 1991). However, Taylor's power

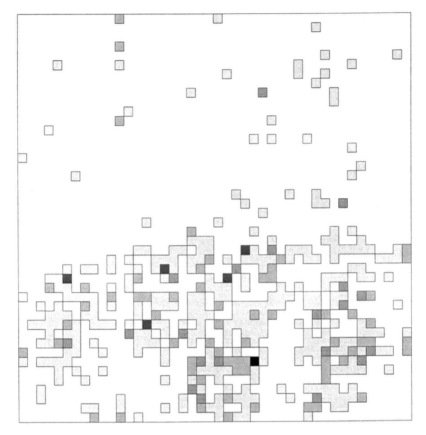

Fig. 12.1. Horizontal distribution of *Helicotylenchus digonicus* in a lucerne field. The 1936 samples were collected on a 44 × 44 grid with spacing between nodes of 6 m. The presence of cells denotes distribution of the nematode population. Increase in density of shading denotes increase in nematode populations: 0, 5–25, 25–75, 125–200, >200 nematodes/200 g soil. (From Goodell and Ferris, 1980.)

law does not have any associated measurement of distance and it produces artefacts at low population densities (Routledge and Swartz, 1992).

To overcome some of these criticisms, more recently geostatistical techniques have been utilized to quantify horizontal nematode aggregation (Webster and Boag, 1992; Wallace and Hawkins, 1994; Robertson and Freckman, 1995; Marshall *et al.*, 1998; Ettema and Yeates, 2003). The advantage of applying geostatistical techniques is that they yield true information on the spatial distribution rather than inferring association from frequency distributions (Williams *et al.*, 1992). Geostatistical techniques also allow the data to be visually inspected (in a three-dimensional fashion as a relief or contour map) (Wallace *et al.*, 1993) and can map errors associated with the displayed distributions. Future research on the hetero-genic nature of nematodes in soil may utilize fractal geometry, as this has been used successfully to describe spatial patterns in plant communities and has been applied to nematode movement in soil (Alexander *et al.*, 1997).

12.2.2. Free-living stages of nematodes parasitic in invertebrates and vertebrates

In explaining the population dynamics of nematode species with one or more vermiform stages outside their host, it is imperative that the more important aspects of
the biology of these free-living stages are understood. To complete its life cycle the
parasite must infect the host. For entomopathogenic nematodes this means finding
an insect in the soil, while for many parasitic nematodes of vertebrates it means that
the infective juvenile must migrate from the host's faeces on to herbage where the
host can ingest it. Research in this area has focused on parasites of domestic animals
and in particular on the spatial distribution of the faeces and infective stages.
However, Hutchinson *et al.* (2002) have studied the distribution of infective juveniles of free-roaming wild Soay sheep on St Kilda, a Scottish island.

12.2.2.1. Vertical distribution

Entomopathogenic nematodes are usually found near the soil surface (Cabanillas
and Raulston, 1994; Campbell *et al.*, 1998), although they have the ability to disperse considerable distances when associated with their host (Kaya, 1990).

The distribution of infective stages of animal parasites in the soil has received
little attention. While the soil can buffer free-living stages from temperature and
moisture extremes, Borgsteede and Boogaard (1983) considered that survival of
Ostertagia ostertagi and *Cooperia onchophora* in soil was negligible. Al Saqur *et al.*
(1982) recorded infective larvae of *Ostertagia* to a depth of 15 cm and suggested
that they may migrate downwards in winter and upwards in spring. The idea that
infective stages could migrate upwards through soil was supported by Gruner *et al.*
(1982). In laboratory experiments, Callinan and Westcott (1986) recorded an
eightfold increase in infective third-stage juveniles in the soil compared with
herbage.

12.2.2.2. Horizontal distribution

Boag *et al.* (1992a) examined the distribution of naturally occurring *Steinernema
feltiae* in a 1 ha area in a grass field and found such a highly aggregated distribution
that work on the population dynamics of this entomopathogenic nematode under
field conditions was suggested to be impracticable. However, Cabanillas and
Raulston (1994) demonstrated that the aggregated horizontal distribution of entomopathogenic nematode populations in the soil varied with depth. Campbell *et al.*
(1996) suggested that the more aggregated distribution of *Heterorhabditis bacteriophora* compared with that of *Steinernema carpocapsae* was due to it being a 'cruise
forager' compared with *S. carpocapsae* being an 'adaptive ambusher'. Campbell *et
al.* (1998) studied the interaction of introduced entomopathogenic nematodes with
naturally occurring species and found that, while the introduced nematode was
released uniformly, the pattern of distribution became aggregated, which was
typical of endemic populations.

Since under field conditions the distribution of faecal material from domestic animals has an aggregated distribution (Donald and Leslie, 1969), the distribution of juvenile stages is also aggregated (Tallis and Donald, 1970). Boag *et al.* (1989), using Taylor's power law and geostatistical techniques, showed that aggregation of infective juveniles of sheep gastrointestinal nematodes was relatively consistent over a range of distances but varied with sampling date. Grazing animals have been shown to actively avoid grazing around faeces (Cooper *et al.*, 2000), probably reducing their exposure to parasitic infection (Hutchinson *et al.*, 2002).

12.2.3. Marine and freshwater nematodes

Freshwater and, especially, marine nematodes can be very numerous, up to 1000 specimens/cm^2 (Fenchel and Riedle, 1970), and they have been used as bioindicators for monitoring pollution incidents (Gourbault, 1987). There has been little research into the spatial distribution of these nematodes.

12.2.3.1. Vertical distribution

Arlt (1973) found most nematodes in the top 2 cm of sediment in shallow areas of the Greifswalder Bodden. This may reflect the fact that nematodes are most numerous in the aerated horizon of sediments, and samples for nematodes are often taken to the depth of redox potential discontinuity (Eskin and Coull, 1987). Eskin and Coull (1987) found that this discontinuity varied between 2 cm and 10 cm for mud and sand, respectively. Other reasons suggested by Heip *et al.* (1982) to be important controlling factors included the distribution of food or the ability to adapt to stress (e.g. elevated temperatures).

12.2.3.2. Horizontal distribution

Arlt (1973) took 300 samples, each 1 cm in diameter, from within a 40 × 40 cm area of the Greifswalder Bodden and demonstrated an aggregated distribution of nematodes at this small scale (Fig. 12.2). Findley (1981) also studied the small-scale horizontal distribution of nematodes and used a contour map to illustrate the aggregated distribution of nematodes in mud-flats.

Small-scale patchiness in the distribution of intertidal nematodes may be explained by nematodes being attracted to hotspots of food (Gerlach, 1977a). Small- and large-scale sediment movements due to currents and wave action in intertidal and marine environments could also have an impact on spatial distribution of nematodes (Gerlach, 1977b). Lambshead and Hodda (1994), using Taylor's power law, investigated the hypothesis that aggregation would be greater in disturbed shallow water than in a deep-sea site but found no significant differences. Neilson *et al.* (1993) also used Taylor's power law to study the distribution of nematodes at a sewage outfall in Scotland, and found that aggregation differed significantly between species.

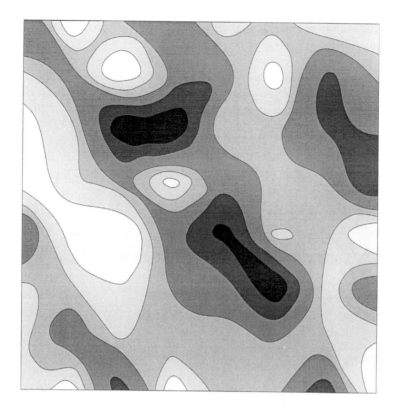

Fig. 12.2. Aggregated distribution of nematodes in a 40 cm × 40 cm area of the Greifwalder Bodden. Increased density of shading denotes increase in total nematode populations: 0–100, 100–125, 125–150, 150–200, 200–250, >250 nematodes/cm². (From Arlt, 1973.)

12.3. Nematode Migration

The role of migration in nematode distribution is not clear, but because of their small size the absolute distances they can travel are also small. Early observations on nematode migration include those of Pitcher (1967), who monitored, from a rhizotron, trichodorid nematodes following the extension of apple roots. Mojtahedi *et al.* (1991) found that *Meloidogyne chitwoodi* migrated upwards to potato roots and Griffiths and Caul (1993) calculated that bacterial-feeding nematodes were attracted to decomposing grass residues and travelled less than 2 cm to feed. Harrison and Winslow (1961) calculated the average horizontal migration of *Xiphinema diversicaudatum* to be 30 cm/year over 75 years. Laboratory experiments (Thomas, 1981) showed that for *X. diversicaudatum, Longidorus elongatus* and *Ditylenchus dipsaci* the 'rate of movement in the vertical plane was broadly similar to that horizontally', i.e. he found the maximum that *L. elongatus* travelled horizontally in 1 month was 10 cm, which was the same as for the vertical migration of *X. diversicaudatum*. However, field observations in Scotland by Taylor *et al.* (1994) showed that mapped patches of *X. diversicaudatum* were virtually unchanged 24–30 years after a previous survey. These findings contrast with those of Steinberger *et al.* (1989) who studied nematode distribution in a 'hot desert' and reported that diurnal migration of 10 cm in 3–6 h may occur due to differences in soil moisture content. These contrasting results illustrate the need to integrate short- and long-term studies. The importance of particle size and moisture in influencing the migration of *Caenorhabditis elegans* were demonstrated by Young *et al.* (1998).

The distances that entomopathogenic nematodes can disperse in different soil types were reviewed by Kaya (1990), who reported distances of 4–90 cm. Many species probably do not move much, with the exception of *Steinernema glaseri* (Schroeder and Beaver, 1987), which moves downwards while other species tend to migrate upwards (Georgis and Poinar, 1983a,b,c). The range in distances that juveniles migrate may be due to some individuals behaving as 'dispersers' while others wait passively for a host (Gaugler *et al.*, 1989). The rate of movement could be up to 5 cm/day (Moyle and Kaya, 1981), with the greatest distance of 90 cm being recorded after 30 days (Schroeder and Beaver, 1987). However, passive dispersal may occur on other soil invertebrates, e.g. mesostigmatid mites can act as phoretic hosts (Epsky *et al.*, 1988).

The migratory behaviour of free-living infective stages of animal-parasitic nematodes has received considerable attention. Crofton (1954a) was able to interpret the distribution of juveniles that occur on grass in contaminated pastures by the juveniles doing a 'random walk'. In 1949, he also showed that the numbers of infective stages available to be eaten by the host increased around midday (Crofton, 1949). The rate at which juveniles migrated from the faecal pats on to herbage was investigated by Rose (1962). Cattle lungworm (*Dictyocaulus viviparus*) juveniles may be assisted in their spread by attachment to the sporangia of a fungus (*Pilobolus*), which shoots the sporangia out of sporangiophores. Duncan *et al.* (1979) explained the epidemiological pattern of *D. viviparus* by suggesting that its juveniles could survive for many months in soil. However, from studies on *Trichostrongylus vitrinus* and a review of the literature, Rose and Small (1985) concluded that relatively few infective juveniles were usually found in the soil and the soil did not act as a large enough overwintering reservoir for the infective stages to cause clinical disease in lambs. Earthworms may play a role in dispersing infective juveniles of *O. ostertagi* from faeces to soil, and hence to the herbage (Grønvold, 1979)

12.4. Seasonal Fluctuations

12.4.1. Soil and plant-parasitic nematodes

Investigations into the seasonal fluctuations of soil and plant-parasitic nematodes have focused on those of economic importance. Some species have a single simple annual generation, apparently controlled by plant growth, root diffusates or temperature via a type of diapause (Sharma and Sharma, 1998). The seasonal fluctuations of *Globodera rostochiensis* and *Globodera pallida*, and factors influencing their fecundity have received much attention. The most important of these factors are considered to be soil type, inter- and intraspecific competition between nematodes and host plant (cultivar resistance and tolerance) (Jones and Kempton, 1978). However, within species of sedentary plant-parasitic nematodes, differences can be observed in diapause, e.g. *Heterodera glycines* enters diapause in North Carolina but not in tropical or subtropical regions (Koenning and Schmitt, 1985).

Most migratory plant-parasitic and free-living nematodes do not have a single annual generation and yet may exhibit strong annual cycles. Yuen (1966) found *Helicotylenchus vulgaris* numbers were lowest in winter. Winslow (1964) monitored a number of plant-parasitic nematodes at two sites in England and found numbers generally decreased between May and July, probably because of drought. Definitive distinct annual cycles are found in the smaller species of nematode with a rapid reproductive cycle and short survival time ('r' strategists). For example, Yeates (1972) found in Denmark that *D. dipsaci* numbers peaked in May while Boag and Alphey (1988) found, after fumigation in a Scottish forest nursery, that *Paratylenchus nanus* numbers rose significantly each spring and decreased each autumn (Fig. 12.3). They also found at the same site that *Rotylenchus uniformis*, *T. primitivus* and *Paratrichodorus pachydermus* acted as 'K' strategists, exhibiting little or no seasonal population cycling. Yeates (1981) examined the population dynamics of nematode genera at five sites in the North Island of New Zealand and found that populations were positively correlated with annual herbage production at the sites but were poorly correlated with the environmental factors measured. Because of the lack of knowledge of the biology of the nematodes concerned he concluded that the 'drivers' of the observed changes in the numbers of various taxa were unknown.

As part of a 25-year investigation of the nematodes in a Swedish forest, Sohlenius and Bostrom (2001) found that during the summer the proportion of fungal- and bacterial-feeding nematodes was similar but that in the wet, cold winters the proportion of bacterial feeders increased. However, within these nematodes, the rapidly multiplying bacteria-feeding species belonging to Rhabditida were common in late summer, while the more slowly multiplying bacterial feeders belonging to Adenophorea were more abundant in winter. In Switzerland, Steiner (1994) found that nematodes in moss exhibited pronounced seasonal fluctuations, numbers being greater in May, while Steinberger *et al.* (1989) studied nematodes of different trophic groups and found that numbers in a desert in Israel were greater in the wet season. A comprehensive study of the interaction between the population dynamics of *Xiphinema americanum* with its spatial distribution in a vineyard was undertaken by Ferris and McKenry (1974), who found its populations greater in the autumn and winter than in spring or summer. Merrifield and Ingham (1996) investigated the seasonal fluctuations of *Pratylenchus* and *Criconemella xenoplax* in peppermint in Oregon and found that numbers of all species were lowest during the winter. Interestingly, Wardle and Yeates (1993) found that numbers of predatory nematodes did not follow the seasonal fluctuations in fungal- and bacterial-feeding nematodes but these were related to the amount of food at the bottom of the detritus food web. However, in a national survey in Great Britain, Boag *et al.*

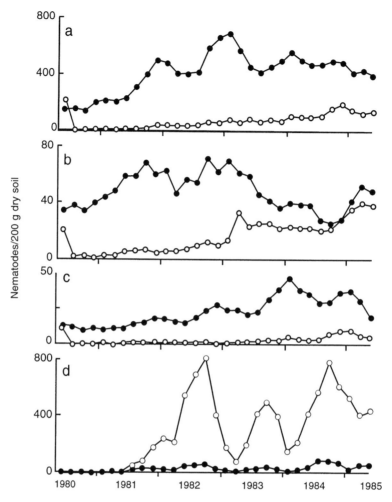

Fig. 12.3. Seasonal fluctuations and population dynamics of *Rotylenchus uniformis* (a), *Trichodorus primitivus* (b), *Paratrichodorus pachydermus* (c) and *Paratylenchus nanus* (d) in fumigated (open circles) and non-fumigated (filled circles) soil. (From Boag and Alphey, 1988.)

(1992b) found distinct seasonal trends in male, female and especially juvenile *Anatonchus tridentatus*, with peak numbers of juveniles occurring in November.

The above examples, from different geographical areas and crops, indicate that soil nematodes show considerable within-season fluctuations in their population behaviour. The reasons for these fluctuations are many and varied, and no unifying set of parameters can be used to explain them. However, plant-parasitic nematode activity tends to be synchronized with that of roots, e.g. Flegg (1968b) found that egg production in *Xiphinema*, *Longidorus profundorum* and *Longidorus macrosoma* was usually greatest in spring, when root growth occurred.

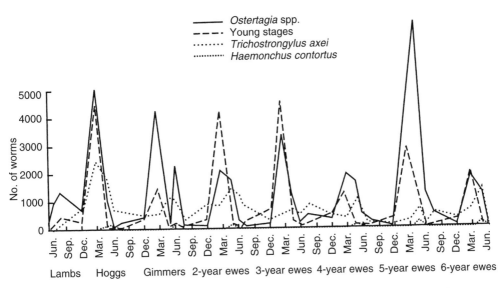

Fig. 12.4. The population dynamics of abomasal worm burdens of hill sheep over a 7-year period. (From Morgan *et al.*, 1951.)

12.4.2. Insect- and animal-parasitic nematodes

No seasonal fluctuations in the numbers of entomopathogenic nematodes in soil were found by Hominick and Briscoe (1990) or Campbell *et al.* (1998) in temperate regions of the world. However, investigations by Glazer *et al.* (1996) in Israel found that numbers tended to be greater in winter when soil moisture increased. One factor to consider when trying to explain the population behaviour of entomopathogenic nematodes is the length of time for which they can survive. To exploit their insect hosts, which only appear intermittently, they must have the ability to survive for long periods. The dehydration and anhydrobiotic potentials of this group of nematodes have received much attention, as understanding this has been a prerequisite to their commercial exploitation as biological control agents (Glazer, 2002; Grewal, 2002).

Seasonal fluctuations in populations of animal parasites in both hosts and soil are common and have long been accepted, e.g. in sheep the regular increase in worm burden in ewes (Tetley, 1949) around the time of parturition was called 'spring rise' (Morgan *et al.*, 1951; Fig. 12.4). The reason for the seasonal fluctuations can involve both the host's immune response and the effect of weather on the free-living stages of the nematodes. The impact of host immunity in determining the size of the parasite burden was demonstrated by Michel (1952a) and initially called 'self-cure'. Subsequently inhibition of nematode development was observed (Michel, 1952b) and found to occur in a range of parasitic species. The impact of host immune response on the morphology and fecundity of nematode parasites was also noted (Michel, 1967a,b).

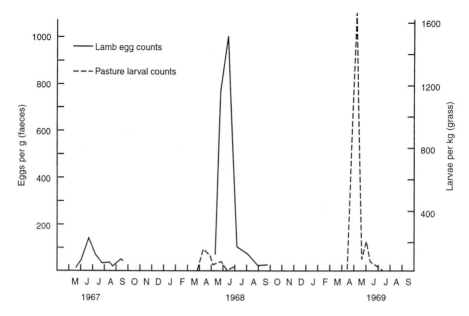

Fig. 12.5. The population build-up of pasture larval counts and lamb faecal egg counts of *Nematodirus battus* over a 3-year period. The initial pasture contamination in 1967 resulted from overwintered eggs in the soil and was undetectable from larval counts taken from the newly reseeded pasture. The pasture in 1969 was not grazed due to the high probability of lamb mortality. (From Boag and Thomas, 1975.)

The seasonal fluctuation in ewe worm burden was found by Crofton (1954b) to be associated with parturition and lactation. Crofton (1957) noted an increase in build-up of parasitic worm burdens in sheep and considered it to be a manifestation of the number of nematode generations occurring within a given season. However, Boag and Thomas (1977) disproved Crofton's hypothesis, explaining the annual epidemiological pattern in the different sheep parasite species in Britain by their differential overwintering capabilities and developmental times.

Some nematode species can be considered to have a type of diapause. *Nematodirus battus*, first reported in Britain in 1951, generally infects lambs and has one generation per year. The infective stage overwinters in the egg and needs a cool spell before hatching once the temperature rises above 10°C (Thomas and Stevens, 1960). This behaviour means that the single seasonal peak in infection occurs over a short period in spring and that contamination can rapidly build up on pasture and cause mortality in lambs within a 3-year period (Boag and Thomas, 1975; Fig. 12.5). However, *N. battus* seems to be evolving rapidly, since by the 1980s there was evidence that it could cause clinical nematodiriasis in calves and that eggs were hatching and causing disease in the autumn, without diapause (Hollands, 1984; Armour *et al.*, 1988).

12.4.3. Marine and freshwater nematodes

Bell (1979) studied nematode populations in a highly saline estuary over 3 years and found no consistent annual pattern in numbers. Fleeger (1985) reported seasonal fluctuations in total nematode numbers in an estuary on the east coast of America, peak numbers being recorded in May. Eskin and Coull (1987) found that the greatest nematode populations in marine mud occurred in March, while peak populations in sand were 2 months later. They suggested that predation by the annual migratory benthic-feeding fish were responsible for the seasonal fluctuations at the mud site and hydrodynamic activity at the sand site.

12.5. Population Cycles

12.5.1. Soil and plant-parasitic nematodes

The year-to-year population dynamics of nematodes has been studied for many habitats but the studies tend to be of short duration, i.e. less than 2–3 years. An excellent data set is given by Sharma (1971), who followed a *Tylenchorhynchus dubius* population over a 5-year period and showed that the annual increase in numbers during the summer period was entirely due to survival of juvenile stages. Boag (1981) monitored trichodorid nematode populations monthly in a forest nursery over 4 years and found gradual changes in nematode depth distribution and community structure. Boag and Geoghegan (1984) followed a population of *Longidorus elongatus* on a monthly basis over 4 years under 49 cropping regimes, and found that *L. elongatus* numbers were only maintained under the grass crop while under all other regimes the population declined slowly due to a reduction in fecundity. Yeates (1982, 1984) described the variation of pasture nematode populations over 3 years and found that populations of many genera differed significantly year to year. He did not find year-to-year differences in *Paratylenchus*. In contrast, Boag and Alphey (1988) in a 6-year study in a forest nursery found that after fumigation *P. nanus* numbers increased significantly for 2 years before a consistent annual cycle was reached. In the same long-term study it was evident that *R. uniformis*, *T. primitivus* and *P. pachydermus* multiplication rates were very low and that these nematodes were 'K' strategists. This study further showed both inter- and intraspecific competition between plant-parasitic nematodes to be important in controlling populations (Fig. 12.3). Jones and Parrott (1969) monitored *G. rostochiensis* populations in small plots growing potatoes annually over a 6-year period and found that numbers declined but oscillated on a 2-year cycle. Wheeler *et al.* (2000) studied *Meloidogyne incognita* at three sites over a 3-year period and found differences among both sites and years. They concluded that, to detect changes in populations, sampling frequency would need to be decided on a field-to-field basis. Johnson *et al.* (2000) followed nematode populations over 6 years in a wheat, cotton and peanut cropping system in the south-eastern USA. They found that no regular seasonal pattern occurred in nematode numbers in untreated control plots, although there was evidence that wheat was a good host for *Helicotylenchus dihystera*. Yeates *et al.* (1999) investigated the impact of different management regimes on soil nematode communities over 7 years. They found that treatment differences could take 3 years to manifest themselves, and therefore, to assess the effect of different crops, sampling should continue until an equilibrium was reached.

12.5.2. Insect- and animal-parasitic nematodes

Long-term information on the population dynamics of nematodes that parasitize insects is not available. However, the persistence of entomopathogenic nematodes was found by Hominick and Reid (1990) to be temperature-dependent.

In domestic animals, the persistence and differential overwintering ability of infective stages can have a significant impact on the behaviour of populations of these nematodes, e.g. Gibson and Everett (1967) and Boag and Thomas (1970) showed that the infective juveniles of *Ostertagia circumcincta* were able to overwinter and persist better than those of *Trichostrongylus colubriformis*. However, because of their requirement for diapause before hatch, *N. battus* and *Nematodirus filicollis* overwinter in the egg, hatch in the spring and infect lambs when they start to graze. If lambs are allowed to graze the same land every spring, this may lead to lamb mortality within 3 years (Boag and Thomas, 1975; Fig. 12.5). The build-up of parasites in free-ranging Soay sheep on the island of St Kilda has been implicated in the population crashes of these animals (Gulland, 1992; Gulland and Fox, 1992). Further evidence that parasitic nematodes are implicated in controlling wild animal populations can be seen in the studies on *Trichostrongylus tenuis,* which infects grouse (*Lagopus lagopus scoticus*) in northern Britain. Hudson *et al.* (1998) demonstrated convincingly that the 'boom and bust' population cycles of grouse could be prevented by use of anthelmintics to remove the parasite burden. In contrast, a 10-year study of the population dynamics of the parasites recovered from 1341 rabbits in eastern Scotland found significant differences in the seasonal patterns exhibited by *Trichostrongylus retortaeformis* (greatest intensity in September–October) and *Graphidium strigosum* (greatest intensity in March) but no evidence of any cyclical trends covering a number of years similar to those found in grouse (Boag, 1988; Fig. 12.6). Presumably factors other than nematodes are controlling rabbit populations.

12.5.3. Marine and freshwater nematodes

Eskin and Coull (1987) studied the abundance of benthic nematodes over a 3-year period and found significant differences among years but failed to correlate differences with environmental conditions. Gourbault (1987) followed the nematode fauna between 1978 and 1984 at the site of the Amoco Cadiz oil spillage. He found that pollution affected the nematodes more in shallow than in deeper areas, but that by 1984 all nematode communities were similar to those in 1978.

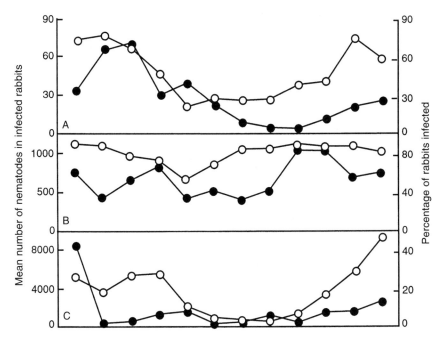

Fig. 12.6. Seasonal fluctuations in nematode populations of parasites collected from wild rabbits (mean of 10 years' data). A, *Graphidium strigosum*; B, *Trichostrongylus retortaeformis*, and C, *Passalurus ambiguous*. Open circles, prevalence; filled circles, intensity of infection. (From Boag, 1989.)

12.6. Nematodes as Bioindicators

Because of their individual affinities for particular ecological niches, nematode species can act as indicators of their environment and can be highly specific. For example, the presence of *G. rostochiensis* in fields in temperate areas of the world would indicate that potatoes had been grown in that soil. Similarly, the presence of *Xenocriconemella macrodora* is associated with oak (*Quercus*) (Bello *et al.*, 1986). Certain nematodes also share a tendency to prefer similar habitats and are often be found together (Yeates, 1984). This has been used to indicate the possible presence of virus-vector nematodes, e.g. *Longidorus caespiticola* (a non-virus-vector species) is significantly associated with *X. diversicaudatum* (a vector of strawberry latent ringspot virus) (Boag and Topham, 1985).

Not only are most animal-parasitic nematodes indicative of the host they parasitize but, due to the interaction with the host, it is sometimes possible to estimate the host age and sex. For example *N. battus* is a parasite of lambs in temperate parts of the world, but because the nematode induces a strong immune reaction it is rarely found in adult sheep (Boag and Thomas, 1975).

Although experimentation with wild fish is difficult, parasites are widely used as indicators of host biology. Nematodes, along with other parasites, have been used as 'tags' to indicate seasonal migration, differentiate populations and investigate marine food-chains (MacKenzie, 1987).

While individual nematodes can act as indicators of the environments in which they live, there is an increasing interest in the use of nematode assemblages to monitor environmental change, as nematodes use a more comprehensive set of niches than do many other groups. Universal ecological indices have been applied to nematode populations to suggest where perturbations have occurred (Warwick, 1988). Bongers (1990) proposed arbitrarily grouping nematode species into five different ecological categories, which could be considered as ranging from 'r' (colonizers) (group 1) to 'K' strategists (persisters) (group 5). Although there have been suggested alterations to this 'maturity index', e.g. the inclusion of plant-parasitic nematodes (Yeates, 1994), the index has detected changes in soil (Ettema and Bongers, 1993) and marine environments (Neilson *et al.*, 1996). However, shortcomings of the original technique led to suggested changes. Bongers and Bongers (1998) proposed integrating life strategies with trophic group classification to obtain a better understanding of nematode biodiversity and soil function. Ferris *et al.* (2001) further 'enhanced the resolution of faunal analyses by providing a weighted system for the indicator importance of the presence and abundance of each functional guild in relation to enrichment and structure of the food web'. This approach has given good resolution in local plot studies (e.g. Hohberg, 2003) but differences in the taxonomic make-up of nematode assemblages between soils and regions make wide comparisons difficult. The use of nematode community structure as a bioindicator in environmental monitoring was reviewed by Bongers and Ferris (1999). Wall *et al.* (2002) found the maturity index a poor indicator of changes along a sand-dune succession, which suggested that nematodes may not be a consistently good indicator of the status of different soils. Further, Yeates *et al.* (2002) provided data that cast doubt on the allocation of nematodes on the 1–5 scale.

12.7. Modelling Populations

Early, simple plant-parasitic nematode models (in which the target nematode had only one generation per year) related populations at the beginning to those at the end of the season, or related yield losses to cyst nematode densities (Seinhorst, 1965, 1967). As knowledge about nematode life cycles grew, more models were developed, covering many aspects of plant and soil nematode biology. These were reviewed by Seinhorst (1970), Jones and Kempton (1978), Duncan and McSorley (1987) and Ferris and Wilson (1987). There are few population dynamic models able to project long-term trends in nematode numbers. Most models have been modified to incorporate local practices, including the effect of different rotations, soil types, planting of resistant varieties, intraspecific competition and judicial use of nematocides to try and develop optimal management regimes (Kinloch, 1986; Been *et al.*, 1995; Elliot *et al.*, 2002).

For domestic animals, Crofton (1971) was one of the first to attempt to model the relationship between hosts and their parasites. In 1979 Gettinby *et al.* published a predictive model for bovine ostertagiasis, which was followed by models for single and mixed infections of sheep (Callinan *et al.*, 1982; Paton *et al.*, 1984; Paton and Boag, 1987; Grenfell, 1988). However, the basis of much of the present modelling of host-parasite relationships was laid down in two 1978 papers. Anderson and May (1978) and May and Anderson (1978) set out theoretical models taking into consideration many of the factors considered responsible for the seasonal fluctuations and population dynamics of parasites and their hosts. These models were subsequently extended to investigate the aggregated distribution patterns of parasites within their hosts and parasite impact on host mortality (Anderson and Gordon, 1982). The application of models to the study of the population dynamics of human parasitic diseases was reviewed in detail by Anderson and May (1985).

Mouse (*Mus domesticus*) outbreaks occur in Australia on a 7–9-year cycle. Use of the nematode *Capillaria hepatica* to control these outbreaks was suggested and was modelled by McCallum and Singleton (1989). Their model suggested that, to have any beneficial economic impact, *C. hepatica* would (due to transmission requiring host death) have to be applied as early in the cycle as possible to achieve reductions in plague intensity. The population dynamics of *T. tenuis* and its impact on grouse numbers were also modelled by Dobson and Hudson (1992). They subsequently showed that this nematode was the 'driver' behind the cycle in grouse numbers (Hudson *et al.*, 1998).

12.8. The Future

The need to understand the population dynamics of nematodes, both parasitic and free-living, is arguably greater now than at any time in the past. Parasitic nematodes, both plant and animal, have been controlled in intensive agricultural systems by the use of nematocides and anthelmintics, respectively. However, the withdrawal of some nematocides from the market in many parts of the world and the development of parasite resistance to anthelmintics have meant that new management methods must be developed. A better understanding of the biology and population dynamics of parasitic nematodes may lead to the development of more environmentally benign control measures. Knowledge of the factors influencing nematode population behaviour is also a prerequisite to any generalized mathematical modelling.

Nematodes may also have a role to play as indicators of environmental 'wellbeing', e.g. as a biological tool to monitor climate change and pollution in soils and the marine environment. However, other phyla have similar claims. In the future, the use of indices using nematodes alone will not advance the future of nematology as much as if nematodes were considered a part of a wider descriptor of soil quality, e.g. in fish biology information from cestodes and nematodes has been combined (MacKenzie, 1987). In the future, the usefulness of indices of soil 'well-being' will be enhanced if they include data on protozoa, earthworms, enchytraeids and collembola as well as nematodes. The development of molecular techniques (Floyd *et al.*, 2002) to aid the identification of nematodes and other soil organisms could make such a comprehensive approach feasible.

Acknowledgements

We thank Ian Pitkethly and Shona Thomson for producing the figures. The Scottish Crop Research Institute is funded by the Scottish Executive Environment and Rural Affairs Department.

References

Abd-Elgawad, M.M. and Hasabo, S.A. (1995) Spatial distribution of the phytonematode community in Egyptian berseem clover (*Trifolium alexandra* L.) fields. *Fundamental and Applied Nematology* 18, 329–334.

Alexander, A.R.A., Young, I.M., Sleeman, B.D., Griffiths, B.S. and Robertson, W.M. (1997) Nematode movement along a chemical gradient in a structurally heterogeneous environment. I. Experiment. *Fundamental and Applied Nematology* 20, 157–163.

Al Saqur, I., Bairden, K., Armour, J. and Gettinby, G. (1982) Population study of bovine *Ostertagia* spp. infective larvae on herbage and in soil. *Research in Veterinary Science* 32, 332–337.

Anderson, R.M. and Gordon, D.M. (1982) Processes influencing the distribution of parasite numbers within host populations with special emphasis on parasite-induced host mortalities. *Parasitology* 85, 373–398.

Anderson, R.M. and May, R.M. (1978) Regulation and stability of host-parasite population interactions I. Regulatory processes. *Journal of Animal Ecology* 37, 219–247.

Anderson, R.M. and May, R.M. (1985) Helminth infections in humans: mathematical models, population dynamics, and control. *Advances in Parasitology* 24, 1–101.

Arlt, G. (1973) Vertical and horizontal microdistribution of the meiofauna in the Greifswalder Bodden. *Oikos* 15, 105–110.

Armour, J., Bairden, K., Dalgleish, R., Ibarra-Silva, A.M. and Salman, S.K. (1988) Clinical nematodiriasis in calves due to *Nematodirus battus* infection. *Veterinary Record* 123, 230–231.

Been, T.H. and Schomaker, C.H. (1996) A new sampling method for the detection of low population densities of potato cyst nematodes (*Globodera pallida* and *G. rostochiensis*). *Crop Protection* 15, 375–382.

Been, T.H., Schomaker, C.H. and Seinhorst, J.W. (1995) An advisory system for the management of potato cyst nematodes (*Globodera* spp.). In: Haverkoert, A.J. and MacKerron, D.K.L. (eds) *Potato Ecology*. Kluwer Academic Publishers, Dordrecht, The Netherlands, pp. 305–321.

Bell, S.S. (1979) Short- and long-term variation in a high marsh meiofauna community. *Estuarine and Coastal Marine Science* 9, 331–350.

Bello, B., Boag, B., Topham, P.B. and Ibanez, J.J. (1986) Geographical distribution of *Xenocriconemella macrodora* (Nematoda: Criconematidae). *Nematologia Mediterranea* 14, 223–229.

Boag, B. (1981) Observations on the population dynamics and vertical distribution of trichodorid nematodes in a Scottish forest nursery. *Annals of Applied Biology* 98, 463–469.

Boag, B. (1988) Population dynamics of parasites of the wild rabbit *Oryctolagus cuniculus* (L.). In: Putman, R.J. (ed.) *Mammals as Pests.* Chapman and Hall, London, pp. 185–195.

Boag, B. and Alphey, T.J.W. (1988) Influence of intra-specific competition on the population dynamics of migratory plant-parasitic nematodes with r and K survival strategies. *Revue de Nématologie* 11, 321–326.

Boag, B. and Geoghegan, I.E. (1984) An evaluation of agricultural and horticultural crops as hosts for the plant-parasitic nematode *Longidorus elongatus* (Nematoda, Dorylaimida). *Crop Research* 24, 85–95.

Boag, B. and Thomas, R.J. (1970) The development and survival of free-living stages of *Trichostrongylus colubriformis* and *Ostertagia circumcincta* on pasture. *Research in Veterinary Science* 11, 380–381.

Boag, B. and Thomas, R.J. (1975) Epidemiological studies on *Nematodirus* species in sheep. *Research in Veterinary Science* 19, 263–268.

Boag, B. and Thomas, R.J. (1977) Epidemiological studies on gastro-intestinal nematode parasites of sheep: the seasonal number of generations and succession of species. *Research in Veterinary Science* 22, 62–67.

Boag, B. and Topham, P.M. (1984) Aggregation of plant-parasitic nematodes and Taylor's Power Law. *Nematologica* 30, 348–357.

Boag, B. and Topham, P.M. (1985) The use of associations of nematode species to aid the detection of small numbers of virus-vector nematodes. *Plant Pathology* 34, 20–24.

Boag, B., Raschke, I.E. and Brown, D.J.F. (1977) Obsevations on the life cycle and pathogenicity of *Paralongidorus maximus* in a forest nursery in Scotland. *Annals of Applied Biology* 85, 389–397.

Boag, B., Brown, D.J.F. and Topham, P.B. (1987) Vertical and horizontal distribution of virus vector nematodes and implications for sampling procedures. *Nematologica* 33, 83–96.

Boag, B., Topham, P.B. and Webster, R. (1989) Spatial distribution of infective larvae of the gastro-intestinal nematode parasites of sheep. *International Journal for Parasitology* 19, 681–685.

Boag, B., Neilson, R. and Gordon, S.G. (1992a) Distribution and prevalence of the entomopathogenic nematode *Steinernema feltiae* in Scotland. *Annals of Applied Biology* 121, 355–360.

Boag, B., Small, R.W., Neilson, R., Gauld, J.H. and Robertson, L. (1992b) The Mononchida of Great Britain: Observations on the distribution and ecology of *Anatonchus tridentatus, Truxonchus dolichurus* and *Miconchoides studerei* (Nematoda). *Nematologica* 38, 502–513.

Bongers, T. (1990) The maturity index: an ecological measure of environmental disturbance based on nematode species composition. *Oecologia* 83, 14–19.

Bongers, T. and Bongers, M. (1998) Functional diversity of nematodes. *Applied Soil Ecology* 10, 239–251.

Bongers, T. and Ferris, H. (1999) Nematode community structure as a bioindicator in environmental monitoring. *TREE* 14, 224–228.

Borgsteede, F.H.M. and Boogaard, G. (1983) Overlevingskansen van Trichostronglideneieren en –larven in de bodem. *Tijdschrift voor Diergeneeskunde* 108, 439–442.

Brodie, B.B. (1976) Vertical distribution of three nematode species in relation to certain soil properties. *Journal of Nematology* 8, 243–247.

Cabanillas, H.E. and Raulston, J.R. (1994) Evaluation of the spatial distribution of *Steinernema riobravis* in corn plots. *Journal of Nematology* 26, 25–31.

Callinan, A.P.L. and Westcott, J.M. (1986) Vertical distribution of trichostrongylid larvae on herbage and in soil. *International Journal for Parasitology* 16, 241–244.

Callinan, A.P.L., Morley, F.H.W., Arundel, J.H. and White, D.H. (1982) A model of the life cycle of sheep nematodes and the epidemiology of nematodiriasis in sheep. *Agricultural Systems* 9, 199–225.

Campbell, J.F., Lewis, E., Yoder, F. and Gaugler, R. (1996) Entomopathogenic nematode (Heterorhabditidae and Steinernematidae) spatial distribution in turfgrass. *Parasitology* 113, 473–482.

Campbell, J.F., Orza, G., Yoder, F., Lewis, E. and Gaugler, R. (1998) Spatial and temporal distribution of endemic and released entomopathogenic nematode populations in turf-grass. *Entomologia Experimementalis et Applicata* 86, 1–11.

Caubel, G., Mungniery, D. and Rivoal, R. (1972) Distribution de l'anguillule des bulbes et des tiges *Ditylenchus dipsaci* (Kuhn) Filipje, dès le sol, à l'interieur d'un foyer d'attaque sur trèfle. *Annales de Zoologie et Ecologie Animales* 4, 385–393.

Cooper, J., Gordon, I.J. and Pike, A.W. (2000) Strategies for the avoidance of faeces by grazing sheep. *Applied Animal Behaviour* 69, 15–33.

Cotton, J. (1979) Effect of annual and perennial cropping regimes on the population density and vertical distribution of *Xiphinema diversicaudatum* and on soil porosity. *Annals of Applied Biology* 86, 397–404.

Crofton, H.D. (1949) The ecology of immature phases of trichostrongyle nematodes. III. Larval populations on hill pastures. *Parasitology* 39, 26–38.

Crofton, H.D. (1954a) Vertical migration of infective larvae of strongylid nematodes. *Journal of Helminthology* 28, 35–52.

Crofton, H.D. (1954b) Nematode parasite populations in sheep on lowland farms. I. Worm egg counts in ewes. *Parasitology* 44, 465–477.

Crofton, H.D. (1957) Nematode parasite populations in sheep on lowland farms. III. The seasonal incidence of species. *Parasitology* 47, 304–318.

Crofton, H.D. (1971) A model of host–parasite relationships. *Parasitology* 63, 343–364.

Dobson, A.P. and Hudson, P.J. (1992) Regulation and stability of a free-living host-parasite system: *Trichostrongylus tenuis* in red grouse. II Population models. *Journal of Animal Ecology* 61, 487–498

Donald, A.D. and Leslie, R.T. (1969) Population studies on the infective stages of some nematode parasites of sheep. II. The distribution of faecal deposits on fields grazed by sheep. *Parasitology* 59, 141–157.

Duncan, J.L., Armour, J., Bairden, K., Urquhart, G.M. and Jørgensen, R.J. (1979) Studies on the epidemiology of bovine parasitic bronchitis. *Veterinary Record* 31, 274–278.

Duncan, L.W. and McSorley, R. (1987) Modeling nematode populations. In: Veech, J.A. and Dickson, D.W. (eds) *Vistas on Nematology*. Society of Nematologists, Hyattsville, Maryland, pp. 377–389.

Elliot, M.J., McNicol, J.W., Phillips, M.S. and Trudgill, D.R. (2002) Computer-based program for modelling integrated control of potato cyst nematodes. In: Heilbron, T.D. (ed.) *Proceedings: Crop Protection in Northern Britain 2002*, Page Bros., Norwich, pp. 257–262.

Epsky, N.D., Walters, D.E. and Akhurst, R.J. (1988) Potential role of nematophagous microarthropods as biotic mortality factors of entomogenous nematodes (Rhabditida: Steinernematidae: Heterorhabditidae). *Journal of Economic Entomology* 81, 821–825.

Eskin, R.A. and Coull, B.C. (1987) Seasonal and three-year variability of meiobenthic nematode populations at two estuarine sites. *Marine Ecology – Progress Series* 41, 295–303.

Ettema, C.H. and Bongers, T. (1993) Characterisation of nematode colonization and succession in disturbed soil using the maturity index. *Biology and Fertility of Soils* 16, 79–85.

Ettema, C.H. and Yeates, G.W. (2003) Nested spatial biodiversity patterns of nematode genera in a New Zealand forest and pasture soil. *Soil Biology and Biochemistry* 35, 339–342.

Feil, H., Westerdahl, B.B., Smith, R.J. and Verdegaal, P. (1997) Effects of seasonal and site factors on *Xiphinema index* populations in two California vineyards. *Journal of Nematology* 29, 491–500.

Fenchel, T.M. and Riedle, R.J. (1970) The sulfide system: a new biotic community underneath the oxidized layer of marine sand bottoms. *Marine Biology* 7, 255–268.

Ferris, H. and McKenry, M.V. (1974) Seasonal fluctuations in the spatial distributions of nematode populations in a California vineyard. *Journal of Nematology* 6, 203–210.

Ferris, H. and Wilson, L.T. (1987) Concepts and principles of population dynamics. In: Veech, J.A. and Dickson, D.A. (eds) *Vistas on Nematology*. Society of Nematologists, Hyattsville, Maryland, pp. 372–376.

Ferris, H., Bongers, T. and de Goede, R.G.M. (2001) A framework for soil food web diagnostics: extension of the nematode faunal analysis concept. *Applied Soil Ecology* 18, 13–29.

Findley, S.E.G. (1981) Small-scale spatial distribution of meiofauna on a mud- and sandflat. *Estuarine, Coastal and Shelf Science* 12, 471–484.

Fleeger, J.W. (1985) Meiofaunal densities and copepod species composition in a Louisiana, USA, estuary. *Transactions of the American Microscopical Society* 104, 321–332.

Flegg, J.J.M. (1968a) The occurrence and depth distribution of *Xiphinema* and *Longidorus* species in southeastern England. *Nematologica* 14, 189–196.

Flegg, J.J.M. (1968b) Life cycle studies of some *Xiphinema* and *Longidorus* species in southeastern England. *Nematologica* 14, 197–210.

Floyd, R., Abebe, E., Papert, A. and Blaxter, M. (2002) Molecular barcodes for soil nematode identification. *Molecular Ecology* 11, 839–850.

Forge, A.T., De Young, R. and Vrain, T.C. (1998) Temporal changes in the vertical distribution of *Pratylenchus penetrans* under raspberry. *Journal of Nematology* 30, 179–183.

Gaugler, R., McGuire, T. and Campbell, J. (1989) Genetic variability among strains of the entomopathogenic nematode *Steinernema feltiae*. *Journal of Nematology* 21, 247–253.

Georgis, R. and Poinar, G.O. Jr (1983a) Effects of soil texture on the distribution and infectivity of *Neoaplectana carpocapsae* (Nematoda: Steinernematidae). *Journal of Nematology* 15, 308–311.

Georgis, R. and Poinar, G.O., Jr (1983b) Effects of soil texture on the distribution and infectivity of *Neoaplectana glaseri* (Nematoda: Steinernematidae). *Journal of Nematology* 15, 329–332.

Georgis, R. and Poinar, G.O., Jr (1983c) Vertical migration of *Heterorhabditis bacteriophora* and *H. heliothidis* (Nematoda: Heterorhabditidae) in sandy loam soil. *Journal of Nematology* 15, 652–654.

Gerlach, S.A. (1977a) Attraction to decaying organisms as a possible cause for the patchy distribution of nematodes in a Bermuda Beach. *Ophelia* 16, 151–165.

Gerlach, S.A. (1977b) Means of meiofaunal dispersal. *Mikrofauna Meeresboden* 61, 89–103.

Gettinby, G., Bairden, K., Armour, J. and Benitez-Usher, C. (1979) A predictive model for bovine ostertagiasis. *Veterinary Record* 105, 57–59.

Gibson, T.E. and Everett, G. (1967) The ecology of the free-living stages of *Trichostrongyulus colubriformis*. *Parasitology* 57, 533–547.

Glazer, I. (2002) Survival biology. In: Gaugler, R. (ed.) *Entomopathogenic Nematology*. CAB International, Wallingford, UK, pp. 169–187.

Glazer, I., Kozodoi, E., Salame, L. and Nestel, D. (1996) Spatial and temporal occurrence of natural populations of spp. (Nematoda: Rhabditida) in a semiarid region. *Biological Control* 6, 130–136.

Goodell, P.B. and Ferris, H. (1980) Graphic techniques for illustrating distribution data. *Journal of Nematology* 12, 151–152.

Goodell, P.B. and Ferris, H. (1981) Sampling optimisation for five plant parasitic nematodes. *Journal of Nematology* 13, 304–313.

Goodman, A. and Payne, E.M.F. (1979) *Longman Dictionary of Scientific Usage*. Longman, Harlow, UK, 684 pp.

Gourbault, N. (1987) Long-term monitoring of marine nematode assemblages in the Morlaix Estuary (France) following the Amoco Cadiz oil spill. *Estuary and Coastal Shelf Science* 24, 657–670.

Gregorich, E.G., Turchenek, L.W., Carter, M.R. and Angers, D.A. (2001) *Soil and Environmental Science*. CRC Press, Boca Raton, Florida, 577 pp.

Grenfell, B.T. (1988) Gastrointestinal nematode parasites and the stability and productivity on intensive ruminant grazing systems. *Philosophical Transactions of the Royal Society, Series B* 321, 541–563.

Grewal, P.S. (2002) Formulation and application technology. In: Gaugler, R. (ed.) *Entomopathogenic Nematology*. CAB International, Wallingford, UK, pp. 265–287.

Griffiths, B.S. and Caul, S. (1993) Migration of bacterial-feeding nematodes, but not protozoa, to decomposing grass residues. *Biology and Fertility of Soils* 15, 201–207.

Griffiths, G.D., Asay, K.H. and Horton, W.H. (1996) Factors affecting population trends of plant-parasitic nematodes on rangeland grasses. *Journal of Nematology* 28, 107–114.

Grønvold, J. (1979) On the possible role of earthworms in the transmission of *Ostertagia ostertagi* third stage larvae from feces to soil. *Journal of Parasitology* 65, 831–832.

Gruner, H., Mauleon, H. and Sauve, C. (1982) Migrations of trichostrongyle infective larvae experiments with ovine parasites in soil. *Annales de Recherches Vétérinaires* 13, 51–59.

Gulland, F.M.D. (1992) The role of nematode parasites in Soay sheep (*Ovis aries* L.) mortality during a population crash. *Parasitology* 105, 493–503.

Gulland, F.M.D. and Fox, M. (1992) Epidemiology of nematode infections of Soay sheep (*Ovis aries* L.) on St Kilda. *Parasitology* 105, 481–492.

Harrison, B.D. and Winslow, R.D. (1961) Laboratory and field studies on the relation of arabis mosaic virus to its nematode vector *Xiphinema diversicaudatum* (Micoletzky). *Annals of Applied Biology* 49, 621–633.

Heip, C., Vincx, M., Smol, M. and Vranken, G. (1982) The systematics and ecology of free-living marine nematodes. *Helminthological Abstracts Series B* 51, 1–31.

Hohberg, K. (2003) Soil nematode fauna of afforested mine sites: general distribution, trophic structure and faunal guilds. *Applied Soil Ecology* 22, 113–126.

Hollands, R.D. (1984) Autumn nematodiriasis. *Veterinary Record* 115, 526–527.

Hominick, W.M. and Biscoe, B.R. (1990) Occurrence of entomopathogenic nematodes (Rhabditida, Steinernematidae and Heterorhabditidae) in British soils. *Parasitology* 100, 295–302.

Hominick, W.M. and Reid, A.P. (1990) Perspectives on entomopathogenic nematology. In: Gaugler, R. and Kaya, H. (eds) *Entomopathogenic Nematodes in Biological Control*. CRC Press, Boca Raton, pp. 327–345.

Hudson, P.J., Dobson, A.P. and Newborn, D. (1998) Prevention of population cycles by parasite removal. *Science* 282, 2256–2258.

Hutchinson, M.R., Milner, J.M., Gordon, I.J., Kyriazakis, I. and Jackson, F. (2002) Grazing decisions of Soay sheep, *Ovis aries*, on St Kilda: a consequence of parasite distribution? *Oikos* 96, 235–244.

Johnson, A.W., Dowler, C.C. and Dandoo, Z.A. (2000) Population dynamics of nematodes and crop yields in rotations of cotton, peanut, and wheat under minimum tillage. *Journal of Nematology* 32, 52–61.

Jones, F.G.W. and Kempton, R.A. (1978) Population dynamics, population models and integrated control. In: Southey, J.F. (ed.) *Plant Nematology*. Her Majesty's Stationery Office, London, pp. 333–361.

Jones, F.G.W. and Parrott, D.M. (1969) Population fluctuations of *Heterodera rostochiensis* Woll. when susceptible potato varieties are grown continuously. *Annals of Applied Biology* 63, 175–181.

Kaya, H.K. (1990) Soil ecology. In: Gaugler, R. and Kaya, H. (eds) *Entomopathogenic Nematodes in Biological Control*. CRC Press, Boca Raton, Florida, pp. 93–105.

Kendall, S.M.G. and Buckland, W.R. (1971) *A Dictionary of Statistical Terms*. Longman, London, 166 pp.

Kinloch, R.A. (1986) Soyabean and maize cropping models for the management of *Meloidogyne incognita* in the coastal plain. *Journal of Nematology* 18, 451–458.

Koenning, S.R. and Schmitt, D.P. (1985) Hatching and diapause of a field population of *Heterodera glycines*. *Journal of Nematology* 17, 502.

Kuzmin, L.L. (1976) Free living nematodes in the tundra of western Taimyr. *Oikos* 27, 501–505.

Lambshead, P.J.D. and Hodda, M. (1994) The impact of disturbance on measurements of variability in marine nematode populations. *Vie Milieu* 44, 21–27.

Lawrence, E. (1995) *Henderson's Dictionary of Biological Terms*. Addison Wesley Longman, Harlow, UK, 693 pp.

McCallum, H.I. and Singleton, G.R. (1989) Models to assess the potential of *Capillaria hepatica* to control population outbreaks of house mice. *Parasitology* 98, 425–437.

MacGuidwin, A.E. and Stanger, B.A. (1991) Changes in vertical distribution of *Pratylenchus scribneri* under potato and corn. *Journal of Nematology* 23, 73–81.

MacKenzie, K. (1987) Parasites as indicators of host populations. *International Journal for Parasitology* 17, 345–352.

McSorley, R. (1982) Simulated sampling strategies for nematode distribution according to the negative binomial model. *Journal of Nematology* 14, 517–522.

McSorley, R. and Dickson, D.W. (1991) Determining consistency of spatial dispersion of nematodes in small plots. *Journal of Nematology* 23, 65–72.

Marshall, B., Boag, B., McNicol, J.W. and Neilson, R. (1998) A comparison of the spatial distributions of plant-parasitic nematode species at three different scales. *Nematologica* 44, 303–320.

May, R.M. and Anderson, R.M. (1978) Regulation and stability of host-parasite population interactions II. Destabilizing processes. *Journal of Animal Ecology* 47, 249–267.

Merrifield, K.J. and Ingham, R.E. (1996) Population dynamics of *Pratylenchus penetrans*, *Pratylenchus* sp., and *Criconemella xenoplax* on western Oregon peppermint. *Journal of Nematology* 28, 557–564.

Michel, J.F. (1952a) Self cure in infections of *Trichostrongylus retortaeformis* and its causation. *Nature* 169, 881.

Michel, J.F. (1952b) Inhibition of development of *Trichostrongylus retortaeformis*. *Nature* 169, 933.

Michel, J.F. (1967a) Morphological changes in a parasitic nematode due to acquired resistance of the host. *Nature* 215, 520–521.

Michel, J.F. (1967b) Regulation of egg output of populations of *Ostertagia ostertagi*. *Nature* 215, 1001–1002.

Mojtahedi, H., Ingham, R.E., Santo, G.S., Pinkerton, J.N., Reed, G.L. and Wilson, J.H. (1991) Seasonal migration of *Meloidogyne chitwoodi* and its role in potato production. *Journal of Nematology* 23, 162–169.

Morgan, D.O., Parnell, I.W. and Rayski, C. (1951) The seasonal variation in the worm burden of sheep. *Journal of Helminthology* 25, 177–212.

Moyle, P.L. and Kaya, H.K. (1981) Dispersal and infectivity of the entomogenous nematode *Neoaplectana carpocapsae* Weiser (Rhabditida: Steinernematidae). *Journal of Nematology* 15, 308–311.

Neilson, R., Boag, B. and Hackett, C.A. (1993) Observations on the use of Taylor's Power Law to describe the horizontal spatial distribution of marine nematodes in an intertidal estuarine environment. *Russian Journal of Nematology* 1, 55–64.

Neilson, R., Boag, B. and Palmer, L.F. (1996) The effect of environment on marine nematode populations as indicated by the Maturity Index. *Nematologica* 42, 233–242.

Noe, J.P. and Campbell, C.L. (1985) Spatial pattern analysis of plant-parasitic nematodes. *Journal of Nematology* 17, 86–93.

Paton, G. and Boag, B. (1987) A model for predicting parasitic gastroenteritis in lambs subject to mixed nematode infections. *Research in Veterinary Science* 43, 67–71.

Paton, G., Thomas, R.J. and Waller, P.J. (1984) A prediction model for parasitic gastroenteritis in lambs. *International Journal for Parasitology* 14, 439–445.

Pitcher, R.S. (1967) The host-parasite relations and ecology of *Trichodorus viruliferus* on apple roots, as observed from an underground laboratory. *Nematologica* 13, 547–557.

Rawsthorne, D. and Brodie, B.B. (1986) Root growth of susceptible and resistant potato cultivars and population dynamics of *Globodera rostochiensis* in the field. *Journal of Nematology* 18, 501–504.

Robertson, G.P. and Freckman, D.W. (1995) The spatial distribution of nematode trophic groups across a cultivated ecosystem. *Ecology* 76, 1425–1432.

Rose, J.H. (1962) Further observations on the free-living stages of *Ostertagia ostertagi* in cattle. *Journal of Comparative Pathology* 72, 11–18.

Rose, J.H. and Small, A.J. (1985) The distribution of the infective larvae of sheep gastrointestinal nematodes in soil and on herbage and the vertical migration of *Trichostrongylus vitrinus* larvae through the soil. *Journal of Helminthology* 59, 127–135.

Rossner, J. (1970) Ein Beitrag zur Vertikalbesiedlung des Bodens durch wandernde Wurzelnematoden. *Nematologica* 16, 556–562.

Rossner, J. (1972) Vertikalbesteilung wandernder Worzelnematoden im Boden in Abhangigkeit von Wassergehalt und Durchwurzelung. *Nematologica* 18, 360–372.

Routledge, R.D. and Swartz, T.B. (1992) Rejoinder to 'Fitting Taylor's power law' by Perry and Woiwod. *Oikos* 65, 543–544.

Ruess, L. (1995) Studies on the nematode fauna of an acid forest soil: spatial distribution and extraction. *Nematologica* 41, 229–239.

Schroeder, W.J. and Beaver, J.B. (1987) Movement of the entomogenous nematodes of the families Heterorhabditidae and Steinernematidae in soil. *Journal of Nematology* 19, 257.

Seinhorst, J.W. (1965) The relationship between nematode density and damage to plants. *Nematologica* 11, 137–154.

Seinhorst, J.W. (1967) The relationship between population increase and population density in plant-parasitic nematodes II. Sedentary nematodes. *Nematologica* 13, 157–171.

Seinhorst, J.W. (1970) Dynamics of populations of plant-parasitic nematodes. *Annual Review of Phytopathology* 8, 131–156.

Shapiro, D.I., Tylka, G.L., Berry, E.C. and Lewis, L.C. (1995) Effects of earthworms on the dispersal of *Steinernema* spp. *Journal of Nematology* 27, 21–28.

Sharma, S.B. (1971) Studies on the plant parasitic nematode *Tylenchorhynchus dubius*. *Mededelingen Landbouwhogeschool Wageningen* 71, 1–154.

Sharma, S.B. and Sharma, R. (1998) Hatch and emergence. In: Sharma, S.B. (ed.) *The Cyst Nematodes*. Kluwer Academic Publishers, Dordrecht, The Netherlands, pp. 191–216.

Sohlenius, B. and Bostrom, S. (2001) Annual and long-term fluctuations of the nematode fauna in a Swedish Scots pine forest soil. *Pedobiologia* 45, 408–429.

Sohlenius, B. and Sandor, A. (1987) Vertical distribution of nematodes in arable soil under grass (*Festuca pratensis*) and barley (*Hordeum distichum*). *Biology and Fertility of Soils* 3, 19–25.

Steinberger, Y., Loboda, I. and Garner, W. (1989) The influence at autumn dewfall on spatial and temporal distribution of nematodes in the desert ecosystem. *Journal of Arid Environments* 16, 177–183.

Steiner, W.A. (1994) The influence of air pollution on moss-dwelling animals. 4. Seasonal and long-term fluctuations of rotifer, nematode and tardigrade populations. *Revue Suisse de Zoologie* 101, 1017–1031.

Tallis, G.M. and Donald, A.D. (1970) Further models for the distribution on pasture of infective larvae the strongyloid nematode parasites of sheep. *Mathematical Biosciences* 7, 179–190.

Taylor, C.E., Brown, D.J.F., Neilson, R. and Jones, A.T. (1994) The persistence and spread of *Xiphinema diversicaudatum* in cultivated and uncultivated biotopes. *Annals of Applied Biology* 124, 469–477.

Taylor, E.L., Woiwod, I.P. and Perry, J.N. (1979) The negative binomial as a dynamic ecological model for aggregation and the density dependence of k. *Journal of Animal Ecology* 48, 289–304.

Taylor, L.R. (1961) Aggregation variance and the mean. *Nature* 189, 732–735.

Taylor, L.R. (1970) Aggregation and the transformation of counts of *Aphis fabae* Scoup. on beans. *Annals of Applied Biology* 65, 181–189.

Taylor, L.R. (1979) Aggregation as a species characteristic. In: Patil, G.P., Pielon, E.C. and Waters, W.E. (eds) *Statistical Ecology*. Pennsylvania Press, University Park, pp. 357–377.

Tetley, J.H. (1949) *Rhythms in Nematode Parasitism in Sheep*. Bulletin No. 96, Department of Scientific and Industrial Research, Wellington, New Zealand.

Thomas, P.R. (1981) Migration of *Longidorus elongatus*, *Xiphinema diversicadatum* and *Ditylenchus dipsaci* in soil. *Nematologia Mediterranea* 9, 75–81.

Thomas, R.J. and Stevens, R.J. (1960) Ecological studies on the development of the pasture stages of *Nematodirus battus* and *N. filicollis* in sheep. *Parasitology* 50, 31–49.

Wall, J.W., Skene, K.R. and Neilson, R. (2002) Nematode community and trophic structure along a sand dune succession. *Biology and Fertility of Soils* 35, 293–301.

Wallace, H.R. (1971) Abiotic influences in the soil environment. In: Zuckerman, B.M., Mai, W.F. and Rohde, R.A. (eds) *Plant Parasitic Nematodes*, Vol. I. Academic Press, New York, pp. 257–280.

Wallace, M.K. and Hawkins, D.M. (1994) Applications of geostatistics in plant nematology. *Journal of Nematology (Supplement)* 26, 626–634.

Wallace, M.K., Rust, R.H., Hawkins, D.M. and MacDonald, D.H. (1993) Correlation of edaphic factors with plant-parasitic nematode population densities in a forage field. *Journal of Nematology* 25, 642–653.

Wardle, D.A. and Yeates, G.W. (1993) The dual importance of competition and predation as regulatory forces in terrestrial ecosystems: evidence from decomposer food-webs. *Oecologia* 93, 303–306.

Warwick, R.M. (1988) The level of taxonomic discrimination required to detect pollution effects on marine benthic communities. *Marine Pollution Bulletin* 19, 259–268.

Wasilewska, L. (1974) Vertical distribution of nematodes in the soil of dunes in the Kampinos forest. *Zeszyty Problemowe Postepow Nauk Rolniczych* 154, 203–212.

Webster, R. and Boag, B. (1992) Geostatistical analysis of cyst nematodes in soil. *Journal of Soil Science* 43, 583–595.

Wheeler, T.A., Kenerley, C.M., Jeger, M.J. and Starr, J.L. (1987) Effects of quadrat and core size on determining the spatial pattern of *Criconemella sphaerocephalus*. *Journal of Nematology* 19, 413–419.

Wheeler, T.A., Baugh, B., Kaufman, H., Schuster, G. and Siders, K. (2000) Variability in time and space of *Meloidogyne incognita* fall population density in cotton fields. *Journal of Nematology* 32, 258–264.

Whitehead, A.G. (1977) Vertical distribution of potato, beet and pea cyst nematodes in some heavily infested soils. *Plant Pathology* 25, 85–90.

Williams, L., III, Schotzko, D.J. and McCaffrey, J.P. (1992) Geostatistical description of the spatial distribution of *Limonius californicus* (Coleoptera: Elateridae) wireworms in Northwestern United States, with comments on sampling. *Environmental Entomology* 21, 983–995.

Winslow, R.D. (1964) Soil nematode population studies. I. The migratory root Tylenchida and other nematodes of the Rothamsted and Woburn six-course rotations. *Pedobiologia* 4, 65–76.

Yeates, G.W. (1967) Studies on the nematodes from dune sands. 9. Quantitative comparisons of the nematode faunas of six localities. *New Zealand Journal of Science* 10, 927–948.

Yeates, G.W. (1972) Population studies on *Ditylenchus dipsaci* (Nematoda: Tylenchida) in a Danish beech forest. *Nematologica* 8, 125–130.

Yeates, G.W. (1980) Populations of nematode genera in soils under pasture. III. Vertical distribution at eleven sites. *New Zealand Journal of Agricultural Research* 23, 117–128.

Yeates, G.W. (1981) Populations of nematodes genera in soils under pasture. IV. Seasonal dynamics at five North Island sites. *New Zealand Journal of Agricultural Research* 24, 107–121.

Yeates, G.W. (1982) Variation of pasture nematode populations over thirty-six months in a summer dry silt loam. *Pedobiologia* 24, 329–346.

Yeates, G.W. (1984) Variation of pasture nematode populations over thirty-six months in a summer moist silt loam. *Pedobiologia* 27, 207–220.

Yeates, G.W. (1994) Modification and qualification of the nematode maturity index. *Pedobiologia* 38, 97–101.

Yeates, G.W., Wardle, D.A. and Watson, R.N. (1999) Response of soil nematode populations, community structure, diversity and temporal variability to agricultural intensification over a seven year period. *Soil Biology and Biochemistry* 31, 1721–1733.

Yeates, G.W., Dando, J.L. and Shepherd, T.G. (2002) Pressure plate studies to determine how moisture affects access of bacterial-feeding nematodes to food in soil. *European Journal of Soil Science* 53, 355–365.

Young, I.M., Griffiths, B.S., Robertson, W.M. and McNicol, J.W. (1998) Nematode (*Caenorhabditis elegans*) movement in sand as affected by particle size, moisture and the presence of bacteria (*Escherichia coli*). *European Journal of Soil Science* 49, 237–241.

Yuen, P.H. (1966) The nematode fauna of the regenerated woodland and grassland of Broadbalk Wilderness. *Nematologica* 12, 195–214.

Zhang, F. and Schmitt, D.P. (1995) Spatial-temporal patterns of *Meloidogyne konaensis* on coffee in Hawaii. *Journal of Nematology* 27, 109–113.

13 Survival Strategies

David A. Wharton

Department of Zoology, University of Otago, PO Box 56, Dunedin, New Zealand

13.1. Introduction

Although the whole biology of an organism is geared towards its survival (and reproduction), 'survival strategies' are usually taken to mean the responses of organisms to the particular biological and physical challenges that may result in their death. Nematodes face a number of such challenges. Biological challenges include: running out of food, predation, pathogens and competition. Parasitic species face the additional problems of the immune, and other, responses of their hosts and the problems of infecting new hosts. Physical challenges include: temperature extremes, desiccation, high pressure, acid or alkaline conditions, osmotic and ionic stress, anoxia and exposure to various types of radiation. This chapter will focus on the behavioural response of nematodes to these physical stresses. The responses of nematodes to a physical challenge, however, also involve morphological, physiological, biochemical and molecular mechanisms. There may also be an interaction between the physical and biological challenges faced by nematodes.

How may a nematode respond to a physical challenge? We might recognize four responses (Wharton, 2002a,b). First, the nematode might move away from the extreme conditions to a less stressful environment that is more favourable to its growth and reproduction (a migratory response). Secondly, the nematode may adapt to the challenge and so be able to grow and reproduce under extreme conditions (capacity adaptation). Thirdly, the life cycle of parasitic nematodes may include timing mechanisms that enable the parasite to avoid seasonally stressful conditions and to synchronize its life cycle with that of its host. Finally, the nematode may survive extremes in a dormant state until conditions suitable for growth and reproduction return (resistance adaptation).

13.2. Migratory Responses to Extreme Conditions

13.2.1. Large-scale migrations

Large-scale migrations are mainly beyond the control of the nematode. Nematodes may be distributed through the effects of human activity (such as on farm machinery), on the bodies of animals (such as in material adhering to the legs of birds) or passively via water currents and air movement. Such mechanisms may only incidentally assist nematodes in avoiding adverse environmental conditions. Migratory mammals and birds will carry their parasitic nematodes with them. This may mean that the free-living stages of their parasitic nematodes are removed from exposure to the harshest of environmental conditions.

13.2.2. Small-scale migrations

Small-scale migrations of nematodes protect against environmental stress by allowing them to move away from harmful conditions. They may also migrate deeper into the soil on a seasonal or a daily basis, providing protection against periodically adverse conditions.

Caenorhabditis elegans moves away from toxic concentrations of cadmium and copper ions but not from nickel ions. This response is mediated via amphidial chemosensory neurones (Sambongi *et al.*, 1999). There are many other examples of chemicals acting as nematode repellents (Croll, 1970; Lee, 2002). Many nematodes migrate away from extreme temperatures to a preferred temperature in a gradient. Infective juveniles of nematode parasites of vertebrates are positively thermotactic and are most active at 35–40°C, which helps them to locate and infect a host (Lee, 2002).

The infective juveniles of trichostrongyle nematodes make diurnal migrations on to vegetation (Saunders *et al.*, 2000). This makes them more available for ingestion by the host but may also help protect them from exposure to adverse conditions, such as avoiding the more exposed parts of plants at a time of day when frost may be present. There is evidence of a seasonal migration of Antarctic nematodes in moss carpets (but not in moss cushions), which involves migration into the deeper layers during autumn and to the surface during spring (Caldwell, 1981; Maslen, 1981). Benthic nematodes in coastal sediment undergo vertical seasonal migrations through the sediment in response to changes in oxygen deficiency and sulphide concentration (Hendelberg and Jensen, 1993).

13.2.3. Phoresis

Nematodes have phoretic relationships with invertebrate and vertebrate hosts. There is no feeding by the nematode upon the phoretic or transport host, and development of the nematode is delayed until it is transported to a fresh habitat. Some nematodes use phoretic hosts as a means of transport from a deteriorating environment to a similar but fresh environment. An example is transport of the dung-inhabiting nematode *Pelodera coarctata* by dung beetles (Poinar, 1983). Phoretic relationships involving nematodes are covered in more detail in Chapter 10 (see Timper and Davies, this volume).

13.3. Capacity Adaptation to Extreme Conditions

If a species of nematode grows and reproduces under environmental conditions that are more extreme than those of a majority of nematode species, we might consider that species to display capacity adaptation (Wharton, 2002a,b). The Antarctic nematode *Plectus antarcticus* lays eggs and completes its life cycle at lower temperatures than nematodes from temperate regions (Caldwell, 1981). Another Antarctic nematode, *Scottnema lindsayae*, had a higher fecundity, faster rates of juvenile development and better survival at 10°C than at 15°C (Overhoff *et al.*, 1993). A nematode from the high Arctic, *Chiloplacus* sp., has faster rates of growth and reproduction at lower temperatures than those of many other nematodes (Procter, 1984). These three polar species thus display capacity adaptation to the cold environments they experience. The Antarctic nematode *Panagrolaimus davidi*, however, grows best at 25°C (Brown, 1993; Wharton, 1997) and thus does not show capacity adaptation to cold.

Three species of nematode (*Chronogaster ethiopica*, *Monhystrella arsiensis* and *Rhabdolaimus aquaticus*) were isolated from an Ethiopian hot spring at a temperature of 41°C. These genera have other species that occur in hot springs (Abebe *et al.*, 2001). Most species reported from hot springs are cosmopolitan in distribution but some are restricted to them (Nicholas, 1984), suggesting capacity adaptation in the latter. Nematodes are abundant in active deep-sea hydrothermal sediments (Vanreusel *et al.*, 1997). It is not known whether they can survive the very hot temperatures (up to 92°C) associated with parts of this habitat. *Turbatrix aceti* shows an extraordinary pH tolerance, maintaining activity from pH 1.6 to 11 and growing in a range from pH 3.5 to 9 (Nicholas, 1984).

13.4. Timing Mechanisms and Survival

Parasitic nematodes can improve their chances of infecting a host and of surviving as a free-living stage outside the host by synchronizing their life cycle with the availability of hosts and by avoiding exposure to adverse seasonal conditions. This involves developmental arrests in the life cycle that retain the nematode as a resistant infective stage until transmission to a host occurs. Environmental conditions may induce a delay in development (diapause) where the nematode is retained as an infective or pre-reproductive parasitic stage until conditions become favourable for transmission. Developmental arrests may also occur in the life cycle of free-living nematodes, enabling them to survive periods of adverse conditions as a resistant stage.

There are two main types of developmental arrests, involving timing mechanisms that can have an impact on nematode survival. Diapause is defined as being triggered by environmental cues (although it can also be intrinsic) and which is temporarily irreversible despite the occurrence of conditions under which development could proceed (Sommerville and Davey, 2002). In other types of arrest (such as infective stages and dauer juveniles, or the simple slowing of development during adverse conditions) development resumes as soon as conditions become favourable – such as infection of a host. Infective juveniles are thus considered to be in a state of dormancy, rather than diapause (Sommerville and Davey, 2002). Diapause is thus temporarily irreversible while other types of dormancy are not.

13.4.1. Resting stages in the life cycle

13.4.1.1. Infective stages

The life cycles of most parasitic nematodes include an infective stage in which development is arrested until infection occurs. The resting stage is usually the infective juvenile and this is often the stage of the life cycle that is most resistant to environmental stress. The infective juvenile does not resume development until it receives a specific stimulus that indicates that infection of the host has occurred. It has survival mechanisms that allow it to persist in the environment until infection occurs and may stay in this state of developmental arrest for many years.

Development to the second-stage juvenile (J2) of *Ascaris lumbricoides* occurs within the egg. The egg does not hatch and development does not resume until it is triggered upon ingestion by the human host. The whole of the free-living phase of the life cycle is thus spent with the embryo or juvenile protected by a tough resistant eggshell. The hatching stimulus is an increase in carbon dioxide concentration, at an appropriate temperature and pH (Rogers, 1960). The eggshell consists of four layers: an outer uterine layer, a vitelline layer, a chitinous layer and an inner lipid layer (Foor, 1967). The chitinous layer is the thickest layer of the eggshell and is composed of a chitin/protein complex that provides structural strength and chemical resistance. Its resistance is thought to be enhanced by a quinone tanning system (Wharton, 1980). The inner lipid layer contains protein and lipid, the lipid fraction consisting of a mixture of glycosides called ascarosides, a group of lipids unique to nematodes (Bartley *et al.*, 1996). The lipid layer is the main permeability barrier of the eggshell, making it extremely impermeable to chemicals and slowing down the rate of water loss when the eggs are exposed to desiccation (Wharton, 1979). Although *Ascaris suum* eggs show some resistance to desiccation, it nevertheless provides a stress that, together with high temperatures, may contribute to the mortality of eggs on pasture (Larsen and Roepstorff, 1999). The eggs of *A. lumbricoides*, however, have been reported to remain viable in soil for as long as 14 years (Crompton, 2001). Many other species of plant- and animal-parasitic nematodes spend their free-living phase within the egg (Lee, 2002).

Trichostrongyle nematodes parasitic in ruminants have a first-stage juvenile (J1) that hatches from the egg and rapidly develops to the infective third-stage juvenile (J3). The infective juvenile retains the J2 cuticle as a sheath. This completely encloses the J3, which is a non-feeding resting stage (Wharton, 1986). The J3 is tolerant of freezing and desiccation. The sheath may play a role in the survival of these stresses (Ellenby, 1968b; Wharton and Allan, 1989; Allan and Wharton, 1990). Upon ingestion by the host, the sheath is shed and development is resumed, in response to a physiological trigger that includes an increase in carbon dioxide concentration (Rogers, 1960). Many other species of animal- and plant-parasitic nematodes have resistant infective juvenile stages, which may or may not have sheaths (Lee, 2002).

Some parasitic nematodes have intermediate hosts in their life cycles (Lee, 2002). This may, at least partially, remove the parasite from the hazards of the external environment and aid transmission by exploiting the food-chain where the intermediate host is the prey of the definitive host. The nematode is encysted, or otherwise in a state of developmental arrest, within the intermediate host, awaiting transmission to the definitive host. Some of these nematodes appear to manipulate the behaviour of their intermediate host in a way that increases the chances of predation by the definitive host, and hence completion of the nematodes' life cycle (Moore and Lasswell, 1986; McCurdy *et al.*, 1999).

13.4.1.2. Dauer juvenile formation

Caenorhabditis elegans produces a special survival stage called a dauer juvenile (dauer larva) in response to deteriorating environmental conditions. This is an alternative developmental pathway that leads to the formation of dauer juveniles, rather than to normal J3 and reproductive development. Dauer juveniles differ from normal J3 in being non-feeding, with a reduced epidermis and a cuticle modified by the addition of a striated basal zone. The excretory glands are inactivated, the intestine is reduced, there are modifications to the sense organs and development to the J4 is delayed (Riddle, 1988). Dauer juveniles have longer lifespans than normal J3 and resist high temperatures and anoxia (Anderson, 1978). They have reduced activity and oesophageal pumping is inhibited, but they will respond to mechanical and chemical stimuli (Cassada and Russell, 1975). They are transported to a fresh environment by phoresy, attaching to an insect (Lee, 2002). This is aided by their ability to mount projections on the substrate and wave their heads in the air, a behaviour that presumably increases their chances of contacting an insect (Cassada and Russell, 1975).

The formation of dauer juveniles in *C. elegans* is triggered primarily by an increase in the concentration of a dauer-inducing pheromone, indicating a rise in population density. This primary signal is influenced by temperature and by a decrease in chemical signals from the bacterial food, indicating exhaustion of the food supply. Development to the J4 is resumed in response to an increase in food supply. Dauer formation is a result of signals acting on the J1 but the developmental decision may be reversed by signals acting on the J2 (Riddle and Albert, 1997). This is mediated by chemosensory neurones (Lee, 2002) and involves over 30 genes (Gems, 2002).

Entomopathogenic nematodes complete two or three generations within the cadaver of their insect host and produce large numbers of dauer juveniles, which act as a survival, dispersal and infective stage (Womersley, 1993). They thus represent an alternative developmental pathway to the formation of normal J3. This may be under the control of a pheromone-based signalling system in a manner analogous to the formation of dauer juveniles by *C. elegans* (Fodor *et al.*, 1990). The infective juvenile (J2) of some anguinid plant-parasitic nematodes have also been referred to as dauer juveniles (Womersley *et al.*, 1998). These do not, however, represent an alternative developmental pathway and are best considered as infective juveniles (Wharton, 2002b).

Control of the developmental decision between dauer juvenile formation and reproductive development in *C. elegans* involves at least two signalling pathways. These are a transforming growth factor beta pathway (Patterson and Padgett, 2000) and an insulin-like pathway (Braeckman *et al.*, 2001). Similar signalling pathways may operate during the resumption of development upon infection of a host by an infective juvenile of a parasitic nematode (Rajan, 1998; Bealla and Pearce, 2002) and in the control of development of the dauer juveniles of entomopathogenic nematodes (Aumann and Ehlers, 2001).

13.4.2. Diapause

Diapause is defined as a period of developmental arrest that is either programmed into the life cycle (obligatory diapause) or is triggered by environmental conditions (facultative diapause). It is often associated with a decrease in metabolism but, unlike quiescence or cryptobiosis, it is not a direct response to unfavourable conditions but rather to stimuli (such as photoperiod) that herald the seasonal onset of such conditions (Nijhout, 1994; Denlinger, 2002).

Diapause in insects is under hormonal control (Nijhout, 1994; Denlinger, 2002). The mechanisms of hormonal control of diapause have yet to be demonstrated in nematodes but the presence of a diapause is inferred from patterns of seasonal development and the responses of particular developmental events to environmental stimuli. Evidence for the existence of diapause in nematodes has been reviewed (Evans and Perry, 1976; Evans, 1987; Antoniou, 1989; Jones *et al.*, 1998; Womersley *et al.*, 1998; Sommerville and Davey, 2002).

13.4.2.1. Egg diapause

Hatching of the eggs of the plant-parasitic nematodes *Heterodera avenae* and *Meloidogyne naasi* is greater after chilling. The response to chilling varies among isolates of these nematodes and may allow overwintering or the survival of a dry season. This suggests adaptation to local environmental conditions (Evans, 1987; Jones *et al.*, 1998; Womersley *et al.*, 1998; Sommerville and Davey, 2002). There may even be a variation in response within a single population, allowing hatching to be spread across two seasons, thus reducing competition for feeding sites on a host (Jones *et al.*, 1998). In some populations of the cereal cyst nematode, *H. avenae*, diapause is induced by high and broken by low temperatures (Mokabli *et al.*, 2001).

In *Nematodirus battus* the infective J3 is retained within the egg, in contrast to the life cycle of most other trichostrongyle nematodes. The eggs hatch in response to a period of chilling over the winter and this ensures a mass hatch in the spring when temperatures are milder and when lambs are available for infection (Evans and Perry, 1976).

Although temperature has been implicated as an environmental trigger for diapause in some nematodes, it is not a reliable indicator of season. Photoperiod is more reliable and evidence of a photoperiod-induced diapause in the eggs of *Globodera rostochiensis* has been presented (Hominick, 1986; Salazar and Ritter, 1993). This is thought to be mediated via the effect of photoperiod on the host plant.

13.4.2.2. Arrested development in trichostrongyle nematodes

As well as having a developmental arrest built into their life cycle (the infective juvenile), some trichostrongyle nematode parasites of ruminants also show a facultative diapause. This involves a seasonal delay in development, where infective juveniles ingested by the host do not immediately develop into adults. This diapause usually occurs at the fourth juvenile stage (J4) but may also involve parasitic J3 and early adult stages (Eysker, 1997). The phenomenon is known as arrested development or hypobiosis (Gibbs, 1986; Eysker, 1997; Sommerville and Davey, 2002). It is distinguished from a developmental arrest due to the immune responses of the host, in that it is triggered by the environmental conditions experienced by the parasite before it is ingested by the host (Sommerville and Davey, 2002).

In temperate regions of the northern hemisphere, *Ostertagia ostertagi* J3

ingested by cattle during autumn become arrested as early J4 in the host aboma-
sum. The parasites resume their development in the spring. Eggs are thus released
from the host at a time when they are likely to complete the free-living phase of the
life cycle and when new hosts (calves) are available for infection. In arid areas, a different pattern of arrested development is observed, associated with survival of a dry
season rather than overwintering (Jacquiet *et al.*, 1996; Gatongi *et al.*, 1998). The
trigger for arrested development involves light and temperature (Fernandez *et al.*,
1999) and is influenced by the host immune response and the management regimes
applied to domestic stock (Eysker, 1997).

Changes in protein profiles occur following exposure to conditions that trigger
arrested development (Kooyman and Eysker, 1995; Dopchiz *et al.*, 2000) and the
identification of a protein diagnostic for inhibited juveniles (Cross *et al.*, 1988)
indicates an underlying molecular mechanism.

13.4.3. Other delays in the life cycle

Deteriorating environmental conditions may result in delays in the life cycle as
growth and development slow or cease altogether. Since this is a direct result of
unfavourable conditions, this is not a diapause but rather a quiescence or cryptobiosis (see Section 13.5 on resistance adaptation). This type of delay can, however,
have significant effects on the dynamics of parasitic nematodes.

First-stage juveniles of protostrongylid nematodes infecting wild Nubian ibex
in the Negev Desert are retained within the faeces of the host during dry conditions,
where they are capable of surviving in an anhydrobiotic state, and they are only
released to infect the snail intermediate host when the faeces are flooded with water
during the rainy season (Solomon *et al.*, 1997). Temperature and moisture are
important factors affecting the transmission of trichostrongyle nematodes of ruminants. Under dry conditions eggs and juveniles are retained with the faeces.
Migration of juveniles on to herbage and infection of the host are delayed until it
rains (Stromberg, 1997).

13.5. Resistance Adaptation

In resistance adaptation, the organism survives environmental stress in a dormant
state until conditions favourable for its growth return (Wharton, 2002a). It is a
common survival strategy in nematodes, particularly in response to low temperatures and desiccation (Wharton, 2002b). The main behavioural response is to cease
activity but there are some specific behaviours associated with desiccation survival.

13.5.1. Survival responses to temperature

All organisms have an optimum temperature at which their metabolism and hence
growth and activity are greatest (Fig. 13.1). As the temperature increases or

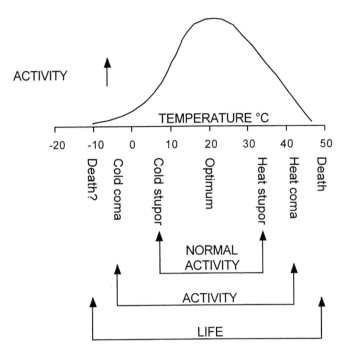

Fig. 13.1. Temperature transitions for a hypothetical nematode. At the optimum temperature, activity is at its maximum. As the temperature increases or decreases from the optimum, activity declines. The nematode enters a state of heat or cold stupor, in which normal activity is disrupted, and then heat or cold coma, in which activity ceases. Death may then ensue. However, nematodes are more likely to survive low than high temperatures (redrawn from Wharton, 2002a).

decreases from the optimum, the rate of metabolism and activity slows. At high and low temperatures, the organism enters a state of heat or cold stupor, where normal activity is disrupted, and then exhibits heat or cold coma, where activity ceases. If temperatures become more extreme, death may ensue (Vannier, 1994). There are no studies that I am aware of which have defined these transitions for any nematode. *Trichostrongylus colubriformis* J3 cease activity upon transfer from high (20°C) to low (3°C) temperature, which may indicate a cold-shock response. Transfer from low (3°C) to high (40°C) temperature produced coiling, which may be a form of heat-shock response (Wharton, 1981a). Activity generally declines with decreasing temperature but nematodes may still be moving at temperatures below 0°C, as long as freezing of the medium does not occur (D.A. Wharton, unpublished observations).

Very low temperatures (< 0°C) are likely to be more survivable than very high temperatures (> 50°C) since their adverse affects are potentially avoidable or reversible. Nematodes may be able to extend the survivable limits of both high and low temperatures by triggering protective responses.

13.5.1.1. Heat tolerance in nematodes

The upper limit for the survival of nematodes at high temperatures is about 47°C, based on nematodes from hot springs (Nicholas, 1984). An acute heat stimulus elicits thermal avoidance behaviour in *C. elegans* (Wittenburg and Baumeister, 1999). Some nematodes can extend their survival via the production of heat-shock proteins (Wharton, 2002b).

13.5.1.2. Cold tolerance in nematodes

Cold-tolerant nematodes are those that survive subzero temperatures in their natural habitat. Such temperatures are likely to be encountered in polar, alpine and temperate regions. Since nematodes are aquatic organisms, they risk external ice seeding through body openings and freezing of their body contents (inoculative freezing). If inoculative freezing is prevented by an eggshell or a sheath, the nematode can supercool to low temperatures in the presence of external ice and survive by freeze avoidance (Wharton and Allan, 1989; Wharton *et al.*, 1993). Some nematodes are freezing tolerant, surviving the formation of ice in their bodies (Wharton, 1995). Freezing-tolerant animals are usually thought to survive only extracellular ice formation. However, the Antarctic nematode *Panagrolaimus davidi* will survive extensive intracellular freezing – the only animal known to do so (Wharton and Ferns, 1995; Wharton *et al.*, 2003). A nematode in a state of anhydrobiosis will also survive low temperatures, since there is no freezable water present (Wharton, 2002b).

A further cold tolerance mechanism has recently been demonstrated in nematodes (Wharton *et al.*, 2003). If the freezing of the medium surrounding *P. davidi* occurs at a high subzero temperature (e.g. −1°C), inoculative freezing does not occur. The nematode then dehydrates, the water loss being driven by the difference in vapour pressure of ice and supercooled water at the same temperature. If the cooling rate is slow, this results in sufficient dehydration for all freezable water to be lost and thus the nematode dehydrates rather than freezes. This process is called 'cryoprotective dehydration'. This cold tolerance mechanism was first described in earthworm cocoons (Holmstrup and Westh, 1994) and may be widespread in soil invertebrates with high cuticular permeabilities (Holmstrup *et al.*, 2002). Forge and MacGuidwin (1992) suggested that nematodes in soil dehydrate rather than freeze but *P. davidi* is the only nematode so far where cryoprotective dehydration has been demonstrated. The cold tolerance mechanism employed by nematodes thus depends upon the microenvironmental conditions experienced during a low-temperature event (Figs 13.2 and 13.3).

The mechanisms underlying cold tolerance in nematodes are not fully understood. In *P. davidi* it may involve the synthesis of trehalose as a cryoprotectant (Wharton *et al.*, 2000a), a recrystallization-inhibiting protein (Ramløv *et al.*, 1996) and/or a substance that inhibits ice nucleation (Wharton and Worland, 1998).

Transmission of *Trichinella* involves ingestion of infective juveniles during predation or scavenging by one host on the carcass of another. Juveniles in the carcasses of Arctic hosts may survive extended periods of freezing (Smith, 1984). Arctic

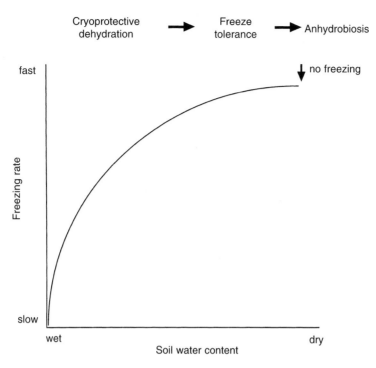

Fig. 13.2. The response of a soil nematode to freezing temperatures will depend upon the temperature of ice nucleation and the volume of soil water (see Wharton *et al.*, 2002). If the water content is high, the nucleation temperature that results in soil freezing will be high (given the large number of soil ice nucleators and the large volume of solution) and the freezing rate will be slow, favouring a cryoprotective dehydration mechanism. As the soil dries out the nematode will be in smaller volumes of fluid with lower nucleation temperatures. Freezing rates will thus be high, favouring inoculative freezing and freezing tolerance. As the soil continues to dry the nematode will become free of surface water, desiccate and survive in a state of anhydrobiosis.

species of *Trichinella* (*T. nativa*) have a greater tolerance to freezing than do those from more temperate latitudes (Pozio *et al.*, 1992; Kapel *et al.*, 1999). Freezing tolerance is enhanced by exposure to low temperatures (Smith, 1984, 1987). Low-temperature exposure inhibits the production of some heat-shock proteins (Hsp90, Hsp60) but enhances the production of others (Hsp70). A heat-shock protein (Hsp50) also acts as a cold-shock protein in *Trichinella* (Martinez *et al.*, 2001) and is induced by both oxidative and cold shock (Martinez *et al.*, 2002). The role these proteins play in the cold tolerance of these nematodes is unknown.

13.5.2. Survival responses to desiccation

For most nematodes, activity ceases once the surrounding water film evaporates. The loss of activity is due to the adherence of nematodes to the substrate in the

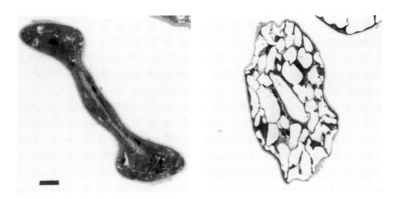

Fig. 13.3. Slow-frozen (left: –1°C 30 min, –4°C 30 min) and fast-frozen (right: –4°C 30 min) *Panagrolaimus davidi*, after then plunging into liquid nitrogen and processing for transmission electron microscopy using a freeze substitution technique. Some slow-frozen nematodes become dehydrated while some fast-frozen nematodes show intracellular ice. Scale bar = 2 μm. (D.A. Wharton, M.P. Downes, G. Goodall and C.J. Marshall, unpublished results.)

absence of water or to the loss of turgor pressure, without which movement cannot occur (Harris and Crofton, 1957). Nematodes may survive partial desiccation with a reduced rate of metabolism (desiccation-induced quiescence) or, after complete desiccation, metabolism may cease altogether (anhydrobiosis).

A slow rate of water loss is important for anhydrobiotic survival (Perry, 1999; Wharton, 2002b). Anhydrobiotic nematodes can be divided into two groups, depending upon the rates of water loss from their environment that they will survive (Womersley, 1987). Slow-dehydration strategists depend upon their surroundings, such as soil or moss, drying slowly; fast-dehydration strategists can survive rapid loss of water from their surroundings but themselves have mechanisms for ensuring the slow loss of water from their bodies. Adaptations for reducing the rate of water loss include both behavioural and physiological mechanisms.

13.5.2.1. Behavioural responses

Various types of aggregation, swarming and synchronized movements occur in nematodes (Croll, 1970), although it is debatable whether these represent distinct classes of behaviours. Some of these behaviours are associated with responses to environmental stress, including desiccation. Croll (1970) suggested that aggregation was associated with quiescence (reduced metabolism) and with survival responses to adverse environmental conditions, while swarming was involved in migration and dispersal. However, aggregation can occur in the absence of an environmental stress and swarming may be associated with such a stress. It would be better to define aggregation as the clumping of nematodes either spontaneously in culture or in water or when they are drawn together by the evaporation of water to form a clump, and to define swarming as a coordinated movement by a mass of nematodes from one location to another. Swarming may be followed by aggregation. Croll (1970) mentions synchronized movements as another category and gives the movement of *Panagrellus redivivus* on to the walls of a culture vessel as an example. It is unclear how this differs from swarming and so I shall mainly consider the phenomena of aggregation and swarming here.

Spontaneous aggregation of nematodes in water or culture media has been noted in a number of species. One example is the formation of dense rosettes by rhabditid nematodes extracted from horse dung (Croll, 1970). Some wild strains of *C. elegans* show a uniform distribution across a bacterial culture plate, while others form dense aggregations or clumps, particularly where the bacteria are thickest (de Bono and Bargmann, 1998). Aggregation is induced by crowding and is detected by specific neurones in the anterior region of the nematode, which presumably detect the accumulation of harmful metabolites (de Bono *et al.*, 2002). This response is regulated via other neurones, which may send an antagonistic signal via the pseudocoelomic fluid (Coates and de Bono, 2002).

Another way by which nematodes form aggregates is where they are drawn together by the evaporation of surrounding water to form a clump. The ability of *Aphelenchus avenae* to survive desiccation depends upon the nematodes being dehydrated in clumps of sufficient size (Crowe and Madin, 1975; Higa and Womersley, 1993). This is a reflection of the need to achieve rates of water loss that mimic those of its natural environment, since it is a slow-dehydration strategist (Higa and Womersley, 1993; Womersley and Higa, 1998). The formation of clumps aids the control of water loss since nematodes on the outside will dry first and form a layer that slows down the rate of water loss from nematodes deeper in the aggregation – the so-called 'eggshell effect' (Ellenby, 1968a). Desiccation-tolerant strains of *Steinernema feltiae* disperse from the clumps that are formed during the initial phases of dehydration to coil and complete desiccation individually. Clumping may nevertheless be important for survival of the initial phases of desiccation. The least desiccation-tolerant strain of this nematode does not show this dispersal effect (Solomon *et al.*, 1999).

Swarming behaviour occurs in some bacteriophagous, mycophagous and plant-parasitic nematodes (McBride and Hollis, 1966; Croll, 1970). This involves migration of nematodes on to the sides and lids of culture vessels (Fig. 13.4) or, in the case of plant-parasitic nematodes, to specific parts of the host, where they form clumps. The swarms consist of nematodes showing synchronized movements. The clumps formed following swarming may aid desiccation survival but swarming is not itself triggered by desiccation. Swarming is probably triggered by the accumulation of toxic waste products, the exhaustion of food supplies and/or the senescence of a host plant (Womersley *et al.*, 1998). In the plant-parasitic nematode *Ditylenchus dipsaci*, J4 swarm from the senescing plant tissue to form clumps known as 'eelworm wool' on infected bulbs or inside bean pods (Ellenby, 1969; Hooper, 1971). Aggregation also occurs in anguinid nematodes, which accumulate in infected host inflorescences and induce them to form galls (Womersley *et al.*, 1998). The formation of clumps by swarming and aggregation will only assist desiccation survival where there are enough nematodes in a restricted space to form

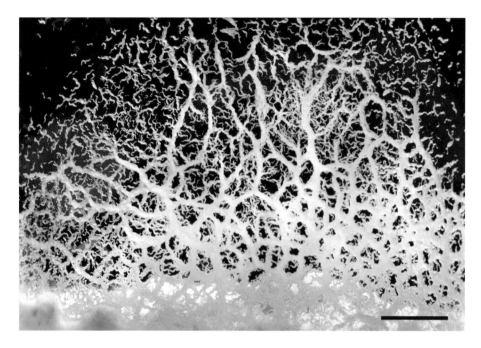

Fig. 13.4. The oatmeal nematode, *Panagrellus redivivus*, swarming on the side of a culture vessel. Scale bar = 4 mm.

clumps of sufficient size. In most situations nematodes are likely to face desiccation stress as individuals (Womersley *et al.*, 1998).

Coiling behaviour has been associated with desiccation survival in a wide range of nematodes (Table 13.1). It may follow or occur in conjunction with swarming and aggregation or the nematodes may coil individually. Coil formation involves the contraction of all the muscles on one side of the body. It appears to first involve the formation of a flat spiral (Fig. 13.5). Slippage of adjacent surfaces over one another then provides the necessary torsion to form a three-dimensional coil, with further slippage being prevented by the lateral alae (Wharton, 1982). The lateral alae thus lie on the inner and outer edges of the coil, limiting its tightness. This is an unstable posture, requiring the active contraction of muscles to maintain the coil in water. However, once water is lost through desiccation, the coil will be maintained by friction or adhesion between adjacent surfaces of the cuticle and with the substrate.

The formation of a coil reduces the surface area of cuticle directly exposed to air and may thus reduce the rate of water loss during desiccation. When desiccating nematodes are immersed in liquid paraffin, they become invisible when their refractive index matches that of the immersion medium (at a water content of about 20%). Using this technique, Womersley (1978) showed that in a partially coiled worm the coiled regions of a desiccated nematode (*Ditylenchus myceliophagus*) had a different water content from that of the uncoiled region. In *Rotylenchus reniformis* coiled nematodes survive desiccation at 80% or 40% relative humidity (RH), but

Table 13.1. Coiling behaviour in some nematodes.

Category	Species	Reference
Infective larvae of animal parasitic nematodes	*Trichostrongylus axei*	Wharton, 1982
	Trichostrongylus colubriformis	Wharton, 1982
	Ostertagia ostertagi	Wharton, 1982
	Ostertagia os circumcincta	Wharton, 1982
	Haemonchus contortus	Wharton, 1982
	Nematodirus filicolis	Wharton, 1982
	Nematodirus spathiger	Wharton, 1982
	Nematodirus battus	Wharton, 1982
	Cooperia onchophora	Wharton, 1982
	Cooperia curticei	Wharton, 1982
	Bunostomum trignocephalum	Belle, 1959
	Dictyocaulus viviparus	Jørgensen, 1980
Entomopathogenic nematodes	*Steinernema glaseri*	Womersley, 1990
	Steinernema carpocapsae	Womersley, 1990
	Steinernema feltiae	Patel *et al.,* 1997
Plant-parasitic nematodes	*Ditylenchus dipsaci*	Evans and Perry, 1976
	Anguina tritici	Bird and Buttrose, 1974
	Anguina amsinckiae	Womersley *et al.,* 1998
	Scutellonema brachyurum	Demure *et al.,* 1979b
	Scutellonema cavenessi	Demure, 1980
	Rotylenchus reniformis	Womersley and Ching, 1989
	Helicotylenchus dihysteria	Womersley and Ching, 1989
	Pratylenchus thornei	Glazer and Orion, 1983
	Pratylenchus penetrans	Townshend, 1984
	Pratylenchus brachyurus	Koenning and Schmitt, 1986
	Hemicriconemoides pseudobrachyurum	Saeed and Roessner, 1984
	Aphelenchoides ritzemabosi	Saeed and Roessner, 1984
	Psilenchus hilarulus	Saeed and Roessner, 1984
Free-living nematodes	*Acrobeloides* sp.	Demure *et al.,* 1979b
	Acrobeloides nanus	Nicholas and Stewart, 1989
	Actinolaimus hintoni	Lee, 1961
	Aphelenchus avenae	Demure *et al.,* 1979b
	Ditylenchus myceliophagus	Cayrol, 1970
	Dorylaimus keilini	Lee, 1961
	Panagrolaimus davidi	Wharton & Barclay, 1993
	Scottnema lindsayae	Treonis *et al.,* 2000
	Eudorylaimus antarcticus	Treonis *et al.,* 2000
	Plectus antarcticus	Treonis *et al.,* 2000
	Plectus sp.	Hendriksen, 1982
	Various species	Freckman *et al.,* 1977

uncoiled nematodes do not (Womersley and Ching, 1989). The survival of coiled *Plectus* sp. is significantly higher than that of uncoiled nematodes (Hendriksen, 1982). Coiling does not, however, guarantee successful induction into anhydrobiosis. *D. myceliophagus, R. reniformis* and some entomopathogenic nematodes can show a coiling response if they are dried slowly, but are still only capable of desiccation-induced quiescence (Womersley and Ching, 1989; Womersley, 1990; Womersley and Higa, 1998). Conversely, not all nematodes need to coil to survive

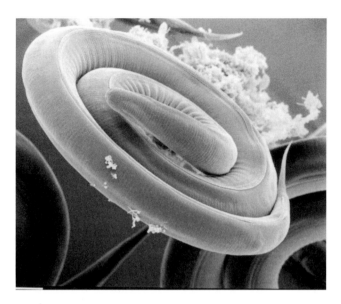

Fig. 13.5. Coiled infective juvenile (J3) of *Trichostrongylus colubriformis*. Scale bar = 20 μm.

desiccation. The galls of *Anguina tritici* are tightly packed with nematodes, which remain uncoiled and yet survive desiccation (Perry, 1999). *Heterorhabditis megidis* and *Heterorhabditis indica* form clumps during desiccation at high RH but neither species shows any tendency to coil nor are they capable of anhydrobiosis (O'Leary *et al.*, 2001).

The primary stimulus for coiling behaviour appears to be the restriction of lateral movement that occurs in an evaporating water film (Wharton, 1981b). The coiling of nematodes in soil increases as the water content of the soil decreases and is induced by the physical forces exerted at six to nine monomolecular layers of water (Demure *et al.*, 1979a). It is unlikely that water films around soil particles would be continuous at such low water contents and therefore restriction of lateral movement would occur.

The relationship between coiling and anhydrobiosis has led some to suggest that the presence of a coiled nematode indicates that it is in a state of anhydrobiosis (Freckman *et al.*, 1977). However, as we have seen, not all coiled nematodes are capable of anhydrobiotic survival and not all anhydrobiotic nematodes coil. Nematodes will coil in response to increases in temperature (Wharton, 1981a) and osmotic stress (Wharton *et al.*, 1983; Charwat *et al.*, 2002), as well as in response to desiccation.

13.5.2.2. Morphological, physiological and biochemical responses

Desiccation-tolerant nematodes tend to have low cuticular permeability. The epicuticle is the most likely permeability barrier, producing the slow rates of water loss necessary for survival (Wharton *et al.*, 1988). In *D. dipsaci* J4, there is a permeability slump, which produces a marked reduction in the rate of water loss, during the initial stages of desiccation (Wharton, 1996). This may involve a narrowing and/or deepening of the grooves of the cuticular annulations (Wharton and Marshall, 2002), although this needs confirmation.

The nematode eggshell has a restricted permeability and hence a slow rate of water loss, which may enable an enclosed juvenile to survive anhydrobiotically (Wharton, 1980). Other structures that are thought to be involved in producing a slow rate of water loss during desiccation include: the cyst wall of plant-parasitic cyst nematodes, the gelatinous matrix surrounding the eggs of root-knot nematodes and the sheaths of the infective juveniles of trichostrongyle and of some other nematodes (Perry, 1999). The presence of a sheath, however, is not always associated with desiccation survival and sheaths may have other functions, such as protection from pathogens (Perry, 1999). The formation of galls in the inflorescence of their host plants by anguinid nematodes also provides a barrier to water loss (Womersley *et al.*, 1998). Desiccation survival of entomopathogenic nematodes is enhanced by retention within the cadaver of their insect host (Koppenhofer *et al.*, 1997). Infective juveniles of *Nippostrongylus brasiliensis* become coated with a thin layer of lipid, obtained from the skin and hairs of their host, which may reduce water loss (Lee, 1972). Extracuticular layers appear to be produced by desiccated *Anguina amsinckiae* (Womersley *et al.*, 1998) and *D. myceliophagus* (Perry, 1999) but the origin, composition and function of these layers is unknown. A negatively charged coat is associated with the surface of the cuticle of a desiccation-tolerant mutant of *H. megidis* (O'Leary *et al.*, 1998). A negative surface charge may assist the desiccation tolerance of nematodes by maintaining a film of water over the surface of the cuticle (Murrell *et al.*, 1983).

A slow rate of water loss ensures the orderly packing and prevents the disruption of internal structures. In *D. dipsaci* this occurs in two phases, corresponding with the permeability slump observed during desiccation (Wharton and Lemmon, 1998). The maintenance of structural integrity during anhydrobiosis has been demonstrated in several species of nematode and appears to be essential for their survival (Wharton, 2002b).

Biochemical changes also occur during entry into anhydrobiosis. Elevated levels of the disaccharide trehalose have been associated with desiccation in a number of species of nematodes and other organisms (Wharton, 2002b). Trehalose is thought to stabilize membranes during desiccation and rehydration by attaching to the polar head groups of phospholipids and preventing phase changes that may cause the membrane to become leaky: 'the water replacement hypothesis' (Crowe and Clegg, 1973; Crowe *et al.*, 1992; Crowe and Crowe, 1999). Tissues may also be stabilized by trehalose via vitrification, which protects them in a sugar glass acting as a high viscosity, low molecular mobility medium (Crowe *et al.*, 1998). Other suggested roles for trehalose during anhydrobiosis include: preventing protein denaturation, oxidative damage and browning reactions, as a free-radical scavenging agent and as an inert energy source (Higa and Womersley, 1993).

There is, however, some doubt that trehalose is the principal adaptation required for anhydrobiosis in nematodes (Perry, 1999; Wharton, 2002a,b). Small aggregates of *A. avenae* exposed to 97% RH do not survive complete desiccation, even though they accumulate similar levels of trehalose to those of large aggregates that do survive anhydrobiotically (Higa and Womersley, 1993). *D. myceliophagus* do not survive anhydrobiosis, even though they may contain up to 16% dry weight of trehalose (Womersley and Higa, 1998). Similar observations have been made on entomopathogenic nematodes (Womersley, 1990).

Attention is now turning to molecular approaches to the study of anhydrobiosis in nematodes. Solomon *et al.* (2000) provided the first report of a desiccation-induced protein (Desc47) in nematodes, from a desiccation-tolerant strain of *S. feltiae* isolated from a desert region in Israel. This protein is heat-stable and has sequence homology to a cold-responsive (COR) protein from wheat and the closest match to late embryogenesis abundant (LEA) homologue proteins from the *C. elegans* predicted proteome database. A desiccation-induced heat-tolerant protein from *A. avenae* has sequence homology with group 3 LEA proteins from plants (Browne *et al.*, 2002). This protein includes several copies of an 11-mer motif that is characteristic of these types of protein and has been found in a wide range of organisms, including *C. elegans* (Dure, 2001).

In plants, LEA and COR are related proteins expressed in response to both desiccation and cold acclimation. Trehalose is also accumulated by *S. feltiae* and *A. avenae* in response to dehydration. The LEA-type proteins may interact with trehalose in these nematodes to form a stable bioglass that aids protection from desiccation-induced damage (Solomon *et al.*, 2000; Browne *et al.*, 2002). Similar associations between LEA proteins and sugars are thought to be important in desiccation survival in plants (Wolkers *et al.*, 2001).

13.5.2.3. Rehydration and the lag phase

When anhydrobiotic nematodes are immersed in water, activity does not start immediately but there is a period of apparent inactivity, called a 'lag phase', before activity commences (Fig. 13.6). Metabolism and water contents rapidly return to normal levels upon immersion in water but it can be as long as several hours before activity resumes (Barrett, 1982, 1991). The length of the lag phase is related to the severity of the desiccation stress during dehydration (Allan and Wharton, 1990; Wharton and Barclay, 1993; Wharton and Aalders, 1999). This may indicate that the lag phase represents a period during which any damage incurred during desiccation is repaired.

An ordered series of morphological changes occurs in *D. dipsaci* J4 during the lag phase (Wharton and Barrett, 1985; Wharton *et al.*, 1985). In *D. dipsaci* J4 and *Anguina agrostis* J2 the cuticle becomes more permeable after desiccation but the permeability barrier is restored upon rehydration (Preston and Bird, 1987; Wharton *et al.*, 1988). The repair of the permeability barrier is disrupted by inhibitors that block post-transcriptional protein synthesis and enzyme activity, suggesting a metabolically dependent repair mechanism (Wharton *et al.*, 1988). In *A. tritici* and *A. avenae* ions leak into the medium during the initial stages of rehydration (Crowe *et al.*, 1979; Womersley, 1981). This leakage ceases as rehydration proceeds, suggesting the restoration of permeability barriers.

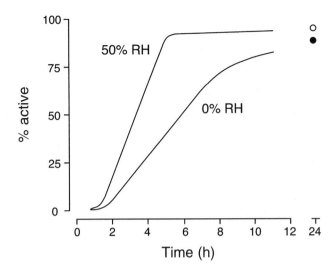

Fig. 13.6. The lag phase in the recovery of activity of *Ditylenchus dipsaci* after exposure to 50% (top line, open symbol) or 0% RH (bottom line, closed symbol) for 6 days, followed by reimmersion in water. It took 6.8 h for 50% of nematodes to recover after exposure to 0% RH and 3.6 h after exposure to 50% RH. (Data from Wharton and Aalders, 1999.)

Changes in the muscle cells of nematodes are likely to be particularly important for the recovery of activity. In *D. dipsaci* a hyaline layer, observed under the light microscope and consisting of the epidermis plus the muscle cells, slowly increases in thickness during the lag phase. The nematodes show a decrease in length during the lag phase before activity resumes. This has been called 'anomalous shrinkage' since one might expect a continuous increase in length during rehydration (Wharton *et al.*, 1985). Similar observations have been made on ensheathed J3 of *T. colubriformis*. In this nematode, anomalous shrinkage is accompanied by the development of birefringence in the hyaline layer under polarized light. This may result from a change in the organization of the muscle filaments of the contractile region of the nematode's muscle cells (Allan and Wharton, 1990). The spacing of the myofilaments of *D. dipsaci* increases during rehydration (Wharton and Barrett, 1985). The nematodes do not commence movement until these morphological changes in the muscle cells occur (Wharton *et al.*, 1985; Allan and Wharton, 1990). This suggests the repair of or restoration of physiological function in the muscle cells.

Rehydrated *D. dipsaci* will respond to electrical and chemical stimuli about 30 min before spontaneous movement occurs (Barrett, 1982), suggesting a partial recovery of nerve and muscle function. Electrical recordings have been made during the lag phase following anhydrobiosis and then rehydration in *D. dipsaci* J4s (Wharton *et al.*, 2000b). The pattern of electrical activity was similar in all nematodes from which recordings were made, although the length of the lag phase varied. Initially there was no electrical activity: this phase lasted up to 40 min after immersion in water. This was followed by bursts of low-amplitude activity and then also of high-amplitude activity (Fig. 13.7). In some specimens, trains of spikes were observed, associated with the movement of the nematode. Spikes were slow (about 100 ms), indicating that they were muscular rather than nervous in origin. The commencement of low-amplitude electrical activity appears to correspond with the time when anomalous shrinkage occurs. This may indicate partial recovery of muscle function. Contraction of both the dorsal and ventral muscle blocks may be responsible for the decrease in length. Recovery of muscles may involve a restoration of the ionic gradients that are essential for their function, but this requires confirmation using techniques that can directly measure the ions involved.

13.5.3. Survival responses to other stresses

Although temperature and desiccation are the major stresses affecting nematodes, there are other physical and chemical stresses whose effects on nematodes are less

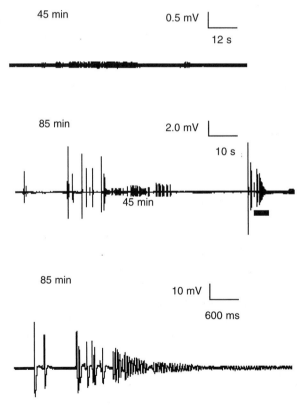

Fig. 13.7. Electrical recordings from a *Ditylenchus dipsaci* J4 after a period of anhydrobiosis and after 45 (top) or 85 min (middle and bottom) reimmersion in water. The bottom trace is an expansion of the section of the recording in the middle trace, indicated by the thick line. (Redrawn from Wharton *et al.*, 2000b.)

well understood. For example, nematodes are the most abundant members of the deep-sea benthic microfauna (Lambshead *et al.*, 2000). Little is known concerning how they survive the high pressures and low temperatures associated with this habitat but this presumably involves capacity adaptation, since they are constantly exposed to these conditions.

Nematodes are exposed to osmotic stress in a variety of habitats. They can survive both hyperosmotic and hypo-osmotic conditions (Thompson and Geary, 2002; see Wright, Chapter 7, this volume). Free-living and plant-parasitic nematodes experience low oxygen tensions where soil is waterlogged or there is a high rate of decomposition of organic material. Animal-parasitic nematodes may face anoxia in the host gut. *Caenorhabditis elegans* has a broad range of oxygen tolerances and is able to maintain a near-normal rate of metabolism at very low oxygen levels. It can, however, survive complete anoxia for only a few days. Given their small size and long, thin shape, the ability to tolerate low oxygen concentrations may be widespread in free-living nematodes (Van Voorhies and Ward, 2000). Some nematodes can survive longer periods of anoxia. Juveniles (but not adults) of the marine benthic nematode *Therisus anoxybioticus* can survive anoxia for more than 15 days (Jensen, 1995). The juveniles of *A. avenae* can survive up to 60 days of anoxia (Cooper and Van Gundy, 1970). Adult ascarids, parasitic in the guts of vertebrates, have a primarily anaerobic metabolism (Barrett, 1981). Other physical and chemical stresses include various types of radiation (especially ultraviolet (UV)), pH extremes and exposure to toxic chemicals.

Animals are rarely exposed to a single physical or chemical stress but to some combination of challenges. Terrestrial Antarctic nematodes, for example, are likely to be exposed to water stress as well as low temperatures. They may also be exposed to high levels of UV radiation and low levels of nutrients and have to survive in transient and unstable substrates (Wharton, 2002a).

13.6. Summary and Conclusions

Nematodes possess a complex suite of survival responses that enable them to complete their life cycles in the face of adverse, and often extreme, environmental stress. In response to such stress they may migrate to avoid the challenge, adapt so they can function under extreme conditions (capacity adaptation), have timing mechanisms (infective juveniles, dauer juveniles, diapause) that allow them to avoid stressful conditions and have resistance adaptations (quiescence and cryptobiosis) that allow them to survive in a dormant state until favourable conditions return.

Each species of nematode has its own set of survival responses and behaviours. Consider, for example, trichostrongyle nematodes of ruminants. They show an obligatory development arrest, an infective juvenile (J3) that requires a stimulus from the host for development to resume and a facultative diapause, the arrested development of the J4. The J3 and, to a lesser extent, the egg are capable of cryptobiosis, surviving both low temperatures and desiccation. The J3 coils in response to desiccation. Migration of infective juveniles from the faeces and on to the herbage is subject to environmental conditions. Further survival mechanisms and behaviours are required upon ingestion by a host in response to the physico-chemical conditions within the vertebrate intestine and the immune response of the host. Nematodes are thus exquisitely adapted to face the challenges of the series of environments through which they pass during the completion of their life cycle.

References

Abebe, E., Mees, J. and Coomans, A. (2001) Nematode communities of Lake Tana and other inland water bodies of Ethiopia. *Hydrobiologia* 462, 41–73.

Allan, G.S. and Wharton, D.A. (1990) Anhydrobiosis in the infective juveniles of *Trichostrongylus colubriformis* (Nematoda: Trichostrongylidae). *International Journal for Parasitology* 20, 183–192.

Anderson, G.L. (1978) Responses of dauer juveniles of *Caenorhabditis elegans* (Nematoda: Rhabditidae) to thermal stress and oxygen deprivation. *Canadian Journal of Zoology* 56, 1786–1791.

Antoniou, M. (1989) Arrested development in plant parasitic nematodes. *Helminthological Abstracts* B58, 1–19.

Aumann, J. and Ehlers, R.U. (2001) Physico-chemical properties and mode of action of a signal from the symbiotic bacterium *Photorhabdus luminescens* inducing dauer juvenile recovery in the entomopathogenic nematode *Heterorhabditis bacteriophora*. *Nematology* 3, 849–853.

Barrett, J. (1981) *Biochemistry of Parasitic Helminths*. Macmillan, London, 308 pp.

Barrett, J. (1982) Metabolic responses to anabiosis in the fourth stage juveniles of *Ditylenchus dipsaci* (Nematoda). *Proceedings of the Royal Society of London* B216, 159–177.

Barrett, J. (1991) Anhydrobiotic nematodes. *Agricultural Zoology Reviews* 4, 161–176.

Bartley, J.P., Bennett, E.A. and Darben, P.A. (1996) Structure of the ascarosides from *Ascaris suum*. *Journal of Natural Products* 59, 921–926.

Bealla, M.J. and Pearce, E.J. (2002) Transforming growth factor-beta and insulin-like signaling pathways in parasitic helminths. *International Journal for Parasitology* 32, 399–404.

Belle, E.A. (1959) The effect of microenvironment on the free-living stages of *Bunostomum trignocephalum*. *Canadian Journal of Zoology* 37, 289–298.

Bird, A.F. and Buttrose, M.S. (1974) Ultrastructural changes in the nematode *Anguina tritici* associated with anhydrobiosis. *Journal of Ultrastructural Research* 48, 177–189.

Braeckman, B.P., Houthoofd, K. and Vanfleteren, J.R. (2001) Insulin-like signaling, metabolism, stress resistance and aging in *Caenorhabditis elegans*. *Mechanisms of Ageing and Development* 122, 673–693.

Brown, I.M. (1993) The influence of low temperature on the Antarctic nematode *Panagrolaimus davidi*. PhD thesis, University of Otago.

Browne, J., Tunnacliffe, A. and Burnell, A. (2002) Anhydrobiosis – Plant desiccation gene found in a nematode. *Nature* 416, 38.

Caldwell, J.R. (1981) The Signy Island terrestrial reference sites: XIII. Population dynamics of the nematode fauna. *British Antarctic Survey Bulletin* 54, 33–46.

Cassada, R.C. and Russell, R.L. (1975) The dauerlarva, a postembryonic developmental variant of the nematode *Caenorhabditis elegans*. *Developmental Biology* 46, 326–342.

Cayrol, J.C. (1970) Action des autres composants de la biocénose du champignon de couche sur le nématode mycophage, *Ditylenchus myceliophagus* J.B. Goodey 1958, et étude de son anabiose: forme de survie en conditions défavorables. *Revue d'Ecologie et de Biologie du Sol* 7, 409–440.

Charwat, S.M., Fisher, J.M. and Wyss, U. (2002) The effect of osmotic stress on desiccation survival and water content of four nematode species. *Nematology* 4, 89–97.

Coates, J.C. and de Bono, M. (2002) Antagonistic pathways in neurons exposed to body fluid regulate social feeding in *Caenorhabditis elegans*. *Nature* 419, 925–929.

Cooper, A.F. and Van Gundy, S.D. (1970) Metabolism of glycogen and neutral lipids by *Aphelenchus avenae* and *Caenorhabditis* sp. in aerobic, microaerobic and anaerobic environments. *Journal of Nematology* 2, 305–315.

Croll, N.A. (1970) *The Behaviour of Nematodes: Their Activity, Senses and Responses*. Edward Arnold, London, 117 pp.

Crompton, D.W.T. (2001) Ascaris and ascariasis. *Advances in Parasitology* 48, 285–375.

Cross, D.A., Klesius, P.H. and Williams, J.C. (1988) Preliminary report: immunodiagnosis of pre-type II ostertagiasis. *Veterinary Parasitology* 27, 151–158.

Crowe, J.H. and Clegg, J.S. (1973) *Anhydrobiosis*. Dowden, Hutchinson and Ross, Stroudsberg, Pennsylvania, 477 pp.

Crowe, J.H. and Crowe, L.M. (1999) Anhydrobiosis: the water replacement hypothesis. In: Glazer, I., Richardson, P., Boemare, N. and Coudert, F. (eds) *Survival of Entomopathogenic Nematodes*. Office for Official Publications of the European Communities, Luxemborg, pp. 15–25.

Crowe, J.H. and Madin, K.A. (1975) Anhydrobiosis in nematodes: evaporative water loss and survival. *Journal of Experimental Zoology* 193, 323–334.

Crowe, J.H., O'Dell, S.J. and Armstrong, D.A. (1979) Anhydrobiosis in nematodes: permeability during rehydration. *Journal of Experimental Zoology* 207, 431–438.

Crowe, J.H., Hoekstra, F. and Crowe, L.M. (1992) Anhydrobiosis. *Annual Review of Physiology* 54, 579–599.

Crowe, J.H., Carpenter, J.F. and Crowe, L.M. (1998) The role of vitrification in anhydrobiosis. *Annual Review of Physiology* 60, 73–103.

de Bono, M. and Bargmann, C.I. (1998) Natural variation in a neuropeptide Y receptor homolog modifies social behavior and food response in *C. elegans*. *Cell* 94, 679–689.

de Bono, M., Tobin, D.M., Davis, W.M., Avery, L. and Bargmann, C.I. (2002) Social feeding in *Caenorhabditis elegans* is induced by neurons that detect aversive stimuli. *Nature* 419, 899–903.

Demure, Y. (1980) Biology of the plant-parasitic nematode, *Scutellonema cavenesii* Sher, 1964: anhydrobiosis. *Revue de Nématologie* 3, 283–289.

Demure, Y., Freckman, D.W. and Van Gundy, S.D. (1979a) *In vitro* response of four species of nematodes to desiccation and discussion of this and related phenomena. *Revue de Nématologie* 2, 203–210.

Demure, Y., Freckman, D.W. and Van Gundy, S.D. (1979b) Anhydrobiotic coiling of nematodes in soil. *Journal of Nematology* 11, 189–195.

Denlinger, D.L. (2002) Regulation of diapause. *Annual Review of Entomology* 47, 93–122.

Dopchiz, M.C., Parma, A.E. and Fiel, C.A. (2000) Hypobiosis induction alters the protein profile of *Ostertagia ostertagi* (Nematoda : Trichostrongylidae). *Folia Parasitologica* 47, 135–140.

Dure, L. (2001) Occurrence of a repeating 11-mer amino acid sequence motif in diverse organisms. *Protein and Peptide Letters* 8, 115–122.

Ellenby, C. (1968a) Desiccation survival in the plant parasitic nematodes, *Heterodera rostochiensis* Wollenweber and *Ditylenchus dipsaci* (Kuhn) Filipjev. *Proceedings of the Royal Society of London* B169, 203–213.

Ellenby, C. (1968b) Desiccation survival of the infective juvenile of *Haemonchus contortus*. *Journal of Experimental Biology* 49, 469–475.

Ellenby, C. (1969) Dormancy and survival in nematodes. *Symposium of the Society for Experimental Biology* 23, 83–97.

Evans, A.A.F. (1987) Diapause in nematodes as a survival strategy. In: Veech, J.A. and Dickson, D.W. (eds) *Vistas on Nematology*. Society of Nematologists, Hyattsville, Maryland, pp. 180–187.

Evans, A.A.F. and Perry, R.N. (1976) Survival strategies in nematodes. In: Croll, N.A. (ed.) *The Organisation of Nematodes*. Academic Press, London, 383–424.

Eysker, M. (1997) Some aspects of inhibited development of trichostrongylids in ruminants. *Veterinary Parasitology* 72, 265–283.

Fernandez, A.S., Fiel, C.A. and Steffan, P.E. (1999) Study on the inductive factors of hypobiosis of *Ostertagia ostertagi* in cattle. *Veterinary Parasitology* 81, 295–307.

Fodor, A., Vecseri, G. and Farkas, T. (1990) *Caenorhabditis elegans* as a model for the study of entomopathogenic nematodes. In: Gaugler, R. and Kaya, H.K. (eds) *Entomopathogenic Nematodes in Biological Control*. CRC Press, Boca Raton, Florida, pp. 249–265.

Foor, W.E. (1967) Ultrastructural aspects of oocyte development and shell formation in *Ascaris lumbricoides*. *Journal of Parasitology* 53, 1245–1261.

Forge, T.A. and MacGuidwin, A.E. (1992) Effects of water potential and temperature on survival of the nematode *Meloidogyne hapla* in frozen soil. *Canadian Journal of Zoology* 70, 1553–1560.

Freckman, D.W., Kaplan, D.T. and Van Gundy, S.D. (1977) A comparison of techniques for extraction and study of anhydrobiotic nematodes from dry soils. *Journal of Nematology* 9, 176–181.

Gatongi, P.M., Prichard, R.K., Ranjan, S., Gathuma, J.M., Munyua, W.K., Cheruiyot, H. and Scott, M.E. (1998) Hypobiosis of *Haemonchus contortus* in natural infections of sheep and goats in a semi-arid area of Kenya. *Veterinary Parasitology* 77, 49–61.

Gems, D. (2002) Ageing. In: Lee, D.L. (ed.) *The Biology of Nematodes*. Taylor and Francis, London, pp. 413–455.

Gibbs, H.C. (1986) Hypobiosis in parasitic nematodes – an update. *Advances in Parasitology* 25, 129–174.

Glazer, I. and Orion, D. (1983) Studies on anhydrobiosis of *Pratylenchus thornei*. *Journal of Nematology* 15, 333–338.

Harris, J.E. and Crofton, H.D. (1957) Internal pressure and cuticular structure in *Ascaris*. *Journal of Experimental Biology* 34, 116–130.

Hendelberg, M. and Jensen, P. (1993) Vertical distribution of the nematode fauna in a coastal sediment influenced by seasonal hypoxia in the bottom water. *Ophelia* 37, 83–94.

Hendriksen, N.B. (1982) Anhydrobiosis in nematodes: studies on *Plectus* sp. In: Lebrun, P., André, H.M., De Medts, A., Grégoire-Wibo, C. and Wauthy, G. (eds) *New Trends in Soil Biology*. Dieu-Brichart, Louvain-la-Neuve, pp. 387–394.

Higa, L.M. and Womersley, C. (1993) New insights into the anhydrobiotic phenomenon: the effects of trehalose content and differential rates of water loss on the survival of *Aphelenchus avenae*. *Journal of Experimental Zoology* 267, 120–129.

Holmstrup, M. and Westh, P. (1994) Dehydration of earthworm cocoons exposed to cold: a novel cold hardiness mechanism. *Journal of Comparative Physiology B* 164, 312–315.

Holmstrup, M., Bayley, M. and Ramløv, H. (2002) Supercool or dehydrate? An experimental analysis of overwintering strategies in small permeable arctic invertebrates. *Proceedings of the National Academy of Sciences, USA* 99, 5716–5720.

Hominick, W.M. (1986) Photoperiod and diapause in the potato cyst-nematode, *Globodera rostochiensis*. *Nematologica* 32, 408–418.

Hooper, D.J. (1971) Stem eelworm (*Ditylenchus dipsaci*), a seed and soil-borne pathogen of field beans (*Vicia fabia*). *Plant Pathology* 20, 25–27.

Jacquiet, P., Cabaret, J., Dia, M.L., Cheikh, D. and Thiam, E. (1996) Adaptation to arid environment – *Haemonchus longistipes* in dromedaries of Saharo-Sahelian areas of Mauritania. *Veterinary Parasitology* 66, 193–204.

Jensen, P. (1995) Life history of the nematode *Theristus anoxybioticus* from sublittoral muddy sediment at methane seepages in the northern Kattegut, Denmark. *Marine Biology* 123, 131–136.

Jones, P.W., Tylka, G.L. and Perry, R.N. (1998) Hatching. In: Perry, R.N. and Wright, D.J. (eds) *The Physiology and Biochemistry of Free-living and Plant-parasitic Nematodes*. CAB International, Wallingford, UK, pp. 181–212

Jørgensen, R.J. (1980) *Dictyocaulus viviparus*: migration in agar of juveniles subjected to a variety of physicochemical exposures. *Experimental Parasitology* 49, 106–115.

Kapel, C.M.O., Pozio, E., Sacchi, L. and Prestrud, P. (1999) Freeze tolerance, morphology, and RAPD-PCR identification of *Trichinella nativa* in naturally infected arctic foxes. *Journal of Parasitology* 85, 144–147.

Koenning, S.A. and Schmitt, D.P. (1986) Reproduction of *Pratylenchus brachyurus* on soybean callus tissue: effects of culture age and observations on anhydrobiosis. *Journal of Nematology* 18, 581–582.

Kooyman, F.N.J. and Eysker, M. (1995) Analysis of proteins related to conditioning for arrested development and differentiation in *Haemonchus contortus* by two-dimensional gel electrophoresis. *International Journal for Parasitology* 25, 561–568.

Koppenhofer, A.M., Baur, M.E., Stock, S.P., Choo, H.Y., Chinnasri, B. and Kaya, H.K. (1997) Survival of entomopathogenic nematodes within host cadavers in dry soil. *Applied Soil Ecology* 6, 231–240.

Lambshead, P.J.D., Tietjen, J., Ferrero, T. and Jensen, P. (2000) Latitudinal diversity gradients in the deep sea with special reference to North Atlantic nematodes. *Marine Ecology Progress Series* 194, 159–167.

Larsen, M.N. and Roepstorff, A. (1999) Seasonal variation in development and survival of *Ascaris suum* and *Trichuris suis* eggs on pasture. *Parasitology* 119, 209–220.

Lee, D.L. (1961) Two new species of cryptobiotic anabiotic fresh water nematodes, *Actinolaimus hintoni* sp. nov. and *Dorylaimus keilini* sp. nov. Dorylaimidae. *Parasitology* 51, 237–240.

Lee, D.L. (1972) Penetration of mammalian skin by the infective juveniles of *Nippostrongylus brasiliensis*. *Parasitology* 65, 499–505.

Lee, D.L. (2002) Life cycles. In: Lee, D.L. (ed.) *The Biology of Nematodes*. Taylor and Francis, London, pp. 61–72.

McBride, J.M. and Hollis, J.P. (1966) The phenomenon of swarming in nematodes. *Nature* 211, 545–546.

McCurdy, D.G., Forbes, M.R. and Boates, J.S. (1999) Evidence that the parasitic nematode *Skrjabinaclava* manipulates host *Corophium* behavior to increase transmission to the sandpiper, *Calidris pusilla*. *Behavioral Ecology* 10, 351–357.

Martinez, J., Perez-Serrano, J., Bernadina, W.E. and Rodriguez-Caabeiro, F. (2001) Stress response to cold in *Trichinella* species. *Cryobiology* 43, 293–302.

Martinez, J., Perez-Serrano, J., Bernadina, W.E. and Rodriguez-Caabeiro, F. (2002) Oxidative and cold shock cause enhanced induction of a 50 kDa stress protein in *Trichinella spiralis*. *Parasitology Research* 88, 427–430.

Maslen, N.R. (1981) The Signy Island terrestrial reference sites: XII. Population ecology of nematodes with additions to the fauna. *British Antarctic Survey Bulletin* 53, 57–75.

Mokabli, A., Valette, S., Gauthier, J.P. and Rivoal, R. (2001) Influence of temperature on the hatch of *Heterodera avenae* Woll. populations from Algeria. *Nematology* 3, 171–178.

Moore, J. and Lasswell, J. (1986) Altered behavior in isopods (*Armadillidium vulgarae*) infected with the nematode *Dispharynx nasuta*. *Journal of Parasitology* 72, 186–189.

Murrell, K.D., Graham, C.E. and McGreevy, M. (1983) *Strongyloides ratti* and *Trichinella spiralis*: net charge of epicuticle. *Experimental Parasitology* 55, 331–339.

Nicholas, W.L. (1984) *The Biology of Free-living Nematodes*, 2nd edn. Clarendon Press, Oxford, 251 pp.

Nicholas, W.L. and Stewart, A.C. (1989) Experiments on anhydrobiosis in *Acrobeloides nanus* (De Man, 1880) Anderson 1986 (Nematoda). *Nematologica* 35, 489–490.

Nijhout, H.F. (1994) *Insect Hormones*. Princeton University Press, Princeton, New Jersey, 267 pp.

O'Leary, S.A., Burnell, A.M. and Kusel, J.R. (1998) Biophysical properties of the surface of desiccation-tolerant mutants and parental strain of the entomopathogenic nematode *Heterorhabditis megidis* (Strain Uk211). *Parasitology* 117, 337–345.

O'Leary, S.A., Power, A.P., Stack, C.M. and Burnell, A.M. (2001) Behavioural and physi-ological responses of infective juveniles of the entomopathogenic nematode *Heterorhabditis* to desiccation. *Biocontrol* 46, 345–362.

Overhoff, A., Freckman, D.W. and Virginia, R.A. (1993) Life cycle of the microbivorous Antarctic Dry Valley nematode *Scottnema lindsayae* (Timm 1971). *Polar Biology* 13, 151–156.

Patel, M.N., Perry, R.N. and Wright, D.J. (1997) Desiccation survival and water contents of entomopathogenic nematodes, *Steinernema* spp. (Rhabditida: Steinernematidae). *International Journal for Parasitology* 27, 61–70.

Patterson, G.I. and Padgett, R.W. (2000) TGF beta-related pathways: roles in *Caenorhabditis elegans* development. *Trends in Genetics* 16, 27–33.

Perry, R.N. (1999) Desiccation survival of parasitic nematodes. *Parasitology (supplement)* 119, S19–S30.

Poinar, G.O., Jr (1983) *The Natural History of Nematodes*. Prentice-Hall, Englewood Cliffs, New Jersey, 323 pp.

Pozio, E., La Rosa, G., Murell, K.D. and Lichtenfels, J. (1992) Taxonomic revision of the genus *Trichinella*. *Journal of Parasitology* 78, 654–659.

Preston, C.M. and Bird, A.F. (1987) Physiological and morphological changes associated with recovery from anabiosis in the dauer juvenile of the nematode *Anguina agrostis*. *Parasitology* 95, 125–133.

Procter, D.L.C. (1984) Population growth and intrinsic rate of natural increase of the high arctic nematode *Chiloplacus* sp. at low and high temperatures. *Oecologia* 62, 138–140.

Rajan, T.V. (1998) A hypothesis for the tissue specificity of nematode parasites. *Experimental Parasitology* 89, 140–142.

Ramløv, H., Wharton, D.A. and Wilson, P.W. (1996) Recrystallization in a freezing toler-ant Antarctic nematode, *Panagrolaimus davidi*, and an alpine weta, *Hemideina maori* (Orthoptera, Stenopelmatidae). *Cryobiology* 33, 607–613.

Riddle, D.L. (1988) The dauer juvenile. In: Wood, W.B. (ed.) *The Nematode Caenorhabditis elegans.* Cold Spring Harbour Laboratory, Cold Spring Harbor, New York, pp. 393–412.

Riddle, D.L. and Albert, P.S. (1997) Genetic and environmental regulation of dauer juvenile development. In: Riddle, D.L., Blumenthal, T., Meyer, B.J. and Priess, J.R. (eds) *C. elegans II.* Cold Spring Harbor Press, Cold Spring Harbor, New York, pp. 739–768.

Rogers, W.P. (1960) The physiology of the infective process of nematode parasites: the stimulus from the animal host. *Proceedings of the Royal Society of London* B152, 367–386.

Saeed, M. and Roessner, J. (1984) Anhydrobiosis in five species of plant associated nematodes. *Journal of Nematology* 16, 119–124.

Salazar, A. and Ritter, E. (1993) Effects of daylength during cyst formation, storage time and temperature of cysts on the *in vitro* hatching of *Globodera rostochiensis* and *G. pallida. Fundamental and Applied Nematology* 16, 567–572.

Sambongi, Y., Nagae, T., Liu, Y., Yoshimizu, T., Takeda, K., Wada, Y. and Futai, M. (1999) Sensing of cadmium and copper ions by externally exposed ADL, ASE, and ASH neurons elicits avoidance response in *Caenorhabditis elegans. NeuroReport* 10, 753–757.

Saunders, L.M., Tompkins, D.M. and Hudson, P.J. (2000) Spatial aggregation and temporal migration of free-living stages of the parasitic nematode *Trichostrongylus tenuis. Functional Ecology* 14, 468–473.

Smith, H.J. (1984) Preconditioning of *Trichinella spiralis nativa* juveniles in musculature to low temperatures. *Veterinary Parasitology* 17, 85–90.

Smith, H.J. (1987) Factors affecting preconditioning of *Trichinella spiralis nativa* juveniles in musculature to low temperatures. *Canadian Journal of Veterinary Research* 51, 169–173.

Solomon, A., Paperna, I., Glazer, I. and Alkon, P.U. (1997) Migratory behaviour and desiccation tolerance of protostrongylid nematode first-stage juveniles. *International Journal for Parasitology* 27, 1517–1522.

Solomon, A., Paperna, I. and Glazer, I. (1999) Desiccation survival of the entomopathogenic nematode *Steinernema feltiae*: induction of anhydrobiosis. *Nematology* 1, 61–68.

Solomon, A., Salomon, R., Paperna, I. and Glazer, I. (2000) Desiccation stress of entomopathogenic nematodes induces the accumulation of a novel heat-stable protein. *Parasitology* 121, 409–416.

Sommerville, R.I. and Davey, K.G. (2002) Diapause in nematodes: a review. *Canadian Journal of Zoology* 80, 1817–1840.

Stromberg, B.E. (1997) Environmental factors influencing transmission. *Veterinary Parasitology* 72, 247–256.

Thompson, D.P. and Geary, T.G. (2002) Excretion/secretion, ionic and osmotic regulation. In: Lee, D.L. (ed.) *The Biology of Nematodes.* Taylor and Francis, London, pp. 291–320.

Townshend, J.L. (1984) Anhydrobiosis in *Pratylenchus penetrans. Journal of Nematology* 16, 282–289.

Treonis, A.M., Wall, D.H. and Virginia, R.A. (2000) The use of anhydrobiosis by soil nematodes in the Antarctic Dry Valleys. *Functional Ecology* 14, 460–467.

Vannier, G. (1994) The thermobiological limits of some freezing intolerant insects: the supercooling and thermostupor points. *Acta Oecologica* 15, 31–42.

Vanreusel, A., Vandenbossche, I. and Thiermann, F. (1997) Free-living marine nematodes from hydrothermal sediments – similarities with communities from diverse reduced habitats. *Marine Ecology Progress Series* 157, 207–219.

Van Voorhies, W.A. and Ward, S. (2000) Broad oxygen tolerance in the nematode *Caenorhabditis elegans. Journal of Experimental Biology* 203, 2467–2478.

Wharton, D.A. (1979) *Ascaris lumbricoides*: water loss during desiccation of embryonating eggs. *Experimental Parasitology* 48, 398–406.

Wharton, D.A. (1980) Nematode egg-shells. *Parasitology* 81, 447–463.

Wharton, D.A. (1981a) The effect of temperature on the behaviour of the infective juveniles of *Trichostrongylus colubriformis*. *Parasitology* 82, 269–276.

Wharton, D.A. (1981b) The initiation of coiling behaviour prior to desiccation in the infective juveniles of *Trichostrongylus colubriformis*. *International Journal for Parasitology* 11, 353–357.

Wharton, D.A. (1982) Observations on the coiled posture of trichostrongyle infective juveniles using a freeze-substitution method and scanning electron microscopy. *International Journal for Parasitology* 12, 335–343.

Wharton, D.A. (1986) The structure of the cuticle and sheath of the infective juvenile of *Trichostrongylus colubriformis*. *Parasitology Research* 72, 779–787.

Wharton, D.A. (1995) Cold tolerance strategies in nematodes. *Biological Reviews* 70, 161–185.

Wharton, D.A. (1996) Water loss and morphological changes during desiccation of the anhydrobiotic nematode *Ditylenchus dipsaci*. *Journal of Experimental Biology* 199, 1085–1093.

Wharton, D.A. (1997) Survival of low temperatures by the Antarctic nematode *Panagrolaimus davidi*. In: Lyons, W.B., Howard-Williams, C. and Hawes, I. (eds) *Ecosystem Processes in Antarctic Ice-free Landscapes*. Balkema Publishers, Rotterdam, pp. 57–60.

Wharton, D.A. (2002a) *Life at the Limits: Organisms in Extreme Environments*. Cambridge University Press, Cambridge, 320 pp.

Wharton, D.A. (2002b) Survival strategies. In: Lee, D.L. (ed.) *The Biology of Nematodes*. Taylor and Francis, London, pp. 389–411.

Wharton, D.A. and Aalders, O. (1999) Desiccation stress and recovery in the anhydrobiotic nematode *Ditylenchus dipsaci* (Nematoda : Anguinidae). *European Journal of Entomology* 96, 199–203.

Wharton, D.A. and Allan, G.S. (1989) Cold tolerance mechanisms of the free-living stages of *Trichostrongylus colubriformis* (Nematoda: Trichostrongylidae). *Journal of Experimental Biology* 145, 353–370.

Wharton, D.A. and Barclay, S. (1993) Anhydrobiosis in the free-living Antarctic nematode *Panagrolaimus davidi*. *Fundamental and Applied Nematology* 16, 17–22.

Wharton, D.A. and Barrett, J. (1985) Ultrastructural changes during recovery from anabiosis in the plant parasitic nematode, *Ditylenchus*. *Tissue and Cell* 17, 79–96.

Wharton, D.A. and Ferns, D.J. (1995) Survival of intracellular freezing by the Antarctic nematode *Panagrolaimus davidi*. *Journal of Experimental Biology* 198, 1381–1387.

Wharton, D.A. and Lemmon, J. (1998) Ultrastructural changes during desiccation of the anhydrobiotic nematode *Ditylenchus dipsaci*. *Tissue and Cell* 30, 312–323.

Wharton, D.A. and Marshall, A.T. (2002) Changes in surface features during desiccation of the anhydrobiotic plant parasitic nematode *Ditylenchus dipsaci*. *Tissue and Cell* 34, 81–87.

Wharton, D.A. and Worland, M.R. (1998) Ice nucleation activity in the freezing tolerant Antarctic nematode *Panagrolaimus davidi*. *Cryobiology* 36, 279–286.

Wharton, D.A., Perry, R.N. and Beane, J. (1983) The effect of osmotic stress on behaviour and water content of the infective juveniles of *Trichostrongylus colubriformis*. *International Journal for Parasitology* 13, 185–190.

Wharton, D.A., Barrett, J. and Perry, R.N. (1985) Water uptake and morphological changes during recovery from anabiosis in the plant parasitic nematode, *Ditylenchus dipsaci*. *Journal of Zoology (London)* 206, 391–402.

Wharton, D.A., Preston, C.M., Barrett, J. and Perry, R.N. (1988) Changes in cuticular permeability associated with recovery from anhydrobiosis in the plant parasitic nematode, *Ditylenchus dipsaci*. *Parasitology* 97, 317–330.

Wharton, D.A., Perry, R.N. and Beane, J. (1993) The role of the eggshell in the cold tolerance mechanisms of the unhatched juveniles of *Globodera rostochiensis*. *Fundamental and Applied Nematology* 16, 425–431.

Wharton, D.A., Judge, K.F. and Worland, M.R. (2000a) Cold acclimation and cryoprotectants in a freeze-tolerant Antarctic nematode, *Panagrolaimus davidi*. *Journal of Comparative Physiology B* 170, 321–327.

Wharton, D.A., Rolfe, R.N. and Perry, R.N. (2000b) Electrophysiological activity during recovery from anhydrobiosis in fourth stage juveniles of *Ditylenchus dipsaci*. *Nematology* 2, 881–886.

Wharton, D.A., Goodall, G. and Marshall, C.J. (2002) Freezing rate affects the survival of a short-term freezing stress in *Panagrolaimus davidi*, an Antarctic nematode that survives intracellular freezing. *CryoLetters* 23, 5–10.

Wharton, D.A., Goodall, G. and Marshall, C.J. (2003) Freezing survival and cryoprotective dehydration as cold tolerance mechanisms in the Antarctic nematode *Panagrolaimus davidi*. *Journal of Experimental Biology* 206, 215–221.

Wittenburg, N. and Baumeister, R. (1999) Thermal avoidance in *Caenorhabditis elegans*: an approach to the study of nociception. *Proceedings of the National Academy of Sciences, USA* 96, 10477–10482.

Wolkers, W.F., McCready, S., Brandt, W.F., Lindsey, G.G. and Hoekstra, F.A. (2001) Isolation and characterization of a D-7 LEA protein from pollen that stabilizes glasses *in vitro*. *Biochemical and Biophysical Acta* 1544, 196–206.

Womersley, C. (1978) A comparison of the rate of drying of four nematode species using a liquid paraffin technique. *Annals of Applied Biology* 90, 401–405.

Womersley, C. (1981) The effect of dehydration and rehydration on salt loss in the second-stage juveniles of *Anguina tritici*. *Parasitology* 82, 411–419.

Womersley, C. (1987) A reevaluation of strategies employed by nematode anhydrobiotes in relation to their natural environment. In: Veech, J.A. and Dickson, D.W. (eds) *Vistas on Nematology*. Society of Nematologists, Hyattsville, Maryland, pp. 165–173.

Womersley, C. (1990) Dehydration survival and anhydrobiotic survival. In: Gaugler, R. and Kaya, H.K. (eds) *Entomopathogenic Nematodes in Biological Control*. CRC Press, Boca Raton, Florida, pp. 117–137.

Womersley, C. (1993) Factors affecting physiological fitness and modes of survival employed by dauer juveniles and their relationship to pathogenicity. In: Bedding, R., Akhurst, R. and Kaya, H. (eds) *Nematodes and the Biological Control of Insect Pests*. CSIRO Publications, East Melbourne, pp. 79–88.

Womersley, C. and Ching, C. (1989) Natural dehydration regimes as a prerequisite for the successful induction of anhydrobiosis in the nematode *Rotylenchus reniformis*. *Journal of Experimental Biology* 143, 359–372.

Womersley, C.Z. and Higa, L.M. (1998) Trehalose – its role in the anhydrobiotic survival of *Ditylenchus myceliophagus*. *Nematologica* 44, 269–291.

Womersley, C.Z., Wharton, D.A. and Higa, L.M. (1998) Survival biology. In: Perry, R.N. and Wright, D.J. (eds) *The Physiology and Biochemistry of Free-living and Plant-parasitic Nematodes*. CAB International, Wallingford, UK, pp. 271–302.

Index

Page numbers in **bold** refer to illustrations and tables

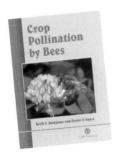
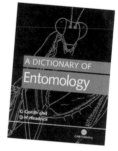